was also applied to FP μ-oxo dimers and quinine and quinidine but will not be discussed here and the reader is referred to the original publication (Leed et al. 2002).

4.2
Resistance to Quinoline-Containing Drugs

One of the most striking characteristics of CQ is its capacity to concentrate in the digestive vacuole of the trophozoite. It is here that it forms complexes with FP and inhibits haemoglobin degradation (see above). It has long been known that CQ-resistant isolates exhibit reduced accumulation of CQ when compared with their CQ-sensitive counterparts (Fitch, 1969, 1970; Verdier et al. 1984, 1985). More recently, we have shown that reduced CQ accumulation is manifest as a reduced apparent affinity of CQ-FP binding in the food vacuole of CQ-resistant isolates (Bray et al. 1998). It is evident that CQ-resistant isolates have evolved a mechanism (most likely a transporter or change in pH gradient) to reduce the access of CQ to FP (Bray et al. 1998). Another well documented property of CQ-resistant isolates is the ability of verapamil to stimulate the accumulation of CQ, make the parasite sensitive to CQ and effectively 'reverse' the resistant phenotype (Martin et al. 1987; Krogstad et al. 1987; Bray and Ward 1998). Verapamil was shown to act by increasing the access of CQ to the FP receptor (Bray et al. 1998). The verapamil effect is widely accepted as an essential phenotypic marker of CQ resistance. Recent groundbreaking work from the Wellems and Fidock laboratories has identified PfCRT (Fidock et al. 2000) an integral digestive vacuole membrane protein that appears to be the determinant of CQ-resistance. Specific polymorphisms of this gene are found in all natural isolates from clinical CQ treatment failures (Wernsdorfer and Noedl 2003) and in vitro in isolates with a CQ-resistant phenotype. Allelic exchange experiments have proven that the characteristic phenotypic markers of CQ resistance (reduced CQ sensitivity, reduced CQ uptake and verapamil effect) can be unambiguously attributed to these specific amino acid changes in PfCRT (Fidock et al. 2000; Sidhu et al. 2002). Being situated on the digestive vacuole membrane, PfCRT is in the right place to influence the access of CQ to FP but at the moment there is no definitive evidence to support any particular mechanism. Two main theories have been proposed: an indirect mechanism in which PfCRT is involved in vacuolar pH homeostasis and a mechanism in which mutant PfCRT directly transports CQ out of the digestive vacuole. The pH hypothesis proposes that reduced vacuolar CQ accumulation arises as a consequence of a primary effect of the PfCRT mutations in reducing the resting pH of the digestive vacuole (Ursos and Roepe 2002). In a mechanistic explanation the authors suggested that reduced vacuolar pH increases the rate of aggregation of FP as it is released from haemoglobin, reducing the amount that is available to bind CQ. However, this scenario would reduce the number of binding sites rather than reducing the apparent affinity of binding and therefore is not consistent with the experimental observations (Bray et al. 1998). Neither is it consistent with the complex patterns of cross-resistance that are observed between CQ and other quinoline antimalarials. Quinoline drugs which are all weak bases and all bind to FP should be influenced in the same qualitative way as CQ by a gross change in vacuolar pH (Cooper et al. 2002). By contrast, the transport hypothesis is consistent with the available data. For instance, it is easy to envisage subtle changes in drug structure conferring very different affinities for a drug transporter and very different patterns of sensitivity. The transporter hypothesis is also consistent with the reduced apparent affinity of FP binding that is observed in CQ-resistant isolates. Thus it is likely that mutant PfCRT is acting as a drug transporter although formal proof will require further experiments.

The original observations of CQ-resistance reversal were reminiscent of the verapamil reversal of multi-drug resistance (MDR) in certain cancer cell lines (Martin et al 1987; Ambudkar et al. 1999). In this case, the MDR phenomenon is attributable to massive amplification of the *mdr1* gene. This leads to overexpression of the protein product, P-glycoprotein, on the plasma membrane of the cell. P-glycoprotein is a primary active transporter of the ATP-binding cassette superfamily. It is polyspecific, being capable of pumping a broad range of hydrophobic, amphiphilic drugs and xenobiotics out of the cell (Ambudkar et al. 1999). A search for P-glycoprotein orthologues in *P. falciparum* identified Pgh1, an integral digestive vacuole membrane protein of the ATP-binding cassette superfamily (Wilson et al. 1989). Specific amino acid changes in Pgh1 undeniably do modulate the susceptibility of parasites to CQ (Reed et al. 2000) but overall, it would be fair to say that this protein probably plays only a minor role in determining parasite resistance to this drug. However, it is becoming abundantly clear that Pgh1 is very important in determining the response of parasites to mefloquine and other quinolines as well as to artemisinins (Reed et al. 2000). Allelic exchange experiments have shown that mutations in Pgh1 can confer resistance to mefloquine, quinine and the structurally related drug halofantrine (Reed et al. 2000). The same mutations were also able to influence the level of susceptibility to artemisinin and CQ (Reed et al. 2000). More recently, it has been shown that a reduced response of clinical isolates to mefloquine and to artemisinin is strongly correlated to the copy number of the *PfMDR1* gene and to expression of Pgh1 (Uhlemann et al. 2004; see the chapter by A.-C. Uhlemann and S. Krishna, this volume).

Amodiaquine (2) is a 4-aminoquinoline antimalarial that is effective against many CQ-resistant strains of *P. falciparum*. However, clinical use has been severely restricted because of associations with hepatotoxicity and agranulocytosis when used in prophylaxis (Jewell et al. 1995). The AQ side-chain contains a 4-aminophenol group; a structural alert for toxicity, because of metabolic oxidation to quinoneimines. We have shown that AQ does indeed readily undergo oxidation to a quinoneimine. (Naisbitt et al. 1998; Tingle et al. 1995).

Figure 5 summarizes the different classes of 4-aminoquinoline that have been investigated over the past 15 years (O'Neill et al. 1998). Because AQ retains antimalarial activity against CQ-resistant parasites our initial studies involved the design and synthesis of fluoroamodiaquine (**4a**) as a safer alter-native to AQ (O'Neill et al. 1994). This analogue cannot form toxic metabolites by P450-mediated processes and retains substantial antimalarial activity ver-sus CQ-resistant parasites. Lead optimization of (**4a**) produced a new lead, compound (**4b**) which is about half as active as AQ against CQ-resistant strains in vitro, but shows equivalent oral in vivo potency versus *P. berghei*. Concern about cost led to the consideration of three other series of synthetically more accessible analogues—the tebuquine series (**5**) (O'Neill et al. 1996, 1997) the bis-Mannich series (**6**) (Barlin et al. 1993; Kotecka et al. 1997) and the 5'-alkyl series class of 4-aminoquinoline (**7**) (Raynes et al. 1999). Compounds in the tebuquine and bis-Mannich series have now been shown to have unacceptable toxicity profiles and extremely long half-lives (Ruscoe et al. 1998). Recently, Delarue et al. have prepared some 4'-dehyroxy analogues, some of which have very good activity profiles and are less toxic than AQ. (Delarue et al. 2001).

Studies on 4-aminoquinoline structure–activity relationships (SAR) have revealed that 2-carbon side-chain CQ analogues retain activity against CQ-resistant *Plasmodium* (De et al. 1997; Ridley et al. 1996). Our own efforts (Stocks et al. 2002) were directed towards compounds less likely to undergo metabolic N-terminal dealkylation, a process that produces N-desalkyl metabolites that are considerably less potent against CQ-resistant strains. Some of these 2-C analogues, e.g., (**9**), display good antiparasitic profiles. Other notable work in the CQ–SAR field has involved the preparation of bisquinoline dimers, some of which possess excellent activity against CQ-resistant parasites (Vennerstrom et al. 1998). Unfortunately, the best candidates (example WR-268, 668) to emerge from this series of analogues were shown to be photosensitizing.

From our SAR studies in the AQ series, it is clear that the presence of a 4-arylamino moiety provides analogues with superior activity against CQ-resistant strains and that the presence of an aromatic hydroxyl function ap-pears to be important for additional levels of antiparasitic activity. We rea-soned that interchange of the 3'-Mannich side chain with the 4'-OH function would provide a new template, chemically incapable of forming potentially toxic quinoneimine metabolites. Furthermore, these compounds should be as cheap to prepare as AQ on an industrial scale. Our initial studies show that

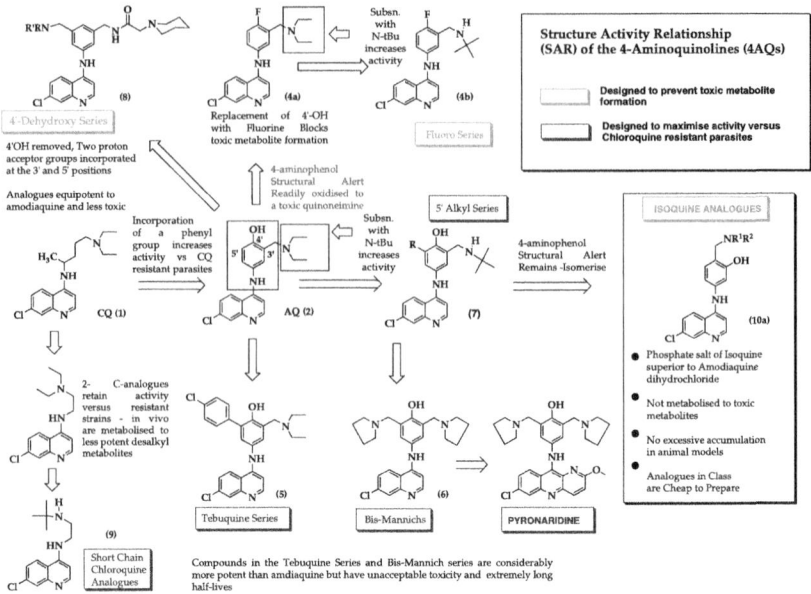

Fig. 5 Summary of 4-aminoquinoline analogues studied in the last 15 years. As shown, chloroquine (*CQ*) and amodiaquine (*AQ*) analogues have been designed to maximize antimalarial activity versus resistant strains. In the case of AQ, analogues have also been designed to reduce potential toxicity based on the proposed mechanism of AQ toxicity

some of the isomeric series of AQ analogues presented in Fig. 5 have potent activity against CQ resistant parasites in vitro and oral activity in rodent models of malaria. Isoquine ISQ-1 (Fig. 5 **10a**, R^1, R^2=ethyl), the direct isomer of AQ, has emerged as the lead candidate and is currently in pre-clinical evaluation in a partnership between the Malaria for Medicines Venture (MMV) and Glaxo Smithkline Pharmaceuticals (O'Neill et al. 2003) Further optimization of the 4-arylamino template is ongoing but it is likely that any new 4-aminoquinoline candidates will be combined with a peroxide-based antimalarial to delay parasite resistance acquisition and prolong the therapeutic life-span of the novel drug entity. Figure 6 summarizes the main SAR observations for (a) CQ and (b) AQ analogues.

To conclude, 4-aminoquinoline-based drug development projects continue; at least three projects are approaching evaluation in man, including short chain CQ analogues, metabolically stable AQ analogues and aza-acridine derivatives (pyronaridine). It is anticipated that successful development of any of these candidates will provide the same sort of therapeutic benefits provided by CQ in its early pre-resistance days.

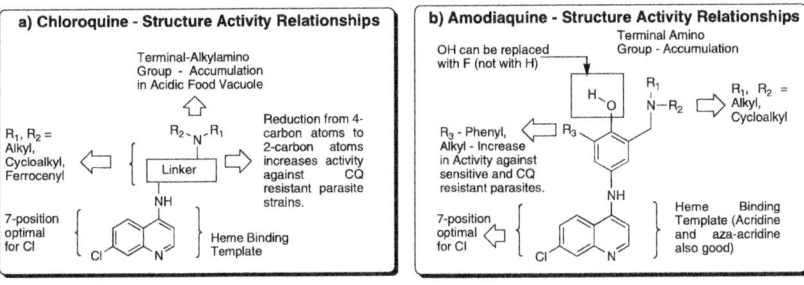

Fig. 6a,b Summary of important SAR observations in the chloroquine and amodiaquine classes

5
Artemisinins

Artemisinin (11a) (qinghaosu) is an unusual 1,2,4-trioxane that has been used in China for the treatment of MDR *P. falciparum* malaria but its therapeutic value is limited by its low solubility in both oil and water (Fig. 7). Consequently, in the search for more effective and soluble drugs, Chinese researchers prepared a number of derivatives of the parent drug. (Klayman 1985; Butler and Wu 1992) Reduction of artemisinin produces dihydroartemisinin (11b), which has in turn led to the preparation of a series of semisynthetic first-generation analogues which include artemether (11c, R=–Me) and arteether (11d, R=–Et) (Fig. 7) (Butler and Wu 1992). Both of these compounds are more potent than artemisinin but have short plasma half-lives and produce fatal central nervous system (CNS) toxicity in chronically dosed rats and dogs (Brewer et al. 1994). Although neurotoxicity is an issue in animal models, recent studies by White and co-workers have shown a lack of neuronal death in patients who had received high doses of artemether by intramuscular injection (Hien et al. 2003). In spite of these observations, there are no

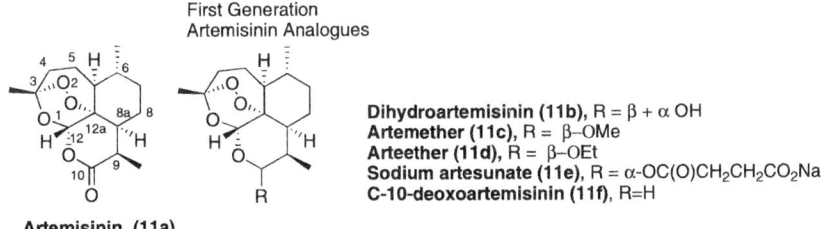

First Generation
Artemisinin Analogues

Dihydroartemisinin (11b), R = β + α OH
Artemether (11c), R = β–OMe
Arteether (11d), R = β–OEt
Sodium artesunate (11e), R = α-OC(O)CH$_2$CH$_2$CO$_2$Na
C-10-deoxoartemisinin (11f), R=H

Artemisinin, (11a)

Fig. 7 Structures of artemisinin and the first-generation 'semi-synthetic peroxides'

comparative data on oral dosing with first generation alkyl ether pro-drugs
(i.e., artemether, arteether) of the neurotoxic dihydroartemisinin.

For treatment of advanced cases of *P. falciparum* malaria, a water-soluble
derivative of artemisinin is desirable. A water-soluble derivative can be injected intravenously (i.v.), (Lin et al. 1989). The sodium salt of artesunic
acid (4e) is such a water-soluble derivative, capable of rapidly diminishing
parasitaemia and restoring consciousness of comatose cerebral malaria patients (Lin et al. 1989). Due to the high recrudescence rate, however, sodium
artesunate (4e) is normally administered in combination therapy, most often with mefloquine for treatment of uncomplicated malaria (Barradell and
Fitton1995). Of the first-generation derivatives, sodium artesunate (4e) is
currently the drug of choice, but the combination of artemether and lumen-
fantrine produced by Novartis is being implemented as part of the artemisinin
combination therapy and is likely to see more widespread use.

5.1
Mechanism of Activation

5.1.1
Carbon Radicals, Open Hydroperoxides and High-Valent Iron-Oxo Species: Reactive Species Implicated in the Mechanism of Action of the Artemisinins

The key pharmacophore in artemisinin is the 1,2,4-trioxane unit and, in
particular, the endoperoxide bond is crucial for expression of antiparasitic
activity. Reduction of the peroxide bridge of artemisinin to give the analogue deoxoartemisinin results in a complete abolition of antimalarial activity (Klayman 1985; O'Neill and Posner 2004). Based on the seminal work
of Posner and co-workers in the early 1990s, the free-radical chemistry of
artemisinin is now very well defined and has been shown to involve an initial chemical decomposition induced by haem Fe(II) (reduced haemin) or
other sources of ferrous iron within the malaria parasite to produce initially an oxy radical that subsequently rearranges into one or both of two
distinctive carbon-centred radical species (Posner and Oh 1992; Posner et
al. 1994; Posner and O'Neill 2004). Figure 8 summarizes the main radical
pathways available for artemisinin following endoperoxide-mediated bioactivation. Since artemisinin is an unsymmetrical endoperoxide, the oxygen
atoms of the peroxide linkage can associate with reducing ferrous ions in two
ways. Association of Fe(II) with oxygen-1 provides an oxy radical that goes on
to produce a primary carbon-centred radical (12a). A surrogate marker for
the intermediacy of this radical species is the ring-contracted tetrahydrofuran
product 12b. Alternatively, association with oxygen 2 provides an oxy radical
species, that, via a 1,5-H shift, can produce a secondary carbon-centred radi-

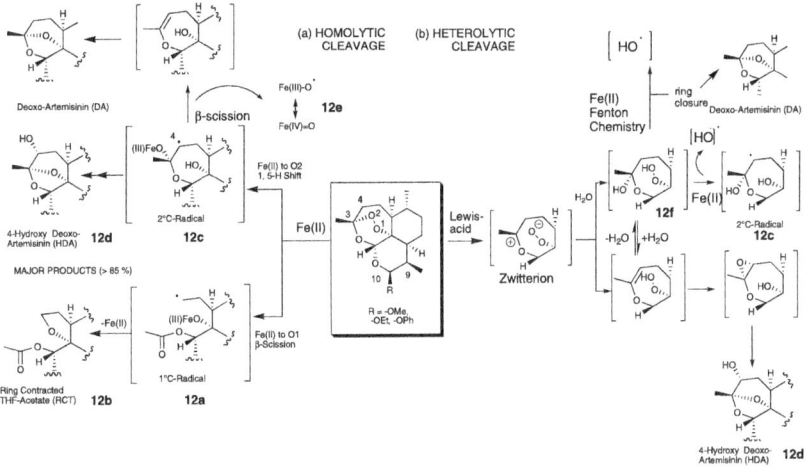

Fig. 8a,b Homolytic and heterolytic mechanisms of bioactivation of the endoperoxide bridge of artemisinin and derivatives

cal (**12c**). Again, like the previous route, a stable end-product, hydroxydeoxo artemisinin (**12d**) functions as a surrogate marker for this secondary carbon centred radical species.

There is evidence to support the roles of each individual carbon radical species as the mediators of antimalarial activity and this subject remains an area of intense debate (Posner and Meshnick 2001; Haynes et al. 2003b; O'Neill and Posner 2004; Haynes 2001; Olliaro et al. 2001b). It has been proposed that final alkylation by these reactive carbon radical intermediates of biomacromolecules such as haem, specific proteins and other targets, result in the death of malaria parasites.

The secondary C-radical intermediate (**12c**) has also been implicated as the precursor to a high-valent iron oxo species (**12e**), and several experimental results support the intermediacy of such a potentially toxic species (Posner et al. 1995, 1996) Although Varotsis has provided Raman spectroscopic support for the generation of a high-valent iron-oxo species during ferrous-mediated endoperoxide decay (Kapetanaki and Varotsis 2000, 2001) the groups of Meunier (Robert and Meunier 1998) and Jefford (Jefford 2001; Jefford et al. 1996) have contested this chemical mechanism.

Definitive evidence for the generation of carbon radical intermediates during ferrous-mediated endoperoxide degradation of both artemisinin (Wu et al. 1998) and arteflene (O'Neill et al. 2000) has been provided by electroparamagnetic resonance spin-trapping techniques (Butler et al. 1998). For artemisinin, both the primary and secondary carbon centered radicals have been efficiently spin-trapped post iron-mediated activation (O'Neill et al. 2000; Wu et al. 1998).

An alternative view to the iron-induced homolytic endoperoxide cleavage hypothesis is that artemisinin acts as a masked source of hydroperoxide (Scheme 2, b). (Haynes et al. 1999; Haynes and Vonwiller 1996a, 1996b) Following specific non-covalent interactions with a given target protein, heterolytic cleavage of the endoperoxide bridge and formation of an unsaturated hydroperoxide is followed by capture by water (or other nucleophile). This process provides a reactive hydroperoxide capable of irreversibly modifying protein residues by direct oxidation.

Fenton like degradation of this hydroperoxide may produce the hydroxyl radical, a species that can go on to oxidize target amino acid residues: this alternative pathway provides a mechanism of producing a whole host of reactive oxygen species that may have an equally important role to play in the antimalarial activity of these compounds. It has been proposed that the heterolytic step is aided by the non-endoperoxide bridging oxygen of the trioxane ring, where the carbocation can be stabilized by resonance (Haynes et al. 1999; Olliaro et al. 2001a). Haynes and co-workers have provided direct chemical evidence for this mechanism of bioactivation by demonstrating that artemisinin can mediate N-oxidation of tertiary alkylamine derivatives via the intermediacy of such a ring opened peroxide form of artemisinin (Haynes et al. 1999).

It is clear from the above discussion that artemisinin and endoperoxide-based drugs have the ability to generate a range of different reactive intermediates, and many of these have been proposed as the mediators of the phenomenal antimalarial activity of this class of drug.

5.2
The Biological Target(s) of Artemisinin Derivatives

Like the quinolines many hypotheses to explain mechanism of artemisinin action have been proposed. Proposed targets of these species are discussed and will conclude with the most recent studies by Eckstein-Ludwig who suggest that artemisinin derivatives target PfATPase6, the sarco/endoplasmic reticulum Ca^{2+}-ATPase (SERCA) of the parasite. Since the O–O bond appears to be crucial, the haemoglobin degradation pathway of malaria parasites would appear to offer the next clue as to the mechanism of action of the artemisinin class of endoperoxide. Since the original proposal by Meshnick and co-workers (Meshnick et al. 1991) it is still believed by many researchers in the field that haem liberated in this proteolysis process (ferrous haem is produced from 1 electron reduction of oxidized ferric FP) is the species responsible for the bioactivation of the endoperoxide bridge to potentially toxic free radicals in the food vacuole of the parasite (see above).

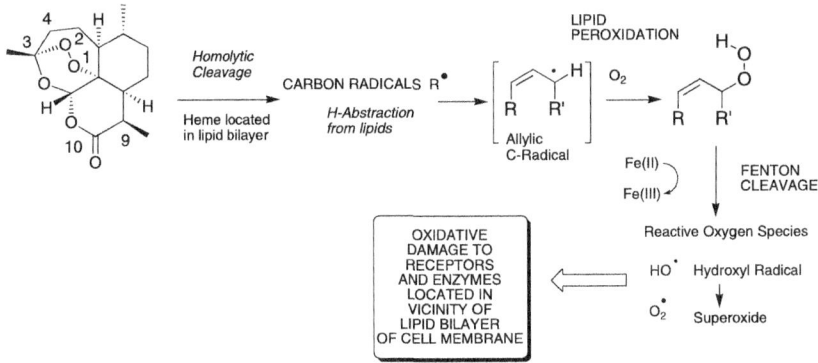

Fig. 9 Proposed chemical mechanism for artemisinin mediated lipid peroxidation of membranes

Consistent with this proposal, biomimetic studies by Berman and Adams have clearly demonstrated that artemisinin can effect a sixfold increase in haem-mediated lipid membrane damage (Berman and Adams 1997). Of importance to this observation are the findings of Fitch and colleagues who have recently demonstrated that unsaturated lipids co-precipitate with FP in the parasite's acidic food vacuole and also dissolve sufficient monomeric FP to allow efficient crystallization (Fitch et al. 1999). A possible mechanism of artemisinin induced lipid peroxidation is depicted in Fig. 9 and provides downstream access to typical reactive oxygen species such as oxyl radicals and the superoxide anion. Interaction of lipid solubilized haem with artemisinin followed by ferrous-mediated generation of oxyl and carbon radicals places these reactive intermediates in the vicinity of target allylic hydrogens of unsaturated lipid bilayers. Hydrogen abstraction and allylic carbon radical formation with subsequent triplet ground state oxygen capture results ultimately in the formation of lipid hydroperoxides. The explicit mechanism depicted in Fig. 9 is supported by the work of Berman and Adams (1997) and others and it was proposed that the damage caused to the parasite's food vacuole membrane leads to vacuolar rupture and parasite autodigestion. The biological significance of hydroperoxides in relation to biological hydroxylation and autoxidation of, for example, lipids and membrane bilayers, is well established. The generation of unsaturated lipid hydroperoxides provides a means of initiation of such processes.

In contrast to these proposals, other workers in the field have suggested that membrane bound haem may have a role to play in reducing the effectiveness of endoperoxides such as dihydroartemisinin (Vattanaviboon et al. 2002) Further work is required to clarify the role of vacuolar membrane bound haem in the mechanism of action of endoperoxide antimalarials.

Although the above scheme would appear to be chemically plausible, several workers have proposed that parasite death in the presence of artemisinin is probably not due to non-specific or random cell damage caused by freely diffusing oxygen radical species, but might involve specific radicals and targets some of which are described below (Robert and Meunier 1998; Robert et al. 2002).

When artemisinin or other active trioxanes were incubated at pharmacologically relevant concentrations within human red blood cells infected by *P. falciparum*, a haem-catalysed cleavage of the peroxide bond was reported to be responsible for the alkylation of haem (Zhang et al. 1992) and a small number of specific parasite proteins, (Yang et al. 1993, 1994) one of which has a molecular size similar to that of a histidine-rich protein (42 kDa). Another possible target protein is the *P. falciparum* translationally controlled tumour protein (TCTP). In vitro, dihydroartemisinin reacts covalently with recombinant TCTP in the presence of haemin. The association between drug and protein increases with increasing drug concentrations until it reaches a stoichiometry of 1 drug molecule/TCTP molecule. The function of TCTP is unknown and thus the role of haemin-mediated artemisinin alkylation in the mode of action of this drug awaits further biochemical definition (Bhisutthibhan et al. 1998; Bhisutthibhan and Meshnick 2001). Artemisinin may also be involved in the specific inhibition of malarial cysteine protease activity (Hong et al. 1994; Pandey et al. 1999).

5.2.1
Haem and Haem Model Alkylation

Alkylation of haem by artemisinin was first reported by Meshnick after identification of haem-drug adducts by mass spectrometry, but no structures were proposed for the resulting covalent adducts (Hong et al. 1994; Zhang et al. 1992). Because of the variety of possible alkylation sites on iron protoporphyrin-IX, Meunier studied the alkylating activity of artemisinin with manganese(II) tetraphenylporphyrin, a synthetic metalloporphyrin having a fourth-order symmetry and only the eight equivalent β-pyrrolic positions as possible alkylation sites. By reacting manganese tetraphenyl porphyrin ($Mn^{II}TPP$) with artemisinin (or artemether and several related synthetic trioxanes) in dichloromethane, a chlorin-type adduct was formed by reaction of the macrocycle with an alkyl radical generated by reductive activation of the drug endoperoxide (Robert et al. 1997).

Further studies involved the investigation of the reactivity of artemisinin toward the pharmacologically more relevant iron(II) containing model of ferriprotoporphyrin IX (Robert et al. 2001). For this purpose, iron(III) protoporphyrin-IX dimethylester was exposed to artemisinin in the presence

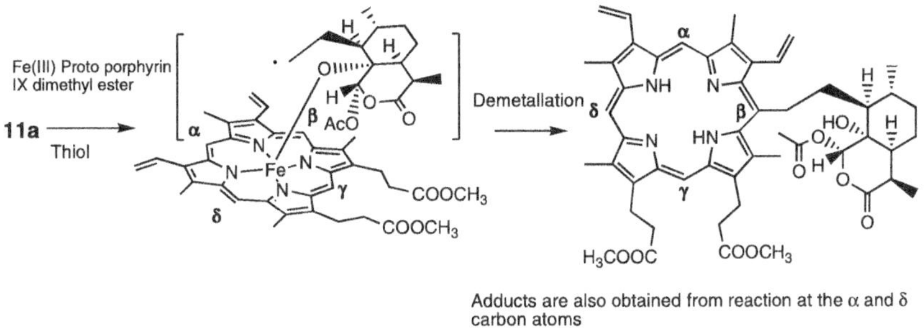

Adducts are also obtained from reaction at the α and δ carbon atoms

Fig. 10 Alkylation of a haem model (the ester of FP) by a primary carbon centred radical derived from bioactivation of artemisinin

of a hydroquinone derivative (or a thiol), as reducing agent, to generate the requisite iron(II) haem species. Haem was readily converted in high yield to haem–artemisinin adducts (Fig. 10; Cazelles et al. 2001). After demetallation of this mixture of three adducts to facilitate the NMR characterization, indications were that the α, β and δ meso carbons were alkylated; such results prompted Meunier to suggest that the low and transient concentration of free haem generated by haemoglobin degradation in vivo may be responsible for the reductive activation of the endoperoxide function of active trioxanes. This pathway generates alkylating species, such as the primary carbon radical (**12a**), which are likely to disrupt vital biochemical processes of the parasite via alkylation of biomolecules located in the close vicinity of the free haem (Robert and Meunier 1998). This proposal is based on the assumption that the primary C-radical has sufficient life-time to migrate from the face of the porphyrin metallocycle and subsequently to interact with its biological target (Olliaro et al. 2001a).

Although it is clear in model systems that artemisinin can efficiently alkylate haem-based models, the role of this event in the mechanism of action of artemisinin has been questioned. For example, it has been proposed that the formation of haem–artemisinin adducts of the type described could result in the prevention of haem crystallization to non-toxic HZ. The resultant build up of redox active alkylated porphyrins could in theory lead to parasite death by a mechanism similar to that proposed for the quinoline-based antimalarials. However, Haynes and co-workers have ruled out this potential mechanism by demonstrating clearly that although artemisinin (**11a**) and dihydroartemisinin (**11b**) have the ability to inhibit beta-FP formation in vitro, the closely related and antimalarially potent C-10 deoxo artemisinin (**11f**) (Fig. 7) has no effect on crystallization (Haynes et al. 2003c) Thus, it was proposed that the observed inhibitory activities in the haem polymerization inhibitory assay (HPIA) for **11a** and **11b** are a reactivity or property not related to the inherent antimalarial mode of action of this class of drugs (Haynes et al. 2003c).

It should be emphasized that virtually all of the above discussion is based on biomimetic chemistry where the Fe(II) source varies from salts such FeSO$_4$ to the more reactive FeCl$_2$.4H$_2$O as well as haem mimetics (TPP) and ester FP variants (O'Neill and Posner 2004). When haem models are used, since porphyrin alkylation is a favoured process, end-product distributions of products can be very different from when a free ferrous ion source is employed. Furthermore, solvent has been shown to have a profound effect on the products obtained in iron-mediated endoperoxide degradation. Thus all of these studies are truly only approximate models of the actual events within the malaria parasites (Posner and Meshnick 2001; Wu 2002). Future work is needed to correlate the results of biomimetic chemistry with the actual situation within the parasite.

5.2.2
Enzymes as Targets

As described earlier, erythrocytic malaria parasites degrade haemoglobin to acquire amino acids for protein synthesis. Falcipain 2 (FP-2; Dua et al. 2000) is a papain family cysteine protease that appears to act in concert with other enzymes including two aspartic proteases (Banerjee et al. 2003; Boss et al. 2003) to degrade haemoglobin. Incubation of erythrocytic parasites with inhibitors of FP-2 blocks haemoglobin degradation and parasite development. Pandey has demonstrated that, in purified digestive vacuoles from *P. yoelii*, cysteine protease activity can be inhibited by artemisinin in a similar manner to the potent cysteine protease inhibitor E-64 (Pandey et al. 1999). Inhibition of falcipain-mediated cleavage of the fluorogenic peptide substrate Z-Phe-Arg-AMC was also demonstrated in a continuous fluorometric assay, and surprisingly protease inhibition was increased in the presence of haem (surprising in the sense that strong arguments have been made that the haem-generated radical species cannot escape the porphyrin macrocycle and hit biological targets, post-reductive endoperoxide cleavage).

To fully validate falcipain 2 and 3 as targets for endoperoxide drugs, it is essential that these studies be expanded to human forms of the parasite. In addition, it would be of great interest to compare the efficiency of falcipain inhibition with the known antimalarial activities of a series of artemisinin analogues of varying potency.

Krishna and co-workers (Eckstein-Ludwig et al. 2003) have very recently provided compelling evidence that artemisinins act by inhibiting PfATPase6, the Sarco/Endoplasmic reticulum Ca^{2+}-ATPase (SERCA) orthologue of *P. falciparum*. When expressed in *Xenopus* oocytes, Ca^{2+}-ATPase activity of PfAT-

Pase6 is inhibited by artemisinin with similar potency to thapsigargin (another sesquiterpene lactone and highly specific SERCA inhibitor), but not by quinine or CQ. As predicted from this observation, thapsigargin antagonizes the parasiticidal activity of artemisinin. Desoxyartemisinin is ineffective as an antimalarial and was shown not to inhibit PfATPase6 activity. Chelation of iron by desferrioxamine abrogates the antiparasitic activity of artemisinins and correspondingly attenuates inhibition of PfATPase6. Single-cell imaging of living parasites with BODIPY-thapsigargin demonstrates cytosolic labelling that is competed by an excess of artemisinin. Furthermore, similar labelling is observed with a novel fluorescent artemisinin derivative. These studies support PfATPase6 as a target of artemisinins operating via an Fe^{2+}-dependent activation mechanism. This information may allow, for the first time, rational biological target-guided drug design efforts to be carried out.

5.3
Semi-synthetic and Synthetic Endoperoxide Analogues

The first generation C-10-acetal derivatives artemether (**11c**) and arteether (**11d**) both have a short half life as a consequence of cytochrome P450 catalysed transformation to dihydroartemisinin (DHA) (**11b**) which in turn is an efficient substrate for Phase II clearance through glucuronidation (Grace et al. 1998; Idowu et al. 1997; Maggs et al. 1997). In addition to metabolism, other first generation analogues such as artesunate are chemically unstable and hydrolyse rapidly to DHA in plasma [studies by Teja-Isavadharm indicate the half-life or artesunate ($t_{1/2}$) is as short as 0.41±0.34 h in man following oral administration; Teja-Isavadharm et al. 2001; Barradell and Fitton 1995). Based on these observations, medicinal chemists have made significant efforts to design more potent and stable analogues of the first-generation semi-synthetic derivatives. The metabolically more robust C-10 carba analogues **13a** and C-10-aryl analogues of DHA **13b** have been the focus of medicinal chemists for 10 years (Haynes 2001). Of note are the C-10 alkyl deoxo analogues prepared by Haynes et al. (2000), Posner et al. (1999), O'Neill et al. (1999, 2001b), Jung (Jung and Lee 1998) and Ziffer (Ma et al. 1999; Pu and Ziffer 1995) and the C-10 aryl or heteroaromatic derivatives **13c** prepared by the groups of Haynes (Haynes et al. 2003a) and Posner (Posner et al. 2003; Fig. 5). Equally impressive are the C-14 modified analogues **13d** prepared by the Avery (Vroman et al. 1999) and Jung groups (Jung et al. 2001).

Recently a C-10 carba analogue (TDR 40292) **13e** has been compared with artemether. This compound cannot form DHA as a metabolite and contains a side chain that can be formulated as a water-soluble salt (Hindley et al. 2002). In addition, this compound has superior activity to artemether and artesunate, both in vitro and in vivo. From initial pharmacokinetic data,

13d has a higher volume of distribution than artemether and is considerably more orally bioavailable (16% versus 1.5% for artemether) (O'Neill et al. unpublished results). (For studies on the bioavailability of artemether in man see Silamut et al. 2003).

A particularly important factor in the design of any new peroxide analogue is the concern about potential neurotoxicity. Any analogue with a higher logP than artemether (3.3–3.5) is likely to cross the blood–brain barrier (Haynes, 2001). Haynes and co-workers have prepared new analogues with reduced neurotoxicity by applying the ADME paradigm for enhancing efficacy through increased drug absorption (coupled with a reduction in the ability of the new analogue to cross the blood–brain barrier). Artemisone (undergoing development by Bayer) is an analogue with much improved properties and represents the success of the ADME approach to drug design (http://www.mmv.org/pages/page_main.htm).

An alternative approach to preventing the formation of dihydroartemisinin by simple P450 metabolism is to replace the methyl function in artemether with an aryl function (O'Neill et al. 2001a). Phenoxy analogues of DHA can easily be prepared in a one-step synthesis from dihydroartemisinin. In addition to having superior in vivo activity to artesunate and artemether, analogues substituted with a p-fluoro (**13f**, R=p-F) or trifluoromethyl group (**13 g**, R=p-CF$_3$) resist metabolism to DHA. In order to improve water solubility, a novel meta carboxylic acid phenoxy derivative (**13i**, R=m-CO$_2$H) has recently been prepared as a metabolically more robust alternative to artesunate and artelinic acid.

With an ever increasing number of artemisinin analogues prepared by semi-syntheses and elegant total synthesis Avery (Avery et al. 1989; Vroman et al. 1999) has developed predictive 3-D quantitative SAR (CoMFA) analyses for the artemisinin class of antimalarial (Avery et al. 1994, 2002). This information coupled with the ADME approach described above should permit highly potent and orally bioavailable semi-synthetic analogues to be designed by a truly rational approach (Haynes, 2001).

The disadvantage of all of the semi-synthetic compounds is that their production requires **11a** as starting material. Artemisinin is extracted from the plant *Artemisinia annua* in low yield (0.01–0.8% yield) (Klayman et al., 1984; Liu et al. 2003; Abdin et al., 2003). To circumvent this problem, a number of groups have produced totally synthetic peroxide analogues, some of which demonstrate remarkable antimalarial activity (Borstnik et al. 2002; Tang et al. 2004). These include the synthetic 1,2,4-trioxane, fenozan B0-7 (**14**) (Peters et al. 1993a, 1993b; Jefford et al. 1995, 2001) the dispiro tetraoxanes (**15, 16**) (Vennerstrom et al. 2000) and the endoperoxide analogues such as arteflene (**17**) (Hofheinz et al. 1994) a synthetic analogue of the naturally occurring yingzhaozu A (**18**) (Zhou and Xu 1994). More recently, tetraoxane (**19**) with

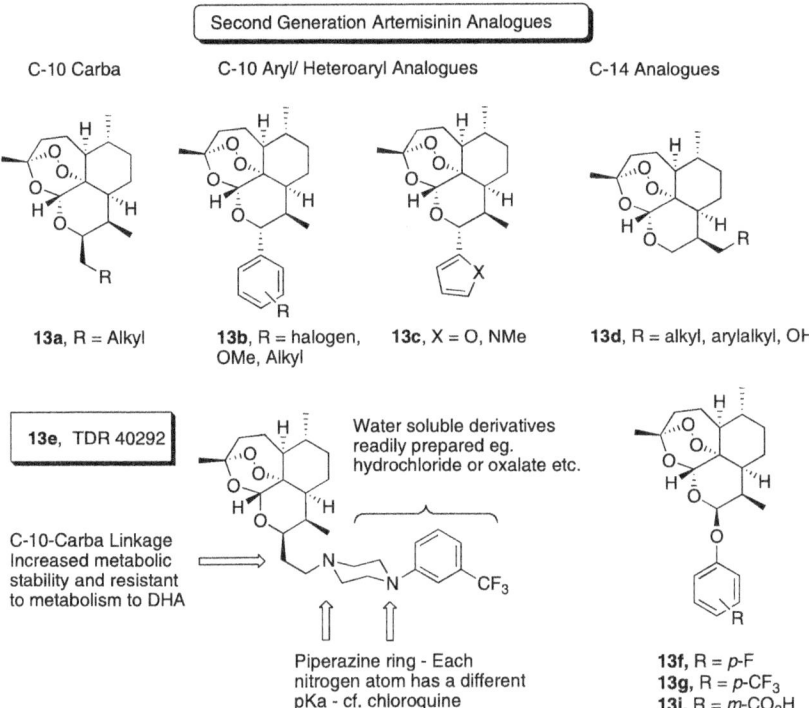

Fig. 11 Selected second-generation artemisinin derivatives

an IC_{50} as low as 3 nM has been discovered (artemisinin $IC_{50}=10$ nM) and analogues in this class have been shown to be effective when given orally in mice infected with *P. berghei* with no observable toxic side-effects (Fig. 11; Kim et al. 2003).

Other synthetic candidates worthy of mention include the C-3 aryl trioxanes (**20a**, R=F and **20b**, R=–COOH) and the endoperoxide analogue (**21**) (Posner et al. 1998; Korshin et al. 2002). These latter compounds have oral activity (ED_{50}) as low as 0.5 mg/kg in mice infected with *P. berghei* (Bachi et al. 2003).

The most significant recent discovery in this area is the discovery that easily synthesized ozonides (1,2,4-trioxolanes) substituted with an adamantane ring (**22a–c**) are not only chemically stable but are active against *P. falciparum* in the low nanomolar range. These compounds are orally active in mice and have a prolonged duration of action when compared with previously available synthetic and semi-synthetic derivatives. This research is supported by the MMV and a 2-year objective is to progress this project from preclinical

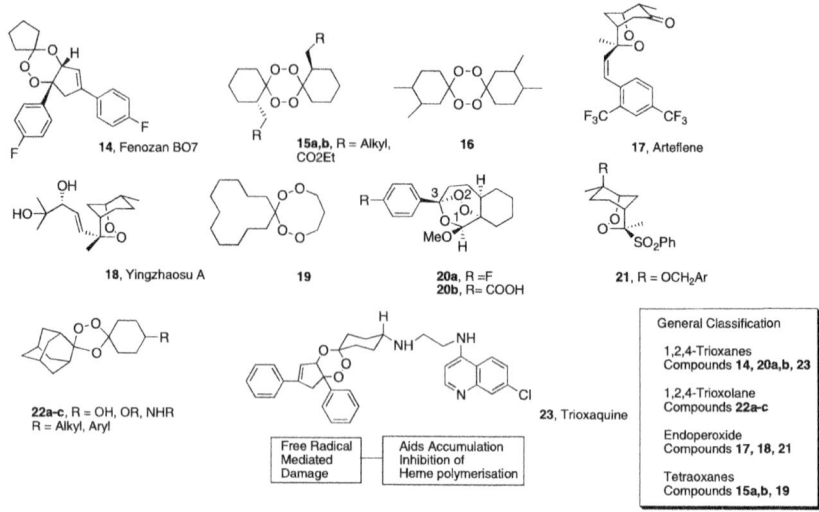

Fig. 12 Synthetic peroxide analogues

development to phase 1 clinical trials. Ranbaxy (an Indian pharmaceutical company) is in partnership with MMV to move the project forward. This project was the MMV project of the year in 2001, and excellent progress has been made since then. Water-soluble compounds that retain good oral activity in the *P. berghei* mouse model have been identified (Vennerstrom et al. 2004). These compounds have longer-lasting activity than current artemisinin derivatives, suggesting that treatment courses of 3 days or less are feasible (see:http://www.mmv.org/pages/page_main.htm).

The final class of analogue of note are the quinoline-peroxide hybrids known as trioxaquines (23) (Fig. 12). These compounds have been designed in order to offset parasite resistance development and to aid parasitized erythrocyte penetration (Dechy-Cabaret et al. 2000, 2001). It is proposed that these hybrids may have the capacity to hit the parasite by two different mechanisms (namely, free radical mediated damage and interference with FP crystallization and detoxication).

6
Summary

In this chapter we have attempted to describe the exciting new advances in our knowledge of the mechanism of action and parasite resistance of two of the most important groups of semisynthetic antimalarials. Clearly we have come a long way since the days of traditional herbal remedies but we are up against a sophisticated foe and we must not allow the drug resistant parasites to gain the upper hand. With that in mind we have presented some of the most exciting recent work on the design and synthesis of analogues of both quinoline and artemisinin/simplified cyclic peroxides. New design strategies encompassing hybrid drugs and identification of chemically and metabolically stable artemisinin derivatives have been explored. These efforts have produced promising candidates, some of which are undergoing preclinical evaluation at present.

References

Abdin MZ, Israr M, Rehman RU, Jain SK (2003) Artemisinin, a novel antimalarial drug: Biochemical and molecular approaches for enhanced production. Planta Medica 69:289–299

Avery MA, Alvim-Gaston M, Vroman JA, Wu B, Ager A, Peters W, Robinson BL, Charman W (2002) Structure-activity relationships of the antimalarial agent artemisinin. 7. Direct modification of (+)-artemisinin and in vivo antimalarial screening of new, potential preclinical antimalarial candidates. J Med Chem 45:4321–4335

Avery MA, Gao F, Mehrotra S, Chong WKM, Milhous WK (1994) Structure-activity-relationships of antimalarial agent artemisinin by comparative molecular-field analysis. Abstracts of Papers of the American Chemical Society 208:43-MEDI

Avery MA, Jenningswhite C, Chong WKM (1989) Synthesis of a C,D-ring fragment of artemisinin. J Org Chem 54:1789–1792

Bachi MD, Korshin EE, Hoos R, Szpilman AM, Ploypradith P, Xie SJ, Shapiro TA, Posner GH (2003) A short synthesis and biological evaluation of potent and nontoxic antimalarial bridged bicyclic beta-sulfonyl- endoperoxides. J Med Chem 46:2516–2533

Banerjee R, Francis SE, Goldberg DE (2003) Food vacuole plasmepsins are processed at a conserved site by an acidic convertase activity in *Plasmodium falciparum*. Mol Biochem Parasitol 129:157–165

Barlin GB, Ireland SJ, Nguyen TMT, Kotecka B, Rieckmann KH (1993) Potential Antimalarials .18. Some mono-mannich and di-mannich bases of 3–7-chloro(and trifluoromethyl)quinolin-4-ylamino phenol. Austra J Chem 46:1685–1693

Barradell LB and Fitton A (1995) Artesunate—a review of its pharmacology and therapeutic efficacy in the treatment of malaria. Drugs 50:714–741

Bendrat K, Berger BJ, Cerami A (1995) Heme polymerization in malaria. Nature 378:138–138

Berger BJ, Bendrat K, Cerami A (1995) High-performance liquid-chromatographic analysis of biological and chemical heme polymerization. Anal Biochem 231:151–156

Berman PA, Adams PA (1997) Artemisinin enhances heme-catalysed oxidation of lipid membranes. Free Rad Biol Med 22:1283–1288

Bhisutthibhan J, Meshnick SR (2001) Immunoprecipitation of H-3 dihydroartemisinin translationally controlled tumor protein (TCTP) adducts from *Plasmodium falciparum*-infected erythrocytes by using anti-TCTP antibodies. Antimicrobial Agents Chemother 45:2397–2399

Bhisutthibhan J, Pan XQ, Hossler PA, Walker DJ, Yowell CA, Carlton J, Dame JB, Meshnick SR (1998) The *Plasmodium falciparum* translationally controlled tumor protein homolog and its reaction with the antimalarial drug artemisinin. J Biol Chem 273:16192–16198

Borstnik K, Paik IH, Shapiro TA, Posner GH (2002) Antimalarial chemotherapeutic peroxides: artemisinin, yingzhaosu A and related compounds. Int J Parasitol 32:1661–1667

Boss C, Richard-Bildstein S, Weller T, Fischli W, Meyer S, Binkert C (2003) Inhibitors of the *Plasmodium falciparum* parasite aspartic protease plasmepsin II as potential antimalarial agents. Curr Med Chem 10:883–907

Bray PG, Janneh O, Raynes KJ, Mungthin M, Ginsburg H, Ward SA (1999) Cellular uptake of chloroquine is dependent on binding to ferriprotoporphyrin IX and is independent of NHE activity in *Plasmodium falciparum*. J Cell Biol 145:363–376

Bray PG, Mungthin M, Ridley RG, Ward SA (1998) Access to hematin: The basis of chloroquine resistance. Mol Pharmacol 54:170–179

Bray PG, Ward SA (1998) A comparison of the phenomenology and genetics of multidrug resistance in cancer cells and quinoline resistance in *Plasmodium falciparum*. Pharmacol Therapeutics 77:1–28

Brewer TG, Grate SJ, Peggins JO, Weina PJ, Petras JM, Levine BS, Heiffer MH, Schuster BG (1994) Fatal neurotoxicity of arteether and artemether. Am J Trop Med Hygiene 51:251–259

Buller R, Peterson ML, Almarsson O, Leiserowitz L (2002) Quinoline binding site on malaria pigment crystal: A rational pathway for antimalaria drug design. Crystal Growth Design 2:553–562

Butler AR, Gilbert BC, Hulme P, Irvine LR, Renton L, Whitwood AC (1998) EPR evidence for the involvement of free radicals in the iron- catalysed decomposition of qinghaosu (Artemisinin) and some derivatives; antimalarial action of some polycyclic endoperoxides. Free Rad Res 28:471–476

Butler AR, Wu YL (1992) Artemisinin (Qinghaosu) – a new type of antimalarial drug. Chem Soc Rev 21:85–90

Cazelles J, Robert A, Meunier B (2001) Alkylation of heme by artemisinin, an antimalarial drug. Comptes Rendus De L Academie Des Sciences Serie Ii Fascicule C-Chimie 4:85–89

Chong CR, Sullivan DJ (2003) Inhibition of heme crystal growth by antimalarials and other compounds: implications for drug discovery. Biochem Pharmacol 66:2201–2212

De DY, Byers LD, Krogstad DJ (1997) Antimalarials: synthesis of 4-aminoquinolines that circumvent drug resistance in malaria parasites. J Heterocyclic Chem 34:315–320

Dechy-Cabaret O, Benoit-Vical F, Robert A, Meunier B (2000) Preparation and antimalarial activities of 'trioxaquines', new modular molecules with a trioxane skeleton linked to a 4- aminoquinoline. Chembiochem 1:281–283

Dechy-Cabaret O, Benoit-Vical F, Robert A, Meunier B (2001) Preparation and antimalarial activities of 4 trioxaquines, new modular molecules with a trioxane skeleton linked to a 4-aminoquinoline. Actualite Chimique 9–11

Delarue S, Girault S, Maes L, Debreu-Fontaine MA, Labaeid M, Grellier P, Sergher-aert C (2001) Synthesis and in vitro and in vivo antimalarial activity of new 4-anilinoquinolines. J Med Chem 44:2827–2833

Diribe CO and Warhurst DC (1985) A study of the uptake of chloroquine in malaria-infected erythrocytes—high and low affinity uptake and the influence of glucose and its analogs. Biochem Pharmacol 34:3019–3027

Dorn A, Vippagunta SR, Matile H, Bubendorf A, Vennerstrom JL, Ridley RG (1998a) A comparison and analysis of several ways to promote haematin (haem) polymeri-sation and an assessment of its initiation in vitro. Biochem Pharmacol 55:737–747

Dorn A, Vippagunta SR, Matile H, Jaquet C, Vennerstrom JL and Ridley RG (1998b) An assessment of drug-haematin binding as a mechanism for inhibition of haematin polymerisation by quinoline antimalarials. Biochem Pharmacol 55:727–736

Dua M, Raphael P, Sijwali PS, Rosenthal PJ, Chishti AH, Hanspal M (2000) Falcipain-2: Molecular cloning and characterization of *Plasmodium falciparum* cysteine protease that cleaves erythrocyte ankyrin and protein 4.1. Blood 96:946

Eckstein-Ludwig U, Webb RJ, van Goethem IDA, East JM, Lee AG, Kimura M, O'Neill PM, Bray PG, Ward SA, Krishna S (2003) Artemisinins target the SERCA of *Plasmodium falciparum*. Nature 424:957–961

Egan TJ (2002) Physico-chemical aspects of hemozoin (malaria pigment) structure and formation. J Inorg Biochem 91:19–26

Egan TJ, Mavuso WW, Ncokazi KK (2001) The mechanism of beta-hematin formation in acetate solution, parallels between hemozoin formation and biomineralization processes. Biochemistry 40:204–213

Famin O, Ginsburg H (2002) Differential effects of 4-aminoquinoline-containing an-timalarial drugs on hemoglobin digestion in *Plasmodium falciparum*-infected erythrocytes. Biochem Pharmacol 63:393–398

Fidock DA, Nomura T, Talley AK, Cooper RA, Dzekunov SM, Ferdig MT, Ursos LMB, Sidhu ABS, Naude B, Deitsch KW, Su XZ, Wootton JC, Roepe PD, Wellems TE (2000) Mutations in the P-falciparum digestive vacuole transmembrane protein PfCRT and evidence for their role in chloroquine resistance. Molecular Cell 6:861–871

Fitch CD (1969) A Mechanism for Chloroquine Resistance in Malaria. J Lab Clin Med 74:872–&

Fitch CD (1970) *Plasmodium falciparum* in owl monkeys—drug resistance and chloro-quine binding capacity. Science 169:289–&

Fitch CD, Cai GZ, Chen YF, Shoemaker JD (1999) Involvement of lipids in ferriproto-porphyrin IX polymerization in malaria. Biochimica Biophysica Acta—Mol Basis Dis 1454:31–37

Fitch CD, Cai GZ, Shoemaker JD (2000) A role for linoleic acid in erythrocytes infected with plasmodium berghei. Biochimica Biophysica Acta—Mol Basis Dis 1535:45–49

Fitch CD, Chevli R, Gonzalez Y (1974) Chloroquine accumulation by erythrocytes—latent capability. Life Sci 14:2441–2446

Fitch CD, Kanjananggulpan P (1987) The state of ferriprotoporphyrin-IX in malaria pigment. J Biol Chem 262:15552–15555

Foley M, Tilley L (1998) Quinoline antimalarials: Mechanisms of action and resistance and prospects for new agents. Pharmacol Ther 79:55–87

Goldberg DE, Slater AFG (1992) The pathway of hemoglobin degradation in malaria parasites. Parasitol Today 8:280–283

Grace JM, Aguilar AJ, Trotman KM, Peggins JO, Brewer TG (1998) Metabolism of beta-arteether to dihydroqinghaosu by human liver microsomes and recombinant cytochrome P450 (vol 26, pg 313, 1998). Drug Metab Disposition 26:704–704

Hawley SR, Bray PG, Mungthin M, Atkinson JD, O'Neill PM, Ward SA (1998) Relationship between antimalarial drug activity, accumulation, and inhibition of heme polymerization in *Plasmodium falciparum* in vitro. Antimicrobial Agents Chemother 42:682–686

Hawley SR, Bray PG, Oneill PM, Park BK, Ward SA (1996) The role of drug accumulation in 4-aminoquinoline antimalarial potency—The influence of structural substitution and physicochemical properties. Biochem Pharmacol 52:723–733

Haynes RK (2001) Artemisinin and derivatives: the future for malaria treatment? Curr Opin Infect Dis 14:719–726

Haynes RK, Chan HW, Cheung MK, Chung ST, La WL, Tsang HW, Voerste A, Williams ID (2003a) Stereoselective preparation of 10 alpha- and 10 beta-aryl derivatives of dihydroartemisinin. Eur J Org Chem :2098–2114

Haynes RK, Chan HW, Cheung MK, Chung ST and Tsang HW (2000) C-10 Halogen, amino and carbon substituted derivatives of artemisinin for treatment of malaria, coccidiosis and neosporosis, in PCT Int. App. WO 2000004024 A1

Haynes RK, Monti D, Taramelli D, Basilico N, Parapini S, Olliaro P (2003b) Artemisinin and heme—Authors' reply. Antimicrob Agents Chemother 47:2712–2713

Haynes RK, Monti D, Taramelli D, Basilico N, Parapini S, Olliaro P (2003c) Artemisinin antimalarials do not inhibit hemozoin formation. Antimicrob Agents Chemother 47:1175–1175

Haynes RK, Pai HHO and Voerste A (1999) Ring opening of artemisinin (qinghaosu) and dihydroartemisinin and interception of the open hydroperoxides with formation of N-oxides—A chemical model for antimalarial mode of action. Tetrahedron Lett 40:4715–4718

Haynes RK, Vonwiller SC (1996a) The behaviour of qinghaosu (artemisinin) in the presence of heme Iron(II) and (III). Tetrahedron Lett 37:253–256

Haynes RK, Vonwiller SC (1996b) The behaviour of qinghaosu (artemisinin) in the presence of non-heme Iron(II) and (III). Tetrahedron Lett 37:257–260

Hempelmann E, Egan TJ (2002) Pigment biocrystallization in *Plasmodium falciparum*. Trends Parasitol 18:11–11

Hempelmann E, Motta C, Hughes R, Ward SA, Bray PG (2003) *Plasmodium falciparum*: sacrificing membrane to grow crystals? Trends Parasitol 19:23–26

Hien TT, Turner GDH, Mai NTH, Phu NH, Bethell D, Blakemore WF, Cavanagh JB, Dayan A, Medana I, Weller RO, Day NPJ, White NJ (2003) Neuropathological assessment of artemether-treated severe malaria. Lancet 362:295–296

Hindley S, Ward SA, Storr RC, Searle NL, Bray PG, Park BK, Davies J, O'Neill PM (2002) Mechanism-based design of parasite-targeted artemisinin derivatives: Synthesis and antimalarial activity of new diamine containing analogues. J Med Chem 45:1052–1063

Hofheinz W, Burgin H, Gocke E, Jaquet C, Masciadri R, Schmid G, Stohler H, Urwyler H (1994) Ro-42-1611 (Arteflene), a new effective antimalarial—chemical-structure and biological-activity. Trop Med Parasitol 45:261–265

Homewood CA, Warhurst DC, Baggaley VC, Peters W (1972) Lysosomes, pH and anti-malarial action of chloroquine. Nature 235:50–

Hong YL, Yang YZ, Meshnick SR (1994) The interaction of artemisinin with malarial hemozoin. Mol Biochem Parasitol 63:121–128

Idowu OR, Lin AJ, Grace JM, Peggins JO (1997) Biomimetic metabolism of artelinic acid by chemical cytochrome P-450 model systems. Pharmaceutical Res 14:1449–1454

Jefford CW (2001) Why artemisinin and certain synthetic peroxides are potent antimalarials. Implications for the mode of action. Curr Med Chem 8:1803–1826

Jefford CW, Kohmoto S, Jaggi D, Timari G, Rossier JC, Rudaz M, Barbuzzi O, Gerard D, Burger U, Kamalaprija P, Mareda J, Bernardinelli G, Manzanares I, Canfield CJ, Fleck SL, Robinson BL, Peters W (1995) Synthesis, structure, and antimalarial activity of some enantiomerically pure, cis-fused cyclopenteno-1,2,4-trioxanes. Helvetica Chimica Acta 78:647–662

Jefford CW, Vicente MGH, Jacquier Y, Favarger F, Mareda J, MillassonSchmidt P, Brunner G, Burger U (1996) The deoxygenation and isomerization of artemisinin and artemether and their relevance to antimalarial action. Helvetica Chimica Acta 79:1475–1487

Jewell H, Maggs JL, Harrison AC, O'Neill PM, Ruscoe JE, Park BK (1995) Role of hepatic-metabolism in the bioactivation and detoxication of amodiaquine. Xenobiotica 25:199–217

Jung M, Lee K, Jung H (2001) First synthesis of (+)-deoxoartemisitene and its novel C-11 derivatives. Tetrahedron Lett 42:3997–4000

Jung M, Lee S (1998) A concise synthesis of novel aromatic analogs of artemisinin. Heterocycles 48:2219–2219

Kapetanaki S, Varotsis G (2000) Ferryl-oxo heme intermediate in the antimalarial mode of action of artemisinin. FEBS Lett 474:238–241

Kapetanaki S, Varotsis C (2001) Fourier transform infrared investigation of non-heme Fe(III) and Fe(II) decomposition of artemisinin and of a simplified trioxane alcohol. J Med Chem 44:3150–3156

Kaschula CH, Egan TJ, Hunter R, Basilico N, Parapini S, Taramelli D, Pasini E, Monti D (2002) Structure-activity relationships in 4-aminoquinoline antiplasmodials. The role of the group at the 7-position. J Med Chem 45:3531–3539

Kim HS, Begum E, Ogura N, Wataya Y, Nonami Y, Ito T, Masuyama A, Nojima M, McCullough KJ (2003) Antimalarial activity of novel 1,2,5,6-tetraoxacycloalkanes and 1,2,5-trioxacycloalkanes. J Med Chem 46:1957–1961

Klayman DL (1985) Qinghaosu (Artemisinin)—an antimalarial drug from China. Science 228:1049–1055

Klayman DL, Lin AJ, Acton N, Scovill JP, Hoch JM, Milhous WK, Theoharides AD (1984) Isolation of Artemisinin Qinghaosu) from *Artemisia annua* growing in the United States. J Natural Prod 47:715–717

Korshin EE, Hoos R, Szpilman AM, Konstantinovski L, Posner GH, Bachi MD (2002) An efficient synthesis of bridged-bicyclic peroxides structurally related to antimalarial yingzhaosu A based on radical co-oxygenation of thiols and monoterpenes. Tetrahedron 58:2449–2469

Kotecka BM, Barlin GB, Edstein MD, Rieckmann KH (1997) New quinoline di-Mannich base compounds with greater antimalarial activity than chloroquine, amodiaquine, or pyronaridine. Antimicrob Agents Chemother 41:1369–1374

Leed A, DuBay K, Ursos LMB, Sears D, de Dios AC, Roepe PD (2002) Solution structures of antimalarial drug-heme complexes. Biochemistry 41:10245–10255

Lin AJ, Lee M, Klayman DL (1989) Antimalarial activity of new water-soluble dihydroartemisinin derivatives. 2. Stereospecificity of the ether side-chain. J Med Chem 32:1249–1252

Liu CZ, Guo C, Wang YC, Fan OY (2003) Factors influencing artemisinin production from shoot cultures of Artemisia annua L. World J Microbiol Biotechnol 19:535–538

Ma JY, Katz E, Ziffer H (1999) A new synthetic route to 10 beta-alkyldeoxoartemisinins. Tetrahedron Lett 40:8543–8545

Maggs JL, Madden S, Bishop LP, Oneill PM, Park BK (1997) The rat biliary metabolites of dihydroartemisinin, an antimalarial endoperoxide. Drug Metab Disposition 25:1200–1204

Martin SK, Oduola AMJ, Milhous WK (1987) Reversal of chloroquine resistance in *Plasmodium falciparum* by Verapamil. Science 235:899–901

McKeage K, Scott LJ (2003) Atovaquone/Proguanil—a review of its use for the prophylaxis of *Plasmodium falciparum* malaria. Drugs 63:597–623

Meshnick SR, Jefford CW, Posner GH, Avery MA, Peters W (1996) Second-generation antimalarial endoperoxides. Parasitol Today 12:79–82

Meshnick SR, Thomas A, Ranz A, Xu CM, Pan HZ (1991) Artemisinin (Qinghaosu) – the role of intracellular hemin in its mechanism of antimalarial action. Mol Biochem Parasitol 49:181–190

Moreau S, Perly B, Biguet J (1982) Interactions between Chloroquine and Ferriprotoporphyrine .9. Nuclear Magnetic-Resonance Study. Biochimie 64:1015–1025

Moreau S, Perly B, Chachaty C, Deleuze C (1985) A nuclear magnetic-resonance study of the interactions of antimalarial-drugs with porphyrins. Biochimica et Biophysica Acta 840:107–116

Mungthin M, Bray PG, Ridley RG, Ward SA (1998) Central role of hemoglobin degradation in mechanisms of action of 4-aminoquinolines, quinoline methanols, and phenanthrene methanols. Antimicrob Agents Chemother 42:2973–2977

Naisbitt DJ, Williams DP, O'Neill PM, Maggs JL, Willock DJ, Pirmohamed M, Park BK (1998) Metabolism-dependent neutrophil cytotoxicity of amodiaquine: A comparison with pyronaridine and related antimalarial drugs. Chem Res Toxicol 11:1586–1595

O'Neill PM, Bishop LPD, Searle NL, Maggs JL, Storr RC, Ward SA, Park BK, Mabbs F (2000) Biomimetic Fe(II)-mediated degradation of arteflene (Ro-42–1611). The first EPR spin-trapping evidence for the previously postulated secondary carbon-centered cyclohexyl radical. J Org Chem 65:1578–1582

O'Neill PM, Bray PG, Hawley SR, Ward SA, Park BK (1998) 4-aminoquinolines—past, present, and future: A chemical perspective. Pharmacol Therapeutics 77:29–58

O'Neill PM, Harrison AC, Storr RC, Hawley SR, Ward SA, Park BK (1994) The effect of fluorine substitution on the metabolism and antimalarial activity of amodiaquine. J Med Chem 37:1362–1370

O'Neill PM, Hawley SR, Storr RC, Ward SA, Park BK (1996) The effect of fluorine substitution on the antimalarial activity of tebuquine. Bioorg Med Chem Lett 6:391–392

O'Neill PM, Miller A, Bishop LPD, Hindley S, Maggs JL, Ward SA, Roberts SM, Scheinmann F, Stachulski AV, Posner GH, Park BK (2001a) Synthesis, antimalarial activity, biomimetic iron(II) chemistry, and in vivo metabolism of novel, potent C-10-phenoxy derivatives of dihydroartemisinin. J Med Chem 44:58–68

O'Neill PM, Mukhtar A, Stocks PA, Randle LE, Hindley S, Ward SA, Storr RC, Bickley JF, O'Neil IA, Maggs JL, Hughes RH, Winstanley PA, Bray PG, Park BK (2003) Isoquine and related amodiaquine analogues: A new generation of improved 4-aminoquinoline antimalarials. J Med Chem 46:4933–4945

O'Neill PM, Posner GH (2004) A medicinal chemistry perspective on artemisinin and related endoperoxides. J Med Chem 47:2945–2964

O'Neill PM, Pugh M, Stachulski AV, Ward SA, Davies J, Park BK (2001b) Optimisation of the allylsilane approach to C-10 deoxo carba analogues of dihydroartemisinin: synthesis and in vitro antimalarial activity of new, metabolically stable C-10 analogues. J Chem Soc-Perkin Transact 1:2682–2689

O'Neill PM, Searle NL, Kan KW, Storr RC, Maggs JL, Ward SA, Raynes K, Park BK (1999) Novel, potent, semisynthetic antimalarial carba analogues of the first-generation 1,2,4-trioxane artemether. J Med Chem 42:5487–5493

O'Neill PM, Willock DJ, Hawley SR, Bray PG, Storr RC, Ward SA, Park BK (1997) Synthesis, antimalarial activity, and molecular modeling of tebuquine analogues. J Med Chem 40:437–448

Olliaro PL, Haynes RK, Meunier B, Yuthavong Y (2001a) Possible modes of action of the artemisinin-type compounds. Trends Parasitol 17:122–126

Olliaro PL, Haynes RK, Meunier RK, Meunier B, Yuthavong Y (2001b) Radical mechanism of action of the artemisinin-type compounds—Response. Trends Parasitol 17:267–268

Pagola S, Stephens PW, Bohle DS, Kosar AD, Madsen SK (2000) The structure of malaria pigment beta-haematin. Nature 404:307–310

Pandey AV, Tekwani BL, Singh RL, Chauhan VS (1999) Artemisinin, an endoperoxide antimalarial, disrupts the hemoglobin catabolism and heme detoxification systems in malarial parasite. J Biol Chem 274:19383–19388

Peters W, Robinson BL, Rossier JC, Misra D, Jefford CW (1993a) The chemotherapy of rodent malaria .49. The activities of some synthetic 1,2,4-trioxanes against chloroquine-sensitive and chloroquine-resistant parasites .2. Structure-activity studies on cis-fused cyclopenteno-1,2,4-trioxanes (fenozans) against drug-sensitive and drug-resistant lines of *Plasmodium berghei* and *P. Yoelii* Ssp Ns in vivo. Annals Trop Med Parasitol 87:221–221

Peters W, Robinson BL, Tovey G, Rossier JC, Jefford CW (1993b) The chemotherapy of rodent malaria .50. The activities of some synthetic 1,2,4-trioxanes against chloroquine-sensitive and chloroquine-resistant parasites .3. Observations on fenozan- 50f, a difluorinated 3,3'-spirocyclopentane 1,2,4-trioxane. Annals Trop Med Parasitol 87:111–123

Posner GH, Cumming JN, Ploypradith P, Chang HO (1995) Evidence for Fe(Iv)=O in the Molecular Mechanism of Action of the Trioxane Antimalarial Artemisinin. J Am Chem Soc 117:5885–5886

Posner GH, Cumming JN, Woo SH, Ploypradith P, Xie SJ, Shapiro TA (1998) Orally active antimalarial 3-substituted trioxanes: New synthetic methodology and biological evaluation. J Med Chem 41:940–951

Posner GH, Meshnick SR (2001) Radical mechanism of action of the artemisinin-type compounds. Trends Parasitol 17:266–267

Posner GH, O'Neill PM (2004) Knowledge of the proposed chemical mechanism of action and cytochrome P450 metabolism of antimalarial trioxanes like artemisinin allows rational design of new antimalarial. Acc Chem Res 37:397–404

Posner GH, Oh CH (1992) A regiospecifically O-18 labeled 1,2,4-trioxane—a simple chemical-model system to probe the mechanism(s) for the antimalarial activity of artemisinin (Qinghaosu). J Am Chem Soc 114:8328–8329

Posner GH, Oh CH, Wang DS, Gerena L, Milhous WK, Meshnick SR, Asawama-hasadka W (1994) Mechanism-based design, synthesis, and in vitro antimalarial testing of new 4-methylated trioxanes structurally related to artemisinin—the importance of a carbon-centered radical for antimalarial activity. J Med Chem 37:1256–1258

Posner GH, Paik IH, Sur S, McRiner AJ, Borstnik K, Xie SJ, Shapiro TA (2003) Orally active, antimalarial, anticancer, artemisinin-derived trioxane dimers with high stability and efficacy. J Med Chem 46:1060–1065

Posner GH, Park SB, Gonzalez L, Wang DS, Cumming JN, Klinedinst D, Shapiro TA, Bachi MD (1996) Evidence for the importance of high-valent Fe=O and of a dike-tone in the molecular mechanism of action of antimalarial trioxane analogs of artemisinin. J Am Chem Soc 118:3537–3538

Posner GH, Parker MH, Northrop J, Elias JS, Ploypradith P, Xie SJ, Shapiro TA (1999) Orally active, hydrolytically stable, semisynthetic, antimalarial trioxanes in the artemisinin family. J Med Chem 42:300–304

Pu YM, Ziffer H (1995) Synthesis and antimalarial activities of 12-beta-allyldeoxo-artemisinin and its derivatives. J Med Chem 38:613–616

Raynes KJ, Stocks PA, O'Neill PM, Park BK, Ward SA (1999) New 4-aminoquinoline mannich base antimalarials. 1. Effect of an alkyl substituent in the 5 '-position of the 4'-hydroxyanilino side chain. J Med Chem 42:2747–2751

Reed MB, Saliba KJ, Caruana SR, Kirk K, Cowman AF (2000) Pgh1 modulates sensi-tivity and resistance to multiple antimalarials in *Plasmodium falciparum*. Nature 403:906–909

Ridley RG, Hofheinz W, Matile H, Jaquet C, Dorn A, Masciadri R, Jolidon S, Richter WF, Guenzi A, Girometta MA, Urwyler H, Huber W, Thaithong S, Peters W (1996) 4-aminoquinoline analogs of chloroquine with shortened side chains retain ac-tivity against chloroquine-resistant *Plasmodium falciparum*. Antimicrob Agents Chemother 40:1846–1854

Robert A, Boularan M, Meunier B (1997) Interaction of artemisinin (qinghaosu) with the tetraphenylporphyrinatomanganese(II) complex. Comptes Rendus De L Academie Des Sciences Serie Ii Fascicule B- Mecanique Physique Chimie As-tronomie 324:59–66

Robert A, Cazelles J, Meunier B (2001) Characterization of the alkylation product of heme by the antimalarial drug artemisinin. Angewandte Chemie-International Edition 40:1954–1957

Robert A, Dechy-Cabaret O, Cazelles J, Meunier B (2002) From mechanistic studies on artemisinin derivatives to new modular antimalarial drugs. Acc Chem Res 35:167–174

Robert A, Meunier B (1998) Is alkylation the main mechanism of action of the anti-malarial drug artemisinin? Chem Soc Rev 27:273–279

Ruscoe JE, Tingle MD, O'Neill PM, Ward SA, Park BK (1998) Effect of disposition of Mannich antimalarial agents on their pharmacology and toxicology. Antimicrob Agents Chemother 42:2410–2416

Sidhu ABS, Verdier-Pinard D, Fidock DA (2002) Chloroquine resistance in *Plasmodium falciparum* malaria parasites conferred by PfCRT mutations. Science 298:210–213

Silamut K, Newton PN, Teja-Isavadharm P, Suputtamongkol Y, Siriyanonda D, Rasameesoraj M, Pukrittayakamee S, White NJ (2003) Artemether bioavailability after oral or intramuscular administration in uncomplicated falciparum malaria. Antimicrob Agents Chemother 47:3795–3798

Slater AFG (1992) Malaria Pigment. Exp Parasitol 74:362–365

Slater AFG, Cerami A (1992) Inhibition by chloroquine of a novel heme polymerase enzyme- activity in malaria trophozoites. Nature 355:167–169

Slater AFG, Swiggard WJ, Orton BR, Flitter WD, Goldberg DE, Cerami A, Henderson GB (1991) An iron carboxylate bond links the heme units of malaria pigment. Proc Natl Acad Sci USA 88:325–329

Stocks PA, Raynes KJ, Bray PG, Park BK, O'Neill PM, Ward SA (2002) Novel short chain chloroquine analogues retain activity against chloroquine resistant K1 *Plasmodium falciparum*. J Med Chem 45:4975–4983

Sullivan DJ (2002) Theories on malarial pigment formation and quinoline action. Int J Parasitol 32:1645–1653

Tang YQ, Dong YX, Vennerstrom JL (2004) Synthetic peroxides as antimalarials. Med Res Rev 24:425–448

Teja-Isavadharm P, Watt G, Eamsila C, Jongsakul K, Li QG, Keeratithakul D, Sirisopana N, Luesutthiviboon L, Brewer TG, Kyle DE (2001) Comparative pharmacokinetics and effect kinetics of orally administered artesunate in healthy volunteers and patients with uncomplicated falciparum malaria. Am J Trop Med Hyg 65:717–721

Tingle MD, Jewell H, Maggs JL, Oneill PM, Park BK (1995) The bioactivation of amodiaquine by human polymorphonuclear leukocytes in vitro—chemical mechanisms and the effects of fluorine substitution. Biochem Pharmacol 50:1113–1119

Ursos LMB, Roepe PD (2002) Chloroquine resistance in the malarial parasite, *Plasmodium falciparum*. Med Res Rev 22:465–491

Vattanaviboon P, Siritanaratkul N, Ketpirune J, Wilairat P, Yuthavong Y (2002) Membrane heme as a host factor in reducing effectiveness of dihydroartemisinin. Biochem Pharmacol 64:91–98

Vennerstrom JL, Ager AL, Dorn A, Andersen SL, Gerena L, Ridley RG, Milhous WK (1998) Bisquinolines. 2. Antimalarial N,N-bis(7-chloroquinolin-4- yl)heteroalkanediamines. J Med Chem 41:4360–4364

Vennerstrom JL, Arbe-Barnes S, Brun R, Charman SA, Chiu FCK, Chollet J, Dong YX, Dorn A, Hunziger D, Matile H, McIntosh K, Padmanilayam M, Thomas JS, Scheurer C, Scorneaux B, Tang YQ, Urwyler H, Wittlin S, Charman WN (2004) Identification of an antimalarial synthetic trioxolane drug development candidate. Nature 430(7002):900–904

Vennerstrom JL, Dong YX, Andersen SL, Ager AL, Fu HN, Miller RE, Wesche DL, Kyle DE, Gerena L, Walters SM, Wood JK, Edwards G, Holme AD, McLean WG, Milhous WK (2000) Synthesis and antimalarial activity of sixteen dispiro-1,2,4,5-tetraoxanes: Alkyl-substituted 7,8,15,16- tetraoxadispiro 5.2.5.2 hexadecanes. J Med Chem 43:2753–2758

Vennerstrom JL, Dong YX, Chollet J, Matile H (2002) Spiro and dispiro 1,2,4-trioxolane antimalarials. United States Patent, Medicines for Malaria Venture (MMV), US 6,486,199 B1

Verdier F, Clavier F, Deloron P, Blayo MC (1984) Distribution of chloroquine and desethyl-chloroquine in blood, plasma and erythrocytes of healthy and malarial subjects—HPLC assay. Pathologie Biologie 32:359–361

Verdier F, Lebras J, Clavier F, Hatin I, Blayo MC (1985) Chloroquine uptake by *Plasmodium falciparum* infected human erythrocytes during in vitro culture and its relationship to chloroquine resistance. Antimicrob Agents Chemother 27:561–564

Vippagunta SR, Dorn A, Matile H, Bhattacharjee AK, Karle JM, Ellis WY, Ridley RG, Vennerstrom JL (1999) Structural specificity of chloroquine-hematin binding related to inhibition of hematin polymerization and parasite growth. J Med Chem 42:4630–4639

Vippagunta SR, Dorn A, Ridley RG, Vennerstrom JL (2000) Characterization of chloroquine-hematin mu-oxo dimer binding by isothermal titration calorimetry. Biochimica Et Biophysica Acta—General Subjects 1475:133–140

Vroman JA, Alvim-Gaston M, Avery MA (1999) Current progress in the chemistry, medicinal chemistry and drug design of artemisinin based antimalarials. Curr Pharmaceutical Des 5:101–138

Warhurst D (2001) New developments: Chloroquine-resistance in *Plasmodium falciparum*. Drug Resistance Updates 4:141–144

Wernsdorfer WH, Noedl H (2003) Molecular markers for drug resistance in malaria: use in treatment, diagnosis and epidemiology. Curr Opin Infect Dis 16:553–558

Wilson CM, Serrano AE, Wasley A, Bogenschutz MP, Shankar AH Wirth DF (1989) Amplification of a gene related to mammalian Mdr genes in drug-resistant *Plasmodium falciparum*. Science 244:1184–1186

Winstanley PA, Ward SA, Snow RW (2002) Clinical status and implications of antimalarial drug resistance. Microbes Infect 4:157–164

Wu WM, Wu YK, Wu YL, Yao ZJ, Zhou CM, Li Y, Shan F (1998) Unified mechanistic framework for the Fe(II)-induced cleavage of qinghaosu and derivatives/analogues. The first spin-trapping evidence for the previously postulated secondary C-4 radical. J Am Chem Soc 120:3316–3325

Wu YK (2002) How might qinghaosu (artemisinin) and related compounds kill the intraerythrocytic malaria parasite? A chemists view. Acc Chem Res 35:255–259

Yang YZ, Asawamahasakda W, Meshnick SR (1993) Alkylation of human albumin by the antimalarial artemisinin. Biochem Pharmacol 46:336–339

Yang YZ, Little B, Meshnick SR (1994) Alkylation of proteins by artemisinin—effects of heme, pH, and drug structure. Biochem Pharmacol 48:569–573

Yayon A, Cabantchik ZI, Ginsburg H (1984) pH-Dependent sensitivity of human malaria parasites to chloroquine. J Protozool 31:A82–A83

Yayon A, Cabantchik ZI, Ginsburg H (1985) Susceptibility of human malaria parasites to chloroquine is pH dependent. Proc Natl Acad Sci USA 82:2784–2788

Yayon A, Ginsburg H (1983) Chloroquine inhibits the degradation of endocytic vesicles in human malaria parasites. Cell Biol Int Rep 7:895–895

Zhang F, Gosser DK, Meshnick SR (1992) Hemin-catalyzed decomposition of artemisinin (Qinghaosu). Biochem Pharmacol 43:1805–1809

Zhang JM, Krugliak M, Ginsburg H (1999) The fate of ferriprotorphyrin IX in malaria infected erythrocytes in conjunction with the mode of action of antimalarial drugs. Mol Biochem Parasitol 99:129–141

Zhou WS, Xu XX (1994) Total synthesis of the antimalarial sesquiterpene peroxide Qinghaosu and Yingzhaosu-A. Acc Chem Res 27:211–216

CTMI (2005) 295:39–53
© Springer-Verlag Berlin Heidelberg 2005

Antimalarial Multi-Drug Resistance in Asia: Mechanisms and Assessment

A.-C. Uhlemann · S. Krishna (✉)

Devision of Cellular and Molecular Medicine, Centre for Infection,
St. George's University of London, Cranmer Terrace, London SW17 0RE, UK
s.krishna@sgul.ac.uk

Abstract The emergence and spread of drug-resistant parasites poses a major problem for management of *Plasmodium falciparum* malaria in endemic areas. Nowhere is this more apparent than in southeast Asia, where multi-drug resistance to chloroquine and sulfadoxine–pyrimethamine was exacerbated when mefloquine monotherapy began failing in the 1980s. A better understanding of mechanisms of (multi-) drug resistance is urgently warranted to monitor and guide antimalarial chemotherapy regimens more efficiently. Here we review recent advances on identification of molecular markers that can be employed in predicting in vitro and in vivo resistance in southeast Asia. Examples include amplification of *PfMDR1* (*P. falciparum* multi-drug resistant gene 1) and mefloquine, K76T *PfCRT* and chloroquine, as well as mutations in the dihydroperoate synthase and dihydrofolate reductase genes and the antifolate class of drugs.

Abbreviations

As	Artesunate
CQ	Chloroquine
DHFR	Dihydrofolate reductase
DHPS	Dihydroperoate synthase
Mfq	Mefloquine
PfCRT	*P. falciparum* chloroquine resistance transporter gene
PfMDR1	*P. falciparum* multi-drug resistance gene 1
PfATP6	*P. falciparum* ATPase 6 gene
SNP	Single nucleotide polymorphism
SERCA	Sarcoplasmic-endoplasmic reticulum Ca^{2+} ATPase

1
Background

Malaria kills between 0.5 and 2.5 million individuals each year, despite decades of attempts to reduce this mortality. These include the use of impregnated bednets (Luxemberger et al. 1994), improvements in treatment by limiting the use of ineffective drugs (Attaran et al. 2004), using combination chemotherapy regimens (Kremsner and Krishna 2004), and developing new drugs. Using existing drugs more effectively [such as rectal artesunate (As) in the initial management of children with malaria unable to receive medical care immediately; Krishna et al. 2001] may also help to reduce malaria-attributable mortality. However, these advances become severely limited when drug resistance in parasites is established rapidly and disseminated widely. Drug resistance often offsets any improvements in mortality, and is now worsening global outcomes from malaria significantly (Trape et al. 1998; Zucker et al. 2003).

Chloroquine (CQ) resistance in *Plasmodium falciparum* was first noticed in the late 1950s in non-immune migrant workers in hyperendemic areas along the Thai–Cambodian border, and was followed by resistance in at least three further geographically separate regions (two in Africa and one in Brazil) (Wellems and Plowe 2001; Wootton et al. 2002; Joy et al. 2003). This marked the beginning of an era of rapid evolution from mono- to multi-drug resistant parasites. Within 20 years CQ resistance has spread to, or developed in, most of the world. These findings should have stopped use of the cheapest antimalarials (4-aminoquinolines) well before now, although they continue to be used inappropriately in many areas (Attaran et al. 2004).

Southeast Asia has continued to be the graveyard of many new antimalarials that have become ineffective (Chongsuphajaisiddhi and Sabchareon 1981; Nosten et al. 1991a, 1991b). In 1973, CQ was replaced by sulphadoxine-pyrimethamine (SP) as first-line treatment in Thailand and Cambodia. After only a few years of extensive use, failing efficacy necessitated increasing doses of SP, and then a change to quinine (combined eventually with tetracycline) as first-line treatment (White 1992).

Then followed the promising introduction of mefloquine (Mfq) as a single-dose treatment in 1980. Cure rates climbed once again to over 95%, but in the early 1990s the failure rate of Mfq mono-therapy had become approximately 50% at Myanmar and Cambodian borders of Thailand (WHO1997b). Similar patterns of resistance soon emerged in adjacent countries.

Since As has been added to Mfq in a 3-day combination regimen a high cure rate (>95%) has been restored, and maintained (Nosten et al. 2000). Other artemisinin-containing combinations therapies are being intensively investigated in areas of multidrug resistance, such as in Vietnam.

Dihydroartemisinin–piperaquine, for example, may be as effective as Mfq given with As, but with the added virtue of greater affordability, providing dosing is optimized (Tran et al. 2004).

The need to monitor antimalarial drug resistance in parasites is obvious from the speed with which it can become established. Several molecular markers of drug resistance have already been developed to monitor changes in drug susceptibilities. For example, discovery of point mutations in the targeted enzymes of parasites resistant to antifolate therapy has been successfully applied to assess in vitro sensitivity to these drugs (Plowe et al. 1995). It took several decades of meticulous work to identify the K76T mutation in the *P. falciparum* CQ resistance transporter gene (*PfCRT*) as being responsible for resistance to 4-aminoquinolines (Wellems 1991; Fidock et al. 2000; Djimde et al. 2001). The high pre-existing frequency of atovaquone resistance is also attributed to a single nucleotide change in the parasite's mitochondrial bc_1 complex (Srivastava et al. 1999). All of these single nucleotide polymorphisms (SNPs) detect parasites that are resistant to antifolates, aminoquinolines or atovaquone with high sensitivity when applied to cultured parasites, but their value in predicting treatment outcomes in vivo has not always been fully explored.

Here we discuss how recent molecular advances in diagnosis of multidrug resistant parasites may guide changes in our treatment and resistance monitoring strategies. Amplification of *P. falciparum* multi-drug resistant gene 1 (*PfMDR1*) has recently proved to be a powerful tool in predicting outcome after Mfq therapy (Price, Uhlemann et al. 2004). Furthermore, the discovery that artemisinins target *P. falciparum* ATPase 6 gene (PfATP6), a parasite calcium-ATPase similar to mammalian sarcoplasmic-endoplasmic reticulum Ca^{2+} ATPase (SERCAs; Eckstein-Ludwig et al. 2003), gives us the chance, for the first time, to monitor changes in this target prospectively and to relate these to shifts in susceptibility to artemisinins. Such monitoring is critical for what is currently our best class of antimalarial.

2
Molecular Aspects of Resistance to CQ and DHFR Inhibitors

CQ was the best treatment for malaria for decades. It was effective for uncomplicated as well as severe malaria (Krishna and White 1996). CQ can be given in three doses once a day by several routes, and is very cheap. But resistance makes it virtually useless in most places. In Southeast Asia CQ is now reserved to treat vivax, malariae or ovale malaria, although it is inappropriately used in many sub-Saharan regions. Initially, it was suspected that SNPs in *PfMDR1* were associated with CQ resistance (Foote et al. 1989; Adagu et al. 1996). However, recently resistance in vitro and in vivo is attributable to *PfCRT,* a putative transporter that modulates intraparasitic drug concentrations. Interestingly, mutations in *PfCRT* orthologues in *P. vivax* and *P. chabaudi* are not associated with the phenotype of CQ resistance (Nomura et al. 2001; Hunt et al. 2004).

Studies to identify *PfCRT* began with the creation of a genetic cross between a CQ-resistant and -sensitive isolate, and have continued using transfection to dissect the contribution of SNPs in *PfCRT* as well as in *PfMDR1* to the CQ-resistance phenotype. After decades of painstaking laboratory research, an SNP in *PfCRT* (encoding a K76T change) in a laboratory cross was found to segregate with resistance to CQ (Wellems et al. 1991; Fidock et al. 2000). This SNP was also observed in parasites from patients in Mali who failed treatment with CQ (Djimde et al. 2001).

Transfection experiments have confirmed K76T as the single most important amino acid exchange causing CQ resistance (Sidhu et al. 2002). Thus other SNPs in *PfCRT* as well as the N86Y change in *PfMDR1* may only serve as bystander mutations, or as modulators of the primary cause of CQ resistance. Indeed, *PfMDR1* N86Y has been a controversial molecular marker of CQ resistance for years (Foote et al. 1990; Adagu and Warhurst 1999; Reed et al. 2000; Djimde et al. 2001; Tinto et al. 2003), perhaps through co-selection with K76T in *PfCRT* (Babiker et al. 2001; Tinto et al. 2003). *PfMDR1* encodes a P-glycoprotein called Pgh-1 ('P' glycoprotein homologue-1), which belongs to the large and diverse family of ABC transporters (Cowman et al. 1991). Many orthologues in different organisms including mammals are associated with multi-drug resistance to an array of structurally unrelated substrates (Marzolini et al. 2004).

Current studies in Southeast Asia on the role of the N86Y mutation are taking place in the genetic context of parasites that are unlikely to be under drug pressure from CQ, as this has long been replaced as first-line treatment for *P. falciparum*. The main drug pressure on falciparum is now from to Mfq, which has an elimination half-time of about 2 weeks (Nosten et al. 2000; Price, Uhlemann et al. 2004). Despite this, the frequencies of K76T in Thailand, Cambodia and some parts of Laos are 100% (Anderson 2003; Berens et al. 2003; Price, Uhlemann et al. 2004). Higher frequencies of wild-type *PfCRT* at about 40% were only observed at the Burma–Bangladesh border and in Southern Laos and Vietnam (Anderson 2003).

The antifolates have failed more rapidly than CQ as antimalarials. Pyrimethamine and sulfadoxine target two enzymes in the folate synthesis pathway of *P. falciparum,* the dihydroperoate synthase (DHPS) and dihydrofolate reductase (DHFR), respectively. When given in combination they act synergistically by disrupting folate synthesis, which eventually kills the parasite. Resistance to this combination has also already evolved worldwide. Experiments in *P. chabaudi* suggested that amplification and rearrangement

Table 1 Important polymorphisms in drug resistance to antimalarials

Gene	Position	Wild type	Mutant
DHPS	436	Ser	Ala
	437	Ala	Gly
	540	Lys	Gln
	581	Ala	Gly
	613	Ala	Ser/Thr
DHFR	16	Ala	Val
	51	Asn	Ile
	59	Cys	Arg
	108	Ser	Asn/Thr
	164	Ile	Leu
PfCRT	76	Lys	Thr
PfMDR1	86	Asn	Tyr
	184	Tyr	Phe
	1034	Ser	Cys
	1042	Asn	Asp
	1246	Asp	Tyr

of the DHFR (TS) gene increases expression and might explain resistance to pyrimethamine (Cowman and Lew 1989, 1990). It is now clear that a few point mutations are sufficient to confer resistance directly (Table 1), S108N being the most important one (Yuvaniyama et al. 2003). The observed fitness advantage of mutant *dhfr* and *dhps* alleles has allowed for rapid selection of resistant parasites under continued drug pressure (see chapter by C. Plowe, this volume). Recent studies suggest a single origin of resistance *dhfr* alleles (Anderson 2003; Nair et al. 2003). The highest frequencies of up to four mutations were observed along the Thai–Burmese border. In contrast in some parts of Laos a significant proportion of *dhfr* (>35%) and *dhps* (>70%) alleles remained wild-type (Anderson 2003; Berens et al. 2003; Nair et al. 2003).

The more rapid emergence of resistance to antifolates compared with much slower emergence of resistance to 4-aminoquinolines in similar geographic areas suggests there may be differences in the molecular mechanisms of resistance. Antifolates select for mutations in their targets, whereas the use of CQ has selected for changes in a transporter protein, which is believed to modulate uptake of CQ. This latter mechanism may have arisen because of the nature of the target for CQ (the process of haem crystallization, see chapter by Scholl et al. and Bray et al., this volume).

3
Molecular Aspects of Resistance to Mfq and Artemisinins

Mfq belongs to the class of antimalarials that includes halofantrine and lume-
fantrine. It was introduced as first-line treatment after quinine, for uncomplicated malaria in Thailand. Amplification of the gene copy number for
PfMDR1 is associated with Mfq but not CQ resistance (Cowman et al. 1994;
Price et al. 1999). This amplification is associated with increased expression
of mRNA for *PfMDR1* when examined in laboratory-adapted isolates (Foote
et al. 1989; Wilson et al. 1989). However, technical difficulties in assessing
PfMDR1 copy number limited this assay in clinical studies to small sample
sizes, and required labour intensive techniques such as ratiometric PCR-based
assays (Price et al. 1997a). More extensive sequencing studies identified non-
synonymous point mutations causing changes in amino acid residues in five
positions (Table 1; Foote et al. 1990).

More recently, *PfMDR1* copy number has become recognized as the best
overall predictor of therapeutic failure after Mfq treatment given either alone
or in combination with As (Fig. 1) (Price, Uhlemann et al. 2004). Furthermore, increased copy number is strongly associated with high IC_{50} values
of structurally related antimalarials: quinine and halofantrine, as well as to
the unrelated sesquiterpene lactones As and dihydroartemisinin. Analysis of
SNPs revealed that point mutations in *PfMDR1* (nucleotide positions 86, 1034
and 1042) previously identified as being associated with Mfq resistance are
only surrogate markers of copy number (Fig. 2). The N1246Y SNP was not
detected in 147 samples and thus may not have relevance in Southeast Asia
(Berens et al. 2003; Ngo et al. 2003). A smaller study also used real-time PCR
(without multiplexing, i.e., without the experimental and reference genes assayed in the same reaction mixture), and observed that increased gene copy
number (three or more) was associated with more resistance to artemisinins
and Mfq (Pickard et al. 2003).

In Southeast Asia, under years of Mfq pressure there was probably a significant decrease in frequency of N86Y despite the very high frequency of the
K76T mutation in *PfCRT*, in effect dissociating the frequency (about 90%)
with which mutations in these two transporters are seen in the African context (Grobusch et al. 1998; Mawili-Mboumba et al. 2002). This observation
highlights the possibility of 'interacting' mutations in different drug transporters in conferring resistance to antimalarials. Such interactions are made
more plausible by the localization of *PfCRT* and *PfMDR1* to the parasite's food
vacuole in separate studies (Cowman et al. 1991; Fidock et al. 2000).

Artemisinin-based combination treatment regimens have been systematically introduced in Southeast Asia over the past decade (Price et al. 1997b;
Adjuik et al. 2004; Tran et al. 2004). They are highly efficacious, well tolerated,
and may provide theoretical protection against the emergence of resistance

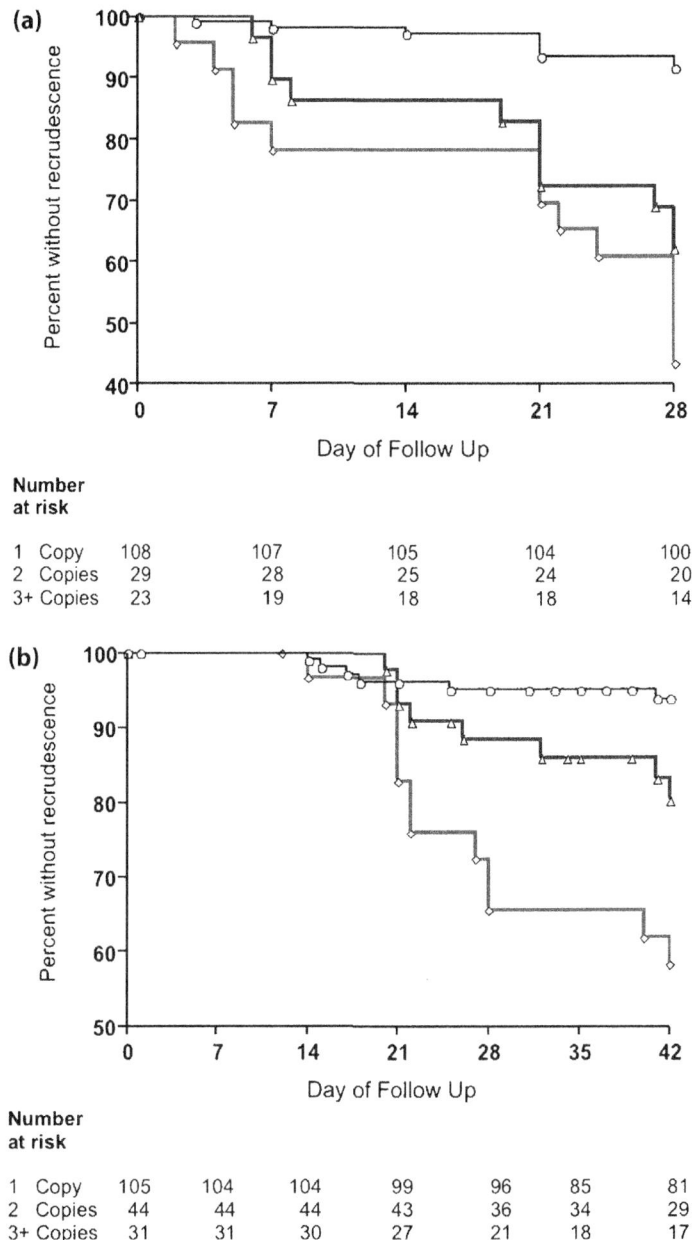

Fig. 1a,b Cumulative percentage of patients free from malaria after treatment with (a) mefloquine monotherapy or (b) mefloquine and 3 days' artesunate (1 copy in black, 2 copies in blue, 3 copies in red)

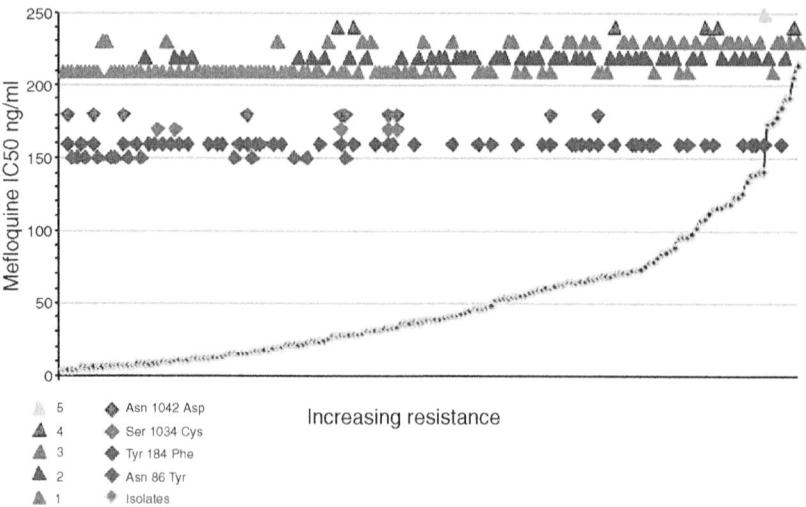

Fig. 2 Distribution of Mfq IC_{50} by copy number (*triangles*) and codon mutations (*diamonds*)

as well as reduced transmissibility by preventing gametocyte development. These features (combined with high treatment efficacy, the principal reason why they are used) may have led to a reduction in incidence of falciparum malaria in areas of Southeast Asia, where malaria transmission is relatively low, although it has not conferred similar benefit in areas of high transmission (Adjuik et al. 2004). The effects of artemisinin combinations on the incidence and on the emergence of drug resistance in areas of higher transmission requires further study.

A target of the artemisinin derivatives has recently been shown to be PfATP6, a Ca^{2+} ATPase. Full-length sequencing of isolates with relatively high As IC_{50} values identified an amino acid (N89Y) polymorphism in the M1–M2 region of the ATPase. This polymorphism was studied further in more than 100 Thai isolates encompassing a wide range (0.26–23.5 ng/ml) of IC_{50} values. There was no linkage between the presence of this polymorphism and higher IC_{50} values for As (Price, Uhlemann et al. 2004). This finding leaves *PfMDR1* copy number as the main modulator of sensitivity to artemisinins in vitro although it should be stressed that there is no clinical evidence of resistance to artemisinins.

The fact that sensitivities to both Mfq and As are modulated by a common molecular mechanism may seem worrying. Paradoxically, however, if *PfMDR1* expression increases the export of antimalarials such as artemisinin, then this could reduce the chances of selection for artemisinin-resistant variants of *pfATP6* by maintaining a low selection pressure on this target. At first sight, Mfq and As targets appear to be localized to different sides of a common membrane: Mfq may interfere with haemoglobin digestion in the food vacuole and As acts on PfATP6 which is (membranes) in the parasite cytosol. Nevertheless, many details are still missing about the membrane connections (both functional and direct) between SERCA-type structures in the parasite cytosol and the membrane components of the food vacuole. These observations clearly point to hypotheses that are amenable to experimental verification.

4
Molecular and In Vitro Techniques to Assess Resistance

Clinical studies are the best way to monitor drug resistance in a given area, because results are the most influential on drug policies. However, they are difficult to conduct, expensive and require follow-up periods of at least 28 days, or longer if long half-life drugs such as Mfq are used. They are also confounded by high rates of reinfection during longer follow-up periods. To complement findings from in vivo studies, in vitro sensitivity testing of parasites has evolved into an important tool despite results being difficult to interpret for some drugs (see chapter by C. Plowe, this volume).

For example, the serum concentrations of folate can affect interpretation of IC_{50} values to antifolates (Wang et al. 1997), and quinine IC_{50} values are influenced by protein binding (Silamut et al. 1985). IC_{50} values determined against fresh parasite isolates for artemisinins may vary by about 100-fold. In Thailand As IC_{50} values range between 0.18 and 23.47 ng/ml (130-fold) between isolates (Price, Uhlemann et al. 2004), perhaps reflecting decomposition of artemisinins by variable amounts of free Fe^{2+} and/or variable genotypes in culture conditions (Kamchonwongpaisan et al. 1994).

Further problems associated with on-site in vitro tests are the need for a relatively sophisticated infrastructure, the dependence on time of transport to the study unit and sufficiently high parasitaemia not to require amplification of numbers by short-term cultures. The schizont maturation test used to be the main test, but results can be influenced strongly by the experience of the microscopist. Furthermore, recent studies using test plates bought from the World Health Organization were inaccurate because plates had been incorrectly dosed (Schwenke et al. 2001; Borrmann et al. 2002). As an alternative, the isotope incorporation assay with tritiated hypoxanthine gives reliable and standardized results (Desjardins et al. 1979), but relies on radioactive material and cannot therefore always be used in field studies. New tests have been developed, for example, the ELISA-based (Histidine Rich Protein2; Hrp2), but these are costly limiting their routine use (Noedl et al. 2002). A colorimetric microtest using double-site enzyme-linked lactate dehydrogenase enzyme immunodetection appears easy to perform, fast and reliable (Moreno et al. 2001). Genotyping for point mutations in resistance-associated genes has now become an essential tool in surveillance of resistance. Assays use PCR and restriction enzyme digest (RFLP) and are relatively easy to perform (Adagu and Warhurst 1999). Only minimal amounts of blood are needed, which can be spotted onto filter paper, allowing for easy storage and transport. Even though RFLP methods are easy to perform they require several steps including nested PCR and gel electrophoresis. Allele-specific fluorescence primer strategies using real-time machines may appear more expensive, but they require less starting material and time making them potentially more useful.

5
Clinical Significance

In Thailand, *PfMDR1* copy number is the best overall predictor of treatment failure after Mfq (sensitivity 81%, specificity 71% and positive predictive value 46%) or Mfq given with As (sensitivity 77%, specificity 65% and positive predictive value 58%). In multivariable analysis, *PfMDR1* copy number was a better predictor of failure than several clinical predictors that had been identified in this population (such as age, parasitaemia, vomiting of drug and mixed infections) (Price, Uhlemann et al. 2004). Genetic analysis also suggests that when combination therapy with Mfq and As fails, this is due to resistance to the Mfq and not the artemisinin component. Increased copy number in field isolates is not seen with mutant *PfMDR1*-86 (N86Y), and only occasionally occurs with mutations at positions 1034 and 1042. The low prevalence of N86Y in *PfMDR1* makes it a reasonable marker for Mfq sensitivity in our study, but use of SNP alone would give an unacceptably low specificity for detecting Mfq resistance.

It may be some time before findings from studies such as these can be replicated in other geographical areas, or that such assays can be applied prospectively to guide treatment of patients at risk of failure, but their potential for increasing understanding of treatment failure is obvious.

In the African context, 76T is strongly associated with treatment outcome, but specificity may be compromised, probably mainly due to the contribution of host immunity to parasite clearance (Djimde et al. 2001). Since 76T is at fixation in many areas in Southeast Asia it is not a suitable molecular marker for CQ response as shown in two recent chemotherapy trials in Laos (Pillai et al. 2001; Berens et al. 2003).

Resistance to anti-malarials has clearly arisen independently in many tropical foci and has been spread by migration of populations (Wootton et al. 2002; Nair et al. 2003). In areas of Africa, where Mfq has not been used extensively resistance does not seem to have spread or widely occurred yet with the possible exception of some cases reporting resistance after quinine and Mfq use (Wichmann et al. 2003).

A novel approach to identify candidate resistance genes uses a genome-wide association screen. Parasites obtained from the field are systematically searched for genetic markers in statistical association with drug resistant phenotypes. Flanking neutral polymorphism will be observed in high frequency but with low variability in strong linkage disequilibrium to yet unknown target genes (Anderson 2004).

6
Conclusions

Recent advances in molecular biology have provided us with tools to study prospectively the emergence of drug-resistance to some of our most important antimalarial regimens. This should improve our understanding of how to monitor and perhaps to limit the spread of resistant parasites. It has already given us the tools to assess the best treatment options in areas where clinical drug resistance is not yet manifest. Two different molecular mechanisms of drug resistance have been discussed: SNPs in the actual drug target as for SP and atovaquone leading to a rapid evolvement of resistance; and mutations or amplification of uptake modulating transporters as for *PfCRT* and CQ or *PfMDR1* and Mfq/As. Furthermore, previous resistance to one drug may provide protective molecular mechanisms against the development of resistance against another unrelated compound.

Acknowledgements This work was funded by The Wellcome Trust (Grant 066201) and by the Medical Research Council of Great Britain (Grant G9800300). We would like to thank our colleagues in collaborative studies over many years, particularly Dr. Richard Price, and Prof. David Walliker and Dr. Tim Anderson for invaluable critical reading of the manuscript.

References

Adagu IS, Dias F, Pinheiro L, Rombo L, do Rosario V, Warhurst DC (1996) Guinea Bissau: association of chloroquine resistance of *Plasmodium falciparum* with the Tyr86 allele of the multiple drug-resistance gene PfMDR1. Trans R Soc Trop Med Hyg 90:90–91
Adagu IS, Warhurst DC (1999a) Allele-specific, nested, one tube PCR: application to PfMDR1 polymorphisms in *Plasmodium falciparum*. Parasitology 119:1–6

Adagu IS, Warhurst DC (1999b) Association of cg2 and PfMDR1 genotype with chloro-quine resistance in field samples of *Plasmodium falciparum* from Nigeria. Parasitology 119:343–348

Adjuik M, Babiker A, Garner P, Olliaro P, Taylor W, White N et al. (2004) Artesunate combinations for treatment of malaria: meta-analysis. Lancet 363: 9–17

Anderson TJ (2004) Mapping drug resistance genes in *Plasmodium falciparum* by genome-wide association. Curr Drug Targets Infect Disord 4:65–78

Anderson TJ, Nair S, Williams JT, Brockman A, Paiphun L, Newton PN, Mayxay M, Guthmann J-P, Simithuis FM, Hien TT, Van den Broek I, Nosten F (2003) Population structure revealed by selected and putatively neutral SNPs in S.E. Asian malaria parasites. American Society of Tropical Medicine and Hygiene, Atlanta

Attaran A, Barnes KI, Curtis C, d'Alessandro U, Fanello CI, Galinski MR, Kokwaro G, Looareesuwan S, Makanga M, Mutabingwa TK, Talisuna A, Trape JF, Watkins WM (2004) WHO, the Global Fund, and medical malpractice in malaria treatment. Lancet 363:237–240

Babiker HA, Pringle SJ, Abdel-Muhsin A, Mackinnon M, Hunt P, Walliker D (2001) High-level chloroquine resistance in Sudanese isolates of *Plasmodium falciparum* is associated with mutations in the chloroquine resistance transporter gene PfCRT and the multidrug resistance Gene PfMDR1. J Infect Dis 183:1535–1538

Berens N, Schwoebel B, Jordan S, Vanisaveth V, Phetsouvanh R, Christophel EM, Phompida S, Jelinek T (2003) *Plasmodium falciparum*: correlation of in vivo resistance to chloroquine and antifolates with genetic polymorphisms in isolates from the south of Lao PDR. Trop Med Int Health 8:775–782

Borrmann S, Binder RK, Adegnika AA, Missinou MA, Issifou S, Ramharter M, Wernsdorfer WH, Kremsner PG (2002) Reassessment of the resistance of *Plasmodium falciparum* to chloroquine in Gabon: implications for the validity of tests in vitro vs. in vivo. Trans R Soc Trop Med Hyg 96:660–663

Chongsuphajaisiddhi T, Sabchareon A (1981) Sulfadoxine-pyrimethamine resistant falciparum malaria in Thai children. Southeast Asian J Trop Med Public Health 12:418–421

Cowman AF, Galatis D, Thompson JK (1994) Selection for mefloquine resistance in *Plasmodium falciaprum* is linked to amplification of the *PfMDR1* gene and cross-resistance to halofantrine and quinine. Proc Natl Acad Sci USA 91:1143–1147

Cowman AF, Karcz S, Galatis D, Culvenor JG (1991) A P-glycoprotein homologue of *Plasmodium falciparum* is localized on the digestive vacuole. J Cell Biol 113:1033–1042

Cowman AF, Lew AM (1989) Antifolate drug selection results in duplication and rearrangement of chromosome 7 in *Plasmodium chabaudi*. Mol Cell Biol 9:5182–5188

Cowman AF, Lew AM (1990) Chromosomal rearrangements and point mutations in the DHFR-TS gene of *Plasmodium chabaudi* under antifolate selection. Mol Biochem Parasitol 42:21–29

Desjardins RE, Canfield CJ et al. (1979) Quantitative assessment of antimalarial activity in vitro by a semiautomated microdilution technique. Antimicrob Agents Chemother 16:710–718

Djimde A, Doumbo OK, Cortese JF, Kayentao K, Doumbo S, Diourte Y, Dicko A, Su XZ, Nomura T, Fidock DA, Wellems TE, Plowe CV Coulibaly D (2001) A molecular marker for chloroquine-resistant falciparum malaria. N Engl J Med 344:257–263

Eckstein-Ludwig U, Webb RJ, Van Goethem ID, East JM, Lee AG, Kimura M, O'Neill PM, Bray PG, Ward SA, Krishna S (2003) Artemisinins target the SERCA of *Plasmodium falciparum*. Nature 424:957–961

Fidock DA, Nomura T, Talley AK, Cooper RA, Dzekunov SM, Ferdig MT, Ursos LM, Sidhu AB, Naude B, Deitsch KW, Su XZ, Wootton JC, Roepe PD, Wellems TE (2000) Mutations in the *P. falciparum* digestive vacuole transmembrane protein PfCRT and evidence for their role in chloroquine resistance. Mol Cell 6:861–871

Foote SJ, Kyle DE, Martin RK, Oduola AM, Forsyth K, Kemp DJ, Cowman AF (1990) Several alleles of the multidrug-resistance gene are closely linked to chloroquine resistance in *Plasmodium falciparum*. Nature 345:255–258

Foote SJ, Thompson JK, Cowman AF, Kemp DJ (1989) Amplification of the multidrug resistance gene in some chloroquine-resistant isolates of *P. falciparum*. Cell 57:921–930

Grobusch MP, Adagu IS, Kremsner PG, Warhurst DC (1998) *Plasmodium falciparum*: in vitro chloroquine susceptibility and allele-specific PCR detection of PfMDR1 Asn86Tyr polymorphism in Lambarene, Gabon. Parasitology 116:211–217

Hunt P, Cravo PV, Donleavy P, Carlton JM, Walliker D (2004) Chloroquine resistance in *Plasmodium chabaudi*: are chloroquine-resistance transporter (crt) and multidrug resistance (mdr1) orthologues involved? Mol Biochem Parasitol 133:27–35

Joy DA, Feng X, Mu J, Furuya T, Chotivanich K, Krettli AU, Ho M, Wang A, White NJ, Suh E, Beerli P, Su XZ (2003) Early origin and recent expansion of *Plasmodium falciparum*. Science 300:318–321

Kamchonwongpaisan S, Chandra-ngam G, et al. (1994) Resistance to artemisinin of malaria parasites (*Plasmodium falciparum*) infecting α-thalassemic erythrocytes in vitro. J Clin Invest 93:467–473

Kremsner PG, Krishna S (2004) Combination chemotherapy. Lancet 364:438–447

Krishna S, Planche T, Agbenyega T, Woodrow C, Agranoff D, Bedu-Addo G, Owusu-Ofori AK, Appiah JA, Ramanathan S, Mansor SM, Navaratnam V (2001) Bioavailability and preliminary clinical efficacy of intrarectal artesunate in Ghanaian children with moderate malaria. Antimicrob Agents Chemother 45:509–516

Krishna S, White NJ (1996) Pharmacokinetics of quinine, chloroquine and amodiaquine. Clinical implications. Clin Pharmacokinetics 30:263–299

Luxemberger C, Perea WA, Delmas G, Pruja C, Pecoul B, Moren A (1994) Permethrin-impregnated bed nets for the prevention of malaria in schoolchildren on the Thai-Burmese border. Trans R Soc Trop Med Hygiene 88:155–159

Marzolini C, Paus E, Buchlin T, Kim RB (2004) Polymorphisms in human MDR1 (P-glycoprotein): recent advances and clinical relevance. Clin Pharmacol Ther 75:13–33

Mawili-Mboumba DP, Kun JF et al. (2002) PfMDR1 alleles and response to ultralow-dose mefloquine treatment in Gabonese patients. Antimicrob Agents Chemother 46:166–170

Moreno A, Brasseur P et al. (2001) Evaluation under field conditions of the colourimetric DELI-microtest for the assessment of *Plasmodium falciparum* drug resistance. Trans R Soc Trop Med Hyg 95:100–103

Nair S, Williams JT, Brockman A, Paiphun L, Mayxay M, Newton PN, Guthmann JP, Smithuis FM, Hien TT, White NJ, Nosten F, Anderson TJ (2003) A selective sweep driven by pyrimethamine treatment in southeast asian malaria parasites. Mol Biol Evol 20:1526–1536

Ngo T, Duraisingh M, Reed M, Hipgrave D, Biggs B, Cowman AF (2003) Analysis of PfCRT, PfMDR1, dhfr, and dhps mutations and drug sensitivities in *Plasmodium falciparum* isolates from patients in Vietnam before and after treatment with artemisinin. Am J Trop Med Hyg 68:350–356

Noedl H, Wernsdorfer WH et al. (2002) Histidine-rich protein II: a novel approach to malaria drug sensitivity testing. Antimicrob Agents Chemother 46:1658–1664

Nomura T, Carlton JM, Baird JK, del Portillo HA, Fryauff DJ, Rathore D, Fidock DA, Su X, Collins WE, McCutchan TF, Wootton JC, Wellems TE (2001) Evidence for different mechanisms of chloroquine resistance in *2 Plasmodium* species that cause human malaria. J Infect Dis 183:1653–1661

Nosten F, ter Kuile F, Chongsuphajaisiddhi T, Luxemburger C, Webster HK, Edstein M, Phaipun L, Thew KL, White NJ (1991a) Mefloquine-resistant falciparum malaria on the Thai–Burmese border. Lancet 337:1140–1143

Nosten F, ter Kuile F, Chongsuphajaisiddhi T, Na Banchang K, Karbwang J, White NJ (1991b) Mefloquine pharmacokinetics and resistance in children with acute falciparum malaria. Br J Clin Pharmacol 31:556–559

Nosten F, van Vugt M, Price R, Luxemburger C, Thway KL, Brockman A, McGready R, ter Kuile F, Looareesuwan S, White NJ (2000). Effects of artesunate-mefloquine combination on incidence of *Plasmodium falciparum* malaria and mefloquine resistance in western Thailand: a prospective study. Lancet 356:297–302

Pickard AL, Wongsrichanalai C, Purfield A, Kamwendo D, Emery K, Zalewski C, Kawamoto F, Miller RS, Meshnick SR (2003). Resistance to antimalarials in Southeast Asia and genetic polymorphisms in PfMDR1. Antimicrob Agents Chemother 47:2418–2423

Pillai DR, Labbe AC et al. (2001). *Plasmodium falciparum* malaria in Laos: chloroquine treatment outcome and predictive value of molecular markers. J Infect Dis 183:789–795

Plowe CV, Djimde A, Bouare M, Doumbo O, Wellems T (1995) Pyrimethamine and proguanil resistance-conferring mutations in *Plasmodium falciparum* dihydrofolate reductase: polymerase chain reaction methods for surveillance in Africa. Am J Trop Med Hygiene 52:656–568

Price R, Robinson G, Brockman A, Cowman A, Krishna S (1997) Assessment of *PfMDR1* gene copy number by tandem competitive polymerase chain reaction. Mol Biochem Parasitol 85:161–169

Price R, Uhlemann A-C, Brockman A, McGready R, Ashley E, Phaipun L, Patel R, Laing K, Looareesuwan S, White N, Nosten F, Krishna S (2004) Mefloquine resistance in *Plasmodium falciparum* results from increased PfMDR1 gene copy number. Lancet 364:438–447

Price RN, Cassar C, et al. (1999) The PfMDR1 gene is associated with a multidrug-resistant phenotype in *Plasmodium falciparum* from the western border of Thailand. Antimicrob Agents Chemother 43:2943–2949

Price RN, Cassar C, Brockman A, Duraisingh M, van Vugt M, White NJ, Nosten F, Krishna S (1997a) Artesunate/mefloquine treatment of multi-drug resistant falciparum malaria. Trans R Soc Trop Med Hyg 91:574–577

Price RN, Nosten F, Luxemburger C, van Vugt M, Phaipun L, Chongsuphajaisiddhi T, White NJ (1997b) Artesunate/mefloquine treatment of multi-drug resistant falciparum malaria. Trans R Soc Trop Med Hyg 91:574–577

Reed MB, Saliba KJ, Caruana SR, Kirk K, Cowman AF (2000) PgH1 modulates sensitivity and reisstance ot multiple antimalarials in *Plasmodium falciparum*. Nature 403:906–909

Schwenke A, Brandts C, Philipps J, Winkler S, Wernsdorfer WH, Kremsner PG (2001) Declining chloroquine resistance of *Plasmodium falciparum* in Lambarene, Gabon from 1992 to 1998. Wien Klin Wochenschr 113:63–64

Sidhu AB, Verdier-Pinard D, Fidock DA (2002) Chloroquine resistance in *Plasmodium falciparum* malaria parasites conferred by PfCRT mutations. Science 298:210–213

Silamut K, White NJ et al. (1985) Binding of quinine to plasma proteins in falciparum malaria. Am J Trop Med Hyg 34:681–686

Srivastava IK, Morrisey JM, Darrouzet E, Daldal F, Vaidya AB (1999) Resistance mutations reveal the atovaquone-binding domain of cytochrome b in malaria parasites. Mol Microbiol 33:704–711

Tinto H, Ouedraogo JB, Erhart A, Van Overmeir C, Dujardin JC, Van Marck E, Guiguemde TR, D'Alessandro U (2003) Relationship between the PfCRT T76 and the PfMDR1 Y86 mutations in *Plasmodium falciparum* and in vitro/in vivo chloroquine resistance in Burkina Faso, West Africa. Infect Genet Evol 3:287–292

Tran TH, Dolecek C, Pham PM, Nguyen TD, Nguyen TT, Le HT, Dong TH, Tran TT, Stepniewska K, White NJ, Farrar J (2004) Dihydroartemisinin-piperaquine against multidrug-resistant *Plasmodium falciparum* malaria in Vietnam: randomised clinical trial. Lancet 363:18–22

Trape JF, Pison G, Preziosi MP, Enel C, Desgrees du Lou A, Delaunay V, Samb B, Lagarde E, Molez JF, Simondon F (1998) Impact of chloroquine resistance on malaria mortality. C R Acad Sci III 321:689–697

Wang P, Sims PF et al. (1997) A modified in vitro sulfadoxine susceptibility assay for *Plasmodium falciparum* suitable for investigating Fansidar resistance. Parasitology 115:223–230

Wellems TE (1991) Molecular genetics of drug resistance in *Plasmodium falciparum* malaria. Parasitol Today 7:110–112

Wellems TE, Plowe CV (2001) Chloroquine-resistant malaria. J Infect Dis 184:770–776

Wellems TE, Walker-Jonah A, Panton LJ (1991) Genetic mapping of the chloroquine-resistance locus on *Plasmodium falciparum* chromosome 7. Proc Natl Acad Sci USA 88:3382–3386

White NJ (1992) Antimalarial drug resistance: the pace quickens. J Antimicrobial Chemother 30:571–585

Wichmann O, Betschart B et al. (2003) Prophylaxis failure due to probable mefloquine resistant P falciparum from Tanzania. Acta Trop 86:63–65

Wilson CM, Serrano AE et al. (1989) Amplification of a gene related to mammalian mdr genes in drug-resistant *Plasmodium falciparum*. Science 244:1184–1186

Wootton JC, Feng X, Ferdig MT, Cooper RA, Mu J, Baruch DI, Magill AJ, Su XZ (2002) Genetic diversity and chloroquine selective sweeps in *Plasmodium falciparum*. Nature 418:320–323

Yuvaniyama J, Chitnumsub P, Kamchongwongpaisan S, Vanichtanankul J, Sirawaraporn W, Taylor P, Walkinshaw MD, Yuthavong Y (2003) Insights into antifolate resistance from malarial DHFR-TS structures. Nat Struct Biol 10:357–365

Zucker JR, Ruebush TK 2nd, Obonyo C, Otieno J, Campbell CC (2003) The mortality consequences of the continued use of chloroquine in Africa: experience in Siaya, western Kenya. Am J Trop Med Hyg 68:386–390

CTMI (2005) 295:55–79

Antimalarial Drug Resistance in Africa: Strategies for Monitoring and Deterrence

C. V. Plowe (✉)

Malaria Section, Center for Vaccine Development, University of Maryland School of Medicine, 685 West Baltimore Street, HSF1-480, Baltimore, MA 21201, USA
cplowe@medicine.umaryland.edu

Abstract Despite the initiation in 1998 by the World Health Organization of a campaign to 'Roll Back Malaria', the rates of disease and death caused by *Plasmodium falciparum* malaria in sub-Saharan Africa are growing. Drug resistance has been implicated as one of the main factors in this disturbing trend. The efforts of international agencies, governments, public health officials, advocacy groups and researchers to devise effective strategies to deter the spread of drug resistant malaria and to ameliorate its heavy burden on the people of Africa have not succeeded. This review will not attempt to describe the regional distribution of drug resistant malaria in Africa in detail, mainly because information on resistance is limited and has been collected using different methods, making it difficult to interpret. Instead, the problems of defining and monitoring resistance and antimalarial drug treatment outcomes will be discussed in hopes of clarifying the issues and identifying ways to move forward in a more coordinated fashion. Strategies to improve measurement of resistance and treatment outcomes, collection and use of information on resistance, and potential approaches to deter and reduce the impact of resistance, will all be considered. The epidemiological setting and the goals of monitoring determine how antimalarial treatment responses should be measured. Longitudinal studies, with incidence of uncomplicated malaria episodes as the primary endpoint, provide the best information on which to base treatment policy changes, while simpler standard in vivo efficacy studies are better suited for ongoing efficacy monitoring. In the absence of an ideal antimalarial combination regimen, different treatment alternatives are appropriate in different settings. But where chloroquine has failed, policy changes are long overdue and action must be taken now.

Abbreviations

DHFR	Dihydrofolate reductase
GRI	Genotype resistance index
GFI	Genotype failure index
mono-DEC	Monodesethylchloroquine

1
The Scope of the Problem

Malaria kills about two million Africans every year, and is the leading cause of disease and death among children and pregnant women in many sub-Saharan African countries. In 1997, at an international workshop convened in Dakar, Sénégal, representatives of funding agencies, public health officials, and scientists from the USA, Europe and Africa addressed the question of how best to bring their combined resources to bear on this enormous and intractable public health problem. The consensus was that existing modalities for the treatment, control and prevention of malaria must be utilized better. However, the participants also concluded that a more substantial and enduring amelioration of the damage caused by malaria would require new tools, most notably effective and inexpensive vaccines. Significant progress has been made toward developing vaccines to prevent malaria infection, disease and transmission, but it will still be many years before an affordable, effective malaria vaccine will be deployed in Africa.

Until a vaccine is available, case management with prompt and effective treatment will remain the mainstay of malaria control (Winstanley 2000), and even if a successful vaccine is developed, drug treatment of malaria will remain a critical pillar of public health in Africa. Because antimalarial drugs will continue to be a primary tool for preventing disease and death, drug resistance will continue to be a major impediment to malaria control on the continent.

Chloroquine is the most important and successful drug ever used against falciparum malaria because of its low cost, low toxicity and high efficacy against susceptible parasites (Wellems et al. 2001). However, resistance has developed, and chloroquine-resistant falciparum malaria has caused large increases in morbidity and mortality in parts of Africa. The introduction of chloroquine resistance into Kinshasa, Zaire was associated with malaria changing from a relatively infrequent admitting diagnosis to the most common diagnosis of pediatric hospital admissions (Greenberg et al. 1989). In Sénégal, as chloroquine resistant falciparum malaria increased over a period of 12 years in different communities it was associated with dramatic increases in malaria-attributable deaths in those communities (Trape et al. 1998). While it is impossible to establish a definite and linear cause-and-effect relationship between the introduction and rise of chloroquine resistance and measures of morbidity and mortality in these ecological studies, it is also difficult to imagine that as chloroquine efficacy falls from nearly 100% to less than 25% in some areas, the result could be anything other than an increase in disease and death caused by malaria. The high-risk groups are relatively non-immune infants and young children in areas of high transmission and persons of all ages in areas with low or epidemic transmission.

In African countries where chloroquine has failed, it has usually been replaced with the antifolate combination sulfadoxine–pyrimethamine. *P. falciparum* resistance to sulfadoxine–pyrimethamine has emerged at various rates in different transmission settings in Africa (Mutabingwa et al. 2001b; Plowe et al. 2004; Roper et al. 2003), and the impact of resistance to sulfadoxine–pyrimethamine is less well documented than that of chloroquine. However, high rates of sulfadoxine–pyrimethamine resistance have been observed in some areas in Africa (Mutabingwa et al. 2001b; Roper et al. 2003) and this development is likely to be accompanied by increases in malaria morbidity and mortality similar to those observed in areas with high chloroquine resistance.

Other antimalarial drugs, usually given in combination in the hope of deterring resistance, have only recently been deployed in Africa in limited areas. The consequences of the eventual failure of these therapies, while not yet known, are predictably dire based on the history of chloroquine and sulfadoxine–pyrimethamine in Africa and of antimalarial combinations in other parts of the world (Nosten et al. 1991).

2
Drug Resistance Versus Therapeutic Efficacy

In discussing drug resistance, it is important to distinguish between classifications of drug-resistant and -sensitive parasites by assays performed in vitro, and those applied to in vivo assessments of antimalarial treatment. In vitro assays are performed under different conditions in different laboratories, and standardized definitions of sensitive versus resistant parasites are not available or agreed upon for all drugs. Even in the case of chloroquine, there is some disagreement as to whether parasites should be classified as strictly sensitive versus resistant in vitro or if there are intermediate ranges of resistance that are meaningful to clinical outcomes. In vitro assays for sulfadoxine–pyrimethamine are notoriously difficult to perform reproducibly. In addition, because the tests must be conducted under nonphysiological conditions, extrapolating from the in vitro results to in vivo treatment outcomes is difficult. For example, decreases in susceptibility can be induced to the artemisinin derivatives in vitro, but these do not reflect true resistance in the sense of being associated with decreased clinical efficacy because in vivo parasitological resistance to the artemisinin derivatives has not been documented (Meshnick 2002).

In vivo outcome assessments are similarly plagued by differences in definition. Traditionally, in vivo resistance has been measured by inspecting peripheral blood smears for the presence or absence of malaria parasites at intervals following treatment (WHO 1973). However, in 1996 the World Health Organization recommended that in vivo outcome definitions also consider the host response, including fever (WHO 1996). Using this approach, an individual found to be parasitemic but afebrile 4–14 days after treatment would be regarded as having had an adequate clinical response. Currently, the WHO is re-defining antimalarial treatment outcomes to incorporate both parasitological and clinical responses. Table 1 provides the older, standard definitions of parasitological resistance, and the latest definition of therapeutic and parasitological outcomes (WHO 2002) are shown in Table 2. The new definitions include a treatment response category called Adequate Clinical and Parasitological Response, which is essentially identical to the traditional sensitive in vivo outcome. This reversion to requiring that detectable parasites be eliminated before an individual can be considered to be adequately treated was based on the recognition that asymptomatic infections, even in areas with high rates of reinfection, are deleterious, and can lead to increased

Table 1 Parasitological resistance outcomes

RIII	No reduction in parasitemia, or reduction to $\geq 25\%$ of day 0 level, by day 3
RII	Reduction in parasitemia to $<25\%$ of day 0 level by day 3 without clearance by day 7, leading to re-treatment or followed by persistent parasitemia
RI	Initial clearance of parasites with subsequent positive thick smear by day 14 or 28
Sensitive	Clearance of parasite by day 6 with no recurrence of parasitemia, without previously meeting criteria for RI, RII or RIII

rates of anemia (Plowe et al. 2004) and even mortality (Zucker et al. 1996).

The definitions of in vivo parasitological resistance and therapeutic efficacy have flaws. Criteria for both high-level parasitological resistance (RIII) and early treatment failure are met if the level of parasite density 3 days after treatment is 25% or more than the parasite density at the time of treatment. However, among semi-immune populations with high initial parasitemias, a substantial proportion of untreated individuals will go on to clear their infections if rescue therapy is not given, indicating that that both the old and current definitions of resistance and efficacy yield a substantial number of 'false positive' RIII and early treatment failure outcomes. These definitions are thus likely to overestimate 'true' parasitological resistance (Plowe et al. 2001; White 2002).

The difficulty in reconciling parasitological and therapeutic outcomes is that post-treatment parasite survival and clinical outcomes depend not only on the ability of malaria parasites to cope with the drug, but also on other factors that can affect the growth of parasites following treatment and the clinical response. Host immunity is probably the most important of these (Djimde et al. 2001a; White 2002) but several other factors can contribute both to parasitological and clinical outcomes.

3
Factors Affecting Treatment Outcome

It has long been known that the level of host immunity acquired through long-term exposure to infection affects the response to drug treatment (Mayxay et al. 2001; White 2002). In Bandiagara, Mali, a site where malaria vaccines are being tested, the median age of persons presenting to clinic with symptomatic uncomplicated falciparum malaria is 10 years (Djimde et al. 2001a). In this setting of moderate malaria transmission, immunity as reflected by age plays a critical role in the host's ability to clear chloroquine-resistant parasites when treated with chloroquine: persons aged less than 10 years cleared chloroquine-resistant parasites at a rate half that of persons aged 10 years or older. The age pattern of the ability to clear chloroquine-resistant parasites mirrors exactly the acquisition of clinical immunity (Fig. 1). Infants aged less than 1 year who are still protected by maternal antibodies or other mechanisms that prevent disease in young infants are able to clear resistant parasites; children aged between 2 and 4 years, the group at highest risk of severe malaria, lose this ability; and then a resumption and retention of the ability to clear chloroquine-resistant parasites in later childhood through to adulthood is observed (Djimde et al. 2003). In contrast, in Blantyre, Malawi, where the median age of persons presenting to clinic with uncomplicated falciparum

Table 2 New World Health Organization classification of response to antimalarial treatment

Intense transmission area	Low to moderate transmission area

Early treatment failure

Intense transmission area	Low to moderate transmission area
Development of danger signs of severe malaria on day 1, day 2 or day 3, in the presence of parasitemia	Development of danger signs of severe malaria on day 1, day 2 or day 3, in the presence of parasitemia
Parasitemia on day 2 higher than day 0 count irrespective of axillary temperature	Parasitemia on day 2 higher than day 0 count irrespective of axillary temperature
Parasitemia on day 3 with axillary temperature $\geq 37.5^\circ$C	Parasitemia on day 3 with axillary temperature $\geq 37.5^\circ$C
Parasitemia on day 3 $\geq 25\%$ of count on day 0	Parasitemia on day 3 $\geq 25\%$ of count on day 0

Late treatment failure

Late clinical failure

Intense transmission area	Low to moderate transmission area
Development of danger signs or severe malaria after day 3 in the presence of parasitemia, without previously meeting any of the criteria of early treatment failure	Development of danger signs or severe malaria after day 3 in the presence of parasitemia, without previously meeting any of the criteria of early treatment failure
Presence of parasitemia and axillary temperature $\geq 37.5^\circ$C on any day from day 4 to day 14, without previously meeting any of the criteria of early treatment failure	Presence of parasitemia and axillary temperature $\geq 37.5^\circ$C (or history of fever) on any day from day 4 to day 28, without previously meeting any of the criteria of early treatment failure

Late parasitological failure

Intense transmission area	Low to moderate transmission area
Presence of parasitemia on day 14 and axillary temperature $<37.5^\circ$C, without previously meeting any of the criteria of early treatment failure or late clinical failure	Presence of parasitemia on any day from day 7 to day 28 and axillary temperature $<37.5^\circ$C, without previously meeting any of the criteria of early treatment failure or late clinical failure

Adequate clinical and parasitological response

Intense transmission area	Low to moderate transmission area
Absence of parasitemia on day 14 irrespective of axillary temperature without previously meeting any of the criteria of early treatment failure or late clinical failure or late treatment failure	Absence of parasitemia on day 28 irrespective of axillary temperature without previously meeting any of the criteria of early treatment failure or late clinical failure or late treatment failure

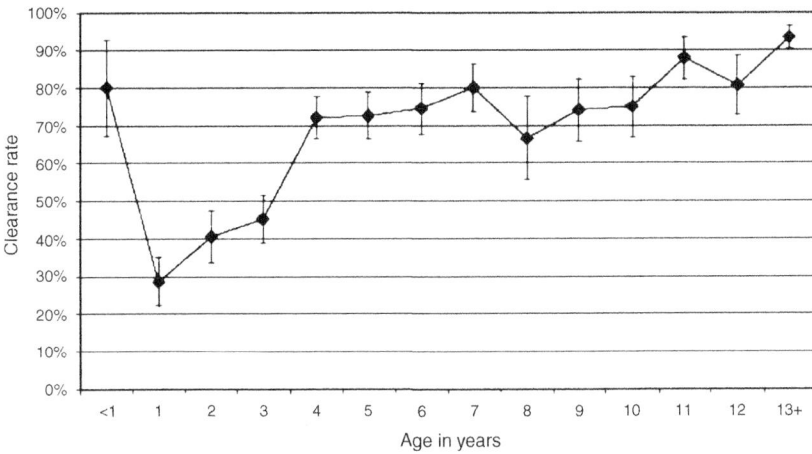

Fig. 1 Relationship between age and clearance of resistant parasites after chloroquine treatment in Bandiagara, Mali. Clearance of chloroquine resistant parasites is strongly associated with age in a pattern that closely follows perinatal and acquired clinical immunity. (Reproduced with permission from Djimde et al. 2003)

malaria was only 2.4 years (Kublin et al. 2002), and in studies that excluded all but the youngest children (Mayor et al. 2001), no association was observed between age and response to sulfadoxine–pyrimethamine or chloroquine treatment, respectively.

Other host factors that are associated with lowered immunity have been implicated in responses to antimalarial treatment and prophylaxis, including malnutrition (Wolday et al. 1995) and infection with HIV (Parise et al. 1998). Given the very high rates of malnutrition and AIDS in much of Africa, if these are determined to be significant factors in response to malaria treatment, the public health impact will be massive (Bloland 2001). Antimalarial treatment failures can even be iatrogenic, e.g. folate supplements given to pregnant women have been shown to decrease the efficacy of sulfadoxine-pyrimethamine (van Hensbroek et al. 1995).

Age itself, irrespective of prior exposure to malaria, may assist with acquisition of the ability to clear parasites following treatment: adult transmigrants from a malaria-free area of Indonesia to a malaria-endemic area acquired immunity significantly more rapidly than did children (Baird et al. 1993). Other forms of innate immunity may assist with clearing resistant parasites, for example HLA types shown to protect from severe malaria might also aid in clearing parasites after treatment (Hill et al. 1991). In addition, other host genetic factors such as sickle cell trait (Miller et al. 1976), hemoglobin C (Agarwal et al. 2000; Modiano et al. 2001), thalassemias (Miller et al. 1976) and glucose-6-phosphase dehydrogenase deficiency (Ruwende et al. 1995) have all been associated with protection from malaria disease and could contribute to parasite clearance.

The bioavailability of a drug formulation and its pharmacokinetics are also likely to be important in antimalarial treatment outcomes. Chloroquine and its principal active metabolite, monodesethylchloroquine (mono-DEC) have comparable activity against chloroquine-sensitive *P. falciparum*, but mono-DEC, which is detected in the plasma in variable ratios to chloroquine among individuals, shows a much greater loss of activity against chloroquine-resistant *P. falciparum* than chloroquine (Hellgren et al. 1989). Mono-DEC is not used in the standard in vitro assays for chloroquine resistance, and given its potentially important in vivo role, it is difficult to interpret differences in IC_{50} among *P. falciparum* strains that fall within the resistant range, and to determine the relevance of these differences to in vivo efficacy (Wellems et al. 2001). Proguanil provides an example of how differences in host metabolism can affect treatment response: a relatively high proportion of East Africans have an impaired ability to metabolize proguanil to its active dihydrofolate reductase (DHFR)-inhibiting metabolite cycloguanil (Watkins et al. 1990).

At the community and regional level, how drugs are distributed and used, e.g., restricting use to dispensation in clinics only following microscopically confirmed diagnosis versus making drugs widely available through programs of home-based therapy by mothers, affect not only individual treatment response but the development and spread of drug resistance. Drug usage patterns and adherence to correct dosing regimens are major determinants of drug pressure and thus of resistance, raising a dilemma in Africa where access to diagnostic facilities and trained health workers is limited. Restriction of drug use has been employed in more economically developed areas but is unlikely to be a viable strategy in most of Africa until primary health care infrastructures are improved. Two recent reviews discuss how such programs and practices affect resistance (Bloland et al. 1999, 2001).

4
Mechanisms of Resistance

The molecular mechanisms of resistance to antimalarial drugs are described in detail in other reviews in this series and will only be briefly summarized here. Chloroquine resistance in *P. falciparum* is conferred by mutations in PfCRT, a transporter protein that has been localized to the parasite digestive vacuole (Fidock et al. 2000). A single Lys→Thr mutation at codon 76 (K76T), always occurring on a variable background of other PfCRT mutations, is the key determinant of chloroquine resistance, as shown in laboratory-based studies of chloroquine-sensitive and resistant parasites (Fidock et al. 2000), genetic transformation experiments (Fidock et al. 2000; Sidhu et al. 2002) and a series of field surveys described in a recent review (Wellems et al. 2001).

Polymorphisms in *PfMDR1* encoding the *P. falciparum* P-glycoprotein homologue 1, modulate chloroquine resistance in mutant *PfCRT*-harboring parasites in vitro (Reed et al. 2000), although their role in chloroquine resistance in vivo, if any, remains unclear (Djimde et al. 2001a). As noted above, the relevance of variations of IC_{50} within the chloroquine-resistant range cannot necessarily be extrapolated to parasite response in vivo, and field studies of the association between these mutations and chloroquine treatment response have been contradictory. *PfMDR1* mutations and other parasite genetic factors may play a secondary role in chloroquine treatment outcomes, but these appear to be neither necessary nor sufficient to cause chloroquine treatment failure, and multivariate analyses have failed to detect statistical interactions between mutations in *PfCRT* and *PfMDR1* (Djimde et al. 2001a; Jelinek et al. 2002), i.e., the presence of mutations in both genes was not more strongly associated with treatment failure than the presence of *PfCRT* T76 alone.

Pyrimethamine and cycloguanil (the active metabolite of proguanil) bind and inhibit DHFR, and the sulfa and sulfone drugs inhibit dihydropteroate synthase (DHPS) (Foote et al. 1994). In vitro resistance of *P. falciparum* to pyrimethamine and to cycloguanil is due to specific point mutations in *P. falciparum* DHFR (Cowman et al. 1988; Foote et al. 1990; Peterson et al. 1988, 1990; Zolg et al. 1989). Mutations in DHPS are similarly associated with in vitro resistance to the sulfa drugs and sulfones (Brooks et al. 1994). Both the DHFR and DHPS mutations occur in a progressive, step-wise fashion, with higher levels of in vitro resistance occurring in the presence of multiple mutations. It is worth noting that a DHFR mutation at codon 164 and a DHPS mutation at codon 581 appear to be the last steps in high-level resistance to the antifolates. Fortunately, the former has yet to be definitively documented in Africa and the latter is rare there.

Mutations and/or changes in copy number of *PfMDR1*, or in the expression of the protein it encodes, Pgh1, have been implicated in resistance to quinine (Duraisingh et al. 2000; Zalis et al. 1998), mefloquine (Price et al. 1999; Reed et al. 2000; Wilson et al. 1993), halofantrine (Peel et al. 1994; Wilson et al. 1993) and to changes in susceptibility to the artemisinin derivatives (Duraisingh et al. 2000; Reed et al. 2000; Walker et al. 2000). However, other studies have failed to find these associations in some cases (Lim et al. 1996; Ritchie et al. 1996) and the role of *PfMDR1* in resistance remains a subject of debate and continued research. Recent genome-wide scanning methods are identifying new transporter genes that may be involved with resistance to quinine and other drugs (Mu et al. 2003) and it is likely that we still have much to learn about transporters and drug resistance in *P. falciparum*.

Resistance to atovaquone when used as a single agent is conferred by mutations in cytochrome b, although this resistance does not appear to be a factor when atovaquone is used in combination with proguanil in the drug Malarone. A pair of DHFR mutations, each rare in nature, is associated with resistance to cycloguanil, the active metabolite of proguanil, but the relevance to potential in vivo atovaquone–proguanil resistance is unclear, in that proguanil appears to act directly against the parasite independently of its metabolite's inhibition of DHFR (Fidock et al. 1997), probably through acting on a mitochondrial pathway (Fidock et al. 1998; Srivastava et al. 1999).

To date, the only drugs for which molecular markers have been consistently and strongly associated with treatment failures are chloroquine, mefloquine (see chapter by P.G. Bray et al. and Uhlemann and Krishna, this volume) and sulfadoxine–pyrimethamine (Plowe 2003).

5
The Spread of Drug Resistance in Africa

Chloroquine resistant *P. falciparum* malaria arrived later in Africa than it did in Asia and South America. Several hypotheses for the later arrival have been put forth, but with the identification of the key molecular determinant of chloroquine resistance it appears that chloroquine-resistant genotypes arose independently in a limited number of foci, none of them in Africa, and spread contiguously and slowly to the continent, arriving in East Africa in the late 1970s (Wellems et al. 2001).

The primary force behind the spread of drug resistance, once it arrives or arises, is drug pressure on the parasite. The relatively slow spread of chloroquine resistance in Africa compared to South East Asia and parts of South America may be explained by models predicting that resistance can only arise and be sustained when a threshold of drug use relative to the overall parasite population in an area is met, particularly if the resistance comes at a cost to parasite survival (Koella et al. 2003). This line of thinking may at first seem counter-intuitive, in that there is much more malaria in Africa than elsewhere and therefore more antimalarial use. However, because high transmission results in semi-immunity, the vast majority of *P. falciparum* infections in Africa are asymptomatic and therefore less likely to come under drug pressure, compared to low transmission areas where infection nearly always leads to disease and therefore to treatment (Luxemburger et al. 1996). Thus in Africa there is always a large reservoir of parasites that are not under drug pressure, permitting more fit, less resistant parasites to regain the upper hand and overall predominance in the population. This phenomenon may explain not only the slow spread of chloroquine resistance in Africa but also the slower development of resistance to sulfadoxine-pyrimethamine resistance in some areas than had been predicted (Plowe et al. 2004).

The pace and patterns of how drug-resistant malaria spreads is no doubt more complex than this. The disparate experiences in Africa and Asia suggest that resistance should take hold most slowly in areas of highest transmission, but within Africa, this model is challenged (Mackinnon 1997; Paul et al. 1995). Chloroquine resistance has remained at relatively low and stable levels in many parts of West Africa where transmission is lower than it is in parts of East Africa where chloroquine has failed. The relative availability of drugs and cultural tendencies to use traditional versus modern pharmaceutical treatments also impacts on the amount of drug pressure applied to the overall parasite population in an area, helping to explain regional variations in development and rates of resistance. Mathematical modeling may help to clarify this picture (Hastings 1997, 1998), but such models will need to incorporate all of the known factors impacting drug pressure, malaria transmission and parasite genetic population dynamics. This will be challenging.

6
Measuring Resistance and Treatment Outcomes

Current in vivo methods for measuring antimalarial drug efficacy in high-transmission areas use a 14-day follow-up period, reflecting a compromise between missing late treatment failures occurring after 14 days and misclassifying reinfections as treatment failures. Predictably, when follow-up is extended to 28 days or longer, more cases of apparent resistance or treatment failure are found (Dorsey et al. 2002; Plowe et al. 2004; White 2002). In high transmission areas, some of these recurrent infections are reinfections, not true recrudescence due to treatment failure. Genotyping post-treatment infections can help to distinguish between reinfections and recrudescence (Cattamanchi et al. 2003; Viriyakosol et al. 1995), but it is impossible to rule out the possibility that apparent new infections actually represent minor subpopulations of the original infections that rose above the threshold of PCR detection after the majority of susceptible parasites were destroyed. Efficacy studies with long follow-up periods and genotyping to distinguish reinfection and recrudescence are also expensive, impractical for most groups conducting drug resistance surveillance, and in some cases may be of questionable relevance to public health policymakers, who are less interested in knowing rates of asymptomatic parasitemia several weeks after treatment than they are in knowing rates of acute recovery from illness and the incidence of uncomplicated malaria treatment episodes, anemia and severe malaria.

These limitations of standard in vivo studies have led some to advocate longitudinal studies of drug efficacy. In addition to measuring efficacy of individual treatments at 14 or 28 days, longitudinal studies measure sustained efficacy with repeated use of the same regimen over time. The primary outcome of interest, the incidence of uncomplicated malaria treatment episodes, as well as the secondary outcomes of incidence of anemia and severe malaria, are highly relevant to public health policy-makers, as they reflect not only the burden of disease but also the utilization of health resources.

Longitudinal studies also permit assessment of how pharmacokinetic properties of drugs affect the incidence of treatment episodes. The first such longitudinal trial of antimalarial drug efficacy was a double-blinded, placebo controlled trial of sulfadoxine–pyrimethamine and chlorproguanil–dapsone in Kenya and Malawi. These otherwise similar antifolate combinations are distinguished by the much shorter half-life of chlorproguanil–dapsone, which was thought to make it less prone to development of resistance, at a cost of losing the post-treatment prophylactic effect offered by longer-acting sulfadoxine-pyrimethamine. Chlorproguanil-dapsone is also more efficacious than sulfadoxine–pyrimethamine in Africa, where it has been shown to clear infections resistant to sulfadoxine–pyrimethamine (Mutabingwa et al. 2001a).

This first longitudinal study randomized children to receive one drug or the other for all uncomplicated falciparum malaria episodes over the course of a year, and compared the cumulative number of treatment episodes and rates of treatment failure and anemia (Sulo et al. 2002). In Malawi, chlorproguanil–dapsone had an efficacy of 95% compared to only 80% for sulfadoxine–pyrimethamine ($p<0.01$). Yet there were no differences between the treatment groups in the incidence of uncomplicated malaria episodes, anemia or severe malaria, indicating that the better efficacy of chlorproguanil-dapsone's short half-life was offset by its lacking the long prophylactic effect offered by sulfadoxine–pyrimethamine.

Although in vivo studies provide the information needed to make evidence-based malaria treatment policies, standard in vivo studies are expensive and time-consuming, and longitudinal clinical efficacy trials are even more so. Longitudinal studies are probably most appropriately done when a country or a region is considering changing its first-line antimalarial, as they provide the most comprehensive evaluation of how a drug will perform over time. Standard in vivo efficacy studies will continue to be the gold standard for monitoring efficacy of drugs currently in use.

In vitro methods for measuring drug resistance (Nguyen-Dinh et al. 1980) have very limited suitability for surveillance because they require that venous blood with a high parasite density be quickly frozen or transported cold to a laboratory for parasite cultivation, the methods are laborious, and failure to establish primary parasite growth is frequent. However, they remain important for confirming and characterizing resistance and for investigating the molecular and cellular mechanisms of resistance.

Although molecular markers for chloroquine and sulfadoxine–pyrimethamine resistance have been advocated as tools for surveillance of resistance for several years, they are still rarely used for this purpose. This is in large part because the prevalence rates of molecular markers for resistance are invariably higher than the prevalence of in vivo drug resistance, almost certainly due to the assistance of the host immune response in clearing parasites following treatment.

To overcome this problem, a simple model was developed that calculates ratios between prevalence rates of the chloroquine-resistant *PfCRT* genotype and therapeutic and parasitological outcomes of in vivo studies (Djimde et al. 2001b). At four sites in Mali over 3 years, a genotype-resistance index (GRI) for each site was calculated by dividing the prevalence of the chloroquine-resistant genotype (*PfCRT* 76T) by the prevalence of in vivo chloroquine resistance (RI, RII and RIII) at that site. A genotype-failure index (GFI) was similarly calculated as the ratio of resistant genotype to chloroquine therapeutic failure (early and late treatment failures) at each site. The GRIs and GFIs increased dramatically with age, presumably reflecting acquired immunity and a higher proportion of older persons who cleared parasites with the chloroquine-resistant genotype when treated with chloroquine. After controlling for age, both GRIs and GFIs were reasonably stable over geography and time, ranging from 1.6 to 2.8 at all study sites over the 3-year period, during which time chloroquine resistance was increasing. This model suggests that if GRIs and GFIs remain stable over time even as rates of resistance and treatment failure change, once these indices are established, molecular surveys could be used as a tool for surveillance to predict in vivo treatment outcomes. The methods are simple: collection of filter-paper blood spots from the same digital punctures used to make malaria smears for diagnosis; drying and transporting the samples at room temperature; and performing simple allele-specific molecular assays. Many laboratories in Africa have had the capacity to perform these assays for several years, and could serve as central laboratories for regional surveillance networks.

Molecular markers will not supplant in vivo surveys for resistance, but they may help to focus limited resources available for in vivo studies to areas where molecular surveys indicate rising rates of resistance markers, and they can also be used to monitor the rates of resistance once a drug has been withdrawn from use (Kublin et al. 2003). GRI and GFI models have thus far been established only for chloroquine, but in principle they should be equally useful for the DHFR and DHPS mutations that serve as markers for sulfadoxine-pyrimethamine resistance (Kublin et al. 2002). Molecular markers for resistance to other antimalarial drugs are not yet well established enough to permit molecular monitoring as a practical tool for surveillance of resistance.

The GRI/GFI model remains to be validated in other settings with wide variations in malaria transmission, and for other drugs such as sulfadoxine-pyrimethamine, and it is not useful where resistance rates are already high, as the prevalence of the molecular markers approaches or reaches 100% in these settings (Dorsey et al. 2001; Pillai et al. 2001). Nevertheless, this approach may provide a way to extend the coverage of drug resistance monitoring beyond a few sentinel sites in areas where chloroquine and sulfadoxine-pyrimethamine are being used.

The new WHO protocol for monitoring antimalarial therapeutic efficacy (WHO 2002) (Table 2) provide revised definitions of clinical and therapeutic failure that can provide the basis for further assessment of the GRI/GFI model. To assure standardized validation of the model, the measures of failure used in the denominator for GFIs should include early treatment failure, late treatment failure and late parasitological failure. Since these incorporate the classical RI, RII and RIII levels of resistance and the category of Adequate Clinical and Parasitological Response is virtually identical to the definition of sensitive in vivo outcome, calculation of GRIs is no longer necessary and GFIs using the new definitions should be used.

7
Deterring Resistance

Combinations of antimalarial drugs targeting different pathways, especially combinations including the rapidly acting and highly efficacious artemisinin derivatives, are being strongly advocated as a strategy for improving efficacy and deterring drug resistant malaria in Africa (White et al. 1999). This strategy holds great promise, not only because the rapid reduction of parasite biomass by artemisinin derivatives is expected to protect other drugs used in combination with them, but because this class of drugs reduces gametocyte production and therefore the transmission of resistant parasites (Price et al. 1996). Nevertheless, several obstacles to successful implementation of combination antimalarial therapy remain; these have been discussed in detail in a recent review (Bloland et al. 2000). As the deployment of these combinations is progressing, it is unfortunately predictable, based on experiences with successful interventions including insecticide-impregnated bednets, childhood vaccines and antiretroviral drugs for HIV, that it will likely take years for these drugs to become available to the majority of Africans, especially those living in areas with little or no primary health care infrastructure.

One novel approach to deterring resistance was suggested by a recent report from Malawi. In 1993, Malawi became the first sub-Saharan African country to switch from chloroquine to sulfadoxine–pyrimethamine as the first-line antimalarial nationwide (Bloland et al. 1993). Since 1993 sulfadoxine–pyrimethamine has been the only treatment for uncomplicated malaria available in government health facilities, where is it is dispensed without prescription. Chloroquine is available only by prescription, a national information campaign was largely successful in convincing health practitioners and the public to accept sulfadoxine–pyrimethamine as the standard treatment for malaria, and chloroquine use was greatly curtailed if not eliminated entirely.

Malawi has thus served as a sentinel site for changing from chloroquine to sulfadoxine–pyrimethamine in Africa. Contrary to predictions that sulfadoxine–pyrimethamine would fail there within 5 years, adequate clinical responses to sulfadoxine–pyrimethamine have been stable at about 80% from 1998 through 2002 in a prospective study that started 5 years after it was introduced (Plowe et al. 2004). Moreover, most of the treatment failures were early treatment failures based solely on parasitological measurements, a substantial proportion of which are likely to be 'false positive' treatment failures, as discussed above. These results demonstrate that the patterns of emergence of resistance in one setting, e.g., Southeast Asia, where sulfadoxine-pyrimethamine failed much more rapidly, cannot be extrapolated to other very different epidemiologic settings. The relatively stable efficacy of sulfadoxine–pyrimethamine in Malawi, combined with the results of the previously cited longitudinal comparison with chlorproguanil-dapsone (Sulo et al. 2002), suggest that African countries with very low chloroquine efficacy, reasonably high sulfadoxine–pyrimethamine efficacy, and no other immediately available alternatives may benefit from interim use of sulfadoxine–pyrimethamine, either alone or in combination with other currently available and affordable drugs (see Sect. 8), while awaiting implementation of superior antimalarial regimens.

This relatively good news of stable clinical efficacy in Malawi is mitigated by the finding that parasitological resistance at 14 and 28 days post-treatment and therapeutic failures at 28 days post-treatment were all increasing modestly but significantly throughout this period. In addition, children with adequate clinical responses but RI parasitological resistant outcomes (i.e., infection but no fever) at 28 days had an increased risk of anemia (Plowe et al. 2004).

The cessation of chloroquine use in Malawi was followed by a dramatic reemergence and predominance of chloroquine-sensitive *P. falciparum*. The prevalence of the chloroquine-resistant *PfCRT* 76T genotype decreased steadily from 85% in 1992 to 13% in 2000 in Blantyre, a large city in central Malawi, and in 2001 and 2002 no mutant *PfCRT* could be found in the area by two independent groups. In 2001, chloroquine cleared 100% of

63 asymptomatic *P. falciparum* infections, and no isolates were resistant to chloroquine in vitro (Kublin et al. 2003). Another study conducted in a district further north in Malawi found a 17% prevalence of *PfCRT* 76T in 1998 and only 2% in 2000 (Mita et al. 2003). The striking disappearance of chloroquine-resistant *P. falciparum* in Malawi at a time when the prevalence of 76T remained over 90% in neighboring Zambia, where it was still the first line drug, is most likely attributable to a cost to parasite fitness caused by the chloroquine resistance-conferring *PfCRT* mutations resulting in selection for parasites with wild type *PfCRT* in the absence of chloroquine drug pressure.

If controlled trials of chloroquine efficacy in areas where chloroquine use has been substantially reduced for a period of years confirm a return of chloroquine's clinical efficacy, countries now using chloroquine may be able to consider withdrawing chloroquine and switching to other drugs on an interim basis, knowing that they may be able to later reintroduce chloroquine, which is unparalleled in its safety and low cost. To preserve chloroquine efficacy it will almost certainly be reintroduced as a component of a combination therapy, ideally with a drug with a similar elimination half-life.

8
Changing Antimalarial Drugs

A full discussion of the scientific, political, economic and social aspects of when and how African countries should make policy decisions on changing first line antimalarial drugs is beyond the scope of this review, and these topics are covered in other recent reviews (Bloland et al. 1999; Bloland 2001). However, in light of the current strong support from some quarters for widespread implementation of artemisinin-derivative based combination therapy (White et al. 1999) across Africa and recent criticisms of the global health community's progress in achieving this (Attaran et al. 2004), some discussion of how drug resistant malaria is defined and monitored is warranted to illuminate this contentious topic. Some of the obstacles that need to be overcome to implement combination therapy successfully in Africa, as has been done in parts of South East Asia, were reviewed in detail by Bloland and Ettling (Bloland et al. 2000), who point out the importance of elimination half-lives when drugs are used in combination. Combinations with mismatched half-lives may be highly efficacious when measured in single-treatment in vivo efficacy studies, but the rapid elimination of one drug will leave the longer-acting drug unprotected, thwarting the goal of deterring development of resistance in areas where reinfection is likely during the period of elimination of the longer-acting drug.

The longitudinal study of chloroproguanil-dapsone versus sulfadoxine-pyrimethamine described above in Section 6 (Sulo et al. 2002) demonstrates another important point that is under-appreciated in the current movement toward high-efficacy, short-acting combination antimalarial drug therapy across Africa. In that study, children treated for all episodes of malaria over the course of a year with sulfadoxine-pyrimethamine had no more episodes of uncomplicated malaria, anemia or severe malaria than those treated for all malaria episodes with chlorproguanil-dapsone, despite the latter's much higher efficacy. Clearly, radical cure, which can be achieved with short-acting, highly efficacious combinations, should be the goal in low transmission settings where the risk of reinfection is low, but this study demonstrates that over time, children in areas of high transmission may do as well or better with a longer acting drug with mediocre efficacy than with a highly curative but short acting drug. The larger point is that one drug or drug combination is not ideal for all settings, and that transmission dynamics and other factors must be considered in choosing optimal antimalarial drug regimens.

Until recently, the WHO recommended that when treatment failure rates (Early and Late Treatment Failures; Table 2) reached 25%, countries should change their first line antimalarial drug. Since post-treatment parasitemia was not considered to be treatment failure unless accompanied with documented fever, this meant that many cases of RI level resistance under the classic definition of resistance were classified as Adequate Clinical Response, and the threshold for recommended change did not occur until clinical resistance rates were very high. Compounded by the fact that many African countries have not made changes despite far exceeding these failure rates, these recommendations have resulted in a large proportion of the sub-Saharan African population being treated for falciparum malaria with a drug, namely chloroquine, that no longer works in many settings. Currently, the WHO is considering lowering the threshold for recommending a change of first line antimalarial drug to 15% or fewer treatment failures (both early and late). This change arose from a sense that despite the poor compliance with the current recommendations for changing drugs, the malaria situation in Africa is becoming increasingly intolerable, and that bold action must be taken to stimulate all of the partners in the malaria control community to work together to reduce malaria-attributable disease and death. As these new recommendations are put in place, it will be important to use carefully defined and standardized methods for monitoring malaria treatment outcomes.

An ideal antimalarial drug for most settings in Africa might be an artemesinin-based combination with two additional long-acting drugs with similar elimination half lives. The artemisinin derivative would rapidly reduce parasite biomass, greatly reducing the probability of survival of resistant parasites, and would block transmission via gametocytes. The other two drugs would offer both extended prophylaxis, reducing incidence of subsequent episodes of malaria and other complications, and also offer some degree of protection for each other against emergence or development of resistance due to either recrudescent or new infections. Inclusion of two long-acting agents in an artemisinin-based combination would likely result in reduced incidence of clinical malaria episodes and anemia, similar to intermittent presumptive treatment regimens giving in pregnancy and being considered for infancy and childhood in Africa (Schellenberg et al. 2001).

Such an ideal combination therapy at the right price and in a dosing regimen that maximizes adherence is not available now and may not be for years, if ever. What, then, should African countries with intolerably low chloroquine efficacy rates who are still using chloroquine do in the meantime? Based on the weight of the available evidence as well as on theoretical considerations of the benefits of artemisinin-based combination therapy, those countries that can afford these drugs now (*e.g.* countries with Global Fund grants), preferably combinations with a long-acting drug offering prophylactic benefit, should use them to replace chloroquine. Countries that will not soon have access to highly effective combinations therapies but where sulfadoxine-pyrimethamine remains highly efficacious should consider switching from chloroquine to sulfadoxine-pyrimethamine as an interim measure that can be implemented immediately while continuing to seek the means to obtain superior drug combinations. Alternatively, the combination of amodiaquine-sulfadoxine-pyrimethamine has been shown to be highly effective even where sulfadoxine-pyrimethamine efficacy is compromised, and this inexpensive combination offers another promising option that is available now (Dorsey et al. 2002). What is indisputable is that no country still using chloroquine in the face of high rates of resistance should continue to use this drug, and that many countries should have switched to sulfadoxine-pyrimethamine several years ago. When chloroquine is withdrawn, its use should be curtailed as strictly as possible, given the evidence that chloroquine sensitive *P. falciparum* may reemerge after several years of its withdrawal (Kublin et al. 2003).

9
Conclusions

The key message of this review is that a "one size fits all" approach is inappropriate both for monitoring antimalarial drug resistance in Africa and for formulating antimalarial treatment and prophylaxis policies on the continent. While strictly standardized definitions and protocols for monitoring resistance should be a high priority, the application and interpretation of these methods must consider the related factors of malaria epidemiology, transmission intensity and host immunity in the area where treatment outcomes are measured. For example, in areas of very high transmission, Adequate Clinical Response may still be the best indicator to use to set thresholds for changing antimalarial drugs; in areas of low to moderate transmission where the benefits of radical cure are more evident, Adequate Clinical and Parasitological Response may be the most appropriate indicator.

It is tempting to argue, as some have done, that artemisinin-based combination therapy resulting in radical cure should be used throughout the sub-Saharan African region. However, a close consideration of the data suggests that highly efficacious combination therapy with short half-life agents or combination therapy with drugs with mismatched elimination times may be the best currently available choice in lower transmission settings, where radical cure or rapid reduction in parasite biomass is likely to deter the emergence of resistance. On the other hand, drugs or drug combinations with a prolonged prophylactic effect, even if they have a somewhat lower efficacy, may provide the most public health benefit over time in higher transmission settings.

Non-pharmacologic interventions such as the use of insecticide-impregnated bednets may change transmission patterns dramatically in an area, impacting malaria epidemiology and immunity, and therefore affecting the rationale for choosing methods for monitoring treatment outcomes and for antimalarial treatment and prophylaxis policies. Ultimately, the goal of antimalarial therapy should not be just to employ the most effective drug as determined in standard in vivo studies that measure efficacy in a single treatment episode, but to reduce the incidence of uncomplicated malaria, anemia, severe malaria, and ultimately malaria mortality when the drug is used to treat consecutive malaria episodes over time. Longitudinal studies may be required to provide the information necessary to make these region-specific choices.

The view that drug resistance monitoring and antimalarial treatment policies must be tailored to different settings across Africa should not be interpreted as pessimism or obstructionism. In fact, with their combined resources, expertise and dedication, international agencies, donors, health officials, drug-resistance monitoring networks, advocacy groups and scientists are up to the task of working together toward a rational, region-specific approach to achieve the goals of reducing malaria disease and deaths in Africa. The key to success will be cooperation and coordination among these partners, who all share the same objectives.

Acknowledgements Thanks to Terrie E. Taylor and Pascal Ringwald for critical reading of the manuscript.

References

Agarwal A, Guindo A, Cissoko, Y, Taylor JG, Coulibaly D, Kone A, Kayentao K, Djimde A, Plowe CV, Doumbo O, Wellems TE, Diallo D (2000) Hemoglobin C associated with protection from severe malaria in the Dogon of Mali, a West African population with a low prevalence of hemoglobin S. Blood 96:2358–2363

Attaran A, Barnes KI, Curtis C, D'Alessandro U, Fanello CI, Galinski MR, Kokwaro G, Looareesuwan S, Makanga M, Mutabingwa TK, Talisuna A, Trape J, Watkins WM (2004) WHO, the Global Fund, and medical malpractice in malaria treatment. Lancet 363:237–240

Baird JK, Purnomo, Basri H, Bangs MJ, Andersen EM, Jones TR, Masbar S, Harjosuwarno S, Subianto B, Arbani PR (1993) Age-specific prevalence of *Plasmodium falciparum* among six populations with limited histories of exposure to endemic malaria. Am J Trop Med Hyg 49:707–719

Bloland PB (2001) Drug resistance in malaria. WHO/CDS/CSR/DRS/2001.4. Geneva, WHO Department of Communicable Disease Surveillance and Response

Bloland PB, Ettling M (1999) Making malaria-treatment policy in the face of drug resistance. Ann Trop Med Parasitol 93:5–23

Bloland PB, Ettling M, Meek S (2000) Combination therapy for malaria in Africa: hype or hope? Bull World Health Organ 78:1378–1388

Bloland PB, Lackritz EM, Kazembe PN, Were JB, Steketee R, Campbell CC (1993) Beyond chloroquine: implications of drug resistance for evaluating malaria therapy efficacy and treatment policy in Africa. J Infect Dis 167:932–937

Brooks DR, Wang P, Read M, Watkins WM, Sims PF, Hyde JE (1994) Sequence variation of the hydroxymethyldihydropterin pyrophosphokinase: dihydropteroate synthase gene in lines of the human malaria parasite, *Plasmodium falciparum*, with differing resistance to sulfadoxine. Eur J Biochem 224:397–405

Cattamanchi A, Kyabayinze D, Hubbard A, Rosenthal PJ, Dorsey G (2003) Distinguishing recrudescence from reinfection in a longitudinal antimalarial drug efficacy study: Comparison of results based on genotyping of MSP-1, MSP-2, and GLURP. Am J Trop Med Hyg 68:133–139

Cowman AF, Morry MJ, Biggs BA, Cross GA, Foote SJ (1988) Amino acid changes linked to pyrimethamine resistance in the dihydrofolate reductase-thymidylate synthase gene of *Plasmodium falciparum*. Proc Natl Acad Sci USA 85:9109–9113

Djimde A, Doumbo OK, Cortese JF, Kayentao K, Doumbo S, Diourte Y, Dicko A, Su XZ, Nomura T, Fidock DA, Wellems TE, Plowe CV (2001a) A molecular marker for chloroquine-resistant falciparum malaria. N Engl J Med 344:257–263

Djimde A, Doumbo OK, Steketee RW, Plowe CV (2001b) Application of a molecular marker for surveillance of chloroquine- resistant falciparum malaria. Lancet 358:890–891

Djimde AA, Doumbo OK, Traore O, Guindo AB, Kayentao K, Diourte Y, Niare-Doumbo S, Coulibaly D, Kone AK, Cissoko Y, Tekete M, Fofana B, Dicko A, Diallo DA, Wellems TE, Kwiatkowski D, Plowe CV (2003) Clearance of drug-resistant parasites as a model for protective immunity in *Plasmodium falciparum* malaria. Am J Trop Med Hyg 69:558–563

Dorsey G, Kamya MR, Singh A, Rosenthal PJ (2001) Polymorphisms in the *Plasmodium falciparum PfCRT* and *pfmdr-1* genes and clinical response to chloroquine in Kampala, Uganda. J Infect Dis 183:1417–1420

Dorsey G, Njama D, Kamya MR, Cattamanchi A, Kyabayinze D, Staedke SG, Gasasira A, Rosenthal PJ (2002) Sulfadoxine/pyrimethamine alone or with amodiaquine or artesunate for treatment of uncomplicated malaria: A longitudinal randomised trial. Lancet 360:2031–2038

Duraisingh MT, Jones P, Sambou I, von Seidlein L, Pinder M, Warhurst DC (2000) The tyrosine-86 allele of the *PfMDR1* gene of *Plasmodium falciparum* is associated with increased sensitivity to the anti-malarials mefloquine and artemisinin. Mol Biochem Parasitol 108:13–23

Fidock DA, Nomura T, Talley AK, Cooper RA, Dzekunov SM, Ferdig MT, Ursos LM, Sihdu AB, Naude B, Deitsch KW, Su X, Wootton JC, Roepe PD, Wellems TE (2000) Mutations in the *P. falciparum* digestive vacuole transmembrane protein PfCRT and evidence for their role in chloroquine resistance. Mol Cell 6:861–871

Fidock DA, Nomura T, Wellems TE (1998) Cycloguanil and its parent compound proguanil demonstrate distinct activities against *Plasmodium falciparum* malaria parasites transformed with human dihydrofolate reductase. Mol Pharmacol 54:1140–1147

Fidock DA, Wellems, TE (1997) Transformation with human dihydrofolate reductase renders malaria parasites insensitive to WR99210 but does not affect the intrinsic activity of proguanil. Proc Natl Acad Sci USA 94:10931–10936

Foote SJ, Cowman, AF (1994) The mode of action and the mechanism of resistance to antimalarial drugs. Acta Trop 56:157–171

Foote SJ, Galatis D, Cowman AF (1990) Amino acids in the dihydrofolate reductase-thymidylate synthase gene of *Plasmodium falciparum* involved in cycloguanil resistance differ from those involved in pyrimethamine resistance. Proc Natl Acad Sci USA 87:3014–3017

Greenberg AE, Ntumbanzondo M, Ntula N, Mawa L, Howell J, Davachi F (1989) Hospital-based surveillance of malaria-related paediatric morbidity and mortality in Kinshasa, Zaire. Bull World Health Organ 67:189–196

Hastings IM (1997) A model for the origins and spread of drug-resistant malaria. Parasitology 115:133–141

Hastings IM Mackinnon MJ (1998) The emergence of drug-resistant malaria. Parasitology 117:411–417

Hellgren U, Kihamia CM, Mahikwano LF, Bjorkman A, Eriksson O, Rombo L (1989) Response of *Plasmodium falciparum* to chloroquine treatment: Relation to whole blood concentrations of chloroquine and desethylchloroquine. Bull World Health Organ 67:197–202

Hill AV, Allsopp CE, Kwiatkowski D, Anstey NM, Twumasi P, Rowe PA, Bennett S, Brewster D, McMichael AJ, Greenwood BM (1991) Common west African HLA antigens are associated with protection from severe malaria. Nature 352:595–600

Jelinek T, Aida AO, Peyerl-Hoffmann G, Jordan S, Mayor A, Heuschkel C, el Valy AO, von Sonnenburg F, Christophel EM (2002) Diagnostic value of molecular markers in chloroquine-resistant falciparum malaria in Southern Mauritania. Am J Trop Med Hyg 67:449–453

Koella JC Antia R (2003) Epidemiological models for the spread of anti-malarial resistance. Malar J 2:3

Kublin JG, Cortese JF, Njunju EM, Mukadam RAG Wirima JJ, Kazembe PN, Djimde AA, Kouriba B, Taylor TE, Plowe CV (2003) Reemergence of chloroquine-sensitive *Plasmodium falciparum* malaria after cessation of chloroquine use in Malawi. J Infect Dis 187:1870–1875

Kublin JG, Dzinjalamala FK, Kamwendo DD, Malkin EM, Cortese JF, Martino LM, Mukadam RA, Rogerson SJ, Lescano AG, Molyneux ME, Winstanley PA, Chimpeni P, Taylor TE, Plowe CV (2002) Molecular markers for failure of sulfadoxine-pyrimethamine and chlorproguanil-dapsone treatment of *Plasmodium falciparum* Malaria. J Infect Dis 185:380–388

Lim AS, Galatis D, Cowman AF. (1996). *Plasmodium falciparum* : amplification and overexpression of PfMDR1 is not necessary for increased mefloquine resistance. Exp Parasitol. 83:295–303

Luxemburger C, Thwai KL, White NJ, Webster HK, Kyle DE, Maelankirri L, Chongsuphajaisiddhi T, Nosten F (1996) The epidemiology of malaria in a Karen population on the western border of Thailand. Trans R Soc Trop Med Hyg 90:105–111

Mackinnon MJ (1997) Survival probability of drug resistant mutants in malaria parasites. Proc Roy Soc Lond—Series B: Biological Sciences 264:53–59

Mayor AG, Gomez-Olive X, Aponte JJ, Casimiro S, Mabunda S, Dgedge M, Barreto A, Alonso PL (2001) Prevalence of the K76T mutation in the putative *Plasmodium falciparum* chloroquine resistance transporter (*PfCRT*) gene and its relation to chloroquine resistance in Mozambique. J Infect Dis 183:1413–1416

Mayxay M, Chotivanich K, Pukrittayakamee S, Newton P, Looareesuwan S, White NJ (2001) Contribution of humoral immunity to the therapeutic response in falciparum malaria. Am J Trop Med Hyg 65:918–923

Meshnick SR (2002) Artemisinin: Mechanisms of action, resistance and toxicity. Int J Parasitol 32:1655–1660

Miller LH, Carter R (1976) A review. Innate resistance in malaria. Exp Parasitol 40:132–146

Mita T, Kaneko A, Lum JK, Bwijo B, Takechi N, Zungu IL, Tsukahara T, Tanabe K, Kobayakawa T, Bjorkman A (2003) Recovery of chloroquine sensitivity and low prevalence of the *Plasmodium falciparum* chloroquine resistance transporter gene mutation K76T following the discontinuance of chloroquine use in Malawi. Am J Trop Med Hyg 68:413–415

Modiano D, Luoni G, Sirima BS, Simpore J, Verra F, Konate A, Rastrelli E, Olivieri A, Calissano C, Paganotti GM, D'Urbano L, Sanou I, Sawadogo A, Modiano G, Coluzzi M (2001) Haemoglobin C protects against clinical *Plasmodium falciparum* malaria. Nature 414:305–308

Mu J, Ferdig MT, Feng X, Joy DA, Duan J, Furuya T, Subramanian G, Aravind L, Cooper RA, Wootton JC, Xiong M, Su XZ. (2003). Multiple transporters associated with malaria parasite responses to chloroquine and quinine. Mol Microbiol. 49:977–989

Mutabingwa T, Nzila A, Mberu E, Nduati E, Winstanley P, Hills E, Watkins W (2001a) Chlorproguanil-dapsone for treatment of drug-resistant falciparum malaria in Tanzania. Lancet 358:1218–1223

Mutabingwa TK, Maxwell CA, Sia IG, Msuya FH, Mkongewa S, Vannithone S, Curtis J, Curtis CF (2001b) A trial of proguanil-dapsone in comparison with sulfadoxine-pyrimethamine for the clearance of *Plasmodium falciparum* infections in Tanzania. Trans R Soc Trop Med Hyg 95:433–438

Nguyen-Dinh, P Payne D (1980) Pyrimethamine sensitivity in *Plasmodium falciparum*: Determination in vitro by a modified 48-hour test. Bull World Health Organ 58:909–912

Nosten F, ter Kuile F, Chongsuphajaisiddhi T, Luxemburger C, Webster HK, Edstein M, Phaipun L, Thew KL, White NJ (1991) Mefloquine-resistant falciparum malaria on the Thai-Burmese border. Lancet 337:1140–1143

Parise ME, Ayisi JG, Nahlen BL, Schultz LJ, Roberts JM, Misore A, Muga R, Oloo AJ, Steketee RW (1998) Efficacy of sulfadoxine-pyrimethamine for prevention of placental malaria in an area of Kenya with a high prevalence of malaria and human immunodeficiency virus infection. Am J Trop Med Hyg 59:813–822

Paul RE, Packer MJ, Walmsley M, Lagog M, Ranford-Cartwright LC, Paru R, Day KP (1995) Mating patterns in malaria parasite populations of Papua New Guinea. Science 269:1709–1711

Peel SA, Bright P, Yount B, Handy J, Baric RS. (1994). A strong association between mefloquine and halofantrine resistance and amplification, overexpression, and mutation in the P-glycoprotein gene homolog (pfmdr) of *Plasmodium falciparum* in vitro. Am J Trop Med Hyg. 51:648–658

Peterson DS, Milhous WK, Wellems TE (1990) Molecular basis of differential resistance to cycloguanil and pyrimethamine in *Plasmodium falciparum* malaria. Proc Natl Acad Sci USA 87:3018–3022

Peterson DS, Walliker D, Wellems TE (1988) Evidence that a point mutation in dihydrofolate reductase- thymidylate synthase confers resistance to pyrimethamine in falciparum malaria. Proc Natl Acad Sci USA 85:9114–9118

Pillai DR, Labbe AC, Vanisaveth V, Hongvangthong B, Pomphida S, Inkathone S, Zhong K, Kain KC (2001) *Plasmodium falciparum* malaria in Laos: Chloroquine treatment outcome and predictive value of molecular markers. J Infect Dis 183:789–795

Plowe CV, Doumbo OK, Djimde A, Kayentao K, Diourte Y, Doumbo SN, Coulibaly D, Thera M, Wellems TE, Diallo DA (2001) Chloroquine treatment of uncomplicated *Plasmodium falciparum* malaria in Mali: Parasitologic resistance versus therapeutic efficacy. Am J Trop Med Hyg 64:242–246

Plowe CV. (2003). Monitoring antimalarial drug resistance: making the most of the tools at hand. J Exp Biol 206:3745–3752.

Plowe CV, Kublin JG, Dzinjalamala FK, Kamwendo DS, Mukadam RA, Chimpeni P, Molyneux ME, Taylor TE (2004) Sustained clinical efficacy of sulfadoxine-pyrimethamine for uncomplicated falciparum malaria in Malawi after 10 years as first line treatment: Five year prospective study. Br Med J 328:545–548

Price RN, Cassar C, Brockman A, Duraisingh M, van Vugt M, White NJ, Nosten F, Krishna S (1999) The *PfMDR1* gene is associated with a multidrug-resistant phenotype in *Plasmodium falciparum* from the western border of Thailand. Antimicrob Agents Chemother 43:2943–2949

Price RN, Nosten F, Luxemburger C, ter Kuile FO, Paiphun L, Chongsuphajaisiddhi T, White NJ (1996) Effects of artemisinin derivatives on malaria transmissibility. Lancet 347:1654–1658

Reed MB, Saliba KJ, Caruana SR, Kirk K, Cowman AF (2000) Pgh1 modulates sensitivity and resistance to multiple antimalarials in *Plasmodium falciparum*. Nature 403:906–909

Ritchie GY, Mungthin M, Green JE, Bray PG, Hawley SR. Ward SA. (1996). In vitro selection of halofantrine resistance in *Plasmodium falciparum* is not associated with increased expression of Pgh1. Mol Biochem Parasitol 83:35–46

Roper C, Pearce R, Bredenkamp B, Gumede J, Drakeley C, Mosha F, Chandramo-
    han D, Sharp B (2003) Antifolate antimalarial resistance in southeast Africa:
    A population-based analysis. Lancet 361:1174–1181
Ruwende C, Khoo SC, Snow RW, Yates SN, Kwiatkowski D, Gupta S, Warn P, Allsopp CE,
    Gilbert SC, Peshu N (1995) Natural selection of hemi- and heterozygotes for G6PD
    deficiency in Africa by resistance to severe malaria. Nature 376:246–249
Sidhu AB, Verdier-Pinard D, Fidock DA (2002) Chloroquine resistance in *Plasmodium
    falciparum* malaria parasites conferred by *PfCRT* mutations. Science 298:210–213
Srivastava IK Vaidya AB (1999) A mechanism for the synergistic antimalarial action
    of atovaquone and proguanil. Antimicrob Agents Chemother 43:1334–1339
Sulo J, Chimpeni P, Hatcher J, Kublin JG, Plowe CV, Molyneux ME, Marsh K, Taylor TE,
    Watkins WM, Winstanley PA (2002) Chlorproguanil-dapsone versus sulfadoxine-
    pyrimethamine for sequential episodes of uncomplicated falciparum malaria in
    Kenya and Malawi: A randomised clinical trial. Lancet 360:1136–1143
Trape JF, Pison G, Preziosi MP, Enel C, Dulou AD, Delaunay V, Samb B, Lagarde E,
    Molez JF, Simondon F (1998) Impact of chloroquine resistance on malaria mortal-
    ity. Comptes Rendus de l'Academie des Sciences de Paris / Life Sciences 321:689–
    697
van Hensbroek MB, Morris-Jones S, Meisner S, Jaffar S, Bayo L, Dackour R, Phillips C,
    Greenwood BM (1995) Iron, but not folic acid, combined with effective anti-
    malarial therapy promotes haematological recovery in African children after
    acute falciparum malaria. Trans R Soc Trop Med Hyg 89:672–676
Viriyakosol S, Siripoon N, Petcharapirat C, Petcharapirat P, Jarra W, Brown KN,
    Snounou G (1995) Genotyping of *Plasmodium falciparum* isolates by the poly-
    merase chain reaction and potential uses in epidemiological studies. Bull World
    Health Organ 73:85–95
Walker DJ, Pitsch JL, Peng MM, Robinson BL, Peters W, Bhisutthibhan J, Meshnick SR
    (2000) Mechanisms of artemisinin resistance in the rodent malaria pathogen
    Plasmodium yoelii. Antimicrob Agents Chemother 44:344–347
Watkins WM, Mberu EK, Nevill CG, Ward SA, Breckenridge AM, Koech DK (1990)
    Variability in the metabolism of proguanil to the active metabolite cycloguanil in
    healthy Kenyan adults. Trans R Soc Trop Med Hyg 84:492–495
Wellems TE, Plowe CV (2001) Chloroquine-resistant malaria. J Infect Dis 184:770–776
White NJ (2002) The assessment of antimalarial drug efficacy. Trends Parasitol 18:458–
    464
White NJ, Nosten F, Looareesuwan S, Watkins WM, Marsh K, Snow RW, Kokwaro G,
    Ouma J, Hien TT, Molyneux ME, Taylor TE, Newbold CI, Ruebush TK, Danis M,
    Greenwood BM, Anderson RM, Olliaro P (1999) Averting a malaria disaster.
    Lancet 353:1965–1967
WHO (1973) Chemotherapy of malaria and resistance to antimalarials. 529. World
    Health Organization, Geneva. WHO Technical Report Series
WHO (1996) Assessment of therapeutic efficacy of antimalarial drugs for uncom-
    plicated falciparum malaria in areas with intense transmission. Geneva, World
    Health Organization, Division of Control of Tropical Diseases
WHO (2002) Monitoring antimalarial drug resistance. Bloland PB, D'Alessandro,
    Ringwald, and Watkins. WHO/CDS/CSR/EPH/2002.17. Geneva, World Health
    Organization

Wilson CM, Volkman SK, Thaithong S, Martin RK, Kyle DE, Milhous WK, Wirth DF (1993) Amplification of *pfmdr 1* associated with mefloquine and halofantrine resistance in *Plasmodium falciparum* from Thailand. Mol Biochem Parasitol 57:151–160

Winstanley PA (2000) Chemotherapy for falciparum malaria: The armoury, the problems and the prospects. Parasitol Today 16:146–153

Wolday D, Kibreab T, Bukenya D, Hodes R (1995) Sensitivity of *Plasmodium falciparum* in vivo to chloroquine and pyrimethamine-sulfadoxine in Rwandan patients in a refugee camp in Zaire. Trans R Soc Trop Med Hyg 89:654–656

Zalis MG, Pang L, Silveira MS, Milhous WK, Wirth DF (1998) Characterization of *Plasmodium falciparum* isolated from the Amazon region of Brazil: Evidence for quinine resistance. Am J Trop Med Hyg 58:630–637

Zolg JW, Plitt JR, Chen GX, Palmer S (1989) Point mutations in the dihydrofolate reductase-thymidylate synthase gene as the molecular basis for pyrimethamine resistance in *Plasmodium falciparum* . Mol Biochem Parasitol 36:253–262

Zucker JR, Lackritz EM, Ruebush TK, Hightower AW, Adungosi JE, Were JB, Metchock B, Patrick E, Campbell CC (1996) Childhood mortality during and after hospitalization in western Kenya: Effect of malaria treatment regimens. Am J Trop Med Hyg 55:655–660

Part II
Malaria the Disease

CTMI (2005) 295:83–104

Uncomplicated Malaria

M. P. Grobusch[1] (✉) · P. G. Kremsner[2]

[1]Infectious Diseases Unit, Faculty of Health Sciences,
University of the Witwatersrand,
7 York Road Parktown, 2196 Johannesburg, South Africa
grobuschmp@pathology.wits.ac.za

[2]Department of Parasitology, Institute of Tropical Medicine, Tübingen University,
Wilhelmstrasse 27, 72074 Tübingen, Germany

Abstract All symptoms and signs of uncomplicated malaria are non-specific, as shared with other febrile conditions, and can occur early or later in the course of the disease. In endemic areas, the presence of hepatosplenomegaly, thrombocytopenia and anaemia is clearly associated with malaria, particularly in children. Fever, cephalgias, fatigue, malaise, and musculoskeletal pain constitute the most frequent clinical features in malaria. Following single exposure to *Plasmodium falciparum* infection, the patient will either die in the acute attack or survive with the development of some immunity. Elderly individuals are prone to a more severe course of disease. The non-fatal *P. vivax* and *P. ovale* cause similar initial illnesses, with bouts of fever relapsing periodically, but irregularly over a period of up to 5 years. Renal involvement of a moderate degree is more common in mild falciparum malaria than initially suspected. The liver is also afflicted in mild disease, but organ damage is limited and fully reversible after parasitological cure. Whereas the cardiotoxic adverse effects of antimalarial chemotherapeutics are well known, clinically relevant cardiac involvement in humans is rare in severe disease and even rarer in uncomplicated falciparum malaria. Co-infection can aggravate malaria. There is a growing body of evidence that there is significant interaction in terms of mutual aggravation of the course of disease between HIV and malaria, particularly in pregnant women. Children with a high level of exposure to *P. falciparum* have a lower risk of developing atopic disorders.

Abbreviations

CD	Complex of differentiation
CI	Confidence interval
ECG	Electrocardiogram
HDL	High-density lipoprotein
HIV	Human immunodeficiency virus
MTCT	Mother-to-child transmission
OR	Odds ratio
PM	Placental malaria
VL	Viral load

1
The Clinical Features of Uncomplicated Malaria

There are four principal *Plasmodium* spp. causing malaria in humans; *P. falciparum*, *P. vivax*, *P. ovale*, and *P. malariae* (Singh 2004). Recently, the simian parasite *P. knowlesi* was also identified as a causative agent of human malaria in the forests of Malaysian Borneo. Infection with each species causes initially similar illness. However, the benign tertian (*P. vivax* and *P. ovale*) and quartan malaria (*P. malariae*) very rarely lead to serious illness or life-threatening complications (although they can cause debilitating disease) whereas falciparum malaria may progress towards severe disease in those not partially protected by acquired immunity or if not treated promptly (for severe disease manifestations in falciparum malaria, see chapters by T. Planche et al. and D. Roberts et al., this volume).

Following an exoerythrocytic, hepatic development phase of a minimum of about 6 days after inoculation in falciparum malaria, 9 days in benign tertian and 15 days in quartan malaria, parasitaemia and disease will develop (the basics of malaria biology are well described in detail in standard textbook chapters on malaria such as that by White 2003). An average incubation period of about 2 weeks between infection and the onset of disease might be considered as a rule of thumb in vivax and falciparum malaria, but much longer incubation periods may occur (Grobusch et al. 2000), particularly in the now rare *P. vivax hibernans* strains from East Asia, and can be found in parts of the Korean peninsula (Ree 2000).

A non-immune individual is infected with about 100 sporozoites on average, of which one or two will be able to invade hepatocytes; post-hepatic infection will become symptomatic in almost all cases. Left untreated, parasites multiply in red blood cells at a rate of about a 10-fold increase every 2 days, leading to a high parasitaemia within a few days and associated clinical complications such as anaemia, metabolic derangements (acidosis and hypoglycaemia) and severe organ impairment (renal or cerebral malaria or pulmonary oedema). The complications may lead to death, which may occur within a week after the onset of disease. However, death will not occur within a day or two after the start of disease. Therefore, the initiation of prompt and effective chemotherapy is the cornerstone of therapeutic intervention.

In general, the onset of malaria might be gradual or fulminant with all four species. All symptoms and signs can occur early or later in the course of the disease. The clinical picture is not uniform and might contain the diverse signs and symptoms in varying proportions. A typical or 'classical' malaria fever paroxysm is a well-described entity, with a 'cold stage' of sharply rising body temperature often leading to frank rigors, a 'hot stage' with flushes, tachycardia that is sustained for a few hours, followed by a defervescence stage of sharply falling body temperature, profuse sweats or diaphoresis and often symptomatic orthostatic hypotension. Fever as the (non-specific) lead symptom is often irregular at onset, which appears to apply in particular to falciparum malaria. Partially immune individuals develop an individual parasitic threshold, or pyrogenic density, and with parasite loads below this threshold, which might roughly be around 1,000 parasites/µl, patients will remain afebrile. Non-immune individuals will develop febrile disease at very low parasite counts (~ 10/µl). In children, as in adults, a regular 'typical' fever pattern seems to occur more frequently in non-falciparum malaria (Steele 1996). In all four species, fever patterns might be very irregular at the beginning prior to synchronization, and again this tends to be most pronounced in falciparum malaria. However, fever in malaria patients follows a circadian rhythm. In a study enrolling 66 children with malaria and fever, 50% had fever at 6 p.m., whereas only 14% had fever at 6 a.m. on the first day of admission (Lell et al. 2000). The use of common antipyretics such as paracetamol cannot reduce fever in all types of malaria (however, Krishna et al. have shown that it can in uncomplicated malaria; Brandts et al. 1997), and fever may be an important innate defence mechanism (Krishna 1995a, 1995b). This is extrapolated by the findings of parasite growth inhibition by elevated temperatures *in vitro* (Long et al. 2001). In endemic areas, the presence of hepatosplenomegaly, thrombocytopenia and anaemia in both febrile and afebrile individuals, particularly children, is clearly associated with malaria (Stein et al. 1985; Olaleye et al. 1998; Schellenberg et al. 1999; Norhayati et al. 2001, D'Acremont et al. 2002). Apart from fever as the lead symptom, cephalgias, fatigue and malaise, musculoskeletal pain, nausea and vomiting constitute the most frequently encountered, though non-specific, symptoms in malaria. In children, abdominal pain and vomiting occur more frequently than in adults. Table 1 lists the main clinical features of malaria as encountered in children compared to the findings in semi-immune and non-immune adults. In semi-immune patients who have experienced several episodes of malaria earlier in life, the onset and progression of disease are often delayed and unprepossessing. Non-immune Europeans appear frequently to exhibit watery, rarely bloody diarrhoea, whereas black Africans more frequently manifest symptoms of the upper gastrointestinal tract with nausea and vomiting. However, data to underpin this clinical observation are lacking. Jelinek et al. (2002) found that 49/790 (6.2%) of adult semi-immune patients with *P. falciparum* parasitaemia were asymptomatic at the time of diagnosis in a large study cohort from Europe, whereas all 869 non-immune Europeans were symptomatic at time point of presentation. Of 660 cases of *P. falciparum* infection observed in a longitudinal study in Gabon, 77% were symptomatic at the time they were identified, and only 7% were preceded by an asymptomatic phase of more than 4 days. In those, sickle cell trait, glucose-6-phosphate dehydrogenase deficiency, and mutation in the promoter region of tumour necrosis factor were significantly associated with asymptomatic *P. falciparum* infection (Missinou et al. 2003).

Table 1 Main clinical features of uncomplicated malaria at time point of diagnosis as encountered in children compared to the findings in semi-immune and non-immune adults[a] (data from Kremsner et al. 1994; Metzger et al. 1995b; Radloff et al. 1996; Brandts et al. 1997; Kun et al. 1998; Jelinek et al. 2002; Borrmann et al. 2003)

Sign or symptom	Semi-immune adults[b]	Non-immune adults[c]	African children[d]
Fever	76%	81%	91%
Headache	49%	50%	68%
Fatigue	24%	35%	50%
Abdominal pain	n.s.[e]	n.s.	40%
Myalgia, arthralgia	17%	23%	22%
Nausea, vomiting	12%	12%	26%
Diarrhoea	10%	14%	7%

[a] Multiple entries are possible.
[b] From various endemic areas ($n = 790$).
[c] Predominantly of German origin ($n = 869$).
[d] From Central Africa (Gabon), age range from 3 to 15 years, ($n = 484$).
[e] Not specified.

Although suggestive of malaria, signs and symptoms might be caused by an underlying or concomitant febrile condition. Depending on the area visited, a large range of infectious diseases might have to be included in the list of differential diagnoses, as well as other non-infectious causes that may explain some aspects of the clinical presentation (autoimmune disorders, neoplasms etc., all of which might present with fever).

The further course of disease depends on the malaria immune status of the patient as well as on the time point of the initiation of therapy. If treatment starts early, a full recovery in terms of clinical and parasitological cure can be achieved within a few days. Four to six days may have to be considered as the critical threshold period between onset of illness and initiation of therapy after which a serious disease progression may be expected in the non-immune malaria patient left untreated. The transition from uncomplicated disease to a life-threatening complicated condition often occurs discretely. In adults, a slight decrease in urinary output or an initially discrete cognitive impairment and behaviour alteration might herald disease progress. The initial phase of disease is often described as resembling a flu-like illness, but one distinct difference is the absence of a 'sore throat' in malaria. Dry cough might occur, but otherwise upper/lower respiratory tract involvement is rare in uncomplicated malaria. However, by the time additional laboratory tests will have been performed, the result of a thick smear should already be available to verify a diagnosis of malaria.

A typical laboratory constellation would include moderately raised C-reactive protein and procalcitonin levels, an often profound thrombo-cytopenia (Ladhani et al. 2002; Hänscheid et al. 2003; Moulin et al. 2003; Erhart et al. 2004) with normal or slightly lowered leukocyte counts. A low haptoglobin level and a raised lactate dehydrogenase level as indicators for haemolysis are typical for early, uncomplicated malaria. It is remarkable that the often pronounced thrombocytopenia does not lead to a higher bleeding tendency, as long as no additional factors become involved, such as a large-scale activation of the coagulation cascade through the intrinsic pathway, or disseminated intravascular haemolysis that occurs in a very small proportion of complicated falciparum malaria cases (Clemens et al. 1994). Thrombocytopenia complicates infection with all four malaria species and is not directly linked to disease severity. It is mainly caused by increased splenic clearance (Skudowitz 1973); excessive splenic platelet pooling and a shortened platelet life-span (Karanikas et al. 2004). Study results regarding the role of platelet-bound antibodies as contributors to malarial thrombocytopenia are controversial (White 2003). Interestingly, in a series of seven patients with uncomplicated malaria, scintigraphy results indicated that the sequestration of thrombocytes appears to be rather diffuse throughout the reticulo-endothelial system than merely splenic (Karanikas et al. 2004). After parasitological cure, reconstitution is usually swift, and frequently a short-lasting overshooting thrombophilia can be observed.

Depending on the duration of illness and the haemoglobin level at onset of malaria, anaemia might be profound and may lead, in the case of falciparum malaria, to grave problems particularly in pregnant women and young children. In this context it seems interesting that non-immune adults with malaria rarely suffer from anaemia. In contrast, non-immune children and especially infants are very susceptible to developing anaemia. This might merely reflect the difference in the number of erythrocytes in adults and children, while the parasite inoculum and the presumed rates of parasite multiplication in both liver and bloodstream are the same in adults as in children. Liver and kidney function tests are initially often unaffected. Later during disease a discrete rise in transaminases and, depending on the hydration status, a rise of retention parameters can be seen.

Interestingly, the plasma lipid profile of malaria patients displays a particular abnormal pattern (Lambrecht et al. 1978; Nilsson-Ehle and Nilsson-Ehle 1990; Kittl et al. 1992; Faucher et al. 2002; Grobusch et al. 2003a). Performed on a fasting blood sample, it is characterized by low total cholesterol, very low high-density lipoprotein (HDL)-cholesterol, apolipoprotein A1 below detection levels, and a raised triglyceride level. Plasma lipid alterations are transient and largely limited to the parasitaemic phase. Temporary HDL depletion has been observed in over 90% in some studies. A decrease in HDL cholesterol levels as part of the acute phase reaction, diminished hepatic synthesis due to liver involvement, and direct and parasite–lipoprotein interactions, have been suggested as possible reasons. Although there is little immediate clinical relevance of this phenomenon, it might offer some deeper insight into the parasite–host interplay. For example, the increasingly complex role of haemozoin formation and its potential for interference with host metabolism might offer a hitherto unnoticed link here, as changes in the host's lipid metabolism might be directly due to processes involved in haemozoin formation. Schwarzer et al. (2003) pointed out that haemozoin formation in the parasitized erythrocytes and its long-term stability appear to constitute two functionally synergistic phenomena in *P. falciparum* malaria. The first one is due to the absence of haem-catabolizing enzymes in the parasite and is a prerequisite for the formation of a series of a whole range of lipid derivatives by haem catalysis. The second is due to the host phagocytes' inability to detoxify haemozoin, and to the very long persistence of haemozoin in macrophages (Metzger et al. 1995a). Therefore, phagocytosed haemozoin may ship vast amounts of diverse and complex bioactive lipid derivatives into host macrophages. Further studies are needed to elucidate whether this mechanism is the driving force behind the lipid metabolism changes observed in falciparum malaria in man (a detailed overview of the current knowledge on haemozoin formation and metabolism is provided in the chapters by Scholl et al., this volume).

Malaria attacks may recrudesce over the next few weeks but die out in the absence of reinfection. The non-fatal *P. vivax* and *P. ovale* cause similar initial illnesses, with bouts of fever relapsing periodically but irregularly over a period of up to 5 years. These are true recrudescences and not simple relapses produced by merozoites developing from hypnozoites in hepatocytes. The relapse pattern is largely determined by geographic origin of infection, as is strain variability particularly of *P. vivax*, where hypnozoites may resist eradication attempts with the 8-aminoquinoline primaquine. Although 'benign' in nature in terms of self-termination, complications similar to those in severe falciparum malaria and afflicting all major organ systems have been described in very rare cases (including pulmonary oedema and acute renal failure) and appear to occur more frequently in vivax than in ovale malaria (Carlini et al. 1999; Modebe and Jain 1999; Mendis et al. 2001; Tanios et al. 2001; Beg et al. 2002; Koibuchi et al. 2003; Prakash et al. 2003). Splenic rupture due to mechanical tear of the enlarged spleen might occur in all human malarias (Facer and Rouse 1991). Following single exposure to the non-fatal *P. malariae* infection, and an incubation period that may extend to several weeks, the patient develops a recurrent fever which occurs at increasing intervals. There may be considerable anaemia, and enlargement of liver and spleen. If left untreated, sometimes recrudescences may occur for more than 30 years, often being iatrogenic following interventions interfering with immune function (Tsuchida et al. 1982; Vinetz et al. 1998; Chadee et al. 2000). As in falciparum malaria, there are no relapses due to hypnozoite activation from hepatocytes. The attack severity diminishes as time goes by.

Mixed infections can occur and are probably often underestimated. In most co-infections involving *P. falciparum*, the other species are suppressed, which might delay the onset of symptoms of non-falciparum malaria for weeks (Looareesuwan et al. 1987; Mason and McKenzie 1999). Conversely, it is possible that *P. vivax* suppresses *P. falciparum* (Mayxay et al. 2001), thus leading to a reduction of disease severity and transmissibility (White 2003).

As adult morbidity and mortality data are lacking for almost the whole of sub-Saharan Africa (Kaufmann et al. 1997), only very limited data on the current situation are available on illnesses and deaths attributable to malaria in the elderly in the world's region with the highest malaria endemicity. As an aggravated course of malaria has been demonstrated in the elderly not exposed before, and therefore lacking immunity, a worsened outcome could be expected in those elderly populations where malaria is of only moderate to low endemicity. In a rural area of Burkina Faso, an area of seasonal malaria transmission, Sankoh et al. (2003) analysed the mortality patterns in adults and older people (≥60 years) based on a demographic surveillance system in a population of 39 villages. Whereas the crude all-cause mortality rate per 1,000 adults was 7.3 in the population overall, and 55.8 in the elderly, malaria and diarrhoea (as recorded through verbal autopsy) accounted for 21% of the total deaths in adults and 22% in the elderly. Meaningful data interpretation is difficult because the data are crude, and are based on retrospective analysis only. However, one might assume that the benefit of sustained and frequently boosted partial immunity is at least counterbalanced by a generally more fragile health status and co-morbidity, together with a decline in the immune system's capacity to control infectious diseases. It has been hypothesized that defects in acquired humoral and T-cell mediated immunity exist and that increased susceptibility to infection may result from defects in the constitutive functioning of macrophages and granulocytes (Khanna and Markham 1999). There is also some evidence for an age-related shift towards a type 2 cytokine profile (Sandmand et al. 2002).

In a transmigrant population in Irian Jaya, Indonesia, Baird et al. (1998) noticed that in a cohort of people with relatively few infants or people of advanced age, the risk of progression from mild to severe falciparum malaria increased with age. More detailed data exist on the manifestations of imported malaria in the elderly as summed up by Schwartz et al. (2001) and Mühlberger et al. (2003). Several smaller studies on non-immune patients suggest that the proportion of more severe or even fatal cases rises with age. In a cross-sectional study from Israel, Schwartz et al. (2001) assessed, amongst other factors, the influence of age on clinical manifestations of P. falciparum malaria in non-immune patients. In most reports a distinction has been made between individuals younger or older than 40 years, but none between those younger or older than 59 years, which appears to be a more reasonable age divide given the average health status of individuals from affluent, non-tropical countries. In their own cohort of 135 falciparum malaria patients, Schwartz et al. found that 95% of the patients younger than 40 years escaped with uncomplicated malaria, whereas this rate dropped to 82% in those individuals aged 40 years or more (odds ratio, 4.29). The mean age of the patients with mild disease was 35±12 (SD) vs. 43±17 (SD) years for patients with severe disease. In addition, all deaths attributable to complications were notified in the older age group. With regard to risk, sex differences were not recorded. The positive correlation between age and disease severity seemed to be unrelated to prophylaxis.

In a cohort of 1181 patients with falciparum malaria imported to Europe analysed by Mühlberger et al. (2003), 78 (6%) were aged over 60 years. The frequency of hospital admission was found to increase with each decade of life [odds ratio (OR), 1.21; 95% confidence interval (CI), 1.06–1.39], and the risk of progression towards severe disease was found to increase by approximately 30% per decade of life (OR, 1.32; 95% CI, 1.14–1.53). However, when elderly patients were compared with younger ones, the elevated risk for elderly patients was non-significant for both aspects.

In summary, all data taken together suggest strongly that the risk of progression beyond moderate levels of illness is elevated in the elderly malaria patient, a clinical finding which is underpinned by immunological studies indicating that a senile decline in humoral and cellular immune function may be the leading underlying cause.

2
The Kidney in Uncomplicated Malaria

P. falciparum and *P. malariae* infections are clearly associated with renal disease. Moderate proteinuria is a common finding (Ehrich and Horstmann 1985) and usually resolves following parasite eradication. In severe disease, a nephrotic syndrome, and less frequently, acute renal failure may develop as a consequence, and arises from a combination of impaired microcirculation due to parasite obstruction of the microvasculature, hypovolaemia, haemolysis, and intravascular coagulation. In a proportion of cases, quartan malaria is associated with nephrotic syndrome as a life-threatening complication in children in endemic areas. A membranoproliferative type of glomerulonephritis with focal and segmental glomerulosclerosis is the most common type of renal damage (Van Velthuysen and Florquin 2000). A combination of endothelial damage and immune complex deposition is observed, but it has been found that immune complexes containing malarial antigens represent only a fraction of the glomerular immune deposits. Polyclonal B-cell activation may contribute to quartan malaria nephritis (Van Velthuysen and Florquin 2000).

Ahmad et al. (1989) assessed renal function in 75 children from India with uncomplicated falciparum or vivax malaria alongside 10 healthy controls. Of the 75 children, 36 showed renal impairment with a decreased endogenous creatinine clearance (<65 ml/min/m^2). In *P. falciparum* infections, renal impairment was twice as frequent as in vivax cases, and reduction in endogenous creatinine clearance was significantly greater. Renal impairment was positively correlated with parasite load as assessed on peripheral blood film. Renal impairment was transient, with the endogenous creatinine clearance returning to normal values within 14 days in all those individuals available for follow-up.

Weber et al. (1999) studied renal function in 80 Gambian children with cerebral malaria and a similarly large group of children with mild malaria. The authors found that renal involvement is common in children with malaria in this part of Africa, with pre-renal, glomerular, and tubulo-interstitial factors contributing. Although more pronounced in children with cerebral malaria, renal dysfunction was found to be mild and fully reversible.

In a study performed in Ghana (Burchard et al. 2003), cystatin C as a novel and sensitive indicator of renal dysfunction was investigated in a cohort of 78 children with uncomplicated falciparum malaria, alongside measurements of albumin, IgG and $\alpha 1$ microglobulin excretion. Eighty-five per cent of the children had proteinuria of glomerular and tubular patterns, indicating that transient, moderate glomerulonephritis and tubular damage occurs frequently in falciparum malaria, but without leading to complicated disease in the majority of cases. Subclinical impairment of the glomerular filtration rate (GFR) was found by elevated cystatin C plasma concentrations in 17% of the children. Correspondingly, Günther et al. (2002) found in a series of 108 adults with falciparum malaria imported to Germany that 55% had a reduced GFR as indicated by elevated cystatin C levels. The proportion of patients with malaria and at least transient and moderately impaired renal function has been underestimated in the past.

Most lesions remain subclinical and resolve soon after successful antimalarial treatment in children in endemic areas. Non-immune adults seem to carry the highest risk of progression towards renal complications in falciparum malaria.

3
The Liver in Uncomplicated Malaria

The liver is afflicted not only in severe malaria but also in mild cases, and this is reflected by the rise in transaminases and a varying degree of hepatomegaly in acute malaria. However, the changes are reversible, and transaminases and organ size revert to normal limits shortly after parasitological cure. Whereas the initial hepatic development and the stage of hypnozoite dormancy in the benign tertian malarias do not lead to traceable signs, liver pathology in the systemic phase of illness is explained by an increased hepatic blood flow whereas it is reduced in severe disease (Pukrittayakamee 1992), and parasite sequestration in the microvasculature (Molyneux et al. 1989; White 2003). The metabolic functions seem to be largely undisturbed in uncomplicated cases. Hyperbilirubinaemia in mild cases and frank jaundice in complicated disease appear to have haemolytic, hepatic, and cholestatic components (Wilairatana et al. 1994; White 2003).

4
The Heart in Uncomplicated Malaria

Whereas the cardiotoxic adverse effects of antimalarial chemotherapeutics are well known, clinically relevant cardiac involvement in humans is rare in severe disease (Bethell et al. 1996) and even rarer in uncomplicated falciparum malaria, particularly if the patient has no underlying cardiac illness. Günther et al. (2003) summed up the scarce available data from case reports and smaller studies, and present the cardiological features of 161 malaria patients treated in Berlin, Germany. Whereas electrocardiogram (ECG) changes were common and observed in 23/161 (14%) of patients, myocardial damage as indicated by an elevated troponin T level was present only in a single individual, whereas creatine kinase muscle–brain levels were normal in all patients. The ECG changes were due to delayed conduction of various kinds in eight patients, and in 15 patients there were non-specific T segment or T wave alterations. A proportion of those could theoretically reflect early signs of myocardial cell damage, as the electrophysiology of cardiomyocytes can be altered before myocytolysis occurs, and as non-specific ECG abnormalities can be found when the affected area of the myocardium is large enough. Only very few studies have used echocardiographic methods to assess cardiac function and morphology in malaria patients presenting without cardiac symptoms. In a series of 22 such patients without a previous history of cardiac disease, two patients with pericardial effusion and one patient with left ventricular hypokinesia were detected, but all patients made an uneventful recovery following malaria therapy and required no cardiological intervention (Franzen et al. 1992). In summary, clinically relevant cardiac involvement appears to be a rare feature in otherwise uncomplicated falciparum malaria.

5
The Role of the Spleen in Limiting the Progression of Disease

Although still poorly understood in its full complexity, it is evident that the spleen plays an essential role in limiting the acute expansion of the infection not only by removing parasitized erythrocytes but also by modulating parasite antigen expression on the surface of infected erythrocytes, as well as by enhancing cellular and humoral immune responses of the host (Garnham 1970; David et al. 1983; Chotivanich et al. 2002). The important role of the spleen becomes dramatically evident in patients enduring a malaria episode following splenectomy (Israeli et al. 1987; Ho et al. 1992; Pongponratn et al. 2000; Grobusch et al. 2003b).

6
Malaria and Concomitant Infection

In a series of studies from Thailand, helminth-infected patients appear to be protected up to a certain level from cerebral malaria. Helminth-infected patients with mild malaria were also more likely to harbour peripheral game-tocytes and there was an increased number of mixed *P. falciparum*/*P. vivax* infections in *Ascaris*-infected patients. The incidence of falciparum malaria was higher in helminth-infected subjects. Nacher (2002) sums up these strik-ing findings with the multiple hypotheses that *P. falciparum* may benefit from helminth-infected hosts, that co-infected hosts may benefit from helminth in-fection by having attenuated courses of malaria, and that, in turn, helminths benefit in terms of protecting their habitat by protecting their hosts.

Coinfection with bacteria or viruses in otherwise uncomplicated disease can aggravate the course of malaria, and vice versa (Ghoshal et al. 2001). It is of interest that not only falciparum malaria but also benign tertian malaria can be significantly aggravated by the presence of additional pathogens (Flatau et al. 2000). Wongsrichanalai et al. (2003) pointed out that the timely recog-nition of dual infections is complicated by their often similar presentations, which applies particularly to the period immediately after onset in which administration of chemotherapeutic agents, if applicable, is most likely to be successful in terms of early and uncomplicated termination of illness. It is well known that patients with severe malaria are prone to secondary bacte-rial infections, particularly non-typhoidal salmonellosis (Mabey et al. 1987). Invasive bacterial disease itself appears to play a role in the progression to-wards a more severe course of malaria (Berkley et al. 1999). In a study of 299 Malawian children with typhoidal *Salmonella* spp. bacteraemia was found to be significantly associated with malarial parasitaemia in children older than 6 months (Graham et al. 2000). Most of the literature is, however, anecdo-tal, and clinical studies of co-infection in otherwise uncomplicated malarial illness are lacking.

7
Malaria and Tuberculosis

An interaction between these two conditions may be likely, but to the best of our knowledge the question of to what extent tuberculosis and malaria interact has so far not been addressed in clinical studies. In a seroepidemio-logical survey from Nigeria, serum samples from 197 patients with malaria, hepatitis B, or tuberculosis and from 166 healthy controls were screened in order to determine any disease interactions. With one patient being human immunodeficiency virus (HIV)-1 seropositive, a statistical association was found between the presence of malaria antibody titres and a diagnosis of tuberculosis ($P<0.05$) (Adebajo et al. 1994). However, the significance of this finding remains unclear.

8
Malaria and HIV

Most of the studies carried out so far on the possible interplay between malaria and HIV have focussed on HIV-1 in relation to uncomplicated falciparum malaria. Early studies investigating a co-morbid influence of HIV and malaria were performed in settings of higher overall population $CD4^+$ cell counts; later studies have shown that in settings of lower $CD4^+$ cell counts malaria and HIV do interact.

In several early cross-sectional studies from Africa, no association was found between HIV-1 infection and falciparum malaria in either paediatric or adult patients (Muller et al. 1990; Allen et al. 1991; Nwanyanwu et al. 1997). In contrast, the risk of malaria in HIV-1 infected patients in a cohort of adults from Uganda was significantly raised (Whitworth et al. 2000). In the same cohort as well as in the placebo arm of a vaccination study, the rate of malaria was inversely correlated with $CD4^+$ cell counts (French et al. 2001).

A major complication of malaria in African children is anaemia, and in a study from Kenya, van Eijk and co-workers (2002) found that HIV-infected infants with malaria parasitaemia had lower mean haemoglobin levels compared with HIV-uninfected infants, or HIV-infected infants without malaria. These results suggested that HIV-infected infants are even more vulnerable to the deleterious effects of malaria on haemoglobin levels than their peers who are unafflicted by HIV infection.

Two cross-sectional studies on pregnant women in Malawi found higher prevalences of malaria parasitaemia at the first prenatal visit amongst HIV-1 infected women compared to seronegative women (32 vs. 19%; and 54 vs. 42%) (Verhoeff et al. 1999; Steketee et al. 1996). In a study by van Eijk et al. (2003) in Kenya, HIV-seropositive pregnant women (no distinction was made between HIV-1 and -2) were more likely than uninfected women to be parasitaemic, to have higher parasite densities, and to be febrile when parasitaemic, and placental infections were more likely to be chronic. Moreover, the typical pregnancy-specific pattern of malaria vanished in seropositive women, and primigravidae had a risk similar to that of multigravidae (Mount 2004). In a study from Malawi, it was shown that HIV impairs antimalarial immunity and, in particular, responses to placental type variant surface antigens with the impairment being greatest in the most immunosupressed women. One recent study investigated the effects of placental malaria on mother-to-child HIV transmission (MTCT) in a prospective community-randomized trial in Uganda (Brahmbhatt et al. 2003). In the 746 HIV-1-positive mother–infant pairs, the MTCT rate was 20.4%, and placental malaria was more common in HIV-positive than in HIV-negative women. The risk for MTCT after multivariate adjustment for viral load (VL) rose to 2.89 in association with placental malaria (PM), and to 2.85 in association with VL. Conflicting evidence arises from a study by Inion et al. (2003) who found no correlation between PM and peripartum transmission in a smaller cohort of 372 infants born to HIV-positive mothers in Mombasa, Kenya (2.9% MTCT in PM-positive vs. 5.6% in PM-negative mothers; $P=0.509$). However, differences in sample size and method of analysis limit the direct comparability of these initial studies regarding this issue.

The epicenter of HIV-2 prevalence lies in West Africa, with a peak prevalence of 8–10% in Guinea-Bissau (Reeves and Doms 2002). HIV-2 seems to be less virulent than HIV-1, infection leads to progression to AIDS after longer latent periods with lower plasma viral loads, slower decline in $CD4^+$ cell counts, and lower heterosexual and vertical transmission rates (Schim van der Loeff and Aaby 1999; Hightower and Kallas 2003). Few studies have been carried out specifically investigating interactions between HIV-2 and malaria. Whereas we know little about a possible increase in frequency and severity of malaria in HIV-2 infected individuals, there is little evidence for major interaction between these two conditions from two studies carried out in rural Guinea–Bissau investigating HIV-2 provirus load and immune stimulation in co-infected individuals (Ariyoshi et al. 1996; N'Gom et al. 1997).

Knowledge about the response to antimalarial therapy in HIV-positive individuals is limited. However, no statistically significant differences were found in the response to antimalarial chemotherapy between HIV-1-positive and HIV-negative children in a large (587 children) prospective, longitudinal cohort study in Kinshasa, Zaire (Greenberg et al. 1991).

With increasing efforts to introduce highly active antiretroviral therapy into those regions in the developing world most in need, early indications of possible interactions between antiretroviral therapy and its influence on the course of malaria episodes deserves attention. Nathoo and colleagues (2003) found that antiretroviral drugs decrease CD36 surface concentrations in vivo. Subsequently, they demonstrated *in vitro* that protease inhibitors impaired CD36-mediated cytoadherence and non-opsonic phagocytosis of parasitized erythrocytes by human macrophages. As binding of CD36 has been postulated to contribute to disease severity, it remains to be shown in further studies whether this interaction might alter malarial co-infection towards milder disease. In contrast, antiretroviral therapy might decrease phagocytosis of parasitized erythrocytes resulting in higher parasite densities and possibly more severe infection (Rogerson 2003).

Lawn (2004) reviewed the impact of co-infectiony on the pathogenesis of HIV-1 infection and came to the general conclusion that the immunologic impact of recurrent co-infections has the potential to increase viral replication, viral genotypic heterogeneity and $CD4^+$ lymphocyte loss, thus accelerating the immunologic decline. With respect to malaria, this view is underpinned by a study by Hoffman et al. (1999) conducted in Malawi in which plasma HIV-1 viral loads were found to be greater in those individuals with symptomatic falciparum malaria compared to controls, and that VL decreased following malaria treatment. These clinical findings have now been reinforced by various *in vitro* studies of the immunologic interplay of both infections (Lawn 2004).

In stark contrast, there is some interest predominantly by Chinese researchers in exploring vivax malaria as supportive treatment in HIV-1 infected patients. This pursuit is based on the assumption that the high levels of cytokine activation as induced in malaria may lead to changes of T-lymphocyte subsets and phenotypes which might be beneficial for HIV-positive individuals in terms of at least transitory $CD4^+$ cell level increases (Chen et al. 2003). For reasons of scientific appropriateness and the unproven benefit of immune-based therapy for HIV, this issue is highly controversial (Nierengarten 2003).

Little is known up to now about the altered disease course of benign tertian and quartan malaria in HIV-positive individuals. Certainly, the effects of interacting co-infections are even more discrete and difficult to assess than with falciparum malaria.

In conclusion, recent studies reveal details of a very delicate and multi-faceted interplay between malaria and HIV, a picture that might become even more complicated through the impact of antiretroviral therapy being introduced in areas where both entities frequently coincide.

9
Malaria and Atopy

A recent study from Gabon (van den Biggelaar et al. 2004) established a direct link between anthelminthic treatment of chronically infected children and increased atopic activity, thus indicating that helminths suppress allergic reactions. Looking at the mutual relationship between malaria and atopy, Lell et al. (2001) performed skin-prick tests with mite antigen in 91 children from Gabon who had been closely followed for an average of 5 years and in whom the exact incidence of malaria attacks was known. Survival analysis yielded a statistically significant lower risk of an atopic skin reaction in those children with a higher incidence of malaria attacks. The authors concluded that a low exposure to malaria parasites contributes to the development of an imbalanced immune system with a subsequent higher reactivity to allergens, and that high exposure to parasite antigens might counterbalance pro-inflammatory immune reactions, amongst other mechanisms involving increased interleukin-10 production, and thus protect against allergic diseases. In their comprehensive review on the complex relationship of exposure to parasites and subsequent remodelling of an individual's immune system with decreased susceptibility to allergens, Yazdanbakhsh et al. (2002) point out that the induction of a robust anti-inflammatory regulatory network by persistent immune challenge offers a unifying explanation for the observed inverse association of many infections with allergic disorders, as instead of Th-1 vs. Th-2, it may be the pro- vs. anti-inflammatory axis with a robust regulatory T-cell network that has to be considered central to the balance and to the prevention of either Th-1 and/or Th-2 diseases.

10
Conclusion

To date, a vast body of knowledge has been accumulated on uncomplicated malaria. Still, many secrets remain to be unveiled, and many puzzling questions are waiting for an answer, as highlighted in this text.

However, the most important area of future research is the transition between an uncomplicated disease episode and a life-threatening condition. Huge efforts in all the relevant fields of epidemiology, pathophysiology, immunology, molecular biology and genetics ought to be made to achieve a more profound insight into the mechanisms underlying the progression of the disease. This will facilitate preventative measures where and whenever possible and lead to optimal care for the individual malaria patient.

References

Adebajo AO, Smith DJ, Hazleman BL, Wreghitt TG (1994) Seroepidemiological associations between tuberculosis, malaria, hepatitis B and AIDS in West Africa. J Med Virol 42:366–368

Ahmad SH, Danish T, Faridi MMA, Ahmad AJ, Fakhir S, Khan AS (1989) Renal function in acute malaria in children. J Trop Pediatr 35:291–294

Allen S, Van de Perre P, Serufilira A, Lepage P, Carael M, De Clercq J, Black D et al (1991) Human deficiency virus and malaria in a representative sample of childbearing women in Kigali, Rwanda. J Infect Dis 164:67–71

Arioshi K, Berry N, Wilkins A, Ricard D, Aaby P, Naucler A, Ngom PT et al (1996) A community-based study of human immunodeficiency virus type 2 provirus load in rural village in West Africa. J Infect Dis 173:245–248

Baird JK, Masbar S, Basri H, Tirtokusumo S, Subianto B, Hoffman SL (1998) Age-dependent susceptibility to severe disease with primary exposure to *Plasmodium falciparum*. J Infect Dis 178:592–595

Beg MA, Khan R, Baig SM, Gulzar Z, Hussain R, Smego RA Jr (2002) Cerebral involvement in benign tertian malaria. Am J Trop Med Hyg 67:230–232

Berkley J, Mwarumba S, Bramham K, Lowe B, Marsh K (1999) Bacteraemia complicating severe malaria in children. Trans R Soc Trop Med Hyg 93:283–286

Bethell DB, Phuong PT, Phuong CX, Nosten F, Waller D, Davis TM, Day NP et al (1996) Electrocardiographic monitoring in severe falciparum malaria. Trans R Soc Trop Med Hyg 90:266–269

Borrmann S, Adegnika AA, Missinou MA, Binder RK, Issifou S, Schindler A, Matsiegui PB, Kun JFJ, Krishna S, Lell B, Kremsner PG (2003) Short-course artesunate treatment of uncomplicated *Plasmodium falciparum* malaria in Gabon. Antimicr Agents Chemother 47:901–904

Brahmbhatt H, Kigozi G, Wabwire-Mangen F, Serwadda D, Sewankambo N, Lutalo T, Wawer MJ (2003) The effects of placental malaria on mother-to-child HIV transmission in Rakai, Uganda. AIDS 17:2539–2541

Brandts CH, Ndjave M, Graninger W, Kremsner PG (1997) Effect of paracetamol on parasite clearance time in *Plasmodium falciparum* malaria. Lancet 350:704–709

Burchard GD, Ehrhardt S, Mockenhaupt FP, Mathieu A, Agana-Nsiire P, Anemana SD, Otchwemah RN et al (2003) Renal dysfunction in children with uncomplicated *Plasmodium falciparum* malaria in Tamale, Ghana. Ann Trop Med Parasitol 4:345–350

Carlini ME, White AC Jr, Atmar RL (1999) Vivax malaria complicated by adult respiratory distress syndrome. Clin Infect Dis 28:1182–1183

Chadee DD, Tilluckdharry CC, Maharaj P, Sinanan C (2000) Reactivation of *Plasmodium malariae* infection in a Trinidad man after neurosurgery. N Engl J Med 342:1924

Chen X, Xiao B, Shi W, Xu H, Gao K, Rao J, Zhang Z (2003) Impact of acute vivax malaria on the immune system and viral load of HIV-positive subjects. Chin Med J 116:1810–1820

Chotivanich K, Udomsangpetch R, McGrady R, Proux S, Newton P, Pukrittayakamee S, Looareesuwan S et al (2002) Central role of spleen in malaria parasite clearance. J Infect Dis 185:1538–1541

Clemens R, Pramoolsinsap C, Lorenz R, Pukrittayakamee S, Bock HL, White NJ (1994) Activation of the coagulation cascade in severe falciparum malaria through the intrinsic pathway. Brit J Haematol 87:100–105

D'Acremont V, Landry P, Mueller I, Pécoud A, Genton B (2002) Clinical and laboratory predictors of imported malaria in an outpatient setting: an aid to medical decision making in returning travellers with fever. Am J Trop Med Hyg 66:481–486

David PH, Hommel M, Miller LH, Udeinya IJ, Olliaro LD (1983) Parasite sequestration in *Plasmodium falciparum* malaria: spleen and antibody modulation of cytoadherence of infected erythrocytes. Proc Natl Acad Sci USA 80:5075–5079

Ehrich JH, Horstmann RD (1985) Origin of proteinuria in human malaria. Trop Med Parasitol 36:39–42

Erhart LM, Yingyuen K, Chuanak N, Buathong N, Laboonchai A, Miller RS, Meshnick SR et al (2004) Hematologic and clinical indices of malaria in a semi-immune population of western Thailand. Am J Trop Med Hyg 70:8–14

Facer CA, Rouse D (1991) Spontaneous splenic rupture due to *Plasmodium ovale* malaria. Lancet 338:896

Faucher JF, Ngou-Milama E, Missinou MA, Ngomo R, Kombila M, Kremsner PG (2002) The impact of malaria on common lipid parameters. Parasitol Res 88:1040–1043

Flatau E, Reichman N, Elias M, Raz R (2000) Malaria and Borrelia co-infection. J Travel Med 7:98–99

Franzen D, Curtius JM, Heitz W, Hopp HW, Diehl V, Hilger HH (1992) Cardiac involvement during and after malaria. Clin Invest 70:670–673

French N, Nakiyingi J, Lugada E, Watera C, Whitworth JAG, Gilks CF (2001) Increasing rates of malarial fever with deteriorating immune status in HIV-1 infected Ugandan adults. AIDS 15:899–906

Garnham PC (1970) The role of the spleen in protozoal infections with special reference to splenectomy. Acta Trop 27:1–14

Ghoshal UC, Somani S, Chetri K, Akhtar P, Aggarwal R, Naik SR (2001) *Plasmodium falciparum* and hepatitis E virus co-infection in fulminant hepatic failure. Indian J Gastroenterol 20:111

Graham SM, Walsh AL, Molyneux EM, Phiri AJ, Molyneux ME (2000) Clinical presentation of non-typhoidal *Salmonella* bacteraemia in Malawian children. Trans R Soc TropMed Hyg 94:310–314

Greenberg AE, Nsa W, Ryder RW, Medi M, Nzeza M, Kitadi N, Baangi M et al (1991) *Plasmodium falciparum* malaria and perinatally acquired human immunodeficiency virus type 1 infection in Kinshasa, Zaire. A prospective, longitudinal cohort study of 587 children. New Engl J Med 325:105–109

Grobusch MP, Wiese A, Teichmann D (2000) Delayed primary attack of vivax malaria. J Travel Med 7:104–105

Grobusch MP, Krüll M, Teichmann D, Gobels K, Suttorp N (2003a) Falciparum malaria and Tangier disease. Intern J Infect Dis 7:74–75

Grobusch MP, Borrmann S, Omva J, Issifou S, Kremsner PG (2003b) Severe malaria in a splenectomized Gabonese woman. Wien Klin Wochenschr/Middle Eur J Med 115:63–65

Günther A, Burchard GD, Slevogt H, Abel C, Grobusch MP (2002) Renal dysfunction in falciparum-malaria is detected more often when assessed by serum concentration of cystatin C instead of creatinine. Trop Med Intern Health 7:931–934

Günther A, Grobusch MP, Slevogt H, Abel W, Burchard GD (2003) Myocardial damage in falciparum malaria detectable by cardiac troponin T is rare. Trop Med Intern Health 8:30–32

Hänscheid T, Grobusch, MP, Melo Cristino J, Pinto BG (2003) Avoiding misdiagnosis of imported malaria: Screening of emergency department samples with thrombocytopenia detects clinically unsuspected cases. J Travel Med 10:155–159

Hightower M, Kallas EG (2003) Diagnosis, antiretroviral therapy, and emergence of resistance of antiretroviral agents in HIV-2 infection: a review. Brazil J Infect Dis 7:7–15

Ho M, Bannister LH, Looareesuwan S, Suntharasamai P (1992) Cytoadherence and ultrastructure of *Plasmodium falciparum*-infected erythrocytes from a splenectomized patient. Infect Immun 60:2225–2228

Hoffman IF, Jere CS, Taylor TE, Munthali P, Dyer JR, Wirima JJ, Rogerson SJ et al (1999) The effect of *Plasmodium falciparum* malaria on HIV-1 RNA blood plasma concentration. AIDS 13:487–494

Inion I, Mwanyumba F, Gaillard P, Chohan V, Verhofstede C, Claeys P, Kishorchandra M et al (2003) Placental malaria and perinatal transmission of human immunodeficiency virus type 1. J Infect Dis 188:1675–1678

Israeli A, Shapiro M, Ephros A (1987) *Plasmodium falciparum* malaria in an asplenic man. Trans R Soc Trop Med Hyg 81:233–234

Jelinek T, Schulte C, Behrens R, Grobusch MP, JP Coulaud, Bisoffi Z, Matteelli A et al (2002) Imported falciparum malaria in Europe: Sentinel surveillance data from the European network on surveillance of imported infectious disease. Clin Infect Dis 34:572–576

Karanikas G, Zedwitz-Liebenstein K, Eidherr H, Schuetz M, Sauerman R, Dudczak R, Winkler S (2004) Platelet kinetics and szintigraphic imaging in thrombocytopenic malaria patients. Thromb Haemost 91:353–357

Kaufmann JS, Asuzu MC, Rotimi CN, Johnson OO, Owoaje EE, Cooper RS (1997) The absence of adult mortality data for sub-Saharan Africa: a practical solution. Bull WHO 75:389–395

Khanna KV, Markham RB (1999) A perspective on cellular immunity in the elderly. Clin Infect Dis 28:710–713

Kittl EM, Diridl G, Lenhart V, Neuwald C, Tomasits J, Pichler H, Bauer K (1992) HDL cholesterol as a sensitive diagnostic parameter in malaria. Wien Klin Wochenschr/Middle Europ J Med 104:21–24

Koibuchi T, Nakamura T, Miura T, Endo T, Nakamura H, Takahashi T, Kim HS et al (2003) Acute disseminated encephalomyelitis following *Plasmodium vivax* malaria. J Infect Chemother 9:254–256

Kremsner PG, Winkler S, Brandts C, Neifer S, Bienzle U, Graninger W (1994) Clindamycin in combination with chloroquine or quinine is an effective therapy for uncomplicated *Plasmodium falciparum* malaria in children from Gabon. J Infect Dis 169:467–470

Krishna S, Sup@@ranond W, Pukrittayakamee S, ter Kuile F, Supputamangkol Y, Attatamsoonthorn K, Ruprah M et al (1995a) Fever in uncomplicated *Plasmodium falciparum* infection: effects of quinine and paracetamol. Trans R Soc Trop Med Hyg 89:197–199

Krishna S, Pukrittayakamee S, Supanaranond W, ter Kuile F, Ruprah M, Sura T, White NJ (1995b) Fever in uncomplicated *Plasmodium falciparum* malaria: randomized double-blind comparison of ibuprofen and paracetamol treatment. Trans R Soc Trop Med Hyg 89:507–509

Kun JFJ, Mordmüller B, Lell B, Lehman LG, Luckner D, Kremsner PG (1998) Polymorphism in promoter region of inducible nitric oxide synthase gene and protection against malaria. Lancet 351:265–266

Ladhani S, Lowe B, Cole AO, Kowuondo K, Newton CR (2002) Changes in white blood cells and platelets in children with falciparum malaria: relationship to disease outcome. Br J Haematol 119:839–847

Lambrecht AJ, Snoeck J, Timmermans U (1978) Transient an-alpha-lipoproteinaemia in man during infection by *Plasmodium vivax*. Lancet 8075:1206

Lawn SD (2004) AIDS in Africa: the impact of co-infections on the pathogenesis of HIV-1 infection. J Infect 48:1–12

Lell B, Brandts CH, Graninger W, Kremsner PG (2000) The circadian rhythm of body temperature is preserved during malarial fever. Wien Klin Wochenschr/Middle Europ J Med 112:1014–1015

Lell B, Borrmann S, Yazdanbakhsh M, Kremsner PG (2001) Atopy and malaria. Wien Klin Wochenschr/Middle Europ J Med 113:927–929

Long HY, Lell B, Dietz K, Kremsner PG (2001) *Plasmodium falciparum*: in vitro growth inhibition by febrile temperatures. Parasitol Res 87:553–555

Looareesuwan S, White NJ, Chittama, Bunnag D, Harinasuta T (1987) High rate of *Plasmodium vivax* relapse following treatment of falciparum malaria in Thailand. Lancet 8567:1052–1055

Mabey DCW, Brown A, Greenwood BM. *Plasmodium falciparum* malaria and Salmonella infections in Gambian children (1987) J Infect Dis 155:1319–1321

Mason DP, McKenzie FE (1999) Blood-stage dynamics and clinical implications of mixed *Plasmodium vivax-Plasmodium falciparum* infections. Am J Trop Med Hyg 61:367–374

Mayxay M, Pukrittayakamee S, Chotivanich K, Imwong M, Looareesuwan S, White NJ (2001) Identification of cryptic co-infection with *Plasmodium falciparum* in patients presenting with vivax malaria. Am J Trop Med Hyg 65:588–592

Mendis K, Sina BJ, Marchesini P, Carter R (2001) The neglected burden of *Plasmodium vivax* malaria. Am J Trop Med Hyg 64:97–106

Metzger WG, Mordmüller BG, Kremsner PG (1995a) Malaria pigment in leucocytes. Trans R Soc Trop Med Hyg 89:637–638

Metzger W, Mordmüller B, Graninger W, Bienzle U, Kremsner PG (1995b) Sulfadoxine/pyrimethamine or chloroquine/clindamycin treatment of Gabonese school children with chloroquine resistant malaria. J Antimicrob Chemother 36:723–728

Missinou MA, Lell B, Kremsner PG (2003) Uncommon asymptomatic *Plasmodium falciparum* infections in Gabonese children. Clin Infect Dis 36:1198–1202

Modebe O, Jain S (1999) Multi-system failure in *Plasmodium vivax* malaria: report of a case. Ann Trop Parasitol 93:409–412

Molyneux ME, Looareesuwan S, Menzies IS, Graniger SL, Phillips RE, Wattanagoon Y, Thomson RP et al (1989) Reduced hepatic blood flow and intestinal malabsorption in severe falciparum malaria. Am J Trop Med Hyg 40:470–476

Moulin F, Lesage F, Legros AH, Maroga C, Moussavou A, Guyon P, Marc E et al (2003) Thrombocytopenia and *Plasmodium falciparum* malaria in children with different exposures. Arch Dis Child 88:540–541

Mühlberger N, Jelinek T, Behrens RH, Gjorup I, Coulaud JP, Clerinx J, Puente S et al (2003) Age as a riks factor for severe manifestations and fatal outcome of falciparum malaria in European patients: observations from TropNetEurop and SIMPID surveillance data. Clin Infect Dis 36:990–995

Muller O, Moser R (1990) The clinical and parasitological presentation of *Plasmodium falciparum* malaria in Uganda is unaffected by HIV-1 infection. Trans R Soc Trop Med Hyg 84:336–338

Nacher M (2002) Worms and malaria: noisy nuisances and silent benefits. Parasite Immunol 24:391–393

Nathoo S, Serghides L, Kain KC (2003) Effect of HIV-1 antiretroviral drugs on cytoadherence and phagocytic clearance of *Plasmodium falciparum*-parasited erythrocytes. Lancet 362:1039–1041

N'gom PT, Jaffar S, Ricard D, Wilkins A, Ariyoshi K, Morgan G, Da Silva AP et al (1997) Immune stimulation by syphilis and malaria in HIV-2-infected and uninfected villagers in West Africa. Br J Biomed Science 54:251–255

Nierengarten MB (2003) Malariotherapy to treat HIV patients? Lancet Infect Dis 2003;3:321

Nilsson-Ehle I, Nilsson-Ehle P (1990) Changes in plasma lipoproteins in acute malaria. J Intern Med 227:151–155

Norhayati M, Rohani AK, Hayati MI, Halimah AS, Sharom MY, Abidin AH, Fatmah MS (2001) Clinical features of malaria in Orang Asli population in Pos Piah, Malaysia. Med J Malaysia 56:271–274

Nwanyanwu OC, Kumwenda N, Kazembe PN, Jemu S, Ziba C, Nkhoma WC, Redd SC (1997) Malaria and human immunodeficiency virus infection among male employees of a sugar estate in Malawi. Trans R Soc Trop Med Hyg 91:567–569

Olaleye BO, Williams LA, d'Alessandro U, Weber MR, Mulholland K, Okorie C, Langerock P et al (1998) Clinical predictors of malaria in Gambian children with fever or a history of fever. Trans R Soc Trop Med Hyg 92:300–304

Pongponratn E, Viriyavejakul P, Wilairatana P, Ferguson D, Chaisri U, Turner G, Looareesuwan S (2000) Absence of knobs on parasitized red blood cells in a splenectomized patient in fatal falciparum malaria. Southeast Asian J Trop Med Public Health 31:829–835

Prakash J, Singh AK, Kumar NS, Saxena RK (2003) Acute renal failure in *Plasmodium vivax* malaria. J Assoc Physicians India 51:265–267

Pukrittayakamee S, White NJ, Davis TM, Looareesuwan S, Supanaranond W, Desakorn V, Chaivisuth B, Williamson DH (1992) Hepatic blood flow and metabolism in severe falciparum malaria: clearance of intravenously administered galactose. Clin Sci (Lond) 82:63–70

Radloff PD, Philipps J, Nkeyi M, Sturchler D, Mittelholzer ML, Kremsner PG (1996) Arteflene compared with mefloquine for treating *Plasmodium falciparum* malaria in children. Am J Trop Med Hyg 55:259–262

Ree HI (2000) Unstable vivax malaria in Korea. Korean J Parasitol 38:119–138

Reeves JD, Doms RW (2002) Human immunodeficiency virus type 2. J Gen Virol 83:1253–1265

Rogerson S (2003) HIV-1, antiretroviral therapy, and malaria. Lancet 362:1008–1009

Sandmand M, Bruunsgaard H, Kemp K, Andersen-Ranberg K, Pedersen AN, Skinhoj P, Pedersen BK (2002) Is ageing associated with a shift in the balance between type 1 and type 2 cytokines in humans? Clin Exp Immunol 127:107–114

Sankoh OA, Kynast-Wolf G, Kouyaté B, Becher H (2003) Patterns of adult and old-age mortality in rural Burkina Faso. J Public Health Med 25:372–376

Schellenberg D, Menendez C, Kahigwa E, Font F, Galindo C, Acosta C, Armstrong Schellenberg J et al (1999) African children with malaria in an area of intense *Plasmodium falciparum* transmission: Features on admission to the hospital and risk factors for death. Am J Trop Med Hyg 61:431–438

Schim van der Loeff M, Aaby P (1999) Towards a better understanding of the epidemiology of HIV-2. AIDS 13:S69–S84

Schwartz E, Sadetzki S, Murad H, Raveh D (2001) Age as a risk factor for severe *Plasmodium falciparum* malaria in non-immune patients. Clin Infect Dis 33:1774–1777

Schwarzer E, Kühn H, Valente E, Arese P (2003) Malaria-parasitized erythrocytes and haemozoin nonenzymatically generate large amounts of hydroxy fatty acids that inhibit monocyte functions. Blood 101:722–728

Skudowitz RB, Katz J, Lurie A, Levin J, Metz J (1973) Mechanisms of thrombocytopenia in malignant tertian malaria. Br Med J 2(5865):515–518

Steele RW (1996) Malaria in children. Adv Pediatr Infect Dis 12:325–349

Stein CM, Gelfand M (1985) The clinical features and laboratory findings in acute *Plasmodium falciparum* malaria in Harare, Zimbabwe. Centr Afr J Med 31:166–170

Steketee RW, Wirima JJ, Bloland PB, Chilima B, Mermin JH, Chitsulo L, Breman JG (1996) Impairment of a pregnant women's acquired ability to limit *Plasmodium falciparum* by infection with human immunodeficiency virus type-1. Am J Trop Med Hyg 55:42–49

Tanios MA, Kogelman L, McGovern B, Hassoun PM (2001) Acute respiratory distress syndrome complicating *Plasmodium vivax* malaria. Crit Care Med 29:665–667

Tsuchida H, Yamaguchi K, Yamamoto S, Ebisawa I (1982) Quartan malaria following splenectomy 36 years after infection. Am J Trop Med Hyg 31:163–165

Van den Biggelaar AHJ, Rodrigues LC, van Ree R, van der Zee JS, Hoeksma-Kruize YCM, Souverijn JHM, Missinou MA et al (2004) Long-term treatment of intestinal helminths increases mite skin-test reactivity in Gabonese schoolchildren. J Infect Dis 189:892–900

Van Eijk AM, Ayisi JG, Ter Kuile FO, Misore AO, Otieno JA, Kolczak MS, Kager PA et al (2002) Malaria and human immunodeficiency virus infection as risk factors for anaemia in infants in Kisumu, Western Kenya. Am J Trop Med Hyg 67:44–53

Van Eijk AM, Ayisi JG, Ter Kuile FO, Misore AO, Otieno JA, Rosen DH, Kager, P et al (2003) HIV increases the risk of malaria in women of all gravidities in Kisumu, Kenya. AIDS 17:595–603

Van Velthuysen ML, Florquin S (2000) Glomerulopathy associated with parasitic infections. Clin Microbiol Rev 13:55–66

Verhoeff F, Brabin BJ, Hart AC, Chimsuku L, Kazembe P, Broadhead RL (1999) Increased prevalence of malaria in HIV-infected pregnant women and its implications for malaria control. Trop Med Intern Health 4:5–12

Vinetz JM, Li J, McCutchan TF, Kaslow DC (1998) *Plasmodium malariae* infection in an asymptomatic 74-year-old Greek woman with splenomegaly. N Engl J Med 338:367–371

Weber MW, Zimmermann U, van Hensbroek MB, Frenkel J, Palmer A, Ehrich JHH, Greenwood BM (1999) Renal involvement in Gambian children with cerebral or mild malaria. Trop Med Intern Health 4:390–394

White NJ (2003) Malaria. In: Cook GC, Zumla A (eds) Manson's Tropical Diseases, 21st ed. Elsevier, Edinburgh, pp 1205–1295

Whitworth J, Morgan D, Quigley M, Smith A, Mayanja B, Eotu H, Omoding N et al (2000) Effect of HIV-1 and increasing immunosuppression on malaria parasitaemia and clinical episodes in adults in rural Uganda: a cohort study. Lancet 356:1051–1056

Wilairatana P, Loareesuwan S, Charoenlarp P (1994) Liver profile changes and complications in jaundiced patients with falciparum malaria. Trop Med Parasitol 45:298–302

Wongsrichanalai C, Murray CK, Gray M, Miller RS, McDaniel P, Liao WJ, Pickard AL et al (2003) Co-infection with malaria and leptospirosis. Am J Trop Med Hyg 68:583–585

Yazdanbakhsh M, Kremsner PG, van Ree R (2002) Allergy, parasites, and the hygiene hypothesis. Science 296:490–494

CTMI (2005) 295:105–136
© Springer-Verlag Berlin Heidelberg 2005

Metabolic Complications of Severe Malaria

T. Planche[1,2] (✉) · A. Dzeing[3] · E. Ngou-Milama[4] · M. Kombila[3] ·
P. W. Stacpoole[5]

[1]Division of Cellular and Molecular Medicine, Centre for Infection,
St. George's University of London, Cranmer Terrace, London SW17 0RE, UK
tim@planche.demon.co.uk

[2]Medical Research Unit, Albert Schweitzer Hospital, Lambaréné, Gabon

[3]Département de Parasitologie, Mycologie et Médecine Tropicale,
Faculté de Médecine, Université des Sciences de la Santé, Libreville, Gabon

[4]Département de Biochimie, Faculté de Médecine,
Université des Sciences de la Santé, Libreville, Gabon

[5]Departments of Medicine (Division of Endocrinology and Metabolism),
Biochemistry and Molecular Biology and the General Clinical Research Center,
University of Florida, Gainsville, FL 32610, USA

Abstract Metabolic complications of malaria are increasingly recognized as contributing to severe and fatal malaria. Disorders of carbohydrate metabolism, including hypoglycaemia and lactic acidosis, are amongst the most important markers of disease severity both in adults and children infected with *Plasmodium falciparum*. Amino acid and lipid metabolism are also altered by malaria. In adults, hypoglycaemia is associated with increased glucose turnover and quinine-induced hyperinsulinaemia, which causes increased peripheral uptake of glucose. Hypoglycaemia in children results from a combination of decreased production and/or increased peripheral uptake of glucose, due to increased anaerobic glycolysis. Patients with severe malaria should be monitored frequently for hypoglycaemia and treated rapidly with intravenous glucose if hypoglycaemia is detected. The most common aetiology of hyperlactataemia in severe malaria is probably increased anaerobic glucose metabolism, caused by generalized microvascular sequestration of parasitized erythrocytes that reduces blood flow to tissues. Several potential treatments for hyperlactataemia have been investigated, but their effect on mortality from severe malaria has not been determined.

Abbreviations

ATP	Adenosine triphosphate
ADP	Adenosine diphosphate
CI	Confidence interval
DCA	Dichloroacetate
EMP	Endothelial microparticle
NAD	Nicotinamide adenine dinucleotide
NADH	Nicotinamide adenine dinucleotide, reduced
$FADH_2$	Flavin adenine dinucleotide, reduced
LDH	Lactate dehydrogenase
PDC	Pyruvate dehydrodenase complex
NAC	*N*-acetyl cysteine
VLDL	Very low-density lipoproteins

1
Introduction

In humans, *Plasmodium falciparum* is the most frequent and important cause of lethal malaria. *P. falciparum* parasitaemia encompasses a wide variety of clinical states that range from asymptomatic carriage to life-threatening complications (Newton and Krishna 1998; White and Ho 1992). Coma and anaemia have been recognized for centuries as frequent complications of severe malaria. However, only in the last 20 years or so have prospective studies examined the relative contributions to mortality of these and other complications. Table 1 summarizes the results of the largest of these studies. Despite inherent demographic variability, these reports illustrate two common themes. First, outcome can be predicted with high accuracy (area under the receiver/operator characteristic curve=0.86) from knowing only a few clinical indices and second, most of these indices are either metabolic in origin or result from metabolic derangements (Marsh et al. 1995; Planche et al. 2003). Current definitions of severe malaria have incorporated many of the findings of these prospective studies.

Severe malaria manifests differently in adults and children. In adult malaria, renal failure, pulmonary oedema, coma and hyperlactataemia are common life-threatening complications, whereas severe anaemia is

Table 1 Clinical predictors of outcome in malaria

	Adults			Children
Reference	Day et al. 2000b	Allen et al. 1996	Marsh et al. 1995	Planche et al. 2003
Number of cases	346	489	1844	854
Country	Vietnam	Papua New Guinea	Kenya	Ghana
Selection	'Unwell' admissions	All admissions	All admissions	'Unwell' admissions
Independent predictors	Bilirubin, base deficit, blood pressure	Hyper-lactataemia, creatinine, bicarbonate	Blantyre coma score, respiratory distress, jaundice, hypogly-caemia	Blantyre coma score, hyperlac-tataemia, body mass index
AUCROC	0.83		Not reported	0.86

AUCROC, area under the receiver/operator not reported characteristics curve.

Table 2 Features of severe malaria

Variables	Adults		Children			
Author	Hien	Day	Van Hensbroek	Marsh	Allen	Planche
Country	Vietnam	Vietnam	Gambia	Kenya	PNG	Ghana
Number of cases	560	346	576	1844	489	854
Coma (BCS ≤2) (%)	52	51	46.9	10	16	20
Convulsions (%)	11	NR	NR	18.3	NR	NR
Hyperlactataemia (%) (lactate ≥5 mmol/l)	NR	35†	NR	NR	20	20
Acidosis (%)	NR	67	NR	NR	NR	NR
Respiratory distress	NR	NR	NR	13.7	NR	28
Hypoglycaemia (%) (glucose ≤2.2 mmol/l)	7	6	24.3	13.2	NR	4.5
PCV (%)	31	31	25	NR	NR	25
Severe anaemia (%) (Hb<50 g/l)	NR	NR	8	18	22	16
Renal failure (%)	NR	27	NR	0.1	NR	0
Jaundice (%)	NR	60	NR	4.5	NR	NR
Case fatality rate (%)	15	15	NR	3.5	3.5	6

[a] Hyperlacataemia in this study defined as >4 mmol/l. PNG, Papua New Guinea; NR, not reported; BCS, Blantyre coma score; PCV, packed cell volume.

relatively uncommon. However, in children, severe anaemia, coma and hyperlactataemia are common but pulmonary oedema and renal failure are rare. Table 2 compares the clinical characteristics of severe malaria in large studies of infected adults and children.

2
Pathogenesis

There are several theories of the pathogenesis of severe malaria. The first states that severe malaria develops as a result of the obstruction or reduction of blood flow caused by the sequestration of parasitized red blood cells in capillary beds. The second suggests that a harmful substance or substances (toxins, either parasite or host derived) are released and these lead to the development of severe malaria. Finally, there is the possibility that severe malaria may result from a combination of the above theories or that there may be a number of unrelated causes for the diverse manifestations of severe malaria.

2.1
Mechanical Theories

First proposed in the mid-nineteenth century (Delafield 1872), this theory states that the principal manifestations of severe malaria are caused by sequestration of parasitized erythrocytes within small vessels. This results in reduced blood flow, hypoxia and altered metabolism of cells distal to the obstruction. Accordingly, microcirculatory obstruction in the splanchnic bed could impair hepatic glucose output and lactate clearance, contributing to hypoglycaemia and lactic acidosis. However, there is no reliable method for quantifying the number of sequestered parasites in vascular beds. Nor is there a strong correlation between the occurrence of cerebral malaria in infected patients and the presence or extent of microcirculatory sequestration identified post-mortem in their brains (Silamut et al. 1999; Taylor et al. 2004). Nevertheless, this does not exclude mechanical obstruction being an early event in the clinical and metabolic derangements of severe malaria, since its resolution ante-mortem could result from specific anti-malarial therapy, general supportive care (e.g. rehydration) or host factors. Some cases may also have been misdiagnosed as cerebral malaria (Taylor et al. 2004).

2.2
Toxin Theories

The notion that manifestations of severe malaria are due to circulating toxins elaborated by the parasite or the host originated in the early twentieth century

Table 3 Proposed toxins or mediators causing severe malaria

Toxin	Reference
Kinins	Maegraith and Fletcher 1972
Endotoxins	Clark 1978
Reactive oxygen species	Clark et al. 1986
Nitric oxide	Clark et al. 1992
Cytokines (particularly tumour necrosis factor)	Clark et al. 1987

(Gaskell and Miller 1920), and continues to be debated. In most cases, the involvement of toxins (summarized in Table 3) in the aetiopathology of severe malaria in humans has been based on experiments conducted in animal models of malaria or on extrapolations from other clinical conditions, with elaborate theoretical justifications. However, to date, there is no clear evidence for an aetiological role for any toxin in the pathogenesis of human malaria caused by *P. falciparum*. In fact:

– Steroids worsen outcome in cerebral malaria (Warrell et al. 1982).

– Anti-tumour necrosis factor monoclonal antibody did not improve survival from cerebral malaria despite its ability to reduce fever (van Hensbroek et al. 1996b) and increased the incidence of delayed neurological sequelae.

– Endotoxin levels do not correlate with disease severity (Aung-Kyaw-Zaw et al. 1988; Usawattanakul et al. 1985).

– Concentrations of nitrates in plasma, urine or cerebrospinal fluid do not correlate with disease severity. However, there is some in vitro evidence that patients with the nitric oxide synthetase 2 gene (*NOS2*) Lambaréné mutation (which is protective against severe malaria) have higher leucocyte NO production in response to stimulation (Kun et al. 2001).

In a heavy infection a 2% peripheral parasitaemia represents about 40 ml of parasitized red cells in the total blood volume of 6 l in adults (even more if sequestered parasites are counted) and contributes tens of grams of malaria parasite material to the circulation. Thus, substances produced by *P. falciparum* are unlikely to cause disease in the same manner as 'classical' bacterial toxins that exert their pathological effects in microgram quantities.

Survival of hospitalized patients with severe malaria has not improved in the last 50 years, despite the advent of new antiparasitic drugs, such as artemether (Hien et al. 1996; van Hensbroek et al. 1996a), or agents designed to modulate the immune response (Warrell et al. 1982), such as steroids or anti-tumour necrosis factor therapy (van Hensbroek et al. 1996b). Because most deaths from *P. falciparum* in hospitalized children occur within the first 24 h, it is possible that interventions aimed at mitigating specific metabolic complications early during the infection might reduce mortality. These complications and their treatment are discussed below.

3
Hypoglycaemia

3.1
Epidemiology and Aetiopathology

Hypoglycaemia, defined as a blood glucose concentration ≤ 2.2 mmol/l, has been known for almost 80 years to be a complication of *P. falciparum* infection (Fitz-Hugh 1944; Peterson 1926). However, the first prospective assessment of the frequency and clinical course of malaria-associated hypoglycaemia was conducted only about 20 years ago, in Thai adults (White et al. 1983). In this study, 12 of 151 subjects with cerebral malaria had hypoglycaemia and the case fatality rate of these patients was significantly higher than of those who were normoglycaemic. Plasma insulin and C peptide levels were inappropriately high during episodes of hypoglycaemia. Measured quinine and insulin concentrations were positively correlated. Quinine infusions in normal volunteers also increased plasma insulin concentrations that resulted in falls in fasting blood glucose levels (White et al. 1983). Pregnant women in Thailand and Sudan are prone to hypoglycaemia without associated hyperinsulinaemia (White et al. 1983; Saeed et al. 1990). [6,6-^{2}H$_2$] glucose turnover studies in pregnant Vietnamese women with uncomplicated malaria and pregnant and non-pregnant controls showed that duration of fasting is a determinant of the fall in blood glucose (van Thien et al. 2004).

Hypoglycaemia is present at the time of hospital admission in 4.5%–30% of children with severe malaria in sub-Saharan Africa (Krishna et al. 1994b; Marsh et al. 1995; Planche et al. 2003; Taylor et al. 1988; White et al. 1987). The risk of iatrogenic hypoglycaemia from quinine administration is approximately 5%–15% of quinine treated children who are normoglycaemic on admission (Agbenyega et al. 2000; English et al. 1998; van Hensbroek et al. 1996a). Children with hypoglycaemia on admission are at highest risk of a subsequent episode of post-admission hypoglycaemia, despite receiving (in most studies) infusions of glucose (~3 mg/kg/min; English et al. 1998; Krishna et al. 2001; Taylor et al. 1988).

On admission, hypoglycaemia was more common in African children (Taylor et al. 1988) than Thai adults (White et al. 1987) but it is often not due to quinine-induced hyperinsulinaemia. In a study of 47 Gambian children with severe malaria who were treated with chloroquine, 15 (32%) had hypogly-caemia, and the case fatality rate of hypoglycaemia was higher than in those children without this complication (33% vs. 3%, $P=0.02$; White et al. 1987). Most hypoglycaemic episodes in these children occurred on admission and insulin concentrations were appropriately low (White et al. 1987). Blood glu-cose concentrations in 95 Malawian children with malaria were negatively correlated with plasma concentrations of the gluconeogenic substrates lac-tate and alanine, but not with the duration of symptoms or with fasting (Taylor et al. 1988). In this study, quinine was infused with a 5% glucose solution. Pre-treatment hypoglycaemia occurred in 17 patients (18%) and was associated with a higher case fatality rate (37% vs. 4%, $P<0.0001$), an increased risk of recurrent episodes of post-admission hypoglycaemia (37% vs. 0%, $P<0.0001$ and an increased risk of neurological sequelae (26% vs. 4%, $P<0.001$).

Hypoglycaemia may be caused by impaired hepatic production of glucose via gluconeogenesis or glycogenolysis, by increased glucose uptake by peri-pheral tissues or a combination of these mechanisms. Isotopic tracer studies in severe malaria indicate that glucose turnover is increased in adults (Davis et al. 1993, 2002) and children (Agbenyega et al. 2000; Dekker et al. 1997a, 1997b, 1996; Singh et al. 1998) with severe malaria. Glucose turnover rates in children with severe malaria are approximately five times higher than in adults with severe malaria (Agbenyega et al. 2000). The blood glucose level at the time of admission is negatively correlated with glucose turnover ($r=0.88$, $n=18$, $P<0.001$; Davis et al. 1993) in adults but not in children (Agbenyega et al. 2000; Dekker et al. 1997a, 1997b, 1996; Singh et al. 1998).

Glucose turnover was measured in seven Vietnamese adults with cerebral malaria using an infusion of $[6,6\ ^2H_2]$ glucose and the contribution of glu-coneogenesis was estimated by ingestion of ^2H_2O and measurement of $[^2H]$ enrichment at the C5 position. In this study glycogenolysis did not contribute to glucose production and hepatic glucose output was solely due to gluconeo-genesis (van Thien et al. 2001). Conversion of galactose (Pukrittayakamee et al. 1992) or glycerol (Pukrittayakamee et al. 1994) to glucose was not significantly impaired in such patients. However liver blood flow, measured by clearance of indocyanine green, was reduced in acute, severe malaria and inversely cor-related with the plasma lactate concentration ($r=0.53$, $P<0.05$). Similarly, the conversion of alanine to glucose was also impaired in severe malaria in adults and is associated with reduced liver blood flow (Pukrittayakamee et al. 2002).

In 20 Kenyan children with uncomplicated malaria, basal glucose produc-tion did not correlate with plasma levels of alanine or lactate, but correlated positively with plasma glucose concentration (Dekker et al. 1996). In uncom-plicated malaria, most (51%–93%) glucose is reported to be derived from gluconeogenesis and glucose turnover increases with an infusion of alanine

(Dekker et al. 1997a). Thus, in uncomplicated childhood malaria, endogenous glucose production is an important determinant of plasma glucose; however, hypoglycaemia does not occur in these children. In contrast, in severe childhood malaria, hypoglycaemia is associated with increased plasma levels of alanine, lactate and glycerol (English et al. 1998), implying that substrate availability is not rate limiting for gluconeogenesis. There was no relationship between glucose turnover and plasma insulin or glucose levels in 21 Ghanaian children with severe malaria (Agbenyega et al. 2000). Children treated with quinine had higher glucose turnover rates than children treated with artesunate (64 vs. 49 µmol/kg/min, $P<0.05$; Agbenyega et al. 2000). Glucose turnover was correlated positively with lactate turnover, suggesting that glucose was mainly consumed by anaerobic glycolysis.

Hyperinsulinaemia causes increased peripheral uptake of glucose in adults with *P. falciparum* infection, so hypoglycaemia is associated with increased glucose turnover. In contrast, glucose turnover and plasma glucose concentration in infected children are not correlated.

A recent microarray hybridization study using the animal model of *P. berghei*-infected mice, showed increased transcription of genes of the host glycolytic pathway (Sexton et al. 2004). Interestingly, the expression of genes encoding components of the Kreb's cycle was not upregulated, whereas pyruvate dehydrogenase kinase (which inhibits the pyruvate dehydrogense complex and consequently mitochondrial removal of lactate) and lactate dehydrogenase were upregulated. These findings imply a potential role for altered gene expression in the perturbations of glucose metabolism associated with this model of severe malaria.

3.2
Therapy

The standard treatment of hypoglycaemia is with intravenous glucose given at 0.5 g/kg (e.g. 25% glucose infusion at 2 mL/kg). Hypoglycaemia should be prevented with a glucose infusion and detected early by regular monitoring of blood glucose, measuring blood glucose every 4 hours, or more frequently if there is any clinical deterioration. Blood glucose monitoring can usually be discontinued when the patient is no longer receiving intravenous quinine.

Analysis of glucose kinetics in children with severe malaria receiving quinine indicate that children who are hypoglycaemic on admission to hospital may require larger amounts of glucose than are conventionally used (up to 6 mg/kg/min) to prevent hypoglycaemia (Agbenyega et al. 2000).

4
Hyperlactataemia and Acidosis

Lactic acidosis defined by a blood lactate ≥ 5 mmol/l, metabolic acidosis or a surrogate marker of these, such as tachypnoea are among the most important independent predictors of death in children or adults with severe malaria (Allen et al. 1996; Day et al. 2000b; Krishna et al. 1994b; Marsh et al. 1995; Planche et al. 2003; Waller et al. 1995; White et al. 1985). Hyperlactatemia and metabolic acidosis are closely related physiologically (Day et al. 2000b; English et al. 1997b; Stacpoole 1993).

4.1
Hyperlactataemia

4.1.1
Epidemiology

Hyperlacatataemia is a common complication of severe malaria in both adults and children. Twenty to thirty-five per cent of cases of severe malaria may be complicated by lactic acidosis and the case fatality rate is 12%–30% (Allen et al. 1996; Day et al. 2000b; Planche et al. 2003; Waller et al. 1995). In childhood malaria lactic acidosis occurs usually between the ages of 2 and 5 years (Allen et al. 1996; Planche et al. 2003; Waller et al. 1995).

4.1.2
Lactate Metabolism

Lactate dehydrogenase (LDH) catalyses the reversible formation and removal of L–lactate in all cells (Eq. 1):

$$CH_3COCOO^- + NAHDH + H^+ \xoverset{LDH}{\longleftrightarrow} CH_3CH(OH)COO^- + NAD^+ \qquad (1)$$
$$\text{(Pyruvate)} \qquad\qquad\qquad \text{(Lactate)}$$

When Eq. 1 is solved for lactate (Eq. 2), it becomes apparent that the immediate determinants of the intracellular concentration of lactate are the concentrations of pyruvate and protons (H^+) and the ratio of reduced: oxidized nicotinamide adenine dinucleotide (NAD):

$$[\text{Lactate}] = k_{eq} \times [\text{Pyruvate}] \times \frac{[\text{NADH}]}{[\text{NAD}^+]} \times [H^+] \qquad (2)$$

Consequently, conditions that give rise to increased levels of pyruvate or to an increased ratio of NADH/NAD$^+$ favour the formation of lactate. The normally high NADH/NAD$^+$ ratio strongly favours lactate formation. In healthy humans at rest who are post-prandial, the whole blood venous lactate concentration is 1 mmol/l (1 m Eq/l; 9 mg/dl) and the pyruvate concentration is 0.1 mmol/l, so the [lactate]/[pyruvate] ratio in the basal state is ~10. The basal concentrations of lactate and pyruvate in arterial blood typically are lower than those in venous blood, while the basal levels of these metabolites in cerebrospinal fluid are similar to those of venous blood.

Lactate is derived exclusively from reduction of pyruvate in the cytoplasm (Eq. 1). In turn, by far the quantitatively most important source of pyruvate is the anaerobic catabolism of glucose by the Embden–Meyerhof pathway of glycolysis. All cells are capable of generating lactate and protons through glycolysis and ATP hydrolysis, respectively. Although the maximum rates by which these ions are generated by individual tissues is not known, it has been estimated that basal rates of lactate and H^+ production are each 1300 mmol/70 kg male/24 h (~0.8 mmol/kg/h).

The metabolic fate of pyruvate determines the manner by which lactate is catabolized. Pyruvate may be transaminated in the cytoplasm to alanine or may undergo reactions catalysed in the mitochondria leading to its conversion to glucose or combustion to CO_2 and water. Pyruvate enters mitochondria via the monocarboxylate transport system and may then undergo either of two irreversible reactions: carboxylation to oxaloacetate by pyruvate carboxylase or decarboxylation to acetyl coenzyme A (acetyl CoA) by the pyruvate dehydrogenase complex (PDC) (Fig. 1).

Pyruvate may be reduced to lactate in the cytoplasm or may be transported into the mitochondria for anabolic reactions, such as gluconeogenesis and lipogenesis, or for oxidation to acetyl CoA by the PDC. Reducing equivalents (NADH, $FADH_2$) are generated by reactions catalysed by the PDC and the tricarboxylic acid (TCA) cycle and donate electrons (e) that enter the respiratory chain at NADH ubiquinone oxidoreductase (complex I) or at succinate ubiquinone oxidoreductase (complex II). Cytochrome c oxidase (complex IV) catalyses the reduction of molecular oxygen to water and ATP synthase (complex V) generates ATP from ADP.

Under aerobic conditions, PDC catalyses the rate-determining reaction for the oxidative removal of glucose, pyruvate and lactate and helps to sustain the TCA cycle through provision of acetyl CoA (Stacpoole 2004). Reducing equivalents (NADH, $FADH_2$) are generated by reactions catalysed by the PDC and by various dehydrogenases in the TCA cycle and provide electrons to the respiratory chain for eventual reduction of molecular oxygen to water and synthesis of ATP. Furthermore, normal functioning of the TCA cycle and respiratory chain is essential for generation of bicarbonate and removal of hydrogen ions. Thus, the PDC participates, albeit indirectly, in cellular acid base homeostasis. The PDC undergoes rapid post-translational changes in catalytic activity by reversible phosphorylation, which renders the complex

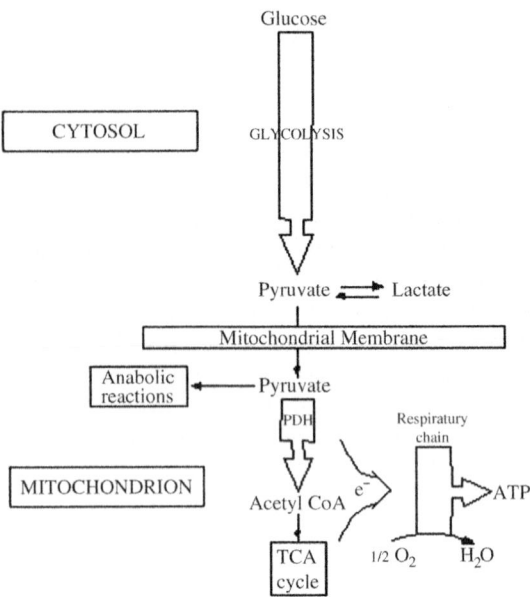

Fig. 1 Pyruvate enters mitochondria via the monocarboxylate transport system and may then undergo either of two irreversible reactions: carboxylation to oxaloacetate by pyruvate carboxylase or decarboxylation to acetyl coenzyme A (*acetyl CoA*) by the pyruvate dehydrogenase complex

inactive. Phosphorylation is mediated by pyruvate dehydrogenase kinase and dephosphorylation is catalysed by pyruvate dehydrogenase phosphatase.

4.2
Etiology of Hyperlactataemia

Hyperlacataemia results from increased lactate production, decreased lactate clearance, or both. Stable isotope lactate turnover studies conducted in 14 adults in Vietnam (Davis et al. 2002) and in 19 children in Ghana (Agbenyega et al. 2000) with severe malaria have shown consistent results. Lactate turnover is elevated in severe disease and plasma lactate concentrations correlate positively with lactate turnover, implying that hyperlactataemia is primarily the result of increased lactate production rather than decreased removal. Glucose and lactate turnover rates are also positively correlated, suggesting that lactate production results from increased anaerobic glycolysis. In both of these studies lactate turnover correlated positively with recovery time from coma, which is consistent with the hypothesis that microvascular obstruction is the underlying mechanism behind both hyperlactataemia and coma in malaria.

Several underlying mechanisms for hyperlactataemia in severe malaria have been proposed, some of which are unlikely to contribute significantly to hyperlactataemia. These are described in detail below.

4.2.1
Lactate Production by Parasites

P. falciparum possesses cytochromes, but is not thought to contain them in a functionally organized respiratory chain capable of undertaking oxidative phosphorylation. Hence, when parasites reside in host erythrocytes (which lack mitochondria), their sole energy source is considered to be anaerobic glycolysis (Mi-Ichi et al. 2003). Parasitized red cells make 20–50 times more lactate than uninfected cells (Vander Jagt et al. 1990). In theory this would roughly double red cell lactate production in a person with 2% parasitaemia. However, this would be unlikely to contribute to the high lactate levels seen in severe malaria, which may occasionally be over 10-fold higher than resting lactate concentrations. Infected red cells produce both D and L-lactate, but concentrations of venous blood D-lactate concentrations are usually low (P. Holloway and S. Krishna unpublished results). This finding does not support the production of lactate by parasites as an important cause of hyperlactataemia in infected hosts.

4.2.2
Convulsions

The intense muscular contractions associated with convulsions may cause a transient lactic acidosis (Orringer et al. 1977). Although convulsions are common in malaria, neither a history of convulsions prior to admission nor convulsions observed in hospital are related to hyperlactataemia (English et al. 1997b; Planche et al. 2003) or increased lactate/pyruvate ratios (Day et al. 2000b). But they are related to larger falls in lactate post-admission (Krishna 1994).

4.2.3
Muscle Damage

Muscle damage, leading to increased proteolysis and conversion of alanine to pyruvate has been proposed as a cause of hyperlactataemia (Davis et al. 1996). However, increased protein breakdown is unlikely to be a major factor in malaria-associated lactic acidosis. Although rhabdomyolysis can occur as muscle damage continues during the first few days of infection the level of blood lactate rapidly declines during this period (Davis et al. 1996). Furthermore, increased production of lactate per se would not be expected to alter significantly the lactate/pyruvate ratio, which is raised in severe malaria (Day et al. 2000b).

4.2.4
Decreased Liver Clearance

Decreased liver blood flow and decreased hepatic clearance of lactate and alanine have been reported in severe malaria (Molyneux et al. 1989; Pukrittayakamee et al. 1994, 2002) and could contribute to hyperlactataemia. However, data from lactate turnover studies in severe malaria demonstrate that increased lactate production, rather than decreased lactate clearance, is the major contributor to hyperlactataemia (Agbenyega et al. 2000; Davis et al. 2002). Thus decreased hepatic clearance is likely to be of secondary importance.

4.2.5
Chronic Salicylate Toxicity

Acute salicylate poisoning may uncouple oxidative phosphorylation and precipitate lactic acidosis (Stacpoole 1993). Salicylate toxicity has been associated with hyperlactataemia and hypoglycaemia in a group of Kenyan children with severe malaria (English et al. 1996). There are great variations in prescribing practice between countries, for example in Kumasi, Ghana none of 50 children presenting with severe malaria had detectable salicylate in plasma (T. Agbenyega, S. Krishna and T. Planche, unpublished results). Although chronic salicylate toxicity may be more prevalent in some geographical areas, salicylates are not widely used in children and are unlikely to be a major cause of hyperlactataemia in malaria worldwide.

4.2.6
Thiamine Deficiency

Thiamine (vitamin B_1), as the pyrophosphate derivative, is a cofactor of the PDC and is required specifically for the decarboxylation of pyruvate (Stacpoole 2004). Chronic thiamine deficiency (beriberi) is associated with lactic acidosis, hypoglycaemia and impaired consciousness. Low blood thiamine levels were detected in 12 of 23 Thai adults (54%) with severe malaria (Krishna et al. 1999). Populations from this region in southeast Asia typically consume a diet of polished rice and fish paste, which contains thiaminase, and thiamine deficiency is common. As with salicylate toxicity, thiamine deficiency is of undertermined importance worldwide as a cause of hyperlactataemia and studies of its prevalence in other geographic areas are awaited.

4.2.7
Increased Resting Energy Expenditure

In a study of 19 Gambian children with uncomplicated malaria, resting energy expenditure was about 30% greater than in uninfected control subjects (Stettler et al. 1992), but fell quickly after the initiation of antimalarial treatment. The increase in resting energy expenditure was positively correlated with oral temperature. Fever and malaria infection, therefore, may increase resting energy expenditure and tissue oxygen requirements. However, this study did not measure lactate levels and there is no direct evidence for a correlation between lactate and resting energy expenditure.

4.2.8
Anaemia

Severe anaemia (haemoglobin ≤ 50 g/l) is common in children and decreases the oxygen delivery capacity of blood. Compensatory changes in the oxygen-dissociation curve of haemoglobin may occur with increased red cell production of 2,3-diphosphoglycerate (Bohr effect) (Jones and MacGregor 1954), though this response may be blunted by acidosis (Krishna et al. 1983). Anaemia per se has a low case fatality rate and is a poor independent predictor of death in children with malaria (Allen et al. 1996; Marsh et al. 1995; Planche et al. 2003; Waller et al. 1995), but it may contribute to raised lactate levels.

Blood lactate levels are highly correlated with the degree of anaemia in infected children (English et al. 1997b; Planche et al. 2003), but not in adults (Day et al. 2000b). A Kenyan study of nine children with severe malaria associated anaemia (English et al. 1997a) showed that the seven cases who had both anaemia and hyperlactataemia had a mean 25% increase in oxygen consumption measured during the course of a blood transfusion. The increase in oxygen consumption was positively correlated with the admission lactate concentration but not with the admission haemoglobin level. The rise in oxygen consumption may reflect the repayment of an 'oxygen debt' that accumulated due to tissue hypoxia. However, due to lack of a control group (no transfusion) in this study it is impossible to determine whether increased oxygen consumption was due primarily to the transfusion or to successful antimalarial treatment. Indeed, the fact that increased oxygen consumption was related to baseline lactate but not to the degree of anaemia makes it less likely the 'oxygen debt' was due to anaemia itself.

4.2.9
Sequestration

P. falciparum is the only malaria parasite infecting humans that causes the sequestration of infected erythrocytes, which occurs in post-capillary venules and capillaries (Silamut et al. 1999). Sequestration may result in decreased blood flow, local competition between host and parasite for nutrients or local release of cytokines or other inflammatory mediators. Parasites adhere to a number of vascular receptors and red cell surface molecules. The most important parasite ligands are a family of proteins called *P. falciparum* erythrocyte protein 1 (Su et al. 1995). Intracellular adhesion molecule, vascular cell adhesion molecule, E-selectin, thrombospondin, CD36 and chondroitin sulphate A are also reported to be important receptors for cytoadherence (Fried and Duffy 1996; Silamut et al. 1999; Turner et al. 1994). Both ring stage parasites and more mature phases may sequester (Silamut et al. 1999). The adhesion ligand on ring stage infected cells is not known. Blood concentrations of endothelial microparticles (EMPs) are elevated in patients with cerebral malaria but not in those with malarial anaemia without cerebral symptoms (Combes et al. 2004). Circulating EMPs are elevated in several other conditions associated with vascular endothelial damage (Combes et al. 1999). These data imply, but do not prove a causal role for circulating EMPs in the pathobiology of cerebral malaria.

Large studies in Thai and Vietnamese adults have found greater numbers of sequestered parasitized red cells in the brains of patients who died from cerebral malaria than in those who died with malaria from other causes, such as renal failure (Silamut et al. 1999; Turner et al. 1994), implying that sequestration is related to the development of cerebral manifestations of disease. Because sequestration is not present in all patients who die with a diagnosis cerebral malaria, it has been suggested that simple mechanical obstruction of the microvasculature may be insufficient to explain the symptoms of severe malaria (Clark and Cowden 2003). However, the clinical diagnosis of cerebral malaria is the presence of coma in a patient with malaria where other causes of coma have been clinically eliminated, which may result from a number of other clinically undetected causes of coma such as hypoglycaemia. Furthermore, treatment may have cleared previously sequestered parasites from the brain of some patients who subsequently died. A recent post-mortem study of 31 children from Malawi who died with a pre-mortem diagnosis of cerebral malaria (Taylor et al. 2004) showed that seven patients [23%; 95% confidence interval (CI), 10%–41%] died of causes other than malaria. All children dying with post-mortem confirmed malaria had evidence of cerebral sequestration. The alternative causes of death in children with a coincidental parasitaemia included Reye's syndrome, intracerebral haemorrhage, meningitis, liver failure and pneumonia. Clearly an important corollary of these findings is that one would expect occasionally to find cases defined pre-mortem as 'cerebral malaria' without evidence of sequestration as they may have died of another cause with a coincidental parasitaemia.

Decreased oxygen availability inhibits the mitochondrial removal of lactate as well as increasing its production. Thus, hypoxia, in theory, is a condition in which the resulting metabolic acidosis may reflects both overproduction and underutilization of lactate and protons (Stacpoole 1993). Therefore, it may not be surprising that the resulting hypoxia would shift the intracellular milieu to a more reduced state, increase the $NADH/NAD^+$ ratio and favour a high lactate/pyruvate ratio. Blood lactate concentrations correlate positively with very high lactate/pyruvate ratios (about 30) in severe malaria (Day et al. 2000a; Krishna et al. 1994a), and which are higher than those reported in patients with sepsis (Dugas et al. 2000; Levy et al. 1997, 1999, 2000; Suistomaa et al. 2000) but similar to ratios measured in patients with congenital lactic acidosis due to defects in respiratory chain enzyme complexes (Robinson 2001)

Coma recovery times are related to lactate turnover (Agbenyega et al. 2000; Davis et al. 2002). Direct measures of tissue blood flows are technically difficult as blood flow increases with fever and anaemia and appropriate corrections need to be made. There is also the possibility of shunting which makes microvascular obstruction difficult to detect with currently available methods. Direct measures of cerebral blood flow in a study of Thai adults using a modification of the Kety–Schmidt method showing reduced cerebral blood flow in cases of cerebral malaria compared to convalescence (Warrell et al. 1988). A study of Kenyan children using trans-cranial Doppler failed to detect a generalized fall in cerebral blood flow compared with convalescence (Newton et al. 1996), though there were differences between blood flows in the right and left cerebral arteries in over half the children. One further cerebral blood flow study was too small to be interpretable ($n=4$) (Clavier et al. 1999). Measures of forearm blood flow in Thai adults ($n=12$) showed no consistent changes in blood flow (Krishna et al. 1994a).

4.2.10
Rheological Factors

Red cell deformability is decreased in both infected and uninfected red cells in malaria and is associated with disease severity and hyperlacataemia (Dondorp et al. 1999, 2002, 2004). Rosetting, in which mature parasitized red cells become surrounded by non-parasitized erythrocytes, has been associated with a poor outcome (Treutiger et al. 1992). Rosetting may increase blood viscosity, and could contribute to microvascular obstruction and poor tissue perfusion in malaria. In vitro models (Shelby et al. 2003; Taylor-Robinson 2004) using artificial microcapillaries confirm the importance of red cell deformability in determining blood flow through the microcirculation.

4.2.11
Circulatory Failure

It has been suggested recently that severe malaria may be the result of shock (Maitland et al. 2003a, 2003b). Shock is defined as 'a clinical state in which there is inadequate tissue perfusion to meet metabolic demands' (Logan 1998). If circulatory failure were responsible for poor tissue perfusion in malaria, then resuscitation with the appropriate intravenous fluids could be a cheap, simple and attractive adjunctive therapy to improve survival. However, hypotension is rare (Day et al. 1996) in adults with severe malaria, in whom cardiac output (Day et al. 1996), intravascular circulating volume (Davis et al. 1992; Sitprija et al. 1967), and total body water (Feldman and Murphy 1945; Flynn et al. 1967; Malloy et al. 1967) are normal or slightly elevated. The renal failure observed in severe adult malaria is not related to hypovolaemia or hypotension (Davis et al. 1992; Day et al. 1996; Sitprija et al. 1967). Thus there is little evidence that circulatory failure contributes importantly in malaria infected adults to poor tissue oxygenation and, hence, to overproduction of lactate and hydrogen ions by underoxygenated tissues.

The relationship between circulatory failure and lactic acidosis in children with severe malaria is less clear. Hypotension is a very late sign of circulatory collapse in children with trauma or sepsis in profound shock. Though hypotension is rare in children with severe malaria this is not evidence of the absence of circulatory failure. In sepsis, systems to diagnose shock exist using a number of validated clinical and laboratory criteria (Association 1997; Sáez-Llorens and McCracken 1993).

Most children with severe malaria have many of the non-specific signs of circulatory failure. For example, a recent Kenyan study found that 212 of 372 (57%) children with severe malaria had 'shock' as defined by two or more of these signs including: tachycardia, fever, hypotension, acidosis, tachypnoea, hypoxia, poor renal output, hypoglycaemia (Maitland et al. 2003a). Children who went on to die had a greater number of these signs than survivors. This was proposed as evidence of circulatory failure in malaria. Signs validated in sepsis to diagnose shock may not apply in severe malaria. For example, if one applied these defining criteria to a series of cases of heart failure or even acute urinary retention these signs would be detected frequently and give the false impression that the symptoms seen in those conditions are also due to shock and will therefore respond to fluid therapy. There are many physiological differences between sepsis and malaria and these are shown in Table 4.

Table 4 Comparison of common clinical features of severe malaria and septic shock

	Severe malaria	Septic shock
Histology		
Sequestration	Yes	No
Amount of infected material	10s of grams	Micrograms
Metabolism Lactate/ pyruvate ratio	~30	~20
Hypoglycaemia	Common	Rare
Haemodynamics		
Hypotension	Rare	Common
Capillary permeability	+	+++
Plasma volume	Normal	Reduced
Total body water	Normal/reduced	Normal/elevated
Extra-cellular water	Normal/reduced	Normal/elevated
Changes in fluids	No change in ECW	ECW elevated for days
Mixed venous O_2	Very low	Low
Haematology		
Leucocytosis (WBC >11/μl)	Rare	Very common
Anaemia (Hb <100 g/l)	Common	Uncommon
Thrombocytopoenia (platelets <140/μl)	Very common	Rare

Measurements of total body water are normal in children with malaria (Planche et al. 2004) and there is no evidence of capillary leak (Hero et al. 1997) or rise in extracellular water in childhood malaria (Planche et al. 2004) that is seen in childhood septic shock. Though no direct comparison has been made, vital signs in children treated with relatively conservative fluid replacement regimens (Planche et al. 2004) return to normal values as quickly as those children in other studies using aggressive fluid resuscitation (Maitland et al. 2003b). In childhood malaria, objective measures of fluid volume status and cardiovascular physiology are needed to demonstrate circulatory failure.

4.2.12
Summary

In summary, the causes of hyperlacataemia in severe malaria are multifactoral. Based on the findings from studies of lactate kinetics and the presumed conse-quences of microcirculatory obstruction by parasitized red blood cells, most cases of malaria-associated hyperlactataemia reflect a condition whereby the rate of lactate production by the host outstrips the rate of clearance.

4.3
Metabolic Acidosis

Metabolic acidosis is common in severe malaria (Taylor et al. 1993), although it has been defined differently by studies. Degree of acidosis is positively correlated with hyperlactataemia, a decreased blood concentration of bicarbonate and a high anion gap (Day et al. 2000b; English et al. 1997b). The serum anion gap is defined by concentration of sodium minus the sum of both chloride and bicarbonate and is 10–12 mEq/l.

A large study of Vietnamese adults with severe malaria showed that acidosis (standardized base deficit >3.3 mEq/l) was more specific than hyperlactataemia in predicting a fatal outcome (Day et al. 2000b). In this study, however, over a quarter of the patients had renal failure and the main predictors of acidosis were the concentration of blood lactate and plasma creatinine. Renal insufficiency causes acidosis by failure to excrete inorganic acids and thus may figure prominently in the pathogenesis of this complication of severe malaria in southeast Asian adults.

Renal insufficiency is extremely rare in children with severe malaria (Marsh et al. 1995; Newton and Krishna 1998). A study of 132 children with malaria and acidosis found that acidosis was related to the plasma concentrations of creatinine and urea, to the blood level of lactate and to age (English et al. 1997b). However, renal impairment was not marked, since the mean (95% CI) plasma creatinine was only 97(80–114) µmol/l (normal range, 33–61 µmol/l) in the most severely acidotic group and would not account for the observed magnitude of acidosis or the raised anion gap.

4.4
Clinical Course and Treatment

Most deaths from severe malaria occur within 24 h after admission to hospital (Krishna et al. 1994b; Marsh et al. 1995; Waller et al. 1995). If hyperlactataemia is present at the time of admission, it usually resolves within a few hours of instituting conservative fluid resuscitation and anti-malarial chemotherapy (Agbenyega et al. 2000, 2003; Krishna et al. 1994a). Whilst blood lactate is an independent predictor of mortality in malaria (Allen et al. 1996; Krishna et al. 1994b; Waller et al. 1995), a prolonged hyperlactaemia ($\geq$4 h) that occurs in a small proportion of cases is an even better indicator of a fatal outcome (Krishna et al. 1994a). It follows that interventions that decrease circulating lactate may improve survival by 'buying time' until anti-malarials can exert their full parasite killing effect. Some of the proposed interventions to improve outcome in malaria are summarized below.

4.4.1
Aggressive Intravenous Infusion of Fluids

The treatment of circulatory failure in sepsis in adults and children has been revolutionized by the use of aggressive fluid resuscitation to treat shock and improve tissue perfusion (Pathan et al. 2003; Rivers et al. 2001, 2004; Welch and Nadel 2003). Were it possible to reproduce this success in severe malaria then tens of thousands of lives might be saved worldwide. However, for the reasons outlined above it is unlikely that circulatory failure plays an important part in the pathophysiology or outcome of severe malaria. Moreover, there are inherent risks with aggressive fluid resuscitation (Maitland et al. 2004; Newton et al. 1997) that are potentially dangerous in settings in which neither mechanical ventilation nor monitoring of infusion rates or serial measurement of plasma electrolyte concentrations are routine. In adults with severe malaria there is unlikely to be any benefit from aggressive fluid management unless hypotension is present (Davis et al. 1992; Day et al. 1996, 2000b). In infected children, in whom hypotension is a less reliable sign of circulatory failure, the situation is less clear. There is no validated evidence of circulatory failure in children with severe malaria (except perhaps immediately before death) and there are potential risks inherent in administering fluid boluses of 20–40 ml/kg to severely ill children whose haemodynamic status cannot be monitored appropriately. Therefore, aggressive fluid resusitation in these patients is probably unwarranted on present evidence.

4.4.2
N-Acetyl Cysteine

N-Acetyl cysteine (NAC) is the acetylated precursor of both L-cysteine and glutathione. It has been used as a mucolytic agent and an antidote for paracetamol overdose (Zimet 1988). Recently NAC has been used for a number of other conditions, such as contrast nephropathy (Kelly 1998). The oral bioavailability of NAC is less than 10% (Olsson et al. 1988) and it is usually given as an infusion in emergency settings. Intravenous preparations of NAC have a shelf life of about 3 years. NAC is safe and well tolerated, although anaphylactic reactions have been reported infrequently (Tenenbein 1984).

NAC has been shown to lower blood lactate levels faster than placebo in Thai adults with severe malaria (Treeprasertsuk et al. 2003; Watt et al. 2002). Its lactate lowering mechanism of action is unknown and its impact on survival has not been tested. The usefulness of NAC is mitigated by the need to infuse it over 20 h and by its cost which is currently about US$ 70 for a 10 kg child (BMA and RPS 2004).

4.4.3
Dichloroacetate

Dichloroacetate (DCA) is an inhibitor of pyruvate dehydrogenase kinase, which reversibly phosphorylates, and inhibits, the activity of the PDC (Stacpoole 1989). Animal experiments indicate that DCA activates the PDC in virtually all tissues within minutes of its parenteral administration and readily crosses the blood–brain barrier. Sodium DCA is inexpensive to manufacture, is stable in solution at room temperature and is safe when administered to children or adults with lactic acidosis (Stacpoole et al. 1998). It has been used extensively in malaria as an investigational drug. DCA lowers lactate more rapidly than placebo in Thai adults (Krishna et al. 1994a, 1996, 2001) and Ghanaian children (Agbenyega et al. 2000, 2003; Krishna et al. 1995) with malaria associated hyperlactataemia. Population pharmacokinetic studies have shown little intersubject variability of phar-macokinetic parameters after intravenous administration of DCA in severe malaria (Agbenyega et al. 2003; Krishna et al. 1994a, 1995, 1996, 2001). In a study of Thai adults single intravenous dose of 46 mg/kg was as effective in lowering circulating lacate concentrations as were two doses (Krishna et al. 1996). DCA has been used with both quinine (Agbenyega et al. 2003; Krishna et al. 1994a, 1995, 1996, 2001) and artesunate (Agbenyega et al. 2000). A large study of Ghanaian children with malaria associated hyperlactataemia (n=124) showed no pharmacokinetic interaction between quinine and DCA and no unexpected adverse events with DCA when administered as a single 10-min infusion of 50 ml/kg body weight (Agbenyega et al. 2003). DCA has been shown to improve outcome in a *P. berghei* infected rodent model of malaria (Holloway et al. 1995) and magnetic resonance imaging studies in *P. berghei* infected mice showed lower brain lactate levels in animals that received DCA (Rae et al. 2000).

In summary, DCA lowers blood lactates in malaria infected humans and animals more rapidly than placebo and improves survival in a rodent model. DCA has been used with both quinine and artesunate and there have been no adverse events attributable to this drug when used in severe malaria. However, the impact of DCA on survival in malaria has not been investigated.

5
Amino Acid Metabolism

Plasma alanine levels are elevated in severe malaria (English et al. 1997b). Pyruvate and alanine are in equilibrium and their interconversion is catalysed in the cytoplasm by alanine aminotransferase. Elevated plasma alanine levels probably result from the increased availability of pyruvate derived from an increased rate of glycolysis and a decreased rate of hepatic gluconeogenesis. Glutamine is a non-essential amino acid, but becomes particularly important in times of metabolic stress, such as during trauma or sepsis (Castell et al. 1994). Low levels of plasma glutamine have been found in a number of critical illnesses and are related to a poor outcome (Oudemans-van Straaten et al. 2001). Glutamine supplementation may improve survival in premature infants (Neu et al. 1997) and critically ill adults (Griffiths 1997).

Plasma glutamine levels are reported to be low in children with moderate malaria (Cowan et al. 1999) but are normal in children with severe malaria (Planche et al. 2002). However, glutamine supplementation is unlikely to be of benefit in patients with severe malaria as plasma glutamine levels in severe malaria are not decreased.

6
Lipid Metabolism

Hypocholesterolaemia is present in *P. falciparum*-infected adults and children in Gabon (Baptista et al. 1996; Cuisinier-Raynal et al. 1990; Djoumessi 1989; Faucher et al. 2002; Mohanty et al. 1992; Ngou-Milama et al. 1995). In these patients the high-density lipoprotein cholesterol is low while the low-density lipoprotein concentration is normal (Faucher et al. 2002). There appears to be no relationship between cholesterol and disease severity (Baptista et al. 1996).

Hypertriglyceridaemia has been observed in Gabonese children (Faucher et al. 2002) and in Indian adults (Mohanty et al. 1992) with acute malaria, but again is unrelated to disease severity (Baptista et al. 1996). Triglyceride levels return to normal within 2 weeks of antimalarial treatment.

Long-chain nonesterified (free) fatty acids are taken up by the liver in proportion to their concentration in the blood. Mitochondrial β-oxidation of free fatty acids provides reducing equivalents for the stimulation of hepatic gluconeogenesis (Chen et al. 1999). Stable isotope ($[6,6\text{-}^2H_2]$ glucose) turnover studies in adults with uncomplicated malaria showed that glucose production and gluconeogenesis were higher in patients with malaria than in uninfected subjects (Van Thien et al. 2004). However, Acipimox, an inhibitor of lipolysis, did not decrease the rate of gluconeogenesis in infected patients despite causing a fall in plasma free fatty acid concentrations.

The causes for the observed changes in plasma lipid concentrations in *P. falciparum* malaria are obscure. However, it is not uncommon to have transient increases in circulating lipids, particularly triglycerides, from infection by bacteria, viruses and protozoa (Edwards and Stacpoole 1999). The hypertriglyceridemic response to infection is associated with an increase in circulating concentrations of very low-density lipoproteins (VLDL) that are the principal triglyceride-rich lipoproteins in the post-absorptive state. Experimental studies indicate that infection-induced hypertriglyceridemia is probably due to both an increase in the hepatic secretion of VLDL and to a decrease in triglyceride clearance from the circulation. This latter effect may be mediated by a decrease in the sensitivity of peripheral tissues to insulin, since systemic infection is a known cause of insulin resistance. Cytokines and endotoxin influence the synthesis and degradation of VLDL and may thus play a central role in infection-induced dyslipidaemia.

It is possible that the lipid abnormalities induced by infectious diseases are beneficial to the host (Edwards and Stacpoole 1999). This intriguing notion has been explored most fully in studies demonstrating that all the major human lipoprotein subclasses can bind endotoxin, block endotoxin-mediated activation of human monocytes and prevent release of various cytokines from these cells. Lipoproteins may also bind and neutralize viruses and may cause lysis of *Trypanosoma brucei* (Vanhamme et al. 2003). It is unknown whether circulating lipoproteins exert protective effects against the biochemical or clinical complications of *P. falciparum* infection.

7
Conclusion

Metabolic disturbances play a major role in deaths from malaria. There are between 1 and 2 million deaths annually from malaria (Breman 2001), making malaria one of the most common causes of fatal metabolic disease worldwide. The identification of hypoglycaemia as a life-threatening complication of severe malaria has led to its widespread recognition and to straightforward measures to prevent or to treat it. Hyperlactataemia is also an attractive metabolic target for intervention because of its independent association with mortality in severe malaria. A few investigational drugs for malaria associated lactic acidosis have been developed, but randomized controlled trials of these agents are required to determine their impact on survival.

References

Agbenyega T, Angus BJ, Bedu-Addo G, Baffoe-Bonnie B, Guyton T, Stacpoole PW, Krishna S (2000) Glucose and lactate kinetics in children with severe malaria. J Clin Endocrinol Metab 85:1569–1576

Agbenyega T, Planche T, Bedu-Addo G, Ansong D, Owusu-Ofori A, Bhattaram VA, Nagaraja NV, Shroads AL, Henderson GN, Hutson AD, Derendorf H, Krishna S, Stacpoole PW (2003) Population kinetics, efficacy, and safety of dichloroacetate for lactic acidosis due to severe malaria in children.J Clin Pharmacol 43:386–396

Allen SJ, O'Donnell A, Alexander ND, Clegg JB (1996) Severe malaria in children in Papua New Guinea. QJM 89:779–788

Association AH (1997) Pediatric Advanced Life Support. Texas

Aung-Kyaw-Zaw, Khin-Maung-U, Myo-Thwe (1988) Endotoxaemia in complicated falciparum malaria. Trans R Soc Trop Med Hyg 82:513–514

Baptista JL, Vervoort T, Van der Stuyft P, Wery M (1996) Changes in plasma lipid levels as a function of *Plasmodium falciparum* infection in Sao Tome.Parasite 3:335–340

BMA and RPS (2004) British National Formulary. Pharmaceutical Press, London

Breman JG (2001) The ears of the hippopotamus: manifestations, determinants, and estimates of the malaria burden. Am J Trop Med Hyg 64:1–11

Castell LM, Bevan SJ, CalderP, Newsholme EA (1994) The role of glutamine in the immune system and in intestinal function in catabolic states. Amino Acids 7:231–243

Chen X, Iqbal N, Boden G (1999) The effects of free fatty acids on gluconeogenesis and glycogenolysis in normal subjects. J Clin Invest 103: 365–372

Clark IA (1978) Does endotoxin cause both the disease and parasite death in acute malaria and babesiosis? Lancet 75–77

Clark IA, Cowden WB (2003) The pathophysiology of falciparum malaria. Pharmacol Therapeutics 99:221–260

Clark IA, Cowden WB, Butcher GA, Hunt NH (1987) Possible roles of tumor necrosis factor in the pathology of malaria. Am J Pathol 127

Clark IA, Hunt NM, Cowden WB (1986) Oxygen derived free radicals in the pathogenesis of parasitic disease. Adv Parasitol 25:1–44

Clark IA, Rockett KA, Cowden WB (1992) Possible central role of nitric oxide in conditions clinically similar to cerebral malaria. Lancet 340, 894–896

Clavier N, Rahimy C, Falanga P, Ayivi B, Payen D (1999) No evidence for cerebral hypoperfusion during cerebral malaria. Crit Care Med 27:628–632

Combes V, Simon AC, Grau GE, Arnoux D, Camoin L, Sabatier F, Mutin M, Sanmarco M, Sampol J, Dignat-George F (1999) In vitro generation of endothelial microparticles and possible prothrombotic activity in patients with lupus anticoagulant. J Clin Invest 104:93–102

Combes V, Taylor TE, Juhan-Vague I, Mege JL, Mwenechanya J, Tembo M, Grau GE, Molyneux ME (2004) Circulating endothelial microparticles in malawian children with severe falciparum malaria complicated with coma. JAMA 291:2542–2544

Cowan G, Planche T, Agbenyega T, Bedu-Addo G, Owusu-Ofori A, Adebe-Appiah J, Agranoff D, Woodrow C, Castell L, Elford B, Krishna S (1999) Plasma glutamine levels and falciparum malaria. Trans R Soc Trop Med Hyg 93:616–618

Cuisinier-Raynal JC, Bire F, Clerc M, Bernard J, Sarrouy J (1990) Paludisme:le syndrome dysglobulinémie-hypocholestérolémie. Med Trop 50:91–95

Davis TME, Benn JJ, Suputtamongkol Y, Weinberg JB, Umpleby AM, Chierakul N, White NJ (1996) Lactate turnover and forearm lactate lactate metabolism in severe falciparum malaria. Endocrinol Metab 3:105–115

Davis TME, Binh TQ, Thu le, TA, Long TT, Johnston W, Robertson K, Barrett PH (2002) Glucose and lactate turnover in adults with falciparum malaria: effect of complications and antimalarial therapy. Trans R Soc Trop Med Hyg 96:411–417

Davis TME, Looareesuwan S, Pukrittayakamee S, Levy JC, Nagachinta B, White NJ (1993) Glucose turnover in severe falciparum malaria. Metabolism 42:334–340

Davis TME, Suputtamongkol Y, Spencer JL, Ford S, Chienkul N, Schulenburg WE, White NJ (1992) Measures of capillary permeability in acute falciparum malaria: relation to severity of infection and treatment. Clin Infect Dis 15:256–266

Day NP, Phu NH, Bethell DP, Mai NT, Chau TT, Hien TT, White NJ (1996) The effects of dopamine and adrenaline infusions on acid-base balance and systemic haemodynamics in severe infection. Lancet 348:219–223

Day NP, Phu NH, Mai NT, Bethell DB, Chau TT, Loc PP, Chuong LV, Sinh DX, Solomon T, Haywood G, Hien TT, White NJ (2000a) Effects of dopamine and epinephrine infusions on renal hemodynamics in severe malaria and severe sepsis. Crit Care Med 28:1353–1362

Day NP, Phu NH, Mai NT, Chau TT, Loc PP, Chuong LV, Sinh DX, Holloway P, Hien TT, White NJ (2000b) The pathophysiologic and prognostic significance of acidosis in severe adult malaria. Crit Care Med 28:1833–1840

Dekker E, Hellerstein MK, Romijn JA, Neese RA, Peshu N, Endert E, Marsh K, Sauerwein HP (1997a) Glucose homeostasis in children with falciparum malaria: precursor supply limits gluconeogenesis and glucose production. J Clin Endocrinol Metab 82:2514–2521

Dekker E, Romijn JA, Moeniralam HS, Waruiru C, Ackermans MT, Timmer JG, Endert E, Peshu N, Marsh K, Sauerwein HP (1997b) The influence of alanine infusion on glucose production in 'malnourished' African children with falciparum malaria. QJM 90:455–460

Dekker E, Romijn JA, Waruiru C, Ackermans MT, Weverling GJ, Sauerwein RW, Endert E, Peshu N, Marsh K, Sauerwein HP (1996) The relationship between glucose production and plasma glucose concentration in children with falciparum malaria. Trans R Soc Trop Med Hyg 90:654–657

Delafield F (1872) A handbook of post-mortem examination and of morbid anatomy. William Wood, New York

Djoumessi S (1989) Serum lipids and lipoproteins during malaria infection. Pathol Biol 37:909–911

Dondorp AM, Angus BJ, Chotivanich K, Silamut K, Ruangveerayuth R, Hardeman MR, Kager PA, Vreeken J, White NJ (1999) Red blood cell deformability as a predictor of anemia in severe falciparum malaria. Am J Trop Med Hyg 60:733–737

Dondorp AM, Nyanoti M, Kager PA, Mithwani S, Vreeken J, Marsh K (2002) The role of reduced red cell deformability in the pathogenesis of severe falciparum malaria and its restoration by blood transfusion. Trans R Soc Trop Med Hyg 96:282–286

Dondorp AM, Pongponratn E, White NJ (2004) Reduced microcirculatory flow in severe falciparum malaria: pathophysiology and electron-microscopic pathology. Acta Trop 89:309–317

Dugas MA, Proulx F, de Jaeger A, Lacroix J, Lambert M (2000) Markers of tissue hypoperfusion in pediatric septic shock. Intensive Care Med 26:75–83

Edwards CM, Stacpoole PW (1999) In: Betheridge DJ, Illingworth DR, Shepherd J (eds) Lipoproteins in health and disease. Oxford University Press, New York

English M, Marsh V, Amukoye E, Lowe B, Murphy S, Marsh K (1996) Chronic salicylate poisoning and severe malaria. Lancet 347:1736–1737

English M, Muambi B, Mithwani S, Marsh K (1997a) Lactic acidosis and oxygen debt in African children with severe anaemia. QJM 90:563–569

English M, Sauerwein R, Waruiru C, Mosobo M, Obiero J, Lowe B, Marsh K (1997b) Acidosis in severe childhood malaria. QJM 90:263–270

English M, Wale S, Binns G, Mwangi I, Sauerwein H, Marsh K (1998) Hypoglycaemia on and after admission in Kenyan children with severe malaria. QJM 91:191–197

Faucher JF, Ngou-Milama E, Missinou MA, Ngomo R, Kombila M, Kremsner PG (2002) The impact of malaria on common lipid parameters. Parasitol Res. 88, 1040–1043

Feldman HA, Murphy FD (1945) The effect of alterations in blood volume on the anemia and hypoproteinemia. J Clin Invest 24:780–792

Fitz-Hugh T (1944) The cerebral form of malaria. Bull US Army Med Dept 83:39–48

Flynn MA, Hanna FM, Lutz RN (1967) Estimation of body water compartments of preschool children I. Normal children. Am J Clin Nutr 20:1125–1128

Fried M, Duffy PE (1996) Adherence of *Plasmodium falciparum* to chondroitin sulfate A in the human placenta. Science 272:1502–1504

Gaskell JF, Miller WL (1920) Studies on malignant malaria in Macedonia. QJM 13:381–426

Griffiths RD (1997) Outcome of critically ill patients after supplementation with glutamine. Nutrition 13:752–754

Hero M, Harding SP, Riva CE, Winstanley PA, Peshu N, Marsh K (1997) Photographic and angiographic characterization of the retina of Kenyan children with severe malaria. Arch Ophthalmol 115:997–1003

Hien TT, Day NPJ, Phu NH, Mai NTH, Chau TTH, Loc PP, Sinh DX, Chuong LV, Vinh H, Waller D, Peto TEA, White NJ (1996) A controlled trial of artemether or quinine in Vietnamese adults with severe falciparum malaria. New Engl J Med 335:76–83

Holloway PA, Knox K, Bajaj N, Chapman D, White NJ, O'Brien R, Stacpoole PW, Krishna S (1995) *Plasmodium berghei* infection: dichloroacetate improves survival in rats with lactic acidosis. Exp Parasitol 80: 624–632

Jones ES, MacGregor IA (1954) Pathological processes in disease. V. Blood physiology of Gambian children infected with *Plasmodium falciparum*. Ann Trop Med Parasitol 48:95–101

Kelly GS (1998)Clinical applications of N-aectylcysteine. Alt Med Rev 3:114–127

Krishna S, Agbenyega T, Angus BJ, Bedu-Addo G, Ofori-Amanfo G, Henderson G, Szwandt IS, O'Brien R, Stacpoole PW (1995) Pharmacokinetics and pharmacodynamics of dichloroacetate in children with lactic acidosis due to severe malaria. QJM 88:341–349

Krishna S, Nagaraja NV, Planche T, Agbenyega T, Bedo-Addo G, Ansong D, Owusu-Ofori A, Shroads AL, Henderson G, Hutson A, Derendorf H, Stacpoole PW (2001) Population pharmacokinetics of intramuscular quinine in children with severe malaria. Antimicrob Agents Chemother 45:1803–1809

Krishna S, Shoubridge EA, White NJ, Weatherall DJ, Radda GK (1983) *Plasmodium yoelii*: blood oxygen and brain perfusion in the infected mouse. Exp Parasitol 56:391–396

Krishna S, Supanaranond W, Pukrittayakamee S, Karter D, Supputamongkol Y, Davis TM, Holloway PA, White NJ (1994a) Dichloroacetate for lactic acidosis in severe malaria: a pharmacokinetic and pharmacodynamic assessment. Metabolism 43:974–981

Krishna S, Supanaranond W, Pukrittayakamee S, Kuile FT, Ruprah M, White NJ (1996) The disposition and effects of two doses of dichloroacetate in adults with severe falciparum malaria. Br J Clin Pharmacol 41:29–34

Krishna S, Taylor AM, Supanaranond W, Pukrittayakamee S, ter Kuile F, Tawfiq KM, Holloway PA, White NJ (1999) Thiamine deficiency and malaria in adults from southeast Asia. Lancet 353:546–549

Krishna S, Waller DW, ter Kuile F, Kwiatkowski D, Crawley J, Craddock CF, Nosten F, Chapman D, Brewster D, Holloway PA, et al. (1994b) Lactic acidosis and hypo-glycaemia in children with severe malaria: pathophysiological and prognostic significance. Trans R Soc Trop Med Hyg 88:67–73

Kun JF, Mordmuller B, Perkins DJ, May J, Mercereau-Puijalon O, Alpers M, Weinberg JB, Kremsner PG (2001) Nitric oxide synthase 2(Lambarene) (G-954C), increased nitric oxide production, and protection against malaria. J Infect Dis 184:330–336

Levy B, Bollaert PE, Charpentier C, Nace L, Audibert G, Bauer P, Nabet P, Larcan A (1997) Comparison of norepinephrine and dobutamine to epinephrine for hemo-dynamics, lactate metabolism, and gastric tonometric variables in septic shock: a prospective, randomized study. Intensive Care Med 23:282–287

Levy B, Nace L, Bollaert PE, Dousset B, Mallie JP, Larcan A (1999) Comparison of systemic and regional effects of dobutamine and dopexamine in norepinephrine-treated septic shock. Intensive Care Med 25:942–948

Levy B, Sadoune LO, Gelot AM, Bollaert PE, Nabet P, Larcan A (2000) Evolution of lac-tate/pyruvate and arterial ketone body ratios in the early course of catecholamine-treated septic shock. Crit Care Med 28:114–119

Logan RW (1998) Forfar and Anneil's Textbook of Pediatrics

Maegraith B, Fletcher A (1972). The pathogenesis of mammalian malaria. Adv Parasitol 10:49–75

Maitland K, Levin M, English M, Mithwani S, Peshu N, Marsh K, Newton CR (2003a) Severe *P. falciparum* malaria in Kenyan children: evidence for hypovolaemia. QJM 96:427–434

Maitland K, Pamba A, Newton CR, Levin M (2003b) Response to volume resuscitation in children with severe malaria. Pediatr Crit Care Med 4:426–431

Maitland K, Pamba A, Newton CR, Lowe B, Levin M (2004) Hypokalemia in children with severe falciparum malaria. Pediatr Crit Care Med 5:81–85

Malloy JP, Brooks MH, Barry KG, Wilt S, McNeil JS (1967) Pathophysiology of acute falciparum malaria. II, fluid compartmentalization. Am J Med 43:745–750

Marsh K, Forster D, Waruiru C, Mwangi I, Winstanley M, Marsh V, Newton C, Winstanley P, Warn P, Peshu N et al. (1995) Indicators of life-threatening malaria in African children. N Engl J Med 332:1399–1404

Mi-Ichi F, Takeo S, Takashima E, Kobayashi T, Kim H, Wataya Y, Matsuda A, Torii M, Tsuboi T, Kita K (2003) In: Marzuki S, Verhoef J, Snippe H (eds) Tropical diseases, from molecule to bedside. Kluwer Academic, New York, pp 117–133

Mohanty S, Mishra SK, Das BS, Satpathy SK, Mohanty D, Patnaik JK, Bose TK (1992) Altered plasma lipid pattern in falciparum malaria. Ann Trop Med Parasitol 86:601–606

Molyneux ME, Looareesuwan S, Menzies IS, Grainger SL, Phillips RE, Wattanagoon Y, Thompson RP, Warrell DA (1989) Reduced hepatic blood flow and intestinal malabsorption in severe falciparum malaria. Am J Trop Med Hyg 40:470–476

Neu J, Roig JC, Meetze WH, Veerman M, Carter C, Millsaps M, Bowling D, Dallas MJ, Sleasman J, Knight T, Auestad N (1997) Enteral glutamine supplementation for very low birth weight infants decreases morbidity. J Pediatr 131:691–699

Newton CR, Crawley J, Sowumni A, Waruiru C, Mwangi I, English M, Murphy S, Winstanley PA, Marsh K, Kirkham FJ (1997) Intracranial hypertension in Africans with cerebral malaria. Arch Dis Child 76:219–226

Newton CR, Krishna S (1998)Severe falciparum malaria in children: current understanding of pathophysiology and supportive treatment. Pharmacol Ther 79:1–53

Newton CR, Marsh K, Peshu N, Kirkham FJ (1996) Perturbations of cerebral hemodynamics in Kenyans with cerebral malaria. Pediatr Neurol 15:41–49 Ngou-Milama E, Duong TH, Minko F, Dufillot D, Kombila K, Richard-Lenoble D, Mouray H (1995) Profil lipidique au cours d'une thérapeutique curative spécifique du paludisme maladie chez l'enfant gabonais. Cahiers Santé 5:95–99

Olsson B, Johansson M, Gabrielsson J, Bolme P (1988). Pharmacokinetics and bioavailability of reduced and oxidised N-actylcysteine. Eur J Clin Pharmacol 34:77–82

Orringer CE, Eustace JC, Wunsch CD, Gardner LB (1977) Natural history of lactic acidosis after grand-mal seizures. A model for the study of an anion-gap acidosis not associated with hyperkalemia. N Engl J Med 297:796–799

Oudemans-van Straaten HM, Bosman RJ, Treskes M, van der Spoel HJ, Zandstra DF (2001) Plasma glutamine depletion and patient outcome in acute ICU admissions. Intensive Care Med 27:84–90

Pathan N, Faust SN, Levin M (2003) Pathophysiology of meningococcal meningitis and septicaemia. Arch Dis Child 88:601–607

Peterson WF (1926) Blood sugar during the crisis of malarial fever. Proc Soc Exp Biol Med 23:753–754

Planche T, Agbenyega T, Bedu-Addo G, Ansong D, Owusu-Ofori A, Micah F, Anakwa C, Asafo-Agyei E, Hutson A, Stacpoole PW, Krishna S (2003) A prospective comparison of malaria with other severe diseases in African children: prognosis and optimization of management. Clin Infect Dis 37:890–897

Planche T, Dzeing A, Emmerson AC, Onanga M, Kremsner PG, Engel K, Kombila M, Ngou-Milama E, Krishna S (2002) Plasma glutamine and glutamate concentrations in Gabonese children with *Plasmodium falciparum* infection. QJM 95:89–97

Planche T, Onanga M, Schwenk A, Dzeing A, Borrmann S, Faucher JF, Wright A, Bluck L, Ward L, Kombila M, Kremsner PG, Krishna S (2004). Assessment of volume depletion in children with malaria. Plos Med 1:e18

Pukrittayakamee S, Krishna S, Ter Kuile F, Wilaiwan O, Williamson DH, White NJ (2002) Alanine metabolism in acute falciparum malaria. Trop Med Int Health 7:911–918

Pukrittayakamee S, White NJ, Davis TM, Looareesuwan S, Supanaranond W, Desakorn V, Chaivisuth B, Williamson DH (1992) Hepatic blood flow and metabolism in severe falciparum malaria: clearance of intravenously administered galactose. Clin Sci (Lond) 82:63–70

Pukrittayakamee S, White NJ, Davis TM, Supanaranond W, Crawley J, Nagachinta B, Williamson DH (1994) Glycerol metabolism in severe falciparum malaria. Metabolism 43:887–892

Rae C, Maitland A, Bubb WA, Hunt NH (2000) Dichloroacetate (DCA) reduces brain lactate but increases brain glutamine in experimental cerebral malaria: a 1H-NMR study. Redox Rep 5:141–143

Rivers E, Nguyen B, Havstad S, Ressler J, Muzzin A, Knoblich B, Peterson E, Tomlanovich M (2001) Early goal-directed therapy in the treatment of severe sepsis and septic shock. N Engl J Med 345:1368–1377

Rivers EP, Nguyen HB, Huang DT, Donnino M (2004) Early goal-directed therapy. Crit Care Med 32:314–315; author reply 315

Saeed BO, Atabani GS, Nawwaf A, Nasr AM, Abdulhadi NH, Abu Zeid YA, Alrasoul MA, Bayoumi RA (1990) Hypoglycaemia in pregnant women with malaria. Trans R Soc Trop Med Hyg 84:349–350

Sexton AC, Good RT, Hansen DS, D'Ombrain MC, Buckingham L, Simpson K, Schofield L (2004) Transcriptional profiling reveals suppressed erythropoiesis, up-regulated glycolysis, and interferon-associated responses in murine malaria. J Infect Dis 189:1245–1256

Shelby JP, White J, Ganesan K, Rathod PK, Chiu DT (2003) A microfluidic model for single-cell capillary obstruction by *Plasmodium falciparum*-infected erythrocytes. Proc Natl Acad Sci USA 100:14618–14622

Silamut K, Phu NH, Whitty C, Turner GD, Louwrier K, Mai NT, Simpson JA, Hien TT, White NJ (1999) A quantitative analysis of the microvascular sequestration of malaria parasites in the human brain. Am J Pathol 155:395–410

Singh B, Choo KE, Ibrahim J, Johnston W, Davis TM (1998) Non-radioisotopic glucose turnover in children with falciparum malaria and enteric fever. Trans R Soc Trop Med Hyg 92:532–537

Sitprija V, Indraprasit S, Pochanugool C, Benyajati C, Piyaratn P (1967) Renal failure in malaria. Lancet:185–188

Stacpoole PW (1989) The pharmacology of dichloroacetate. Metabolism 38:1124–1144

Stacpoole PW (1993) Lactic acidosis. Endocrinol Metab Clin North Am 22:221–245

Stacpoole PW (2004) In: Clinical studies in medical biochemistry. Oxford University Press, New York

Stacpoole PW, Henderson GN, Yan Z, Cornett R, James MO (1998) Pharmacokinetics, metabolism and toxicology of dichloroacetate. Drug Metab Rev 30: 499–539

Stettler N, Schutz Y, Whitehead R, Jequier E (1992) Effect of malaria and fever on energy metabolism in Gambian children. Pediatr Res 31:102–106

Su XZ, Heatwole VM, Wertheimer SP, Guinet F, Herrfeldt JA, Peterson DS, Ravetch JA, Wellems TE (1995) The large diverse gene family var encodes proteins involved in cytoadherence and antigenic variation of *Plasmodium falciparum*-infected erythrocytes. Cell 82:89–100

Suistomaa M, Ruokonen E, Kari A, Takala J (2000) Time-pattern of lactate and lactate to pyruvate ratio in the first 24 hours of intensive care emergency admissions. Shock 14:8–12

Sáez-Llorens X, McCracken GH (1993) Sepsis syndrome and septic shock in pediatrics: Current concepts of terminology, pathophysiology, and management. J Pediatri 123:497–508

Taylor TE, Borgstein A, Molyneux ME (1993). Acid-base status in paediatric *Plasmodium falciparum* malaria. QJM 86:99–109

Taylor TE, Fu WJ, Carr RA, Whitten RO, Mueller JG, Fosiko NG, Lewallen S, Liomba NG, Molyneux ME (2004) Differentiating the pathologies of cerebral malaria by post-mortem parasite counts. Nat Med 10:143–145

Taylor TE, Molyneux ME, Wirima JJ, Fletcher A, Morris K (1988) Blood glucose levels in Malawian children before and during the adminstration of intravenous quinine for severe falciparum malaria. N Engl J Med 319:1040–1047

Taylor-Robinson A (2004) In-vitro model offers insight into the pathophysiology of severe malaria. Lancet 363:1661–1663

Tenenbein M (1984) Hypersensitivity-like reactions to N-acetylcysteine. Vet Human Toxicol 26:S3–S5

Treeprasertsuk S, Krudsood S, Tosukhowong T, Maek-A-Nantawat W, Vannaphan S, Saengnetswang T, Looareesuwan S, Kuhn WF, Brittenham G, Carroll J (2003) N-acetylcysteine in severe falciparum malaria in Thailand. Southeast Asian J Trop Med Public Health 34:37–42

Treutiger CJ, Hedlund I, Helmby H, Carlson J, Jepson A, Twumasi P, Kwiatkowski D, Greenwood B, Wahlgren M (1992) Rosette formation in *Plasmodium falciparum* isolates and anti-rosette activity of sera from Gambians with cerebral or uncomplicated malaria. Am J Trop Med Hyg 46:503–510

Turner GD, Morrison H, Jones M, Davis TM, Looareesuwan S, Buley ID, Gatter KC, Newbold CI, Pukritayakamee S, Nagachinta B et al. (1994) An immunohistochemical study of the pathology of fatal malaria. Evidence for widespread endothelial activation and a potential role for intercellular adhesion molecule-1 in cerebral sequestration. Am J Pathol 145:1057–1069

Usawattanakul W, Tharavanij S, Warrell DA, Looareesuwan S, White NJ, Supavej S, Soikratoke S (1985) Factors contributing to the development of cerebral malaria. II. Endotoxin Clin Exp Immunol 61:562–568

van Hensbroek MB, Onyiorah E, Jaffar S, Schneider G, Palmer A, Frenkel J, Enwere G, Forck S, Nusmeijer A, Bennett S, Greenwood B, Kwiatkowski D (1996a) A trial of artemether or quinine in children with cerebral malaria. N Engl J Med 335:69–75

van Hensbroek MB, Palmer A, Onyiorah E, Schneider G, Jaffar S, Dolan G, Memming H, Frenkel J, Enwere G, Bennett S, Kwiatkowski D, Greenwood B (1996b) The effect of a monoclonal antibody to tumor necrosis factor on survival from childhood cerebral malaria. J Infect Dis 174:1091–1097

van Thien H, Ackermans MT, Dekker E, Thanh Chien VO, Le T, Endert E, Kager PA, Romijn JA, Sauerwein HP (2001) Glucose production and gluconeogenesis in adults with cerebral malaria. QJM 94:709–715

van Thien H, Ackermans MT, Weverling GJ, Thanh Chien VO, Endert E, Kager PA, Sauerwein HP (2004a) Influence of prolonged starvation on glucose kinetics in pregnant patients infected with *Plasmodium falciparum*. Clin Nutr 23:59–67

Van Thien H, Weverling G, Ackermans MT, Canh Hung N, Endert E, Kager PA, Sauerwein HP (2004b) Free fatty acids are not involved in the regulation of gluconeogenesis and glycogenolysis in adults with uncomplicated *P. falciparum* malaria. Am J Physiol Endocrinol Metab

Vander Jagt DL, Hunsaker LA, Campos NM, Baack BR (1990) D-lactate production in erythrocytes infected with *P. falciparum*. Mol Biochem Parasitol 42:277–284

Vanhamme L, Paturiaux-Hanocq F, Poelvoorde P, Nolan DP, Lins L, Van Den Abbeele J, Pays A, Tebabi P, Van Xong H, Jacquet A, Moguilevsky N, Dieu M, Kane JP, De Baetselier P, Brasseur R, Pays E (2003) Apolipoprotein L-I is the trypanosome lytic factor of human serum. Nature 422:83–87

Waller D, Krishna S, Crawley J, Miller K, Nosten F, Chapman D, ter Kuile FO, Craddock C, Berry C, Holloway PA et al. (1995). Clinical features and outcome of severe malaria in Gambian children. Clin Infect Dis 21:577–587

Warrell DA, Looareesuwan S, Warrell MJ, Kasemsarn P, Intaraprasert R, Bunnag D, Harinasuta T (1982) Dexamethasone proves deleterious in cerebral malaria. A double-blind trial in 100 comatose patients. N Engl J Med 306:313–319

Warrell DA, White NJ, Veall N, Looareesuwan S, Chanthavanich P, Phillips RE, Karbwang J, Pongpaew P, Krishna S (1988) Cerebral anaerobic glycolysis and reduced cerebral oxygen transport in human cerebral malaria. Lancet 2:534–538

Watt G, Jongsakul K, Ruangvirayuth R (2002) A pilot study of N-acetylcysteine as adjunctive therapy for severe malaria. QJM 95:285–290

Welch SB, Nadel S (2003)Treatment of meningococcal infection. Arch Dis Child 88:608–614

White NJ, Ho M (1992) The pathophysiology of malaria. Adv Parasitol 31:134–173

White NJ, Miller KD, Marsh K, Berry CD, Turner RC, Williamson DH, Brown J (1987) Hypoglycaemia in African children with severe malaria. Lancet i:708–711

White NJ, Warrell DA, Chanthavanich P, Looareesuwan S, Warrell MJ, Krishna S, Williamson DH, Turner RC (1983) Severe hypoglycemia and hyperinsulinemia in falciparum. N Engl J Med 309:61–66

White NJ, Warrell DA, Looareesuwan S, Chanthavanich P, Phillips RE, Pongpaew P (1985) Pathophysiological and prognostic significance of cerebrospinal-fluid lactate in cerebral malaria. Lancet 1:776–778

Zimet I (1988) Acetylcysteine: A drug that is much more than a mucokinetic. Biomed Phamacother 42:513–520

CTMI (2005) 295:137–167

The Clinical and Pathophysiological Features of Malarial Anaemia

D. J. Roberts (✉) · C. Casals-Pascual · D. J. Weatherall

Nuffield Department of Clinical Laboratory Sciences, University of Oxford and Blood Research Laboratory, John Radcliffe Hospital, Headington, Oxford OX3 9DU, UK
david.roberts@ndcls.ox.ac.uk

Abstract This review will focus on the principal clinical and pathophysiological features of the anaemia of falciparum malaria, including the problems of treating malarial anaemia, and also will suggest how recent advances in genomics may help our understanding of cellular and molecular mechanisms underlying this syndrome.

Abbreviations

DAT	Direct Coombs' test
Epo	Erythropoietin
TNF-α	Tumour necrosis factor-alpha
IFN-γ	Interferon-gamma
IL-10	Interleukin-10
MIF	Macrophage inhibitory factor
HNE	Hydroxynoenal
GPI	Glycerol phophatidyl inositol

1
Introduction

Malaria is a major public health problem in tropical areas and it is estimated that it is responsible for 1–3 million deaths annually and 300–500 million infections. The vast majority of morbidity and mortality from malaria is caused by infection with *Plasmodium falciparum,* although *P. vivax, P. ovale* and *P. malariae* also are responsible for human infections. A significant proportion of the mortality and morbidity is due to anaemia. The salient features that make this clinical problem a major public health concern are the very large numbers of children affected and the difficulty of satisfactory treatment by blood transfusion outside specialist centres. These concerns have been enhanced by recent data from primate studies, which suggest that monkeys vaccinated with erythrocytic-stage antigens and that have acquired protection from acute infection may succumb to severe anaemia during a subacute or chronic phase of infection (Egan et al. 2002). However, ever since the first descriptions of malaria by Greek and Roman authors the study of malarial anaemia has been relatively neglected compared with those of other syndromes of severe malaria.

2
Prevalence and Aetiology of Anaemia in the Developing World

It is difficult to be sure how many children in the developing world suffer from anaemia. A precise synthesis of the numerous surveys of the prevalence of anaemia across the developing world is impossible. Different studies have used different definitions, demographic design and methodology. Nevertheless, using the World Health Organization definitions of anaemia (Table 1), it is quite clear that the overall prevalence is very high and in many parts of the world the majority of young children are anaemic (Table 2).

Table 1 World Health Organization Criteria for haemoglobin concentrations below which anaemia is considered to be present in populations at sea level (from World Health Organization 1972)

Age	Haemoglobin concentration (g/dl)
Children (6 months–6 years)	11
Children (6–14 years)	12
Adult males	13
Adult females (non-pregnant)	12
Adult females (pregnant)	11

Table 2 Prevalence (%) of anaemia (World Health Organization criteria) in different populations (from Weatherall and Ledingham 1987)

Geographic area	Preschool children	Non-pregnant women	Pregnant women	Adult men
Latin America (seven countries)	–	17	24	4
Chile	35	–	–	–
Nigeria	63	46	52	36
Northern India	90	84	80	48
Southern India	76	81	88	56
Burma	3–27	5–15	82	1–5
Philippines	42	37	72	7

It is equally difficult to determine the relative importance of different causes of anaemia in many tropical populations. Clinical studies that have examined the aetiology of severe anaemia in children presenting to hospital have concluded that acute or chronic malaria infection is a major precipitating factor in children with severe anaemia causing admission to hospital (Newton et al. 1997). Many community surveys have concentrated only on one mechanism of anaemia, for example iron or folate deficiency. These studies serve to emphasise the multifactorial aetiology of anaemia. Nevertheless, it is clear that iron deficiency, which affects at least 20% of the world's population, is a major factor in the seasonal surge of anaemia in the tropical rainy season in many, but not all, regions and that the numerous other diseases that exacerbate anaemia are often operating in the setting of low or inadequate body iron stores.

The iron and folate stores of the newborn reflect the respective maternal stores and folate content of breast milk, which may be reduced by maternal deficiency and maternal malaria. After weaning, iron deficiency is common in those communities eating mainly cereals, legumes and/or vegetables. The iron content of these diets is low and non-haem iron intake is further diminished by the presence of fibre, polyphenols and phosphates. In many regions, iron stores are reduced by blood losses following hookworm or schistosomal infection.

In many populations anaemia may be associated with folate deficiency. The dietary intake of folate is highly variable and depends not only on the presence of green leafed vegetables in the diet but also on methods of food preparation. Moreover, chronic illness may reduce intake further due to anorexia and/or malabsorption. Finally, the demand for folate may be increased by chronic haemolysis due to thalassaemia or sickle cell disease or indeed malaria. Nutritional vitamin B_{12} deficiency is rare in children, with the notable exception of breast fed children born to mothers with sprue. Common haemoglobinopathies provide a further cause of anaemia after the first 6 months of life.

The aetiology of anaemia in children in developing countries is therefore multifactorial; children have low haematinic reserves and increased demand for haematinics, and they are drawn into a vicious circle of infection, reduced immunity and malnutrition, particularly in the rainy season. In malaria endemic areas, anaemia secondary to acute and/or chronic malaria infection is superimposed on this complex epidemiological background.

Therefore, to obtain a true picture of the cause of anaemia it is necessary to use data from longitudinal studies carried out over a considerable period. For example, studies in The Gambia have stated that mean haemoglobin levels in children vary significantly at different times of the year; anaemia is much more common in the rainy season when malaria transmission is its highest (Brewster and Greenwood 1993) (Fig. 1). However, the rains also have profound effects on sanitation measures and at this time the water supply frequently becomes contaminated, with a concomitant increase in water-borne diarrhoeal disease. Such illnesses are accompanied by malabsorption and compounded by an impoverished food supply as food stores reach a nadir. So the seasonal increase in anaemia in malaria endemic areas arises on a background of low or frankly deficient stores of haematinics and/or other micronutrients. Recent studies have shown that a low baseline haemoglobin level (average 9.7 g/dl) at the start of the season for malaria transmission is a major risk factor for developing severe malarial anaemia (Dicko et al. 2004).

The exact role of borderline stores of haematinics, other nutrients and malaria per se in malarial anaemia would be expected to vary from region to region. Reliable determination of contribution of these individual factors to anaemia requires intervention studies rather than descriptive surveys. However, there has been considerable debate over the risk that iron sup-

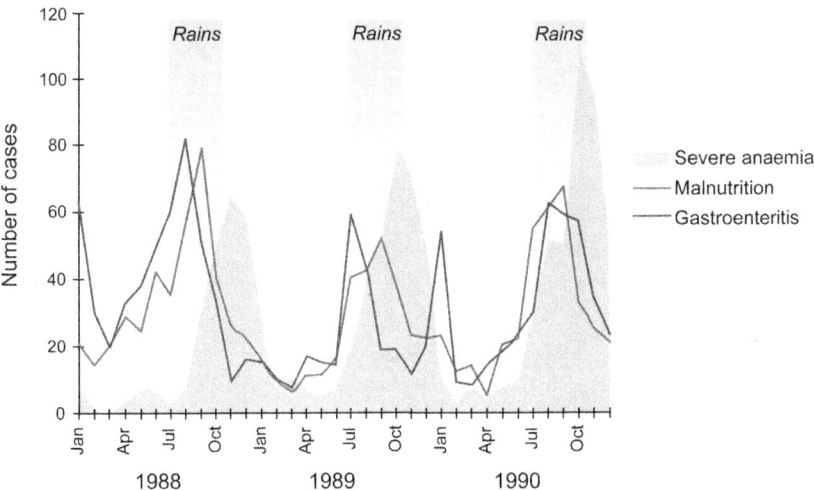

Fig. 1 Number of admissions to a children's ward in a hospital where malaria transmission is confined to the rainy season. The peaks of severe malarial anaemia correspond to seasonal epidemics of fever and cerebral malaria. Gastroenteritis and malnutrition peak incidence occurs in the early part of the rainy season, consistent with a multifactorial aetiology of the anaemia. (Redrawn from Brewster and Greenwood 1993)

plementation may improve anaemia but increase the prevalence of malaria disease or precipitate clinical episodes. In Tanzania, Menendez and Alonso and their colleagues have shown that iron supplementation and anti-malaria prophylaxis prevented 30% and 60%, respectively of all cases of moderate anaemia presenting during the malaria transmission season (Menendez et al. 1997). However, after the end of the intervention period, children who had received malaria chemoprophylaxis had approximately double the rates of severe anaemia and malaria compared to non-chemoprophylaxis groups. This study suggested that iron supplementation of infants could prevent iron-deficiency anaemia, even in malaria-endemic areas. Here, malaria chemoprophylaxis during the first year of life was more effective in prevention of malaria and anaemia, but apparently impaired the development of natural immunity. This trial provoked considerable interest in using chemoprophylaxis and iron supplementation to prevent malarial anaemia in early childhood. A recent meta-analysis of many different trials suggested that individuals with severe anaemia below 7 g/dl should receive malaria chemoprophylaxis as well as iron supplementation. Individuals with mild to moderate anaemia will not have an adverse effect of malaria disease with iron supplementation.

Menendez and colleagues performed another trial, giving sulphadoxine-pyrimethamine or placebo at the same time as routine vaccinations delivered through WHO's Expanded Program on Immunization (EPI). All children received iron supplementation between 2 and 6 months of age. This simple intervention reduced severe anaemia by 50% compared to the placebo group, while serological responses to EPI vaccines were not affected by the intervention (Schellenberg et al. 2001).

There is now widespread interest in repeating these trials elsewhere, using such powerful longitudinal studies to understand the other significant factors that contribute to anaemia and so define strategies to prevent anaemia in children living in areas with different patterns of malaria endemicity.

3
Features of Malarial Anaemia

3.1
The Spectrum of Disease Caused by Malarial Infection

The signs and symptoms of malaria infection in humans are caused by the asexual blood stage of the parasite. Infection with these stages may result in a wide range of outcomes and pathologies. Indeed, the spectrum of severity ranges from asymptomatic infection to rapidly progressive, fatal illness. The clinical presentation of malaria infection is also wide and influenced by age, immune status, genotype and pregnancy status of the host and by the species, genotype and, perhaps, the geographical origin of the parasite. In endemic areas many infections present as an uncomplicated febrile illness. In more severe forms of the disease children may present with prostration or inability to take oral fluids, or in younger children inability to suckle. Alternatively, they may exhibit a number of syndrome(s) of severe disease including anaemia, coma, respiratory distress and hypoglycaemia, and may also have a high rate of bacteraemia (Marsh et al. 1995; Berkley et al. 1999). The salient features and pathophysiology of many of these syndromes are described elsewhere in this book (see the chapters on uncomplicated malaria by M.P. Grobusch and P.G. Kremsner, severe malaria by T. Planche et al. and placental malaria by P.E. Duffy and M. Fried, this volume).

In most age groups, anaemia is frequently accompanied by more than one syndrome of severe disease. This was clearly illustrated in the studies of Marsh and colleagues (Fig. 2). This large series also demonstrates how the already substantial case fatality rate of 15%–20% for severe malaria in African children rises significantly when multiple syndromes of severe disease are present (Fig. 2; Marsh et al. 1995).

The age distribution of anaemia and other syndromes of severe disease are a consistent but puzzling feature of the epidemiology of clinical malaria.

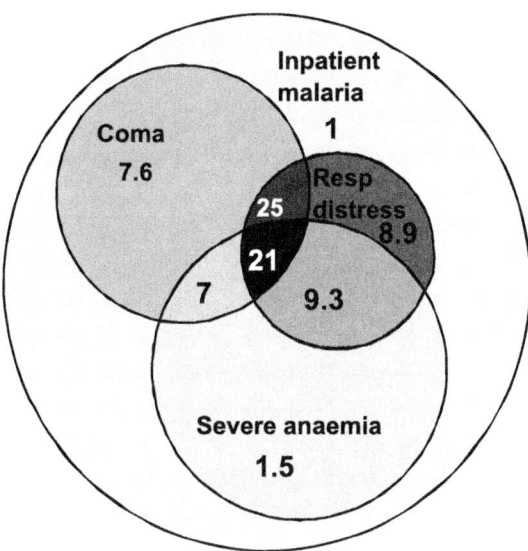

Fig. 2 The spectrum of clinical malaria in Kilifi, Kenya. The spectrum of clinical malaria in a malaria endemic area from a series of 1,800 children admitted to hospital. The *areas of the circles* in this Venn diagram are roughly proportional to the number of children in each group. The case fatality rate is given as a percentage. (Adapted from Marsh et al. 1995)

Children born in endemic areas are protected from severe malaria in the first 6 months of life by the passive transfer of maternal immunoglobulins and by foetal haemoglobin. Beyond infancy, the presentation of disease changes from severe anaemia in children aged between 1 and 3 years in areas of high transmission to cerebral malaria in older children in areas of lower transmission (Snow et al. 1997). As transmission intensity declines, severe malaria is most frequently found in older age groups. For example, in Papua New Guinea, malaria presents with multi-system disease in older children and young adults (Table 3).

The reasons for this pattern are unknown but it is often suggested that age-specific or cross-reactive responses activated during infection in older children may predispose to or precipitate strong pro-inflammatory reactions that have been implicated in the pathogenesis of cerebral malaria. It is also conceivable that anaemia may in some way confer protection from other syndromes of severe disease (see Sect. 6 for further discussion).

Table 3 The presenting features of malaria in African children and Papua New Guinean adults admitted to hospital, by WHO Criteria for Severe Falciparum Malaria (adapted from Newton and Krishna 1996)

	African children Marsh et al. 1995 Prevalence (%)	Non-immune adults Lalloo et al. 1996 Prevalence (%)
Defining criteria		
Coma	10.0	17.1
Severe anaemia	17.6	10.0
Respiratory distress	13.7	–
Hypoglycaemia	13.2	5.7
Circulatory collapse	0.4	0
Renal failure	0.1	22.9
Spontaneous bleeding	0.1	0.1
Haemoglobinaemia	0.1	0.1
Acidosis	63.6	NA
Repeated convulsions	18.3	0.3
Supporting criteria		
Impaired consciousness	8.2	37.1
Jaundice	4.7	45.7
Prostration	12.2	–
Hyperpyrexia	10.6	20
Hyperparasitaemia	8.9	40

NA, not applicable. The distribution of overlapping syndromes for children with severe malaria in African children and the case fatality rates are given in Fig. 1. The case fatality rates for severely ill non-immunes depend on the available supporting facilities.

3.2
The Clinical Features of Malarial Anaemia

The blood stage of falciparum malaria may cause life-threatening anaemia; a haemoglobin level of less than 5 g/dl is considered to represent severe disease. Children with anaemia may present with malaise, fatigue and dyspnoea or respiratory distress, defined by tachypnoea, deep gasping breathing and by use of the secondary muscles of respiration. In acute malaria respiratory distress usually represents metabolic acidosis, but in an ill child acute respiratory infection must be carefully excluded (Krishna et al. 1994; Marsh et al. 1995; English et al. 1997).

Acidosis is largely due to excessive lactic acid and other anions. Salicylate toxicity and dehydration may also play a role in studies from East Africa

(Maitland et al. 2003), but others have recently shown insignificant depletion of total water in children presenting with severe malaria (Planche et al. 2004; for a review see chapter by T. Planche et al., this volume). However, the majority of children presenting with respiratory distress are severely anaemic, have a metabolic acidosis secondary to reduced oxygen carrying capacity and respond to rapid transfusion of fresh blood (for review see English et al. 1996). A minority of those with respiratory distress do not respond to appropriate resuscitation; they probably represent a heterogeneous clinical group and may have renal failure, systemic bacterial infection or a more profound syndrome of systemic disturbance due to malaria parasites.

3.3
Haemolytic Syndromes Including Blackwater Fever

No discussion of the pathology of malarial anaemia is complete without consideration of 'Blackwater Fever'. The sudden appearance of haemoglobin in the urine indicating severe intravascular haemolysis leading to haemoglobinemia and haemoglobinuria received particular attention in early studies of anaemia in expatriates living in endemic areas (Stephens 1937). There was an association between Blackwater Fever and the irregular use of quinine for chemoprophylaxis. This drug can act as a hapten and stimulate production of a drug-dependent complement-fixing antibody. Recent studies of sudden intravascular haemolysis have shown it is rare in Africa, but more common in Southeast Asia and Papua New Guinea, where some cases are associated with glucose-6-phosphate dehydrogenase deficiency and treatment with variety of drugs including quinine, mefloquine and artesunate (Price et al. 1999). However, in the majority of cases the cause of sudden haemolysis cannot be identified.

3.4
The Haematological Features of Malaria Infection

The anaemia of falciparum malaria is typically normocytic and normochromic, with a notable absence of reticulocytes, although microcytosis and hypochromia may be present due to the very high frequency of alpha and beta thalassaemia traits and/or iron deficiency in many endemic areas (Newton et al. 1997; Yeats et al. 1999).

The anaemia of malaria may be accompanied by changes in the white cell and platelet counts and in clotting parameters but these changes are not in themselves diagnostic nor guide management. Malaria is accompanied by a modest leucocytosis, although leucopoenia may also occur. Occasionally, leukemoid reactions have been observed. Leucocytosis has been associated with severe disease (Molyneux et al. 1989; Ladhani et al. 2002). A high neutrophil count may also suggest intercurrent bacterial infection. Monocytosis and increased numbers of circulating lymphocytes are also seen in acute infection although the significance of these changes is not established (Abdalla 1988). However, malarial pigment is often seen in neutrophils and in monocytes and has been associated with severe disease and unfavourable outcome (Amodu et al. 1998).

Thrombocytopoenia is almost invariable in malaria and so may be helpful as a sensitive, but non-specific, marker of active infection. However, severe thrombocytopoenia ($<50\times10^9$/l) is rare. Increased removal of platelets may follow absorption of immune complexes, but there is no evidence for platelet-specific alloantibodies (Kelton et al. 1983). By analogy with erythropoiesis, there may be a defect in thrombopoiesis but this has not been established. Thrombocytopoenia is not associated with disease severity, although somewhat paradoxically platelets have been shown to contribute to disease pathology in animal and in human malaria (Lou et al. 2001). Moreover, in human infections, platelets may form clumps with infected erythrocytes (Pain et al. 2001). One explanation of this paradox could be that low levels of platelets might not only be a marker of parasite burden but also be protective from severe disease.

Disordered coagulation with clinical evidence of bleeding is seen in less that 5% of non-immune adults contracting malaria and presenting with severe disease. Patients may present with bleeding at injection sites, gums or epistaxis. Abnormalities of laboratory tests of haemostasis, suggesting activation of the coagulation cascades, occur in acute infection. However, histological evidence of intravascular fibrin deposition is notably absent in those dying from severe malaria (MacPherson et al. 1985). Factor XIII, normally responsible for cross-linking fibrin, is inactivated during malaria infection, and these data may explain low levels of fibrin deposition in the face of increased procoagulant activity (Holst et al. 1999).

The bone marrow is typically hypercellular. The most striking findings are of grossly abnormal development of erythroid precursors, or dyserythropoiesis. The developing erythroid cells typically demonstrate cytoplasmic and nuclear bridging and irregular nuclear outline (Abdalla et al. 1980; Abdalla and Wickramasinghe 1988). These changes are probably central to the pathophysiology of malarial anaemia and are discussed in detail below. Children who present with either acute or chronic malaria have significant suppression of reticulocytes (Casals-Pascual and Roberts, unpublished results). However, it has been firmly established that the proportion of abnormal erythroid precursors and the degree of dyserythropoiesis are much greater in chronic compared to acute infection (Abdalla et al. 1980). These data suggest that the inhibition and abnormal maturation of erythroid precursors may have somewhat different aetiologies in acute and chronic infection and/or that the factors that suppress erythropoiesis in acute malaria cause different morphological and functional abnormalities in chronic infection. There is no lack of candidates for factors that suppress bone marrow function during acute infection (see Sects. 4.7 and 4.8) but understanding their relative significance and the functional abnormalities they provoke requires detailed studies of bone marrow function in children presenting with anaemia at different stages of malaria infection.

The role of haematinic deficiency in children presenting with malaria and anaemia may be difficult to assess. First, the diagnosis of iron deficiency is complicated by the increase in serum ferritin as an acute phase reactant. Second, even in apparently iron replete patients, the demand for iron by the erythron can exceed the available supply. In such circumstances, the only diagnostic test of iron deficiency may be the response to iron supplementation. In this context, the studies that show an increase in haemoglobin level after iron supplementation during convalescence from an acute attack of malaria can be interpreted as correcting not only low absolute iron stores but also bypassing functional iron deficiency (Finch 1982). The relative importance of absolute and functional iron deficiency have not been defined. Nevertheless, many hospitals give a course of iron supplementation after an episode of acute malaria, although no general guidelines have been established. Chronic haemolysis may increase folate requirements, but frank deficiency is uncommon in children presenting with acute malaria, at least in East Africa (Newton et al. 1997). Folate deficiency may be more common in West Africa and protocols for folate supplementation after malaria must reflect local experience.

3.5
Malarial Anaemia in Pregnancy

Pregnancy is accompanied by a series of physiological changes that predispose not only to anaemia but also to malaria. The total blood volume increases during pregnancy but there is a disproportionate increase in plasma volume with a commensurate decrease in the normal haemoglobin level. Moreover, the demand for both iron and folate increases as the foetus grows and often precipitates frank folate or iron deficiency, particularly in multigravid women.

During and after pregnancy women are more susceptible to malaria infection and its consequences. Occult malaria infection, often without fever, may cause anaemia and placental dysfunction This effect is greatest in primigravidae and has been attributed to the adhesion of parasitized erythrocytes to chondroitin sulfate A and hyaluronic acid in the placenta (Fried and Duffy 1996; Beeson et al. 2000). Foetal growth is impaired and babies born to

women with placental malaria are on average 100 g lighter than controls born
to women without malaria. The subsequent contribution of malaria to infant
mortality is substantial (Guyatt and Snow 2001). Furthermore, the increase
in haematopoiesis demanded by haemolysis during malaria infection may
precipitate frank folate deficiency. Finally, women who are not immune to
malaria are more likely to develop hypoglycaemia and pulmonary oedema
during pregnancy.

The increase in maternal and foetal morbidity and mortality secondary
to malaria may be prevented by routine haematinic supplementation and
by intermittent treatment with anti-malarials during the second and third
trimesters with Fansidar (sulphadoxine–pyrimethamine). These simple, successful strategies are threatened by the spread of drug resistant malaria parasites; devising successful and safe alternatives is a major challenge.

3.6
Anaemia and *P. vivax*, *P. ovale* and *P. malariae* Infection

In *P. vivax* and *P. ovale* malaria, high parasitaemias are rare as invasion of
erythrocytes is limited to reticulocytes. The parasites rarely sequester in the
peripheral circulation or cause organ-specific syndromes, but may occasionally cause severe haemolysis and pulmonary oedema. Mortality is limited
to occasional deaths from splenic rupture and from intercurrent illnesses in
already weakened individuals.

Nevertheless *P. vivax* malaria has been clearly associated with anaemia
during pregnancy and with low birth weight of children of affected mothers.
Here cytokines or other inflammatory mediators appear to cause placental
dysfunction (Nosten et al. 1999). *P. malariae* infection is also rarely fatal but
is distinguished by the persistence of blood stage parasites for up to 40 years.
It can however cause a progressive and fatal nephrotic syndrome.

4
The Pathophysiology of Malarial Anaemia

The pathophysiology of severe anaemia is complex and is a relatively neglected
subject of study. In principle, all anaemias are caused by disturbance of the
balance of red cell production and loss and malaria is no exception

4.1
Loss of Infected Erythrocytes

Destruction of red blood cells is inevitable as parasites complete their 48-h
growth cycle and lyse their temporary host cell. It is noteworthy that some

parasites may be removed from erythrocytes as immature ring forms by phagocytic cells, leaving the red blood cells with residual parasite antigens to continue to circulate, albeit with reduced survival (Angus et al. 1997). These observations were made in patients receiving artemisinin treatment and the general significance of these interesting observations remains to be established. Infected erythrocytes may also be phagocytosed by macrophages following opsonization by immunoglobulins and/or complement components (Groux and Gysin 1990; Scholander et al. 1998). Other signals for recognition of infected erythrocytes by macrophages and other effector cells are less well defined, but may include abnormally rigid membranes, exposure of phosphatidylserine and other altered host antigens (Dondorp et al. 1999; McGilvray et al. 2000; Waitumbi et al. 2000). Infected erythrocytes may also be destroyed by antibody-dependent cytotoxicity and by NK cells, although the precise mechanisms are unclear. These effector mechanisms destroy erythrocytes and thus may control peripheral parasitaemia and therefore, indirectly, ameliorate anaemia.

4.2
Loss of Uninfected Erythrocytes

Several studies have shown that survival of uninfected erythrocytes in malaria infection is reduced due not only to changes to the uninfected erythrocytes themselves but also to the activity and number of phagocytes. Undoubtedly, the signals for recognition of uninfected erythrocytes for removal by macrophages are enhanced. Uninfected erythrocytes bind increased amounts of immunoglobulin and/or complement as detected in the direct antiglobulin test (DAT or Coomb's Test). The increased frequency of positive DAT tests was documented in clinical studies in the 1970s and thereafter confirmed by others in East Africa (Facer et al. 1979; Facer 1980; Abdalla et al. 1983). However, these studies were of insufficient power to detect an association between a positive DAT and severe anaemia.

More recently quantitative studies of immunoglobulin (Ig)G and complement regulatory proteins CR1, CD55, and CD59 on the surface of red blood cells from patients with severe anaemia and controls have shown that red cells from these patients were not only more susceptible to phagocytosis but also showed increased surface IgG and deficiencies in CR1 and CD55 compared with controls. These changes could contribute to the accelerated destruction of red cells in these patients by mechanisms such as phagocytosis or complement-mediated lysis (Waitumbi et al. 2000).

The specificity of the immunoglobulins on the surface of the red cells has remained controversial. Careful studies by Facer and our own work in Kenya suggest that these antibodies do not have a particular specificity, but are more likely to represent immune complexes absorbed onto the surface of red blood cells. Immune complexes are bound to the surface of red blood cells by complement receptor 1 (CD35) and also other receptors; this pathway for clearance of immune complexes by erythrocytes probably represents a physiological mechanism to limit systemic complement activation (Ahvazi et al.; Rowe et al. 1997)

4.3
Role of the Spleen

Some degree of splenomegaly is a normal feature of malarial infection, and the prevalence of splenomegaly in regions of malarial transmission is used as a major indicator of the level of malarial endemicity. The importance of the spleen in host defence against malaria has been demonstrated in experimental systems and non-immune individuals and are more susceptible to severe infection post-splenctomy. However, splenectomy does not abrogate the protective, acquired immune responses in those living in endemic areas (for review see Looareesuwan et al. 1993)

The phenomenon of parasitic sequestration, discussed earlier, is thought to have evolved primarily as an immune evasion strategy whereby the mature parasite can avoid passing through the spleen (Allred 1995).

Active erythrophagocytosis is a conspicuous feature within the bone marrow during *P. vivax* and *P. falciparum* malaria (Wickramasinghe et al. 1987, 1989), and it is highly probable that this also occurs within the spleen. Several studies have attempted to define the pathophysiological changes in the spleen during acute malaria. In animal models, malaria is accompanied by increased intravascular clearance of infected or rigid, heat-treated cells by the spleen (Wyler 1983), alterations in the splenic microcirculation (Weiss et al. 1986), and myelomonocytic recruitment to the spleen and liver (Lee et al. 1986). In studies of human malaria it has been found that increased splenic clearance of heat damaged red cells occurs during acute attacks (Looareesuwan et al. 1987).

Both clinical and experimental studies suggest that cytokines may be responsible for activating macrophages during malaria infection. Children with acute *P. falciparum* malaria have high circulating levels of interferon (IFN)-γ and tumour necrosis factor (TNF)-α(Kwiatkowski et al. 1990), a synergistic combination of cytokines that activate macrophages. Transgenic mice that constitutively overexpress human TNF are anaemic and show enhanced clearance of autologous erythrocytes, which is presumed (although not proven) to be due to erythrophagocytosis (Taverne et al. 1994). When infected with *P. yoelii* or *P. berghei*, transgenic mice suppress parasite density much more effectively than their littermates.

Changes to the uninfected red blood cells during infection also contribute to their own enhanced clearance by phagocytes. Uninfected red cells in children and adults with severe disease are less deformable and this is a significant predictor of the severity of anaemia and indeed outcome, consistent with the notion that these cells are being removed by the spleen (Dondorp et al. 1999). It has also been found that IgG-sensitized red cells are rapidly removed from the circulation by the spleen and that unusually rapid clearance persists well into the convalescent phase (Lee et al. 1989).

Thus, all the available evidence points to increased reticuloendothelial clearance in *P. falciparum* malaria, persisting long after recovery. These changes are presumably a host defence mechanism, maximising the clearance of parasitized erythrocytes.

4.4
The Role of Ineffective Erythropoiesis in Severe Malarial Anaemia

Reticulocytopoenia has been confirmed in numerous clinical studies of malarial anaemia (Vryonis 1939; Thonnard-Neumann; 1944; Srichaikul et al. 1967; Dormer et al. 1983). The implication that erythropoiesis is ineffective has become accepted by clinical scientists and by biologists modelling erythrocyte loss in acute malaria as a typical but somewhat unremarkable feature of the haematological response to malaria infection.

However, in general haematological practice, reticulocytopoenia in the face of rapid haemolysis is unusual. Increases in red cell clearance are normally compensated for by a swift and effective increase in red cell production. In auto-immune haemolytic anaemia it is commonplace for the relative reticulocyte count to rise from 1.5% to >20%, and thus the absolute reticulocyte count to rise from 10^{10} to 2×10^{11}/l as the appropriate bone marrow response in the presence of severe intravascular or extravascular haemolysis.

It is therefore surprising that many studies of erythropoiesis in malaria have concluded that there is no statistical and therefore biological relationship between reduced red cell production and degree of anaemia during acute malarial infection. These conclusions may be based on erroneous comparisons. The degree of reticulocytopoenia in malarial anaemia of 3%–4% should be compared not with the reticulocyte counts seen in normal non-anaemic healthy controls, but in subjects with a similar degree of anaemia due to intra- or extra-vascular haemolysis, where the reticulocyte count rises up to 10 times normal. This perspective of normal erythrocyte homeostasis suggests that destruction and ineffective production have at least an equal role in causing malarial anaemia.

4.5
Normal Erythropoiesis

The least mature committed erythroid progenitor is known as a primitive erythroid burst-forming unit or pBFU-E. These colonies derive from single cells and have in vitro a burst-like appearance because cells from the first few divisions migrate or burst, into clumps, prior to forming a colony composed of multiple subclusters of erythroid cells. BFU-Es mature and lose their ability to divide and migrate in vitro but increase their sensitivity to erythropoietin (Epo) until the stage where they are known as erythroid colony-forming units (CFU-E). Haematopoietic differentiation also requires the appropriate microenvironment. The bone marrow cells that play a key role in determining the fate of haematopoietic stem cells and their progeny include amongst others: macrophages, endothelial cells, stromal reticular cells and adipocytes. Our understanding of how this normal process of development is disturbed in those suffering from malarial anaemia is far from complete.

4.6
Previous Studies of Ineffective Erythropoiesis

The histopathological study of the bone marrow of children with malarial anaemia shows erythroid hyperplasia with increased numbers of erythroid precursors (Fig. 3). However, the maturation is abnormal by light and electron microscopy. Abdalla and Weatherall described the hallmark characteristics of such abnormal maturation, namely cytoplasmic and nuclear bridging and irregular nuclear outline (Abdalla et al. 1980; Abdalla and Wickramasinghe

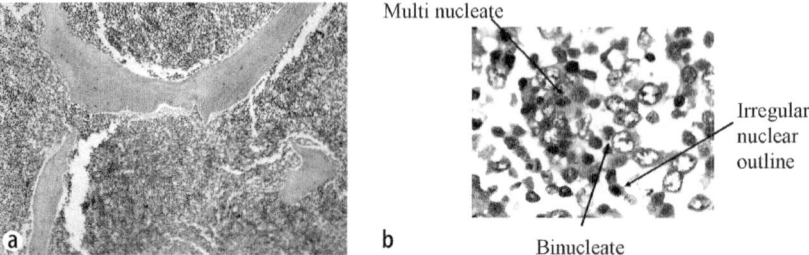

Fig. 3 Histopathology of bone marrow from a child dying with severe malarial anaemia. **a** A low-power view of a bone marrow trephine sample taken from a child dying from severe malaria and anaemia. The marrow is hypercellular and shows plentiful erythroid precursors, although reticulocytopoenia during acute infection indicated that erythropoiesis is ineffective. **b** A high-power view of the same section showing marked dyserythropoiesis, including multinucleate erythroid precursors and cytoplasmic bridging (*arrowed*)

1988). They later confirmed that the distribution of the erythroid progenitors through the cell cycle is abnormal in malarial anaemia with an increased proportion of cells in the G_2 phase compared with normal controls (Abdalla et al. 1984) Horstmann and colleagues confirmed these findings using the same criteria as Abdalla: they studied six patients with falciparum malaria before and after treatment (Dormer et al. 1983).

Ineffective erythropoieis also contributes to anaemia in animal models of malaria. A recent study has shown that vaccinated *Aotus* monkeys, after a challenge infection, may develop moderate to severe anaemia following rapid clearance of uninfected erythrocytes but with low reticulocyte counts indicating bone marrow dysfunction (Egan et al. 2002). Erythropoiesis is disrupted in mice during malaria infection. Mice infected with *P. berghei* show reduced bone marrow cellularity, erythroblasts, BFU-E and CFU-E as early as 24 h post-infection (Maggio-Price et al. 1985). The cellularity and CFU-S content of the femoral marrow of BALB/c mice infected with *P. chabaudi adami* decreases as the parasitaemia increases (Silverman et al. 1987).

4.7
Modulation of Erythropoiesis by Cytokines

Given the importance of Epo to erythropoiesis, attention has been focussed on the levels of this crucial cytokine in malarial infection. Serum Epo was appropriately raised in a single study of African children suffering from malarial anaemia (Kurtzhals et al. 1997; Newton et al. 1997). However, other studies in adults from Thailand and Sudan have suggested that the Epo concentration, although raised, was inappropriate for the degree of anaemia (Burgmann et al. 1996; el Hassan et al. 1997). It is possible that ineffective or inadequate Epo synthesis does contribute to malarial anaemia. Reduced production of Epo may be secondary to inflammatory mediators such as TNF-α, the level of which is known to be raised in acute malaria (Tonelli et al. 2001). Without further studies of the level of Epo in African children, the significance of these studies in adults is unclear.

The prime candidates for the host factors mediating dyserythropoiesis are imbalances of TNF-α, IFN-γ and interleukin (IL)-10. The concentrations of TNF-α and IFN-γ have been correlated with the severity of the disease (Kurtzhals et al. 1998; Othoro et al. 1999; Zamai et al. 2000). While low concentrations of TNF-α (<1 ng/ml) stimulate erythropoiesis, higher levels of TNF-α have been shown to suppress erythropoiesis (Zamai et al. 2000). Furthermore, it is possible that high levels of these inflammatory cytokines may contribute to reduced and abnormal production of erythrocytes, and also to increased erythrophagocytosis.

Recent evidence has suggested that the release of macrophage inhibitory factor (MIF) inhibits the growth of early erythroid and myeloid progenitors during murine malaria infection (Martiney et al. 2000). However, the role of MIF in human malaria infection has not been established

The possibility has been raised that high levels of the Th2 type cytokine, IL-10, might prevent the development of severe malarial anaemia. Low levels of IL-10 have been described in African children with severe malarial anaemia (Kurtzhals et al. 1998). However, other studies and our own results do not confirm these findings. Similarly, defective IL-12 production has been shown experimentally to be associated with increased severity of malaria in a rodent model and low IL-12 levels have been associated with severe malaria in African children (Mohan and Stevenson 1998, 1998; Luty et al. 2000; Perkins et al. 2000).

4.8
Modulation of Erythropoiesis by Infected Erythrocytes and Their Products

The hypothesis that parasite products directly stimulate the production of inflammatory cytokines, including TNF-α, has been widely promoted. In humans, peripheral blood mononuclear cells do secrete TNF-α and other cytokines when incubated with intact or lysed parasite cells or malaria pigment in vitro (Kwiatkowski et al. 1989; Pichyangkul et al. 1994). However, interpretation of the older literature on this topic has been severely compromised by the finding that malarial cultures have been contaminated worldwide by mycoplasma species (Turrini 1997; Rowe et al. 1998).

There is also substantial evidence that lysate of infected erythrocytes may directly modulate the function of host cells. During its blood stage, the malaria parasite proteolyses host haemoglobin in an acidic vacuole to obtain amino acids, releasing haem as a by-product, which is autoxidized to potentially toxic haematin [aquaferriprotoporphyrin IX, $H2O-Fe(III)PPIX$]. β-haematin forms as a crystalline cyclic dimer of $Fe(III)PPIX$ and is complexed with protein and lipid products as malarial pigment or haemozoin. Arese and colleagues showed that the function of monocytes and of monocyte-derived macrophages is severely inhibited after ingestion of malaria pigment or haemozoin. These cells were unable to repeat phagocytosis and to generate oxidative burst when appropriately stimulated (Schwarzer et al. 1992). Membrane-bound protein kinase C and NADPH oxidase, and thus the oxidative burst, were severely impaired (Schwarzer et al. 1993, 1996). Furthermore, after phagocytosis of haemozoin human and murine myeloid cells were unable to kill ingested fungi, bacteria and tumour cells (Fiori et al. 1993) and to respond to IFN-γ stimulation, but instead responded by increased release of IL-1β and TNF-α (Pichyangkul et al. 1994), macrophage inflammatory protein (MIP)-1α and MIP-1β (Sherry et al. 1995).

The haemozoin crystal of haem moieties may be complexed with biologically active compounds. The oxidation of membrane lipids catalysed by the ferric haem produces the lipoperoxides (Schwarzer et al. 1996, 1999). A key feature in lipid peroxidation is the breakdown of polyunsaturated fatty acids to yield a broad array of smaller fragments, 3 to 9 carbons in length, including aldehydes, such as 4-hydroxy-2-alkenals (HNE). HNE or hydroxynoenal is known to be the major aldehyde formed during lipid peroxidation of omega-6 polyunsaturated fatty acids, such as linoleic and arachidonic acid. HNE accumulates in membranes at concentrations of 10 µM–5 µM in response to oxidative insults.

There is accumulating evidence that HNE and other lipoperoxides including 15-hydroxy-arachidonic acid [15 (R,S) HETE] may play a role in the pathophysiology of malaria. Arese and colleagues showed that protein kinase C was inhibited at >1 µM in haemozoin-fed monocytes (Schwarzer et al. 1996). Most recently they have shown that HNE and HETE are generated in parasitized erythrocytes and that the toxicity of haemozoin was reduced after removal of lipids. HNE also inhibits the up-regulation of surface expression of intercellular adhesion molecule (ICAM)-1, vascular cell adhesion molecule (VCAM)-1 and E-selectin following TNF-stimulation (Minekura et al. 2001). Moreover, endoperoxides produced in pigment-containing monocytes or macrophages may impair erythroid growth (Giribaldi et al. 2004). These data suggest one possible mechanism for inhibition of erythropoiesis by haemozoin, although many other potential routes of oxidative damage exist and the role of haemozoin itself as an inhibitor of erythropoiesis remains to be established.

4.9
Modulation of Erythropoiesis by Intercurrent Infection

In children presenting with acute malaria bacteraemia is associated not only with anaemia but also with excess mortality (Berkley et al. 1999). Parvovirus B19 may cause a transient reticulocytopoenia and thus severe and sudden anaemia in those suffering from a haemolytic anaemia or foetal anaemia and intrauterine death (Brown 2000). Most children in Africa have serological evidence of infection early in life. However, acute infection with this virus is uncommon in those presenting with severe malarial anaemia (Newton et al. 1997; Yeats et al. 1999).

4.10
Summary

Anaemia in falciparum malaria is clearly multifactorial. Considering the dynamics of erythropoiesis there is a powerful argument that erythrocyte destruction and ineffective erythropoiesis play equal parts in the aetiology of malarial anaemia. The causes of ineffective erythropoiesis are themselves manifold and may include not only host cytokines but also inflammatory mediators produced directly or indirectly by parasite products.

5
Treatment

Malaria requires urgent effective chemotherapy to prevent progression of disease and this may be the most crucial public health intervention to reduce global mortality from malaria. The drug treatment of malaria must take account of the expected pattern of drug resistance in the area where infection was contracted, the severity of clinical disease and the species of parasite. The spread of drug resistant parasites and the optimal use of affordable, effective drugs are of continual concern and these have recently been reviewed (Winstanley 2001; Wongsrichanalai et al. 2002). In severely ill patients, good nursing care is vital. Monitoring and treatment of fits and hypoglycaemia is essential and anti-pyretics should be given (Njuguna and Newton 2004), although it has been shown that paracetamol may prolong the parasite clearance time (see the chapter by M.P. Grobusch and P.G. Kremsner, this volume).

Certainly, blood transfusion is in principle a straightforward solution to the treatment of severe malarial anaemia, although controversy exists over the trigger for transfusion and the rate of administration of blood. The standard regimes of cautious and slow delivery of blood have been challenged by the demonstration that rapid initial flow rates may correct lactic acidosis and hypovolaemia (Maitland et al. 2003). However, the fluid management of severe malaria is controversial and Planche and colleagues have recently published convincing evidence that significant volume depletion is uncommon in severely ill children in Gabon (Planche et al. 2004). At present fluid management must depend on careful clinical assessment of each case. There remains general agreement that in non-immunes and in pregnant women blood transfusion must be accompanied by careful haemodynamic monitoring to avoid precipitating or exacerbating pulmonary oedema.

No formal controlled trials for the transfusion of patients with malaria have been performed. Whatever clinical guidelines emerge, in reality blood transfusion in the heartland of malaria endemic areas is beset by many practical and theoretical problems. First, the absence of well-characterized donor panels (and thus systematic blood collection) frequently jeopardizes the supply of blood. Secondly, even when standard screening for HIV is in place, the residual risk of HIV transmission in the serological window of infectivity remains at 1 in 2000 (Fleming 1997). At a practical level, positive indirect antiglobulin tests in acute infection may make the exclusion of alloantibodies difficult. Depending on the clinical urgency and transfusion history the least serologically incompatible blood may have to be given.

One therapeutic option available in North America and in Europe for the urgent treatment of non-immune patients with severe disease would be an exchange blood transfusion. This procedure removes non-sequestered, infected erythrocytes and possibly circulating 'toxins'. In the absence of evidence from trials for the use of exchange transfusion in malaria, some have suggested that this treatment could be given for hyperparasitaemia (>20%) in severely ill non-immune patients by CDC criteria (Newton and Krishna 1998; Riddle et al. 2002).

6
Relation of Anaemia to Parasite Virulence

The clinical features of malarial anaemia can be readily described, and the pathophysiology of this condition is steadily being revealed. However, a real appreciation of the role and significance of malarial anaemia in the overall dynamic host–parasite relationships is far from clear.

At first sight, malarial anaemia appears to be potentially harmful to the host and offers little or no advantages to the parasite. However, it is conceivable that anaemia may limit host pathology and also be of some benefit for the transmission and thus survival of the parasite. The overall picture of disordered physiology in malarial anaemia suggests that poor erythropoietic response in the face of haemolytic anaemia leads to a fall in the haemoglobin concentration. Anaemia limits the total parasite burden in two ways. First, most malaria parasites, including *P. falciparum*, have a higher growth rate in reticulocytes or young erythrocytes than in normocytes or older erythrocytes. It is true that this reticulocyte preference is not as marked for *P. falciparum* as it is for *P. vivax*, but it does exist. Reducing or limiting the rise in the proportion of reticulocytes in the blood will therefore reduce the parasite growth rate. Second, reducing the number of total erythrocytes in the circulation would limit the total parasite burden. These two factors are therefore likely to reduce the number of parasitized erythrocytes, the number of adherent and sequestered parasitized erythrocytes and also the total quantity of toxic parasite products released into the circulation after rupture of schizonts.

There is little direct evident to support such a relationship between susceptibility to malarial anaemia and resulting protection from cerebral malaria or other end-organ damage. One piece of circumstantial evidence from clinical malaria is the clear separation of age-related susceptibility to difference syndromes of malarial infection. Over a wide range of transmission rates, the incidence of malarial anaemia peaks at a younger age than the incidence of cerebral malaria. The difference is most obvious at lower transmission rates, i.e., low entomological inoculation rates (EIR). At the highest EIRs, of several hundred infective mosquito bites each year, the predominant syndrome of severe malaria is malarial anaemia (Snow et al. 1997). This age-specific incidence of syndromes of severe disease is not understood. One possible explanation is that younger children are susceptible to anaemia, but those developing anaemia are less likely to suffer from cerebral malaria.

This hypothesis cannot of course be tested experimentally in human infections but would be supported by recent data from Stevenson and colleagues studying *P. chabaudi* infections in mice. In this model, anaemia of the murine host, is associated with inadequate reticulocytosis and suppressed proliferation, differentiation and maturation of erythroid precursors (Chang et al. 2004). They were able to demonstrate that Epo-induced reticulocytosis was important for recovery from malarial anaemia and survival of mice. However, pre-treatment of mice with Epo, resulting in a significant reticulocytosis during the pre-patent period of erythrocytic growth, increased the rates of parasite growth and increased mortality in otherwise resistant mice (Chang et al. 2004). These observations are consistent with the mathematical model of Hetzel and Anderson, suggesting that the growth rates during the initial phase of infection depend on the number and type of host erythrocytes (Hetzel and Anderson 1996). Therefore, in this model of mouse malaria, anaemia can be a protective factor for outcome in malaria.

Similar observations cannot be made in humans. However, it may be possible to gain indirect supporting evidence for anaemia as a protective factor for other syndromes of severe malaria when the mechanism of suppression of erythropoiesis is more fully understood. One would hypothesize that a group of human polymorphisms, that reduce the erythropoietic response during malaria infection, confer susceptibility to malarial anaemia but give protection from cerebral malaria. At present, the only polymorphisms with this characteristic appears to be Melanisian ovalocytosis (Allen et al. 1998), although the relationship of this trait to altered erythropoiesis is not obvious. Other candidate loci and polymorphisms of this type may be revealed as the cellular and molecular pathogenesis of anaemia is described (see Fig. 4).

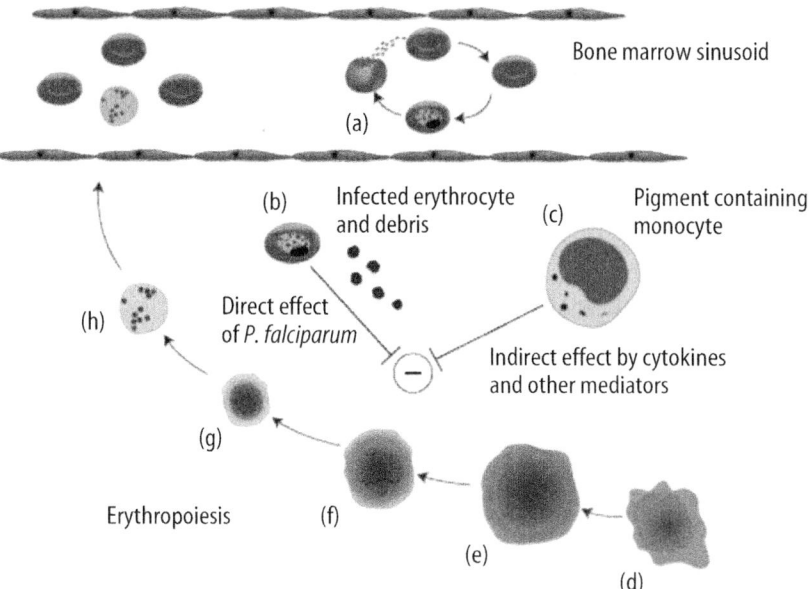

Fig. 4 Schematic diagram of the pathogenesis of malarial anaemia. Malaria parasites multiply by invasion of uninfected erythrocytes. Not only lysis of infected erythrocytes (*a*) but also splenic clearance of infected (*b*) and uninfected erythrocytes contribute to haemolysis during infection. Pigment-containing macrophages may release inflammatory cytokines and other biologically active mediators (*c*). It is also possible that malarial pigment or other parasite products have a direct inhibitory effect on erythropoiesis (*e*). Inhibition of erythropoiesis may be at one or more sites in the growth and differentiation of haematopoietic progenitor cells (*f*), through erythroblasts, early and late normoblasts (*g*), to reticulocytes (*h*). Both indirect and direct effects may cause bone marrow suppression and result in inadequate reticulocyte counts for the degree of anaemia

Until recently, anaemia had no significance for the biology of the malaria parasite during infection. However, some elegant studies now suggest that gametocytogenesis or the switch to the development of sexual forms of the parasite from asexual erythrocytic parasites may be influenced by host anaemia. In two models of malaria infection, with avian malarial parasite *P. gallinacium* and the murine malarial parasite *P. vinckei*, Brey and colleagues showed that the gametocyte sex ratio was strongly associated with the presence of reticulocytes (Paul et al. 2000). Moreover, male to female ratio could be increased by stimulating erythropoietic activity, by phlebotomy, hypoxia or addition of exogenous Epo. All of these manipulations reduce the infectivity of feeding mosquitoes to malaria infection, suggesting that the sex ratio of gametocytes is not only determined by the environment with the host during the course of infection, but is optimized during natural infection to maintain maximum infectivity for feeding mosquitoes. Similar mechanisms may exist to regulate the development of gametocytes in human infections (Paul et al. 2002).

In summary, the blunted erythropoietic response that contributes to anaemia and significant morbidity and mortality may ameliorate other manifestations of severe disease. Furthermore, the erythropoietic response influences the sex ratio of gametocytes, in avian, murine and—by extrapolation—probably in human infections.

7
Future Directions and Conclusion

After many years of relative obscurity, the study of the clinical, experimental and animal models of malarial anaemia is attracting renewed interest. This is primarily fuelled by the realization that the problem of malarial anaemia will be with us for some time to come and may even increase in relative importance as drug resistance increases and if vaccines giving partial immunity but not complete immunity promote chronic infections. In addition, safe treatment of severe malaria remains problematic in the absence of a ready supply of safe blood. At the same time, the research infrastructure in Africa has improved for linked epidemiological–clinical studies and, more generally, the tools to study molecular and cellular pathology have become much more powerful

Several clinical questions remain unanswered. What is the natural history of severe malarial anaemia? In particular, what is the role of pre-existing anaemia, pre-existing malaria infection and micronutrient status for the development of severe anaemia? Can some immune responses give specific protection and what are they? Do children who appear to recover from an episode of severe malarial anaemia suffer medium- and/or long-term malaria-specific or other morbidity and mortality? Several centres now have well-developed facilities for the longitudinal study of large numbers of children to answer these questions.

The relative roles of ineffective erythropoiesis and increased red cell clearance and the specific mechanisms that underline these processes have not been defined. Experimental studies suggest the cytokines including MIF and parasite products (including GPI and haemozoin) may influence erythropoiesis. However, further linked clinical and experimental studies are needed to determine if these factors are significant during the development of severe anaemia in children with malaria. These studies will require, not only the measurement of the levels of these putative endogenous and exogenous 'toxins', but a more precise definition of the cellular and molecular basis of their effect. Ideally, one should not only be able to establish an association between the levels of these factors and outcome but also show that the putative molecular and cellular pathology caused by these toxins developing erythroid cells in vitro is present in bone marrow samples taken from children with malaria.

Genomics and genetic epidemiology may be able to establish the significance of particular pathogenic mechanisms. Modern genomic techniques, in particular the ability to study the transcription of several thousand genes in a chosen tissue using micro-arrays enhances the descriptive power by several orders of magnitude compared to existing techniques measuring change of a handful of variables.

Schofield and colleagues have used this technology to outline the transcriptional profile of genes in the brain and spleen of mice infected with *P. berghei* ANKA. In murine splenocytes, enzymes of the glycolytic pathway and IFN-inducible genes were up regulated while genes associated with erythropoiesis were suppressed (Sexton et al. 2004). Significant up-regulation of genes associated with proteosome, cytokines including MIP-1, C-C chemoreceptor 5, chemokine receptor CXCR3, collagen genes, anti-oxidative genes and a wide variety of genes expressed in the immune system was seen. Obviously, the equivalent human studies, comparative studies in both human and animal hosts and the respective in vitro studies with these technologies are awaited with great interest.

In clinical studies, further refinement of the pathways of inhibition of erythropoiesis and/or enhanced clearance of erythrocytes will point to candidate genes and their functional polymorphisms. The robust association of these polymorphisms with protection or susceptibility to malarial anaemia may be a reliable method to distinguish the biological significance of putative pathogenic mechanisms.

Finally, we must ask what is the purpose of ever more powerful descriptive studies? Some have suggested that descriptive studies are low priority in the race to define new vaccines and find new drugs. However, for malarial anaemia, it is becoming increasingly plain that these descriptive studies are the necessary prelude to the development of novel preventative or therapeutic strategies as the fundamental mechanisms of disease are so poorly defined. Given the technological armoury at our disposal, it provides to be a fascinating and fruitful time for the study of this important manifestation of malaria.

Acknowledgements DJR is supported by the National Blood Service (NBS) and the Howard Hughes Medical Institute. CCP was supported by the University of Oxford and now by the NBS. We thank Profs. Marsh, Snow, Warrell, White and Krishna and Drs. Newton, English, Crawley and colleagues in Oxford for very many helpful discussions.

References

Abdalla S, Weatherall DJ, Wickramasinghe SN, Hughes M (1980) The anaemia of
    P. falciparum malaria. Br J Haematol 46:171–183
Abdalla SH (1988) Peripheral blood and bone marrow leucocytes in Gambian chil-
    dren with malaria: numerical changes and evaluation of phagocytosis. Ann Trop
    Paediatr 8:250–258
Abdalla SH, Kasili FG, Weatherall DJ (1983) The coombs direct antiglobulin test in
    Kenyans. Trans R Soc Trop Med Hyg 77:99–102
Abdalla SH, Wickramasinghe SN (1988) A study of erythroid progenitor cells in the
    bone marrow of Gambian children with falciparum malaria. Clin Lab Haematol
    10:33–40
Abdalla SH, Wickramasinghe SN, Weatherall DJ (1984) The deoxyuridine suppression
    test in severe anaemia following Plasmodium falciparum malaria. Trans R Soc
    Trop Med Hyg 78:60–63
Ahvazi BC, Jacobs P, Stevenson MM (1995) Role of macrophage-derived nitric oxide
    in suppression of lymphocyte proliferation during blood-stage malaria. J Leukoc
    Biol 58:23–31
Allred DR (1995) Immune evasion by Babesia bovis and Plasmodium falciparum:
    cliff-dwellers of the parasite world. Parasitol Today 11:100–105
Amodu OK, Adeyemo AA, Olumese PE, Gbadegesin RA (1998) Intraleucocytic malaria
    pigment and clinical severity of malaria in children. Trans R Soc Trop Med Hyg
    92:54–56
Angus BJ, Chotivanich K, Udomsangpetch R, White NJ (1997) In vivo removal of
    malaria parasites from red blood cells without their destruction in acute falci-
    parum malaria. Blood 90:2037–2040
Beeson JG, Rogerson SJ, Cooke BM, Reeder JC, Chai W, Lawson AM, et al. (2000)
    Adhesion of Plasmodium falciparum -infected erythrocytes to hyaluronic acid in
    placental malaria. Nat Med 6:86–90
Berkley J, Mwarumba S, Bramham K, Lowe B, Marsh K (1999) Bacteraemia compli-
    cating severe malaria in children. Trans R Soc Trop Med Hyg 93:283–286
Brewster DR, Greenwood BM (1993) Seasonal variation of paediatric diseases in The
    Gambia, west Africa. Ann Trop Paediatr 13:133–146
Brown KE (2000) Haematological consequences of parvovirus B19 infection. Baillieres
    Best Pract Res Clin Haematol 13:245–259
Burgmann H, Looareesuwan S, Kapiotis S, Viravan C, Vanijanonta S, Hollenstein U,
    et al. (1996) Serum levels of erythropoietin in acute Plasmodium falciparum
    malaria. Am J Trop Med Hyg 54:280–283
Chang KH, Tam M, Stevenson MM (2004) Inappropriately low reticulocytosis in severe
    malarial anemia correlates with suppression in the development of late erythroid
    precursors. Blood
Chang KH, Tam M, Stevenson MM (2004) Modulation of the course and outcome
    of blood-stage malaria by erythropoietin-induced reticulocytosis. J Infect Dis
    189:735–743
Dicko A, Klion AD, Thera MA, Sagara I, Yalcouye D, Niambele MB, et al. (2004)
    The etiology of severe anemia in a village and a periurban area in Mali. Blood
    104:1198–1200

Dondorp AM, Angus BJ, Chotivanich K, Silamut K, Ruangveerayuth R, Hardeman MR, et al. (1999) Red blood cell deformability as a predictor of anemia in severe falciparum malaria. Am J Trop Med Hyg 60:733–737

Dormer P, Dietrich M, Kern P, Horstmann RD (1983). Ineffective erythropoiesis in acute human *P. falciparum* malaria. Blut 46:279–288

Egan AF, Fabucci ME, Saul A, Kaslow DC,Miller LH (2002) Aotus New World monkeys: model for studying malaria-induced anemia. Blood 99:3863–3866

el Hassan AM, Saeed AM, Fandrey J, Jelkmann W (1997) Decreased erythropoietin response in *Plasmodium falciparum* malaria- associated anaemia. Eur J Haematol 59:299–304

English M, Muambi B, Mithwani S, Marsh K (1997) Lactic acidosis and oxygen debt in African children with severe anaemia. QJM 90563–569

English M, Waruiru C, Marsh K (1996) Transfusion for respiratory distress in life-threatening childhood malaria. Am J Trop Med Hyg 55:525–530

Facer CA (1980) Direct Coombs antiglobulin reactions in Gambian children with *Plasmodium falciparum* malaria. II. Specificity of erythrocyte-bound IgG. Clin Exp Immunol 39:279–288

Facer CA, Bray RS, Brown J (1979) Direct Coombs antiglobulin reactions in Gambian children with *Plasmodium falciparum* malaria. I. Incidence and class specificity. Clin Exp Immunol 35:119–127

Finch CA (1982) Erythropoiesis, erythropoietin, and iron. Blood 60:1241–1246

Fiori PL, Rappelli P, Mirkarimi SN, Ginsburg H, Cappuccinelli P, Turrini F (1993) Reduced microbicidal and anti-tumour activities of human monocytes after ingestion of *Plasmodium falciparum*-infected red blood cells. Parasite Immunol 15:647–655

Fleming AF (1997) HIV and blood transfusion in sub-Saharan Africa. Transfus Sci 18:167–179

Fried M, Duffy PE (1996) Adherence of *Plasmodium falciparum* to chondroitin sulfate A in the human placenta. Science 272:1502–1504

Giribaldi G, Ulliers D, Schwarzer E, Roberts I, Piacibello W, Arese P (2004) Hemozoin- and 4-hydroxynonenal-mediated inhibition of erythropoiesis. Possible role in malarial dyserythropoiesis and anemia. Haematologica 89:492–493

Groux H, Gysin J (1990) Opsonization as an effector mechanism in human protection against asexual blood stages of *Plasmodium falciparum* : functional role of IgG subclasses. Res Immunol 141:529–542

Guyatt HL, Snow RW (2001) Malaria in pregnancy as an indirect cause of infant mortality in sub-Saharan Africa. Trans R Soc Trop Med Hyg 95:569–576

Hetzel C, Anderson RM (1996) The within host cellular dynamics of blood-stage malaria. Parasitology 113:25–28

Holst FG, Hemmer CJ, Foth C, Seitz R, Egbring R, Dietrich M (1999) Low levels of fibrin-stabilizing factor (factor XIII) in human *Plasmodium falciparum* malaria: correlation with clinical severity. Am J Trop Med Hyg 60:99–104

Kelton JG, Keystone J, Moore J, Denomme G, Tozman E, Glynn M, et al. (1983) Immune-mediated thrombocytopenia of malaria. J Clin Invest 71:832–836

Krishna S, Waller DW, ter Kuile F, Kwiatkowski D, Crawley J, Craddock CF, et al. (1994) Lactic acidosis and hypoglycaemia in children with severe malaria: pathophysiological and prognostic significance. Trans R Soc Trop Med Hyg 88:67–73

Kurtzhals JA, Adabayeri V, Goka BQ, Akanmori BD, Oliver-Commey JO, Nkrumah FK, et al. (1998) Low plasma concentrations of interleukin 10 in severe malarial anaemia compared with cerebral and uncomplicated malaria. Lancet 351:1768–1772

Kurtzhals JA, Rodrigues O, Addae M, Commey JO, Nkrumah FK, Hviid L (1997) Reversible suppression of bone marrow response to erythropoietin in Plasmodium falciparum malaria. Br J Haematol 97:169–174

Kwiatkowski D, Cannon JG, Manogue KR, Cerami A, Dinarello CA, Greenwood BM (1989) Tumour necrosis factor production in falciparum malaria and its association with schizont rupture. Clin Exp Immunol 77:361–366

Kwiatkowski D, Hill AV, Sambou I, Twumasi P, Castracane J, Manogue JR, et al. (1990) TNF concentration in fatal cerebral, non-fatal cerebral, and uncomplicated Plasmodium falciparum malaria. Lancet 336:1201–1204

Ladhani S, Lowe B, Cole AO, Kowuondo K, Newton CR (2002) Changes in white blood cells and platelets in children with falciparum malaria: relationship to disease outcome. Br J Haematol 119:839–847

Lee SH, Crocker P, Gordon S (1986) Macrophage plasma membrane and secretory properties in murine malaria. Effects of Plasmodium yoelii blood-stage infection on macrophages in liver, spleen, and blood. J Exp Med 163:54–74

Lee SH, Looareesuwan S, Wattanagoon Y, Ho M, Wuthiekanun V, Vilaiwanna N, et al. (1989) Antibody-dependent red cell removal during P. falciparum malaria: the clearance of red cells sensitized with an IgG anti-D. Br J Haematol 73:396–402

Looareesuwan S, Suntharasamai P, Webster HK, Ho M (1993) Malaria in splenectomized patients: report of four cases and review. Clin Infect Dis 16:361–366

Looareesuwan S, Ho M, Wattanagoon Y, White NJ, Warrell DA, Bunnag D, et al. (1987) Dynamic alteration in splenic function during acute falciparum malaria. N Engl J Med 317:675–679

Lou J, Lucas R, Grau GE (2001) Pathogenesis of cerebral malaria: recent experimental data and possible applications for humans. Clin Microbiol Rev 14:810–820

Luty AJ, Perkins DJ, Lell B, Schmidt-Ott R, Lehman LG, Luckner D, et al. (2000) Low interleukin-12 activity in severe Plasmodium falciparum malaria. Infect Immun 68:3909–3915

MacPherson GG, Warrell MJ, White NJ, Looareesuwan S, Warrell DA (1985) Human cerebral malaria. A quantitative ultrastructural analysis of parasitized erythrocyte sequestration. Am J Pathol 119:385–401

Maggio-Price L, Brookoff D, Weiss L (1985) Changes in hematopoietic stem cells in bone marrow of mice with Plasmodium berghei malaria. Blood 66:1080–1085

Marsh K, Forster D, Waruiru C, Mwangi I, Winstanley M, Marsh V, et al. (1995) Indicators of life-threatening malaria in African children. N Engl J Med 332:1399–1404

Maitland K, Levin M, English M, Mithwani S, Peshu N, et al. (2003) Severe P. falciparum malaria in Kenyan children: Evidence for hypovolaemia. QJM 96:427–434

Martiney JA, Sherry B, Metz CN, Espinoza M, Ferrer AS, Calandra T, et al. (2000) Macrophage migration inhibitory factor release by macrophages after ingestion of Plasmodium chabaudi-infected erythrocytes: possible role in the pathogenesis of malarial anemia. Infect Immun 68:2259–2267

McGilvray ID, Serghides L, Kapus A, Rotstein OD, Kain KC (2000) Nonopsonic mono-cyte/macrophage phagocytosis of *Plasmodium falciparum*-parasitized erythro-cytes: a role for CD36 in malarial clearance. Blood 96:3231–3240

Menendez C, Kahigwa E, Hirt R, Vounatsou P, Aponte JJ, Font F, et al. (1997) Ran-domised placebo-controlled trial of iron supplementation and malaria chemo-prophylaxis for prevention of severe anaemia and malaria in Tanzanian infants. Lancet 350:844–850

Minekura H, Kumagai T, Kawamoto Y, Nara F, Uchida K (2001) 4-Hydroxy-2-nonenal is a powerful endogenous inhibitor of endothelial response. Biochem Biophys Res Commun 282:557–561

Mohan K, Stevenson MM (1998) Dyserythropoiesis and severe anaemia associated with malaria correlate with deficient interleukin-12 production. Br J Haematol 103:942–949

Mohan K, Stevenson MM (1998) Interleukin-12 corrects severe anemia during blood-stage *Plasmodium chabaudi* AS in susceptible A/J mice. Exp Hematol 26:45–52

Molyneux ME, Taylor TE, Wirima JJ, Borgstein A (1989) Clinical features and prognos-tic indicators in paediatric cerebral malaria: a study of 131 comatose Malawian children. QJM 71:441–459

Newton CR, Krishna S (1998) Severe falciparum malaria in children: current under-standing of pathophysiology and supportive treatment. Pharmacol Ther 79:1–53

Newton CR, Warn PA, Winstanley PA, Peshu N, Snow RW, Pasvol G, et al. (1997) Severe anaemia in children living in a malaria endemic area of Kenya. Trop Med Int Health 2:165–178

Njuguna P, Newton C (2004) Management of severe falciparum malaria. J Postgrad Med 50:45–50

Nosten F, McGready R, Simpson JA, Thwai KL, Balkan S, Cho T, et al. (1999) Effects of *Plasmodium vivax* malaria in pregnancy. Lancet 354:546–549

Othoro C, Lal AA, Nahlen B, Koech D, Orago AS, Udhayakumar V (1999) A low interleukin-10 tumor necrosis factor-alpha ratio is associated with malaria anemia in children residing in a holoendemic malaria region in western Kenya. J Infect Dis 179:279–282

Pain A, Ferguson DJ, Kai O, Urban BC, Lowe B, Marsh K, et al. (2001) Platelet-mediated clumping of *Plasmodium falciparum* -infected erythrocytes is a common adhesive phenotype and is associated with severe malaria. Proc Natl Acad Sci USA 98:1805–1810

Paul REL, Brey PT, Robert V (2002) *Plasmodium* sex determination and transmission in mosquitos. Trends Parasitol 18:32–38

Paul REL, Coulson TN, Raibaud A, Brey PT (2000) Sex determination in malaria parasites. Science 287:128–131

Perkins DJ, Weinberg JB, Kremsner PG (2000) Reduced interleukin-12 and transform-ing growth factor-beta1 in severe childhood malaria: relationship of cytokine balance with disease severity. J Infect Dis 182:988–992

Pichyangkul S, Saengkrai P, Webster HK (1994) *Plasmodium falciparum* pigment induces monocytes to release high levels of tumor necrosis factor-alpha and interleukin-1 beta. Am J Trop Med Hyg 51:430–435

Planche T, Onanga M, Schwenk A, Dzeing A, Borrmann S, et al. (2004) Assessment of volume depletion in children with malaria. PLoS Med 1:e18

Price R, van Vugt M, Phaipun L, Luxemburger C, Simpson J, McGready R, et al. (1999) Adverse effects in patients with acute falciparum malaria treated with artemisinin derivatives. Am J Trop Med Hyg 60:547–555

Riddle MS, Jackson JL, Sanders JW, Blazes DL (2002) Exchange transfusion as an adjunct therapy in severe *Plasmodium falciparum* malaria: a meta-analysis. Clin Infect Dis 34:1192–1198

Rowe JA, Moulds JM, Newbold CI, Miller LH (1997) *P. falciparum* rosetting mediated by a parasite-variant erythrocyte membrane protein and complement-receptor 1. Nature 388:292–295

Rowe JA, Scragg IG, Kwiatkowski D, Ferguson DJ, Carucci DJ, Newbold CI (1998) Implications of mycoplasma contamination in *Plasmodium falciparum* cultures and methods for its detection and eradication. Mol Biochem Parasitol 92:177–180

Schellenberg D, Menendez C, Kahigwa E, Aponte J, Vidal J, Tanner M, et al. (2001) Intermittent treatment for malaria and anaemia control at time of routine vaccinations in Tanzanian infants: a randomised, placebo-controlled trial. Lancet 357:1471–1477

Scholander C, Carlson J, Kremsner PG, Wahlgren M (1998) Extensive immunoglobulin binding of *Plasmodium falciparum*-infected erythrocytes in a group of children with moderate anemia. Infect Immun 66:361–363

Schwarzer E, Ludwig P, Valente E, Arese P (1999) 15(S)-hydroxyeicosatetraenoic acid (15-HETE), a product of arachidonic acid peroxidation, is an active component of hemozoin toxicity to monocytes. Parassitologia 41:199–202

Schwarzer E, Muller O, Arese P, Siems WG, Grune T (1996) Increased levels of 4-hydroxynonenal in human monocytes fed with malarial pigment hemozoin. A possible clue for hemozoin toxicity. FEBS Lett 388:119–122

Schwarzer E, Turrini F, Giribaldi G, Cappadoro M, Arese P (1993) Phagocytosis of *P. falciparum* malarial pigment hemozoin by human monocytes inactivates monocyte protein kinase C. Biochim Biophys Acta 1181:51–54

Schwarzer E, Turrini F, Ulliers D, Giribaldi G, Ginsburg H, Arese P (1992) Impairment of macrophage functions after ingestion of *Plasmodium falciparum*-infected erythrocytes or isolated malarial pigment. J Exp Med 176:1033–1041

Sexton AC, Good RT, Hansen DS, D'Ombrain MC, Buckingham L, Simpson K, et al. (2004) Transcriptional profiling reveals suppressed erythropoiesis, up-regulated glycolysis, and interferon-associated responses in murine malaria. J Infect Dis 189:1245–1256

Sherry BA, Alava G, Tracey KJ, Martiney J, Cerami A, Slater AF (1995) Malaria-specific metabolite hemozoin mediates the release of several potent endogenous pyrogens (TNF, MIP-1 alpha, and MIP-1 beta) in vitro, and altered thermoregulation in vivo. J Inflamm 45:85–96

Silverman PH, Schooley JC, Mahlmann LJ (1987) Murine malaria decreases hemopoietic stem cells. Blood 69:408–413

Snow RW, Omumbo JA, Lowe B, Molyneux CS, Obiero JO, Palmer A, et al. (1997) Relation between severe malaria morbidity in children and level of *Plasmodium falciparum* transmission in Africa. Lancet 349:1650–1654

Srichaikul T, Panikbutr N, Jeumtrakul P (1967) Bone-marrow changes in human malaria. Ann Trop Med Parasitol 61:40–51

Stephens JWW (1937). Blackwater Fever. Liverpool, Univerisity of Liverpool Press

Taverne J, Sheikh N, de Souza JB, Playfair JH, Probert L, Kollias G (1994) Anaemia and resistance to malaria in transgenic mice expressing human tumour necrosis factor. Immunology 82:397–403

Tonelli M, Blake PG, Muirhead N (2001) Predictors of erythropoietin responsiveness in chronic hemodialysis patients. Asaio J 47:82–85

Turrini FGG, Valente E, Arese P (1997) Mycoplasma contamination of Plasmodium cultures—A case of parasite parasitism. Parasitol Today 13:367–368

Vryonis G (1939) Observations on the parasitization of erythrocytes by Plasmodium vivax, with special reference to reticulocytes. Am J Hygiene 30:41

Waitumbi JN, Opollo MO, Muga RO, Misore AO, Stoute JA (2000) Red cell surface changes and erythrophagocytosis in children with severe Plasmodium falciparum anemia. Blood 95:1481–1486

Weatherall DJ, Ledingham (eds) The Oxford textbook of medicine. Oxford, Oxford University Press, 1987. Copyright DJ Weatherall, JGG Ledingham and DA Warrell

Weiss L, Geduldig U, Weidanz W (1986) Mechanisms of splenic control of murine malaria: reticular cell activation and the development of a blood-spleen barrier. Am J Anat 176:251–285

Wickramasinghe SN, Looareesuwan S, Nagachinta B, White NJ (1989) Dyserythropoiesis and ineffective erythropoiesis in Plasmodium vivax malaria. Br J Haematol 72:91–99

Wickramasinghe SN, Phillips RE, Looareesuwan S, Warrell DA, Hughes M (1987) The bone marrow in human cerebral malaria: parasite sequestration within sinusoids. Br J Haematol 66:295–306

Winstanley P (2001) Modern chemotherapeutic options for malaria. Lancet Infect Dis 1:242–250

Wongsrichanalai C, Pickard AL, Wernsdorfer WH, Meshnick SR (2002) Epidemiology of drug-resistant malaria. Lancet Infect Dis 2:209–218

World Health Organization (1972) Nutritional Anaemia: World Health Organ Tech Rep Serv 503:1

Wyler DJ (1983) The spleen in malaria. Ciba Found Symp 94:98–116

Yeats J, Daley H, Hardie D (1999) Parvovirus B19 infection does not contribute significantly to severe anaemia in children with malaria in Malawi. Eur J Haematol 63:276–277

Zamai L, Secchiero P, Pierpaoli S, Bassini A, Papa S, Alnemri ES, et al. (2000) TNF-related apoptosis-inducing ligand (TRAIL) as a negative regulator of normal human erythropoiesis. Blood 95:3716–3724

CTMI (2005) 295:169–200
© Springer-Verlag Berlin Heidelberg 2005

Malaria in the Pregnant Woman

P. E. Duffy[1,2] (✉) · M. Fried[1]

[1]Seattle Biomedical Research Institute, 307 Westlake Avenue, Seattle, WA, USA
patrick.duffy@sbri.org
[2]Walter Reed Army Institute of Research, 503 Robert Grant Avenue,
Silver Spring, MD, USA

Abstract Women become more susceptible to *Plasmodium falciparum* malaria during pregnancy, and the risk of disease and death is high for both the mother and her fetus. In low transmission areas, women of all parities are at risk for severe syndromes like cerebral malaria, and maternal and fetal mortality are high. In high transmission areas, where women are most susceptible during their first pregnancies, severe syndromes like cerebral malaria are uncommon, but severe maternal anemia and low birth weight are frequent sequelae and account for an enormous loss of life. *P. falciparum*-infected red cells sequester in the intervillous space of the placenta, where they adhere to chondroitin sulfate A but not to receptors like CD36 that commonly support adhesion of parasites infecting nonpregnant hosts. Poor pregnancy outcomes due to malaria are related to the macrophage-rich infiltrates and pro-inflammatory cytokines such as tumor necrosis factor-α that accumulate in the intervillous space. Women who acquire antibodies against chrondroitin sulfate A (CSA)-binding parasites are less likely to have placental malaria, and are more likely to deliver healthy babies. In areas of stable transmission, women acquire antibodies against CSA-binding parasites over successive pregnancies, explaining the high susceptibility to malaria during first pregnancy, and suggesting that a vaccine to prevent pregnancy malaria should target placental parasites. Prevention and treatment of malaria are essential components of antenatal care in endemic areas, but require special considerations during pregnancy. Recrudescence after drug treatment is more common during pregnancy, and the spread of drug-resistant parasites has eroded the usefulness of the few drugs known to be safe for the woman and her fetus. Determining the safety and effectiveness of newer antimalarials in pregnant women is an urgent priority. A vaccine that prevents pregnancy malaria due to *P. falciparum* could be delivered before first pregnancy, and would have an enormous impact on mother–child health in tropical areas.

Abbreviations

LBW	Low birth weight
IUGR	Intrauterine growth retardation
PTD	Preterm delivery
IE's	Parasite-infected erythrocytes
CSA	Chrondroitin sulfate A
CQ	Chloroquine
SP	Sulfadoxine–pyrimethamine

1
Introduction

Pregnant women are uniquely susceptible to *Plasmodium falciparum* malaria. The pathogenesis, epidemiology, clinical sequelae, prevention, and treatment of malaria all have unique features during pregnancy. For example, the issues of teratogenicity or embryotoxicity complicate the effort to identify new treatments and prophylaxis strategies for pregnant women against drug-resistant parasites. Conversely, a new understanding of the pathogenesis of pregnancy malaria suggests that vaccines or anti-adhesion treatments against pregnancy malaria are feasible. This chapter will discuss the distinct features of malaria in pregnancy, new knowledge regarding pathogenesis, the status of available treatments, and research priorities for the future.

Pregnancy malaria has many names, including maternal malaria, pregnancy-associated malaria, or malaria of, in or during pregnancy. Placental malaria refers to the accumulation of parasites, inflammatory cells and pigment in the placenta that occurs during *P. falciparum* infection (of the four human malaria parasite species, only *P. falciparum* can adhere to endothelium and sequester in deep vascular beds).

2
Epidemiology

In general, pregnancy malaria due to *P. falciparum* has two distinct clinical presentations based on malaria epidemiology: one for women in areas of unstable or low transmission (less than one infected mosquito bite per pregnancy) who have little or no immunity; and one for women residing in areas of stable transmission who have acquired substantial immunity. Women with little or no immunity are highly likely to develop severe syndromes like cerebral malaria and respiratory distress syndrome, and are at high risk of death unless effective therapy is provided promptly. Women with high pre-existing immunity are unlikely to develop cerebral malaria or respiratory distress syndrome during pregnancy malaria. However, these women have increased susceptibility to parasitemia and febrile episodes during pregnancy, and commonly suffer severe anemia and low birth weight (LBW; <2,500 g) deliveries that predispose to maternal and neonatal death, respectively. The scope of the problem in high transmission is further demonstrated by the prevalence of parasitemia in pregnant mothers, which among primigravidas in high transmission areas can exceed 80% in the second trimester (Brabin 1983).

2.1
Pregnancy Malaria in Areas of Low, Unstable or Epidemic Malaria, Where Immunity Is Low

For many centuries, malaria was known to be more severe in pregnant victims (Duffy and Desowitz 2001), although few studies have examined outcomes in cohorts of women with low immunity. The modern literature on severe malaria in pregnant women begins with the detailed studies of Wickramasuriya (1935, 1936) during the 1934–35 epidemic in Ceylon (Sri Lanka). Cerebral malaria was relatively frequent (4.8%) among infected pregnant women, and, compared to other adult cases, pregnancy increased the case fatality rate of cerebral malaria threefold to 89.5% (Wickramasuriya 1936). Malaria during pregnancy commonly caused severe anemia, cachexia, and anasarca (a severe, generalized form of edema), and the majority of women with this complex of symptoms died during or shortly after labor. Wickramasuriya observed that pregnancy malaria was often complicated by (or predisposed to) eclampsia, cardiac failure, and puerperal sepsis, contributing to the high mortality rate. Overall, the case fatality rate observed by Wickramasuriya in 358 pregnant women with malaria was 13.1% (Wickramasuriya 1936).

More recent studies have also highlighted the risks of severe disease and death in pregnant women. In an area of Thailand with low malaria transmission, the risk of severe malaria is threefold greater in pregnant versus nonpregnant women (Luxemburger et al. 1997), but mortality is prevented by early detection and prompt treatment (Nosten and McGready 2003). In Gujarat, India, fatal complications of pregnancy malaria included cerebral malaria, postpartum hemorrhage, acute pulmonary edema, and hypoglycemia (Maitra et al. 1993).

In Africa, severe syndromes are common among pregnant women in areas of unstable or epidemic malaria transmission. In a highland area of Rwanda, malaria-related maternal deaths increased fivefold and total maternal mortality tripled during the epidemic year of 1998 (Hammerich et al. 2002). Pregnant women were estimated to be at 1.6–4.9 times higher risk of admission for malaria than other adults in this area of Rwanda. In Ethiopia, women hospitalized for malaria are significantly more likely to be pregnant in areas of unstable malaria transmission, but not in areas of stable transmission (Newman et al. 2003). In a highland area of northern Tanzania with low malaria transmission, 44% of maternal deaths in 1995–96 were due to cerebral malaria, and half of these died postpartum (Hinderaker et al. 2003). The postpartum period has been noted in several studies to be a particularly high-risk period for maternal mortality related to malaria (Duffy and Desowitz 2001).

Separate from maternal mortality, fetal wastage and neonatal death are also frequent. Wickramasuriya in Ceylon reported the combined rate of neonatal and fetal deaths to be 66.9% in cases of pregnancy malaria (Wickramasuriya 1936). In South Vietnam, 135 of 1,255 (10.4%) parturients admitted to a Saigon hospital from July to September 1950 were malaria infected. Of the infected women, 63 (48%) delivered premature babies that had a 36.6% mortality rate, and 10 (8%) ended in abortion (van Hung 1951).

Despite early detection and prompt treatment in Thailand, malaria or other febrile conditions occurring in the week before delivery increased the risk of neonatal death 2.5-fold, exclusive of the effect of malaria-related anemia, LBW or premature delivery to predispose to neonatal mortality (Luxemburger et al. 1997). In Gujarat, India, fetal loss among 445 pregnant women with malaria was 31.1%, twice the rate in the general population (Maitra et al. 1993). In a low transmission area of northern Tanzania, malaria was a cause in 20 of 60 stillbirths (33.3%) and 21 of 67 neonatal deaths (31.3%) (Hinderaker et al. 2003). In Ethiopia, women with placental parasitemia had a sevenfold increased risk of stillbirth (19.1%) compared to aparasitemic women (2.7%) in areas of unstable malaria transmission, but not in areas of stable transmission (Newman et al. 2003).

2.2
Pregnancy Malaria in Areas of High Stable Transmission Where Immunity Is High

In areas of high and stable malaria transmission, pregnancy malaria is a major cause of severe maternal anemia and LBW babies. Pregnant women in high transmission areas are less likely to develop other severe malaria syndromes, presumably because their acquired immunity prevents disease. In high transmission areas, malaria has an unusual epidemiology—susceptibility is greatest in first pregnancies, and diminishes over successive pregnancies (Cannon 1958; McGregor 1984). This is unlike women with little or no immunity, in whom susceptibility is similar regardless of parity (Duffy and Desowitz 2001; McGregor 1984; Newman et al. 2003).

The pathogenesis of anemia in pregnant women is multifactorial, making it difficult to estimate the total contribution of malaria to severe anemia. Maternal hemoglobin levels are substantially lower in pregnant African women in endemic areas compared to nonendemic areas. However, the level of malaria transmission in endemic areas is not strongly related to maternal hemoglobin levels (Guyatt and Snow 2001). In high transmission areas where malaria susceptibility is greatest during first pregnancies, the risk of severe anemia is at least 50% higher in primigravidae than multigravidae (Brabin and Rogerson 2001). Cross-sectional studies throughout Africa yield a combined estimate that 7.3% of severe anemia cases in women of all parities result from malaria (Guyatt and Snow 2001). In a meta-analysis of four chemoprophylaxis studies (Garner and Gulmezoglu 2003), drugs given to prevent malaria in pregnant women substantially reduced the relative risk of severe maternal anemia to 0.62 [95% confidence interval (CI), 0.50–0.78], with the benefits primarily limited to women of low parity.

Severe maternal anemia increases the risk for maternal mortality, particularly in women who suffer postpartum hemorrhage. Malaria and anemia may conspire to increase mortality: severe anemia increases the risk of prolonged labor and operative or Cesarean deliveries (Malhotra et al. 2002), and the risk of postpartum hemorrhage is increased in malarious areas of Papua New Guinea (Piper et al. 2001). Anemia-related maternal mortality is largely preventable through proper antenatal and obstetric care. Maternal case fatality rates therefore vary substantially, from <1% to >50% (Brabin et al. 2001), based on the availability of proper facilities. The overall relative risk of mortality in women with severe anemia has been estimated at 3.51 (95% CI, 2.05–6.00) (Brabin et al. 2001), but this will be higher in areas with limited health care resources. In west Africa, anemia may be responsible for 20% of maternal deaths (van den Broek 1996).

In high transmission areas, pregnancy malaria also causes disease and death of the fetus. LBW has long been known to be a consequence of pregnancy malaria (Duffy and Desowitz 2001), including among immune populations in areas of stable transmission (Archibald 1956; Bruce-Chwatt 1952). The prevalence of LBW in primiparae varies with intensity of malaria transmission, ranging from 10% in low transmission areas to over 50% in high transmission areas (Brabin and Rogerson 2001). The true impact of pregnancy malaria on LBW may be much greater, because other factors that contribute to LBW are themselves caused by or related to malaria, such as maternal anemia and first births in areas of stable transmission. For example, in rural Malawi, primiparity, severe anemia, and malaria at delivery were all independent risk factors for intrauterine growth retardation (IUGR)(Verhoeff et al. 2001).

LBW may result from IUGR or preterm delivery (PTD), and malaria has been associated with both IUGR and PTD in high transmission areas. LBW increases mortality during the neonatal period and infancy. Based on its prevalence and the associated risk of mortality, malaria-induced LBW is estimated to kill 62,000–363,000 newborns in Africa each year, or 3–17 deaths per 1,000 live births (Murphy and Breman 2001). In a study from Malawi, PTD accounted for 65% of perinatal mortality and 68% of neonatal mortality, or 42.5 and 25.2 deaths per 1000 live births, respectively (Kulmala et al. 2000).

In Zaire, evidence of active parasitemia or chronic placental malaria were related to different patterns of IUGR, and together increased the prevalence of LBW fourfold, to 25.9% (Meuris et al. 1993). The timing of malaria during pregnancy determined the relative contribution of IUGR and PTD to LBW in Malawi: malaria during the antenatal period was associated with IUGR, whereas placental malaria at delivery was associated with PTD (Sullivan et al. 1999). IUGR tends to be more common than PTD in high transmission areas (Brabin and Rogerson 2001), possibly due to the greater frequency of antenatal infections that cause IUGR, and immunity preventing the malaria fevers that induce PTD.

2.3
Pregnancy Malaria and the Nonfalciparum Malaria Parasites

Pregnant women can be infected by any of the four human malaria parasite species. Susceptibility to *P. ovale* or *P. malariae* does not increase during pregnancy (Diagne et al. 1997). Susceptibility to *P. vivax* increases during pregnancy. Paradoxically, the risk of *P. vivax* parasitemia is greatest in first pregnancy, while the poor outcomes related to *P. vivax* malaria are greatest in later pregnancies (Nosten et al. 1999a; Singh et al. 1998, 1999). The basis for the *P. vivax* epidemiology is unknown. *P. vivax* exclusively invades reticulocytes, and susceptibility to vivax malaria could increase due to the reticulocytosis of pregnancy and diminish as women develop progressive micronutrient deficiencies (Duffy 2001). More research is needed to understand the pathogenesis and epidemiology of pregnancy malaria due to *P. vivax*.

All the malaria parasites can precipitate poor pregnancy outcomes. In Vietnamese women with presumably low immunity, abortion and premature delivery were caused by the different parasite species (*P. falciparum*, *P. vivax*, and *P. malariae*) roughly in proportion to the prevalence of the parasites in peripheral blood (van Hung 1951). *P. vivax* is related to maternal anemia and LBW newborns, although these sequelae are less severe in women infected with *P. vivax* than in those infected with *P. falciparum* (Nosten et al. 1999a; Singh et al. 1998). Programs to improve pregnancy outcomes in endemic areas must include strategies to prevent and treat infection with any malaria parasite species.

3
Parasite Adhesion and Sequestration in the Placenta

P. falciparum parasites that infect the human placenta differ from other parasites. Parasite-infected erythrocytes (IEs) that sequester in the placenta adhere to chondroitin sulfate A (CSA) but not to receptors like CD36 or intercellular adhesion molecule (ICAM)-1 that support binding of IEs collected from nonpregnant donors. Unlike many other isolates, placental IEs do not rosette (aggregate uninfected red cells) and agglutinate poorly in the presence of immune sera. Thus, adhesion to CSA and possibly other novel receptors in the placenta selects for a distinct form of parasite that causes pregnancy malaria.

P. falciparum IEs sequester in deep vascular beds, and this property of the parasite has long been thought to account for parasite virulence (Laveran 1882). For example, the mass of IEs sequestered in brain vessels is increased in fatal cases of cerebral malaria compared to other fatal malaria cases (Pongponratn et al. 1991). Similarly, IEs sequester in abundance in the placenta, and the accumulation of parasites and inflammatory cells in the placenta is related to poor pregnancy outcomes (Duffy 2001).

The earliest histologic studies of cerebral malaria demonstrated that IEs sequestering in the brain were adhering to vascular walls (Marchiafava and Bignami 1894), and electron microscopic (EM) studies subsequently demonstrated that electron-dense 'knobs' on the IE surface served as points of attachment to cerebral endothelium (MacPherson et al. 1985). In the placenta, IEs accumulate throughout the maternal vascular spaces (called intervillous spaces), and the initial EM studies of infected placentas found few IEs adhering to the cellular syncytium (called syncytiotrophoblast) that lines the wall of the intervillous spaces (Bray and Sinden 1979). Placental IEs displayed surface knobs typical of adherent parasites, but on the basis of the EM studies, sequestration in the placenta was attributed to rheologic properties of the IE and 'sluggish' blood flow in the placenta, and not specific adhesion (Bray and Sinden 1979).

This view was radically altered by adhesion studies using freshly collected samples of placental IEs. In a seminal study from Kenya, IEs extracted from infected placentas specifically adhered to the glycosaminoglycan CSA, but not to CD36 (Fried and Duffy 1996). IEs collected from nonpregnant adults in the same area commonly bound to CD36, but did not bind to CSA. In ex vivo assays using placental cryosections, placental IEs bound to the surface of syncytiotrophoblast (Fig. 1), and this binding was inhibited by more than 95% by soluble CSA or by pre-treatment of the placental cryosection with chondroitinases (Fried and Duffy 1996). Adhesion to CSA has been observed in IEs collected from placentas or pregnant women at several geographically distant sites, including Kenya, Thailand, Malawi, and Cameroon (Beeson et al.

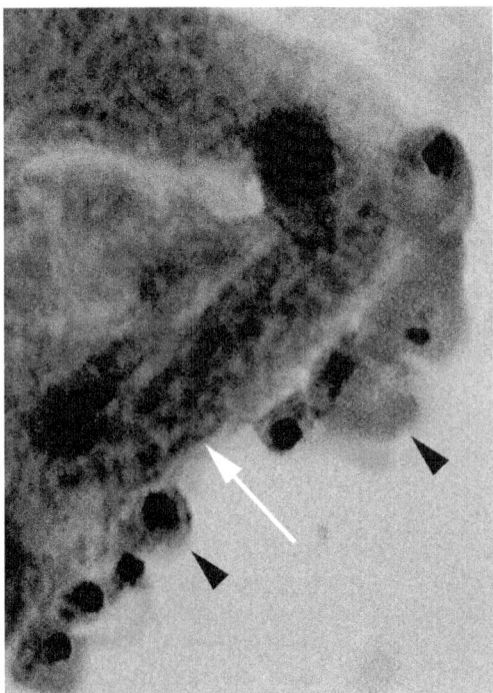

Fig. 1 Placental IEs adhere to syncytiotrophoblast in an ex vivo assay. IEs (*black arrowheads*) collected from an infected placenta are allowed to bind to a cryosection of uninfected placenta tissue. IEs bind specifically along the surface of the syncytiotrophoblast (*white arrow*), and this binding is inhibited by soluble CSA or by pre-treatment of the tissue with chondroitinase AC

1999; Fried and Duffy 1996; , 1998; Maubert et al. 2000). The results suggest that pregnant women throughout the world are infected with a distinct adhesive form of the malaria parasite, selected by adhesion to CSA in the placenta.

Placental or CSA-binding IEs have additional distinct features. Unlike parasites collected from many nonpregnant individuals, placental IEs do not form rosettes with uninfected red cells (Maubert et al. 1998; Rogerson et al. 2000). Placental IEs also agglutinate only poorly in the presence of immune sera collected in endemic areas (Beeson et al. 1999; Fried and Duffy 1998). CSA-binding IEs may adsorb IgM to their surface (Creasey et al. 2003). These various distinct features of placental IEs may contribute to their sequestration in the placenta, or possibly to their evasion of antibodies formed against other parasites that rosette or agglutinate.

Research is underway to identify additional receptors for IE adhesion in the placenta. Hyaluronic acid (HA), a nonsulfated GAG, supported binding of 12 of 15 placental IE samples in Malawi (Beeson et al. 2000), but none of four isolates from Kenya (Fried et al. 2000). Detailed studies have highlighted several factors that can influence IE binding to HA preparations, including CSA contamination of commercial HA preparations (Fried et al. 2000; Valiyaveettil et al. 2001) and the length of HA molecules (Beeson and Brown 2004). Hypothetically, antibodies adsorbed to the IE surface may also play a role in parasite adhesion, either by attaching to Ig receptors on the surface of syncytiotrophoblast (Flick et al. 2001) or 'cross-linking with syncytial knots' (Creasey et al. 2003). Laboratory parasites selected to bind CSA will adsorb IgM on the IE surface (Creasey et al. 2003), and IgG is frequently detected on the surface of placental IEs (Flick et al. 2001), but these studies have made conflicting claims that have not yet been resolved. Of note, no Ig receptors have been detected on the surface of syncytiotrophoblast: the neonatal Fc receptor that mediates transplacental antibody transfer localizes to intracellular vesicles in syncytiotrophoblast, and binds IgG at acidic but not neutral pH (Simister 2003). Additional studies are required to determine the contribution of receptors other than CSA to IE adhesion and sequestration in the placenta.

Placental parasites clearly display a distinct repertoire of surface proteins, and serologic studies suggest that the CSA-binding protein(s) of *P. falciparum* are conserved worldwide (Fried et al. 1998b). Until recently, only two proteins, called PfEMP1 and rifin, were known to be displayed on the IE surface. PfEMP1 and rifin are variant antigens encoded by multicopy gene families: PfEMP1 by the ~60-member *var* gene family, and rifin by the ~150-member *rif* gene family. PfEMP1 has previously been shown to bind endothelial receptors in vitro (Baruch et al. 1996), and therefore efforts to identify CSA-binding IE surface proteins have focused on this molecule. Recently completed proteomic studies suggest that additional parasite proteins encoded by single copy genes may be expressed on the IE surface (Florens et al. 2004; Fried et al. 2004), but these await further characterization to understand their function, such as a role in adhesion.

PfEMP1 is a large antigen comprising multiple extracellular domains, with substantial sequence diversity among the alleles in the prototype 3D7 strain (Gardner et al. 2002), and also with extensive diversity between isolates. However, some PfEMP1 variants appear to be present in many parasite isolates, and their sequence is highly conserved (Fried and Duffy 2002; Rowe et al. 2002; Salanti et al. 2002), a property consistent with the conserved serologic reactivity of the CSA-binding protein. The extensive variation in PfEMP1 is a formidable obstacle for designing reagents that can be used for differential expression studies in field isolates (Taylor et al. 2000), including PCR primers, microarray probes or antibodies. Efforts to identify the PfEMP1 of placental parasites have used degenerate or specific primers to amplify transcripts from either CSA-binding laboratory isolates or from placental parasite samples, and these have yielded several candidate surface proteins or CSA-binding molecules.

The first transcript identified by this approach, called *varCS2*, was amplified with degenerate primers from ITG strain parasites selected to bind CSA. Recombinant forms of two extracellular domains (CIDRα and DBLγ domains) encoded by *varCS2* elicited antibodies that inhibited adhesion of parasites to CSA (Reeder et al. 1999), and bound to CSA and other sulfated glycans in soluble binding assays (Reeder et al. 2000). However, detailed studies of this gene in laboratory isolates (Duffy et al. 2002) and placental samples (Fried and Duffy 2002; Fried et al. 2004) indicate that it is transcribed or expressed at low or undetectable levels that do not increase in CSA-binding parasites.

The second candidate *var* transcript, called *FCR3varCSA*, was amplified with degenerate primers from FCR3 strain parasites selected to bind CSA (Buffet et al. 1999). *FCR3varCSA* is highly homologous to *var* alleles identified in other parasites (Fried and Duffy 2002; Rowe et al. 2002), and together these are referred to as *var1csa* (Salanti et al. 2003). Recombinant forms of the DBLγ domain from *FCR3varCSA* support binding of soluble CSA (Buffet et al. 1999), and elicit monoclonal antibodies that reacted with ~50% of IEs in cryosections of six infected placentas collected in Cameroon (Lekana Douki et al. 2002). However, studies of laboratory isolates (Duffy et al. 2002; Kyes et al. 2003) and field isolates (Fried and Duffy 2002; Fried et al. 2004; Winter et al. 2003) indicate that transcription or expression of *FCR3varCSA* or *var1csa* is not upregulated in CSA-binding or placental parasites. Further, although serologic recognition of the target protein is expected to increase over successive pregnancies, sera collected from pregnant women in endemic areas do not react with recombinant DBLγ domains (Winter et al. 2004) or other *FCR3varCSA* domains (Jensen et al. 2003) in a parity-specific fashion.

A third *var* gene implicated in pregnancy malaria, called *var2csa*, has been identified by quantitative RT–PCR from NF54 strain parasites using specific primers against all possible 3D7 strain *var* genes (Salanti et al. 2003). Transcription of *var2csa* is upregulated in parasites selected to bind CSA, and is increased in placental parasites compared to parasites infecting children (Salanti et al. 2003). Further studies are needed to better understand whether *var2csa* (Salanti et al. 2003) or other novel IE surface proteins (Florens et al. 2004; Fried et al. 2004) play a role in pregnancy malaria.

Tandem mass spectrometry has recently been used to define the IE surface proteins of placental or CSA-binding parasites (Fried et al. 2004). In these studies, no single PfEMP1 was shared between all (or even most) of the isolates tested. Proteins encoded by *var1csa* and *var2csa* were detected in IEs that did not bind to CSA, and were not detected in placental parasites. Peptides encoded by *varCS2* were not detected in any samples. Two novel PfEMP1 forms, encoded by *PF07_0051* and *PFI0005w*, were detected in four and two of six placental parasite samples, respectively. Further work is needed to determine whether these novel variants react with sera in a parity-specific fashion, elicit antibodies against placental IEs, and possibly bind to CSA, as would be expected of the IE surface proteins involved in placental malaria (Fried et al. 2004).

4
Parasite-Specific Immune Responses

A distinct CSA-binding form of *P. falciparum* sequesters in the human placenta (Fried and Duffy 1996). In endemic areas, women do not display immune responses specific for CSA-binding parasites prior to their first pregnancy, but develop specific humoral and cellular responses against these parasites over successive pregnancies. Antibody responses against CSA-binding parasites have been associated with reduced likelihood of placental malaria and improved fetal and maternal outcomes. These findings explain the parity-specific epidemiology of malaria, in which the high susceptibility during first pregnancies diminishes in subsequent pregnancies as women acquire specific immunity against placental parasites.

In endemic areas, antibodies develop over successive pregnancies that inhibit IE adhesion to the placental receptor CSA (Fried et al. 1998b) or that react with the surface of CSA-binding IEs in flow cytometry assays (Ricke et al. 2000). Antibodies acquired by Asian women inhibit the adhesion of placental IEs collected in Africa, and antibodies acquired by African women react with IEs collected from pregnant women in Asia, suggesting that surface antigens of placental IEs are conserved around the world (Fried et al. 1998b). Serum levels of the antibody that inhibits placental parasite binding are related to resistance to placental malaria (Fried et al. 1998b), as well as increased birth weight and gestational age of the newborn (Duffy and Fried 2003). In a high transmission area, Kenyan secundigravidae with serum antiadhesion antibodies delivered babies that were on average 398 g heavier and 2 weeks more mature than babies born to secundigravidas without antibodies (Duffy and Fried 2003; Staalsoe et al. 2004). In an area of seasonal transmission, the subset of Kenyan women with chronic placental malaria had higher hemoglobins (by 17 g/l) and delivered heavier babies (by 260 g) if their antisera reacted to CSA-binding parasites in flow cytometry assays (Staalsoe et al. 2004).

Cellular immune responses against CSA-binding parasites are also acquired over successive pregnancies, but it is unknown whether cellular responses correlate with resistance. Immune cells collected 3 months postpartum from multigravid Cameroonian women had significantly higher proliferative and interleukin (IL)-4 responses to CSA-binding parasites (but not other parasites) than those collected from primigravid women (Fievet et al. 2002). Interestingly, proliferative responses to CSA-binding parasites (but not other parasites) were significantly higher in 3-month postpartum samples than samples collected at delivery. The lower responses at delivery could reflect pregnancy-related immunomodulation (Fievet et al. 2002), but this explanation is unsatisfying because responses to a non-CSA-binding parasite were not lower at delivery. Alternatively, active infection with CSA-binding parasites (as often occurs at delivery) may suppress specific immune responses to homologous parasites; this would explain why responses against CSA-binding parasites but not non-CSA-binding parasites are lower at the time of delivery than in the postpartum period. IFNγ responses of immune cells collected from placentas of multigravid women are higher in uninfected women than infected women (Moore et al. 1999), but further studies are needed to determine whether these responses target the unique antigens of placental parasites.

The epidemiology of malaria in pregnant women infected with the human immunodeficiency virus (HIV) suggests that HIV delays but does not prevent the acquisition of immunity (Duffy 2001). Studies in Malawi and Kenya have shown that both HIV-infected and uninfected women acquire resistance over successive pregnancies, but HIV-infected women acquire resistance more slowly (van Eijk et al. 2003; Verhoeff et al. 1999). A study from Rwanda suggests that the susceptibility of HIV-infected women persists through the postpartum period (Ladner et al. 2002). By impairing humoral immune responses to novel antigens, HIV could delay the acquisition of specific immunity against placental parasites (Duffy 2001), and HIV-infected parturients in Kenya and Malawi had lower levels of antibody directed against placental parasites (and other parasites) than uninfected parturients, although the difference did not achieve significance (Staalsoe et al. 2004). Further studies are required to examine this question that has important implications for risk of transmission to the newborn. The frequency of congenital malaria remains controversial but may be increased in women who have heavier burdens of placental parasitemia (Redd et al. 1996). The effect of placental malaria on perinatal mother-to-child transmission (MTCT) of HIV is also controversial: one study found no effect of placental malaria on MTCT (Inion et al. 2003), one found an increased risk associated with placental malaria (Brahmbhatt et al. 2003), and a third study found the risk of MTCT depended on placental parasite density with low parasite density reducing risk and high parasite density increasing risk (Ayisi 2004).

Taken together, the studies of placental parasites and acquired immune responses suggest a model to explain malaria susceptibility and resistance in pregnant women. Women lack immunity to placental IEs before their first pregnancy. This would account for the susceptibility of primigravid women to malaria (Fried and Duffy 1996). Exposure during first and subsequent pregnancies elicits specific immunity that confers protection to multigravid women. Antibodies and possibly cellular responses against placental parasites prevent infection (Fried et al. 1998b) and the poor pregnancy outcomes associated with malaria (Duffy and Fried 2003; Staalsoe et al. 2004). This model of protective immunity suggests that a vaccine against placental parasites could be delivered before a woman's first pregnancy, and exposure during pregnancy would boost her protective responses and prevent poor outcomes for both mother and fetus.

5
Immunopathology

The inflammatory infiltrate and pro-inflammatory cytokines elicited by placental malaria have been associated with poor fetal and maternal outcomes. After *P. falciparum* parasites accumulate in the placenta, a massive cellular infiltrate may appear in the intervillous spaces (Garnham 1938), with elevated numbers of monocytes/macrophages (Fig. 2) and lymphocytes and the selective absence of natural killer (NK) cells (Ordi et al. 2001; Rasheed et al. 1992). Elevated levels of β and α chemokines in malaria-infected placentas are likely to play a role in attracting and activating leukocyte populations. Placental macrophage activity, and the inflammatory mediators tumor necrosis factor (TNF)-α, interferon (IFN)-γ and IL-8, are related to LBW and severe maternal anemia.

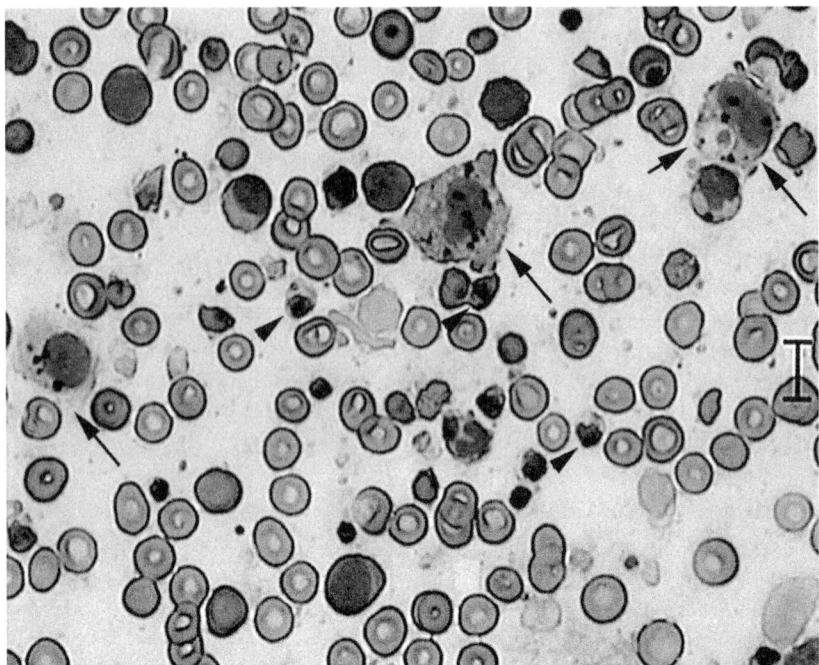

Fig. 2 Macrophages are a prominent feature of a placental bloodsmear obtained from an infected woman in Kenya. Large, distorted macrophages containing pigment (*large arrows*) are common in the cellular infiltrate that occurs after sequestration of IEs (*arrowheads*) in the placenta. An intact IE (*small arrow*) has been ingested by a macrophage

The earliest pathologic studies of *P. falciparum*-infected placentas observed the accumulation of phagocytic cells containing parasite 'pigment' (Bignami 1898; Sereni 1902). Also called hemozoin, pigment is the heme crystal produced by intraerythrocytic malaria parasites as they digest hemoglobin. The accumulation of inflammatory cells occurs following sequestration of parasites in the placenta (Garnham 1938, and has been related to the poor pregnancy outcomes induced by malaria (Duffy 2001; Leopardi et al. 1996). Placental macrophages are associated with severe maternal anemia (Jilly 1969) and LBW (Leopardi et al. 1996; Ordi et al. 1998) in malaria-infected women. Hemozoin is a marker of placental malaria, but is not associated with poor birth outcomes (Sullivan et al. 2000).

Healthy pregnancy is characterized by a bias toward anti-inflammatory or Type 2 cytokines in mice (Wegmann et al., 1993) and humans (Fried et al., 1998a; Marzi et al., 1996). Exposure to malaria during pregnancy elicits inflammatory cytokines in the placenta, including TNF-α, IFN-γ, and IL-2 (Fried et al., 1998a). TNF-α (Fried et al., 1998a; Rogerson et al., 2003), IFN-γ (Fried et al. 1998a), and IL-8 (Moormann et al. 1999) have been related to severe maternal anemia and LBW deliveries in malaria-exposed women. Histologic (Moormann et al. 1999) and tissue explant studies (Suguitan et al. 2003) suggest that infiltrating macrophages may be the source of TNF-α and IL-8, while the chorionic villi of the placenta may in part contribute to IFN-γ during pregnancy malaria. Chemokines, including α and β chemokines, are elevated during placental malaria (Abrams et al. 2003; Suguitan et al. 2003), and the infiltrating macrophages express CC chemokine receptor 5 (Tkachuk et al. 2001), and these are likely to enhance the inflammatory infiltrate.

In areas of stable transmission, the relationship between inflammatory cytokines and poor pregnancy outcomes appears to be strongest among primigravid women (Fried et al. 1998a) who suffer disproportionately from malaria and malaria-related complications. This may be due to the chronicity of infection and the prolonged cytokine responses observed in this parity group (Fried et al. 1998a). The mass of placental parasites and macrophages that accumulate in the placenta could theoretically impair the transplacental exchange of oxygen and nutrients, but studies to test this hypothesis have failed to yield confirmatory evidence (Duffy 2001).

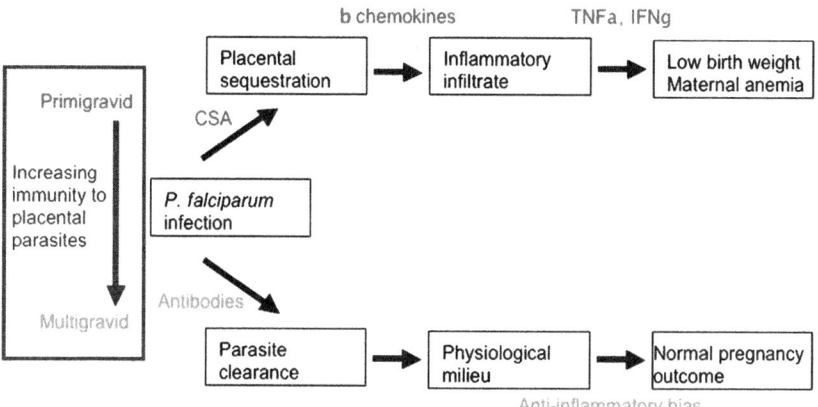

Fig. 3 A model of protective immunity against *P. falciparum* during pregnancy. Parasites that adhere to the placental receptor CSA are selected for growth during pregnancy, and antibodies against these parasites are acquired over successive pregnancies. Primigravid women lack immunity to CSA-binding parasites, and suffer heavy placental infections and inflammation, leading to maternal anemia and low birth weight deliveries. In multigravid women, antibodies against placental parasites limit the infection and the inflammatory response, thereby preventing poor pregnancy outcomes

The available evidence suggests that primigravid women lack specific immunity against the CSA-binding forms of *P. falciparum* that are selected for adhesion in the placenta. In the absence of specific immunity, these parasites accumulate in large numbers in the placenta, inducing an infiltrate of inflammatory cells whose accumulation may be accelerated by expression of chemokines. The inflammatory cells may reduce parasite multiplication, but are inefficient for clearing the parasite, allowing a prolonged inflammatory response associated with poor pregnancy outcomes including LBW and maternal anemia. The acquisition of specific immunity against CSA-binding parasites may facilitate the clearance of parasites prior to the influx of inflammatory cells, thereby avoiding the cascade of events leading to disease and death (Fig. 3).

6
Drugs to Prevent Pregnancy Malaria

Two chemotherapeutic approaches have been successfully used to protect women from the consequences of pregnancy malaria. Chemoprophylaxis entails continuous provision of antimalarials during pregnancy in order to prevent parasitemia. Intermittent presumptive treatment (IPT) provides full treatment doses at specific times during pregnancy in order to reduce disease. Both strategies are effective in areas where parasites remain sensitive to the drugs, but drug-resistant parasites are spreading rapidly throughout the world. New drugs are urgently needed to prevent and treat malaria in pregnant women, but safety issues including teratogenicity and embryotoxicity are paramount.

6.1
Chemoprophylaxis and Pregnancy Outcome

Malaria chemoprophylaxis for pregnant women aims to reduce the risks of LBW and severe maternal anemia. Several drug combinations are effective for chemoprophylaxis during pregnancy, although the rapid development of drug resistance has undermined the usefulness of some.

The selection of prophylactic drugs for pregnant women is a challenge. Drugs such as amodiaquine cannot be used as prophylaxis for safety reasons, while the safety of several newer drugs in pregnant women is unknown. The success of prophylaxis in pregnant women is measured as reduced parasitemia, increased maternal hemoglobin level, increased birth weight and reduced rate of LBW. The improvement in birth weight is perhaps the most important measure of benefit due to its impact on mortality during infancy.

In a high transmission area in Tanzania, chloroquine (CQ) prophylaxis was inferior to proguanil prophylaxis for reducing clinical malaria episodes in pregnant women (Mutabingwa et al. 1993a), and those receiving CQ alone had the highest rates of placental malaria and LBW (Mutabingwa et al. 1993b). The benefits of prophylaxis were greatest in primigravid women. The studies confirmed the emergence of CQ-resistant parasites in the study area, and established proguanil as a suitable agent for prophylaxis in areas of CQ resistance (Mutabingwa et al. 1993a). In Cameroon, CQ continues to be recommended as the drug of choice for prophylaxis in pregnant women, despite the emergence of CQ-resistant parasites. In an evaluation of CQ prophylaxis, the rate of anemia was 20.7% in the treatment group compared to 32% in the untreated group, and the rates did not vary significantly with parity [relative risk, (RR) for treated group: 0.57 in primigravidas, 0.67 in multigravidas] (Salihu et al. 2002).

In The Gambia, chemoprophylaxis with pyrimethamine/dapsone (Maloprim®) provided significant benefits, primarily to primigravid women (Greenwood et al. 1992) and grandemultigravid (parity >7) women (Greenwood et al. 1989; Menendez et al. 1994). Compared to those receiving placebo, primigravidas receiving pyrimethamine/dapsone had lower rates of parasitemia and LBW deliveries, as well as higher hematocrits. Although the rate of parasitemia was reduced in multigravid women, birth weights and maternal anemia rates were not significantly different. Pyrimethamine/dapsone chemoprophylaxis reduced the mortality of infants born to primigravid women by 18%, and that of infants born to multigravid women by 5%, although this benefit was only significant for first pregnancies (Greenwood et al. 1992). In Malawi, mefloquine prophylaxis during pregnancy also significantly reduced the frequency of LBW, and this effect was again most pronounced in first pregnancies (Steketee et al. 1996a, 1996c).

Chemoprophylaxis during first pregnancy did not reduce immunity in subsequent pregnancies. Secundigravid women in The Gambia who had received pyrimethamine/dapsone during their first pregnancy did not differ from untreated women according to the prevalence of peripheral or placental malaria or birth weight. Because most trial participants started their prophylaxis during the third trimester of their first pregnancy, the results do not exclude the possibility that prophylaxis provided throughout pregnancy may hinder the acquisition of natural immunity (Greenwood et al. 1994).

6.2
Intermittent Presumptive Treatment

During intermittent presumptive treatment (IPT), women periodically take full treatment doses of antimalarial drug, without regard to presence or absence of parasitemia. The goal of IPT is to reduce severe disease or chronic illness without affecting the acquisition of protective immunity. The World Health Organization currently recommends that women should receive IPT at two scheduled visits after the first trimester (www.who.org).

Several studies examined the effect of IPT using sulfadoxine–pyrimethamine (SP) to reduce maternal parasitemia, LBW, and maternal anemia, with varying results. In a study of 575 women in Malawi, primigravidas and multigravidas who received two or three doses of SP delivered significantly heavier babies compared to mothers who received one dose of SP, although parasite densities and hemoglobin levels did not differ between groups (Verhoeff et al. 1998). In studies in Kenya and Malawi, IPT with either two doses or monthly doses of SP reduced maternal parasitemia (Parise et al. 1998; Schultz et al. 1994). Compared to fever case management, IPT yielded heavier babies, but did not reduce rates of severe anemia, premature deliveries, abortion or stillbirth (Parise et al. 1998). In Kenya, primigravidas receiving IPT with SP had significantly lower rates of parasitemia and severe maternal anemia, although women receiving one, two or three doses of SP achieved similar benefits (Shulman et al. 1999). None of these studies detected an adverse effect of SP on the mother or the newborn.

SP resistance has emerged quickly in areas where it has been introduced as first-line therapy, and therefore combinations of SP with artesunate may prolong their effectiveness. Gambian women who received one dose of artesunate–SP during the third trimester delivered babies that were significantly heavier than babies of unprotected women, indicating effectiveness, and the rates of abortion, stillbirth and infant death were similar, indicating safety (Deen et al. 2001).

Although the different studies yielded varying magnitudes of results, all studies have observed benefits of IPT. Pregnant women with placental parasitemia commonly lack symptoms, and have low or undetectable peripheral parasite densities. These women are unlikely to seek treatment, and this may explain the advantage of IPT over treatment for symptomatic malaria in pregnant women. Similar to the benefits of chemoprophylaxis, the benefits of IPT are greatest in primigravid women.

7
Drugs to Treat Pregnancy Malaria

The drugs traditionally used to treat pregnancy malaria are losing their efficacy, and experience in pregnant women with newer antimalarial drugs is limited. Pregnant women are more likely to fail treatment than nonpregnant women (Keuter et al. 1990), but the urgent need to provide more effective treatments to pregnant women is hindered by safety concerns.

Intravenous quinine is the mainstay of therapy for severe or complicated malaria in pregnant women (Looareesuwan et al. 1985). Traditionally, CQ was the mainstay of therapy for uncomplicated malaria in pregnant women. In areas of CQ resistance, SP or amodiaquine are recommended for nonpregnant individuals (Keuter et al. 1990; Winstanley 2003), but amodiaquine is not recommended in pregnant women due to a lack of safety data (Taylor and White 2004). In areas with chloroquine and SP resistance, artesunate combinations (McGready et al. 2003) or oral quinine in combination with clindamycin (McGready et al. 2001b) have been shown to be effective for outpatient treatment of pregnant women. Quinine is generally well tolerated, although hypoglycemia is a potentially fatal side effect (Kochar et al. 1995; McGready et al. 2001b), and has been reported after oral administration (Taylor and White 2004).

8
Drug Safety During Pregnancy

Drug availability is an urgent issue for pregnant women exposed to malaria. Parasites that are resistant to traditional drugs are spreading rapidly, but there is not enough experience with most new drugs to know whether they are safe in pregnant women (Table 1).

Numerous studies have examined the safety of CQ, SP and mefloquine given to pregnant women at treatment or prophylaxis dosages. CQ is considered safe during all three trimesters, although mild side effects (itching, dizziness, and gastrointestinal discomfort) are frequent in pregnant women after treatment or prophylactic doses (Steketee et al. 1996b). Treatment with SP is thought to be safe during the last two trimesters (Keuter et al. 1990; Taylor and White 2004). SP is an antifolate, and the use of antifolates during the first trimester increases the risks of neural tube defects, cardiovascular defects, oral clefts, and urinary tract defects, and the use of folate supplements reduces the risks (Hernandez-Diaz et al. 2000).

Among 4,187 women in Malawi, the rate of stillbirths (3.9%) and abortions (1.2%) were similar in groups that received mefloquine or CQ as treatment or prophylaxis (Steketee et al. 1996b). Among 208 women in Thailand, stillbirths were more frequent among those treated with mefloquine (4.5%) than those treated with quinine (1.6%), but the rates of other severe outcomes (abortion, LBW, congenital malformation, neurologic deficits) were similar (Nosten et al. 1999b). Among 53 female US military personnel in Somalia who inadvertantly used mefloquine early in pregnancy, there were 17 elective abortions, 12 spontaneous abortions, one molar pregnancy, and 23 live births (Smoak et al.

Table 1 Current status of antimalarial drugs that may be considered for treatment, prophylaxis, or intermittent presumptive treatment (IPT) of pregnant women

Drug (and regimen)	Studies	Status and safety	References
Chloroquine, CQ (Treatment, prophylaxis)	Tanzania Cameroon	Safe in all trimesters Resistant parasites widespread	Mutabingwa et al. 1993 Salihu et al. 2002
Pyrimethamine/dapsone (Prophylaxis)	Gambia	Safe in second and third trimesters Limited first trimester experience Resistant parasites spreading rapidly	Greenwood et al. 1989 Greenwood et al. 1992
Mefloquine (Treatment, prophylaxis)	Malawi Thailand	Prophylaxis safe in second and third trimesters Treatment dose requires further evaluation	Steketee et al. 1996 Nosten et al. 1994, 1999
Sulfadoxine/ pyrimethamine, SP (Treatment, IPT)	Malawi Kenya	Safe in second and third trimesters Limited first trimester experience	Verhoeff et al. 1998 Schultz et al. 1994 Parise et al. 1998 Shulman et al. 1999
Quinine, quinine with clindamycin (Treatment)	Thailand	Safe in second and third trimesters Limited first trimester experience Risk of hypoglycemia	Kochar et al. 1995 McGready et al. 2001
Artemisinin derivatives (Treatment)	Gambia Thailand	Safe in a limited number of studies WHO recommends first trimester use be limited to life-saving situations when no alternative drugs are available	Deen et al. 2001 McGready et al. 2001 WHO 1998
Atovaquone/proguanil (Treatment)	Thailand	Safety requires further evaluation Limited experience with no adverse effects Proguanil alone is safe in pregnancy May be used when no alternatives are available	McGready et al. 2003 Mutabingwa et al. 1993 Taylor and White 2004

1997), suggesting that mefloquine may be embryotoxic and teratogenic when used just prior to or during the first trimester (Taylor and White 2004).

Quinine at treatment dosage is generally safe in pregnant women, although it causes hypoglycemia in 50% of pregnant women with severe malaria compared to 10% of nonpregnant patients (Kochar et al. 1995; Looareesuwan et al. 1985). Quinine is commonly administrated with tetracyclines in nonpregnant patients, but tetracycline is contraindicated during pregnancy. The combination of quinine and clindamycin is safe to use in pregnant women (McGready et al. 2001b). Atovaquone/proguanil is used for both treatment and prophylaxis, and no adverse outcomes in pregnant women have been reported in a preliminary study in Thailand, but further evaluation is needed (McGready et al. 2003).

Pregnant women with acute malaria who were treated with artesunate or artemether had rates of abortion, stillbirth and congenital abnormalities that were comparable to rates in the local population (Deen et al. 2001; Mc-Gready et al. 2001a). Artemisinins appear to be effective in the treatment of quinine-resistant parasites, and unlike quinine do not cause hypoglycemia (Taylor and White 2004). Halofantrine, amodiaquine and the combination of artemether/lumefantrine cannot currently be recommended for use in pregnant women due to a lack of information on their safety (Taylor and White 2004).

The inexorable spread of drug-resistant parasites has rendered traditional antimalarials useless in many areas. Conversely, experience with newer drugs in pregnant women is limited, and therefore concerns about their safety remain. The use of insect repellents, such as DEET (N,N-diethyl-M-toluamide), is therefore an attractive alternative for preventing malaria. In a trial at the Thai–Burmese border, women who applied DEET daily experienced birth outcomes similar to other women, measured by gestational age, birth weight, and neurological parameters, including child development during the first year of life (McGready et al. 2001c). The safety of DEET during the first trimester has not been studied. Insecticide-treated bed nets (ITNs) offer another means of protecting pregnant women. Studies of ITNs in Ghana and coastal Kenya did not detect benefits for mothers using ITNs, whereas birth weight was increased (77–130 g) in western Kenya and Gambia, and the incidence of anemia and parasitemia was reduced among women in Thailand, Tanzania, and western Kenya (Browne et al. 2001; D'Alessandro et al. 1996; Dolan et al. 1993; Marchant et al. 2002; Shulman et al. 1998; ter Kuile et al. 2003). In western Kenya, women (gravida one to four) using ITNs had a 24.1% rate of poor fetal outcomes (abortion, stillbirth, LBW due to PTD or IUGR) versus 32.1% in women without ITNs (ter Kuile et al. 2003). In The Gambia, rates of abortions or stillbirths were similar in women who did or did not use ITNs (D'Alessandro et al. 1996).

In Asia, *P. falciparum* parasites have become increasingly resistant to quinine, and therefore artemisinin derivatives have been evaluated in pregnant women. In a prospective study from Thailand, parasites reappeared after treatment in 6.8% of women receiving artesunate alone as primary treatment, 15.9% in women receiving artesunate as a re-treatment, and 37% in women receiving quinine as a re-treatment (McGready et al. 2001a). The rate of abortions, stillbirths and congenital abnormalities was similar among women who received artemisinins and the general population (McGready et al. 2001a).

Artesunate combined with mefloquine cleared fever and parasites more rapidly than quinine in a study of 50 Thai women (Bounyasong 2001), however mefloquine has been linked to the risk of stillbirth (McGready et al. 2003).

Atovaquone-proguanil (Malarone™) is a new antimalarial drug, but may rapidly select for resistant parasites (Nosten and McGready 2001), and therefore has been studied in combination with artemisinins (McGready et al. 2003). Proguanil is thought to be safe during pregnancy (Mutabingwa et al. 1993a), and fetal toxicity due to atovaquone in rabbits was observed only at doses that caused maternal toxicity (McGready et al. 2003). Atovaquone-proguanil in combination with artesunate was well-tolerated by Thai women who received it during the second or third trimester. No premature deliveries or congenital abnormalities were observed, and birth weights were similar to women who received other standard therapies (McGready et al. 2003). The cure rate was 100%, but the plasma concentrations of both atovaquone and proguanil were lower in pregnant women compared to other adults with uncomplicated malaria, warranting additional pharmacokinetics studies with higher doses of atovaquone-proguanil (McGready et al. 2003).

Other drugs that have been evaluated in pregnant women include clindamycin and dapsone-chlorproguanil (LapDap™). Clindamycin in combination with quinine achieved 100% cure rates in pregnant women with uncomplicated malaria in southeast Asia, and has therefore been recommended as a backup to artesunate combinations for the treatment of quinine-resistant parasites (McGready et al. 2001b). Dapsone-chlorproguanil initially cleared parasites in 81 pregnant women in Kenya, but parasitemia reappeared in 67% of primigravidas within 28 days of treatment. Pregnancy outcomes were not reported in this study, so the safety of dapsone-chlorproguanil in pregnant women remains unclear (Keuter et al. 1990).

9
Areas for Future Research

Today, the most urgent issue requiring research is the availability of safe, effective and affordable drugs to prevent and treat pregnancy malaria. Research priorities should reflect this. Several new antimalarial drugs are known to be effective in other populations, and research should focus first on determining whether these drugs are safe and effective in pregnant women. Progress has been made in this area, particularly in Thailand, where multidrug-resistant parasites are widespread. However, drug-resistant parasites are rapidly spreading throughout Africa, and drugs that are effective for prevention are a necessary element of antenatal health care throughout most of the continent.

Barring the eradication of malaria, the long-term solution to pregnancy malaria will require vaccines. Studies performed in several countries support a model of protective immunity—pregnant women who acquire antibodies against placental parasites are resistant to infection and disease. This knowledge will guide the identification of the antigens required for a pregnancy malaria vaccine, but several questions remain. Are antibodies sufficient for protection? We need a better understanding of cellular immune responses against placental parasites. Are there receptors other than CSA that support parasite sequestration in the placenta? Vaccines may need to target more than one parasite subpopulation. How many IE surface antigens mediate binding to CSA, or elicit protective antibodies? The best vaccine may comprise several surface antigens of placental parasites.

Research into pregnancy malaria will also yield important insights that illuminate other research areas, such as the normal physiology of pregnancy or general immunity against blood stage parasites. Parasites multiply to abundance in the placenta despite high titer antibodies against many antigens of blood stage parasites. What does this tell us about developing vaccines that are intended to control blood stage malaria in children? In high transmission areas, placental malaria causes severe maternal anemia and poor fetal outcomes, although the infection generally causes only low peripheral parasitemia in the mother and no infection of the fetus. What are the roles of the placenta in sustaining the health of the mother and fetus, and how can placental malaria shed light on other important diseases of pregnancy, such as eclampsia?

Pregnancy malaria is an enormous public health problem in the tropical world. Research should emphasize the development and delivery of drugs and vaccines that prevent and treat infection. Basic research is needed to understand protective immunity and to identify targets for new therapies. Translational research and clinical trials are needed to deliver the products that women desperately need. Research into the host–parasite interactions that underlie pregnancy malaria will also yield insights into the remarkable processes of placentation, pregnancy, parturition, and human development in utero and during early life.

Acknowledgements Atis Muehlenbachs reviewed the manuscript. The authors are supported by the Bill & Melinda Gates Foundation, the US National Institutes of Health (grants AI43680 and AI52059 to P.E.D.), and the US Military Infectious Disease Research Program. The views expressed in this article do not necessarily reflect those of the US Department of Defense.

References

Abrams ET, Brown H, Chensue SW, Turner GD, Tadesse E, Lema VM, Molyneux ME, Rochford R, Meshnick SR, Rogerson SJ (2003) Host response to malaria during pregnancy: placental monocyte recruitment is associated with elevated beta chemokine expression. J Immunol 170:2759–2764

Archibald HM (1956) The influence of malarial infection of the placenta on the incidence of prematurity. Bull WHO 15:842–845

Ayisi JG, Newman RD, ter Kuile FO, Shi YP, Yang C, et al. (2004) Maternal malaria and perinatal HIV transmission, western Kenya. Emerg Infect Dis 10:643–652

Baruch DI, Gormely JA, Ma C, Howard RJ, Pasloske BL (1996) *Plasmodium falciparum* erythrocyte membrane protein 1 is a parasitized erythrocyte receptor for adherence to CD36, thrombospondin, and intercellular adhesion molecule 1. Proc Natl Acad Sci USA 93:3497–3502

Beeson JG, Brown GV (2004) *Plasmodium falciparum*-infected erythrocytes demonstrate dual specificity for adhesion to hyaluronic acid and chondroitin sulfate A and have distinct adhesive properties. J Infect Dis 189:169–179

Beeson JG, Brown GV, Molyneux ME, Mhango C, Dzinjalamala F, Rogerson SJ (1999) *Plasmodium falciparum* isolates from infected pregnant women and children are associated with distinct adhesive and antigenic properties. J Infect Dis 180:464–472

Beeson JG, Rogerson SJ, Cooke BM, Reeder JC, Chai W, Lawson AM, Molyneux ME, Brown GV (2000) Adhesion of *Plasmodium falciparum*-infected erythrocytes to hyaluronic acid in placental malaria. Nat Med 6:86–90

Bignami A (1898) Al Policlinico (Supplemento) 4:763–767

Bounyasong S (2001) Randomized trial of artesunate and mefloquine in comparison with quinine sulfate to treat *P. falciparum* malaria pregnant women. J Med Assoc Thai 84:1289–1299

Brabin BJ (1983) An analysis of malaria in pregnancy in Africa. Bull WHO 61:1005–1016

Brabin BJ, Hakimi M, Pelletier D (2001) An analysis of anemia and pregnancy-related maternal mortality. J Nutr 131:604S–614S; discussion 614S–615S

Brabin BJ, Rogerson SJ (2001) The epidemiology and outcomes of maternal malaria. In: Duffy PE, Fried M (eds) Malaria in pregnancy: deadly parasite, susceptible host. Taylor & Francis, New York

Brahmbhatt H, Kigozi G, Wabwire-Mangen F, Serwadda D, Sewankambo N, Lutalo T, Wawer MJ, Abramowsky C, Sullivan D, Gray R (2003) The effects of placental malaria on mother-to-child HIV transmission in Rakai, Uganda. AIDS 17:2539–2541

Bray RS, Sinden RE (1979) The sequestration of *Plasmodium falciparum* infected erythrocytes in the placenta. Trans R Soc Trop Med Hyg 73:716–719

Browne EN, Maude GH, Binka FN (2001) The impact of insecticide-treated bednets on malaria and anaemia in pregnancy in Kassena-Nankana district, Ghana: a randomized controlled trial. Trop Med Int Health 6:667–676

Bruce-Chwatt LJ (1952) Malaria in African infants and children in Southern Nigeria. Ann Trop Med Parasitol 46:173–200

Buffet PA, Gamain B, Scheidig C, Baruch D, Smith JD, Hernandez-Rivas R, Pouvelle B, Oishi S, Fujii N, Fusai T, Parzy D, Miller LH, Gysin J, Scherf A (1999) *Plasmodium falciparum* domain mediating adhesion to chondroitin sulfate A: a receptor for human placental infection. Proc Natl Acad Sci USA 96:12743–12748

Cannon DSH (1958) Malaria and prematurity in the western region of Nigeria. BMJ ii:877–878

Creasey AM, Staalsoe T, Raza A, Arnot DE, Rowe JA (2003) Nonspecific immunoglobulin M binding and chondroitin sulfate A binding are linked phenotypes of *Plasmodium falciparum* isolates implicated in malaria during pregnancy. Infect Immun 71:4767–4771

D'Alessandro U, Langerock P, Bennett S, Francis N, Cham K, Greenwood BM (1996) The impact of a national impregnated bed net programme on the outcome of pregnancy in primigravidae in The Gambia. Trans R Soc Trop Med Hyg 90:487–492

Deen JL, von Seidlein L, Pinder M, Walraven GE, Greenwood BM (2001) The safety of the combination artesunate and pyrimethamine-sulfadoxine given during pregnancy. Trans R Soc Trop Med Hyg 95:424–428

Diagne N, Rogier C, Cisse B, Trape JF (1997) Incidence of clinical malaria in pregnant women exposed to intense perennial transmission. Trans R Soc Trop Med Hyg 91:166–170

Dolan G, ter Kuile FO, Jacoutot V, White NJ, Luxemburger C, Malankirii L, Chongsuphajaisiddhi T, Nosten F (1993) Bed nets for the prevention of malaria and anaemia in pregnancy. Trans R Soc Trop Med Hyg 87:620–626

Duffy MF, Brown GV, Basuki W, Krejany EO, Noviyanti R, Cowman AF, Reeder JC (2002) Transcription of multiple var genes by individual, trophozoite-stage *Plasmodium falciparum* cells expressing a chondroitin sulphate A binding phenotype. Mol Microbiol 43:1285–1293

Duffy PE (2001) Immunity to malaria during pregnancy: Different host, different parasite. In: Duffy PE, Fried M (eds) Malaria in pregnancy: deadly parasite, susceptible host. Taylor & Francis, New York, pp 71–126

Duffy PE, Desowitz RS (2001) Pregnancy malaria throughout history: Dangerous labors. In: Duffy PE, Fried M (eds) Malaria in pregnancy: deadly parasite, susceptible host. Taylor & Francis, New York, pp 1–25

Duffy PE, Fried M (2003) Antibodies that inhibit *Plasmodium falciparum* adhesion to chondroitin sulfate A are associated with increased birth weight and the gestational age of newborns. Infect Immun 71:6620–6623

Fievet N, Tami G, Maubert B, Moussa M, Shaw IK, Cot M, Holder AA, Chaouat G, Deloron P (2002) Cellular immune response to *Plasmodium falciparum* after pregnancy is related to previous placental infection and parity. Malaria J 1:16

Flick K, Scholander C, Chen Q, Fernandez V, Pouvelle B, Gysin J, Wahlgren M (2001) Role of nonimmune IgG bound to PfEMP1 in placental malaria. Science 293:2098–2100

Florens L, Liu X, Wang Y, Yang S, Schwartz O, Peglar M, Carucci DJ, Yates JR, Wub Y (2004) Proteomics approach reveals novel proteins on the surface of malaria-infected erythrocytes. Mol Biochem Parasitol 135:1–11

Fried M, Duffy PE (1996) Adherence of *Plasmodium falciparum* to chondroitin sulfate A in the human placenta. Science 272:1502–1504

Fried M, Duffy PE (1998) Maternal malaria and parasite adhesion. J Mol Med 76:162–171

Fried M, Duffy PE (2002) Two DBLgamma subtypes are commonly expressed by placental isolates of *Plasmodium falciparum*. Mol Biochem Parasitol 122:201–210

Fried M, Lauder RM, Duffy PE (2000) *Plasmodium falciparum*: adhesion of placental isolates modulated by the sulfation characteristics of the glycosaminoglycan receptor. Exp Parasitol 95:75–78

Fried M, Muga RO, Misore AO, Duffy PE (1998a) Malaria elicits type 1 cytokines in the human placenta: IFN-gamma and TNF-alpha associated with pregnancy outcomes. J Immunol 160:2523–2530

Fried M, Nosten F, Brockman A, Brabin BJ, Duffy PE (1998b) Maternal antibodies block malaria. Nature 395:851–852

Fried M, Wendler JP, Mutabingwa TK, Duffy PE (2004) Mass spectrometric analysis of *Plasmodium falciparum* erythrocyte membrane protein-1 variants expressed by placental malaria parasites. Proteomics 4:1086–1093

Gardner MJ, Hall N, Fung E, White O, Berriman M, Hyman RW, Carlton JM, Pain A, Nelson KE, Bowman S, Paulsen IT, James K, Eisen JA, Rutherford K, Salzberg SL, Craig A, Kyes S, Chan MS, Nene V, Shallom SJ, Suh B, Peterson J, Angiuoli S, Pertea M, Allen J, Selengut J, Haft D, Mather MW, Vaidya AB, Martin DM, Fairlamb AH, Fraunholz MJ, Roos DS, Ralph SA, McFadden GI, Cummings LM, Subramanian GM, Mungall C, Venter JC, Carucci DJ, Hoffman SL, Newbold C, Davis RW, Fraser CM, Barrell B (2002) Genome sequence of the human malaria parasite *Plasmodium falciparum*. Nature 419:498–511

Garner P, Gulmezoglu AM (2003) Drugs for preventing malaria-related illness in pregnant women and death in the newborn. Cochrane Database Syst Rev CD000169

Garnham PCC (1938) Transact R Soc Trop Med Hyg 32:13–48

Greenwood AM, Armstrong JR, Byass P, Snow RW, Greenwood BM (1992) Malaria chemoprophylaxis, birth weight and child survival. Trans R Soc Trop Med Hyg 86:483–485

Greenwood AM, Menendez C, Todd J, Greenwood BM (1994) The distribution of birth weights in Gambian women who received malaria chemoprophylaxis during their first pregnancy and in control women. Trans R Soc Trop Med Hyg 88:311–312

Greenwood BM, Greenwood AM, Snow RW, Byass P, Bennett S, Hatib-N'Jie AB (1989) The effects of malaria chemoprophylaxis given by traditional birth attendants on the course and outcome of pregnancy. Trans R Soc Trop Med Hyg 83:589–594

Guyatt HL, Snow RW (2001) The epidemiology and burden of *Plasmodium falciparum*-related anemia among pregnant women in sub-Saharan Africa. Am J Trop Med Hyg 64:36–44

Hammerich A, Campbell OM, Chandramohan D (2002) Unstable malaria transmission and maternal mortality—experiences from Rwanda. Trop Med Int Health 7:573–576

Hernandez-Diaz S, Werler MM, Walker AM, Mitchell AA (2000) Folic acid antagonists during pregnancy and the risk of birth defects. N Engl J Med 343:1608–1614

Hinderaker SG, Olsen BE, Bergsjo PB, Gasheka P, Lie RT, Havnen J, Kvale G (2003) Avoidable stillbirths and neonatal deaths in rural Tanzania. BJOG 110:616–623

Inion I, Mwanyumba F, Gaillard P, Chohan V, Verhofstede C, Claeys P, Mandaliya K, Van Marck E, Temmerman M (2003) Placental malaria and perinatal transmission of human immunodeficiency virus type 1. J Infect Dis 188:1675–1678

Jensen AT, Zornig HD, Buhmann C, Salanti A, Koram KA, Riley EM, Theander TG, Hviid L, Staalsoe T (2003) Lack of gender-specific antibody recognition of products from domains of a var gene implicated in pregnancy-associated *Plasmodium falciparum* malaria. Infect Immun 71:4193–4196

Jilly P (1969) Anaemia in parturient women, with special reference to malaria infection of the placenta. Ann Trop Med Parasitol 63:109–116

Keuter M, van Eijk A, Hoogstrate M, Raasveld M, van de Ree M, Ngwawe WA, Watkins WM, Were JB, Brandling-Bennett AD (1990) Comparison of chloroquine, pyrimethamine and sulfadoxine, and chlorproguanil and dapsone as treatment for falciparum malaria in pregnant and non-pregnant women, Kakamega District, Kenya. BMJ 301:466–470

Kochar DK, Kumawat BL, Kochar SK, Sanwal V (1995) Hypoglycemia after oral quinine administration. J Assoc Physicians India 43:654, 657

Kulmala T, Vaahtera M, Ndekha M, Koivisto AM, Cullinan T, Salin ML, Ashorn P (2000) The importance of preterm births for peri- and neonatal mortality in rural Malawi. Paediatr Perinat Epidemiol 14:219–226

Kyes SA, Christodoulou Z, Raza A, Horrocks P, Pinches R, Rowe JA, Newbold CI (2003) A well-conserved *Plasmodium falciparum var* gene shows an unusual stage-specific transcript pattern. Mol Microbiol 48:1339–1348

Ladner J, Leroy V, Simonon A, Karita E, Bogaerts J, De Clercq A, Van De Perre P, Dabis F (2002) HIV infection, malaria, and pregnancy: a prospective cohort study in Kigali, Rwanda. Am J Trop Med Hyg 66:56–60

Laveran A (1882) Bulletins et Memoires de la Societe Medicale de la hopitaux de Paris 18:168–176

Lekana Douki JB, Traore B, Costa FT, Fusai T, Pouvelle B, Sterkers Y, Scherf A, Gysin J (2002) Sequestration of *Plasmodium falciparum*-infected erythrocytes to chondroitin sulfate A, a receptor for maternal malaria: monoclonal antibodies against the native parasite ligand reveal pan-reactive epitopes in placental isolates. Blood 100:1478–1483

Leopardi O, Naughten W, Salvia L, Colecchia M, Matteelli A, Zucchi A, Shein A, Muchi JA, Carosi G, Ghione M (1996) Malaric placentas. A quantitative study and clinico-pathological correlations. Pathol Res Pract 192, 892–898; discussion 899–900

Looareesuwan S, Phillips RE, White NJ, Kietinun S, Karbwang J, Rackow C, Turner RC, Warrell DA (1985) Quinine and severe falciparum malaria in late pregnancy. Lancet 2:4–8

Luxemburger C, Ricci F, Nosten F, Raimond D, Bathet S, White NJ (1997) The epidemiology of severe malaria in an area of low transmission in Thailand. Trans R Soc Trop Med Hyg 91:256–262

MacPherson GG, Warrell MJ, White NJ, Looareesuwan S, Warrell DA (1985) Human cerebral malaria. A quantitative ultrastructural analysis of parasitized erythrocyte sequestration. Am J Pathol 119:385–401

Maitra N, Joshi M, Hazra M (1993) Maternal manifestations of malaria in pregnancy: a review. Indian J Matern Child Health 4:98–101

Malhotra M, Sharma JB, Batra S, Sharma S, Murthy NS, Arora R (2002) Maternal and perinatal outcome in varying degrees of anemia. Int J Gynaecol Obstet 79:93–100

Marchant T, Schellenberg JA, Edgar T, Nathan R, Abdulla S, Mukasa O, Mponda H, Lengeler C (2002) Socially marketed insecticide-treated nets improve malaria and anaemia in pregnancy in southern Tanzania. Trop Med Int Health 7:149–158

Marchiafava E, Bignami A (1894) On summer-autmnal fevers. In: Charles TE (ed.) Two monographs on malaria and the parasites of malarial fevers. The New Sydenham Society, London

Marzi M, Vigano A, Trabattoni D, Villa ML, Salvaggio A, Clerici E, Clerici M (1996) Characterization of type 1 and type 2 cytokine production profile in physiologic and pathologic human pregnancy. Clin Exp Immunol 106:127–133

Maubert B, Fievet N, Tami G, Boudin C, Deloron P (1998) *Plasmodium falciparum*-isolates from Cameroonian pregnant women do not rosette. Parasite 5:281–283

Maubert B, Fievet N, Tami G, Boudin C, Deloron P (2000) Cytoadherence of *Plasmodium falciparum*-infected erythrocytes in the human placenta. Parasite Immunol 22:191–199

McGready R, Cho T, Keo NK, Thwai KL, Villegas L, Looareesuwan S, White NJ, Nosten F (2001a) Artemisinin antimalarials in pregnancy: a prospective treatment study of 539 episodes of multidrug-resistant *Plasmodium falciparum*. Clin Infect Dis 33:2009–2016

McGready R, Cho T, Samuel D, Villegas L, Brockman A, van Vugt M, Looareesuwan S, White NJ, Nosten F (2001b) Trans R Soc Trop Med Hyg 95:651–656

McGready R, Hamilton KA, Simpson JA, Cho T, Luxemburger C, Edwards R, Looareesuwan S, White NJ, Nosten F, Lindsay SW (2001c) Safety of the insect repellent N,N-diethyl-M-toluamide (DEET) in pregnancy. Am J Trop Med Hyg 65:285–289

McGready R, Stepniewska K, Edstein MD, Cho T, Gilveray G, Looareesuwan S, White NJ, Nosten F (2003) The pharmacokinetics of atovaquone and proguanil in pregnant women with acute falciparum malaria. Eur J Clin Pharmacol 59:545–552

McGregor IA (1984) Epidemiology, malaria and pregnancy. Am J Trop Med Hyg 33:517–525

Menendez C, Todd J, Alonso PL, Lulat S, Francis N, Greenwood BM (1994) Malaria chemoprophylaxis, infection of the placenta and birth weight in Gambian primigravidae. J Trop Med Hyg 97:244–248

Meuris S, Piko BB, Eerens P, Vanbellinghen AM, Dramaix M, Hennart P (1993) Gestational malaria: assessment of its consequences on fetal growth. Am J Trop Med Hyg 48:603–609

Moore JM, Nahlen BL, Misore A, Lal AA, Udhayakumar V (1999) Immunity to placental malaria. I. Elevated production of interferon-gamma by placental blood mononuclear cells is associated with protection in an area with high transmission of malaria. J Infect Dis 179:1218–1225

Moormann AM, Sullivan AD, Rochford RA, Chensue SW, Bock PJ, Nyirenda T, Meshnick SR (1999) Malaria and pregnancy: placental cytokine expression and its relationship to intrauterine growth retardation. J Infect Dis 180:1987–1993

Murphy SC, Breman JG (2001) Gaps in the childhood malaria burden in Africa: cerebral malaria, neurological sequelae, anemia, respiratory distress, hypoglycemia, and complications of pregnancy. Am J Trop Med Hyg 64:57–67

Mutabingwa TK, Malle LN, de Geus A, Oosting J (1993a) Malaria chemosuppression in pregnancy. I. The effect of chemosuppressive drugs on maternal parasitaemia. Trop Geogr Med 45:6–14

Mutabingwa TK, Malle LN, de Geus A, Oosting J (1993b) Malaria chemosuppression in pregnancy. II. Its effect on maternal haemoglobin levels, placental malaria and birth weight. Trop Geogr Med 45:49–55

Newman RD, Hailemariam A, Jimma D, Degifie A, Kebede D, Rietveld AE, Nahlen BL, Barnwell JW, Steketee RW, Parise ME (2003) Burden of malaria during pregnancy in areas of stable and unstable transmission in Ethiopia during a nonepidemic year. J Infect Dis 187:1765–1772

Nosten F, McGready R (2001) The treatment of malaria in pregnancy. In: Duffy PE, Fried M (eds) Malaria in pregnancy: deadly parasite, susceptible host. Taylor & Francis, New York, pp

Nosten F, McGready R (2003) Burden of malaria during pregnancy in areas of stable and unstable transmission in Ethiopia during a nonepidemic year. J Infect Dis 188:1259–1261; author reply 1561–1562

Nosten F, McGready R, Simpson JA, Thwai KL, Balkan S, Cho T, Hkirijaroen L, Looareesuwan S, White NJ (1999a) Effects of Plasmodium vivax malaria in pregnancy. Lancet 354:546–549

Nosten F, Vincenti M, Simpson J, Yei P, Thwai KL, de Vries A, Chongsuphajaisiddhi T, White NJ (1999b) The effects of mefloquine treatment in pregnancy. Clin Infect Dis 28:808–815

Ordi J, Ismail MR, Ventura PJ, Kahigwa E, Hirt R, Cardesa A, Alonso PL, Menendez C (1998) Massive chronic intervillositis of the placenta associated with malaria infection. Am J Surg Pathol 22:1006–1011

Ordi J, Menendez C, Ismail MR, Ventura PJ, Palacin A, Kahigwa E, Ferrer B, Cardesa A, Alonso PL (2001) Placental malaria is associated with cell-mediated inflammatory responses with selective absence of natural killer cells. J Infect Dis 183:1100–1107

Parise ME, Ayisi JG, Nahlen BL, Schultz LJ, Roberts JM, Misore A, Muga R, Oloo AJ, Steketee RW (1998) Efficacy of sulfadoxine-pyrimethamine for prevention of placental malaria in an area of Kenya with a high prevalence of malaria and human immunodeficiency virus infection. Am J Trop Med Hyg 59:813–822

Piper C, Brabin BJ, Alpers MP (2001) Higher risk of post-partum hemorrhage in malarious than in non-malarious areas of Papua New Guinea. Int J Gynaecol Obstet 72:77–78

Pongponratn E, Riganti M, Punpoowong B, Aikawa M (1991) Microvascular sequestration of parasitized erythrocytes in human falciparum malaria: a pathological study. Am J Trop Med Hyg 44:168–175

Rasheed FN, Bulmer JN, Morrison L, Jawla MF, Hassan-King M, Riley EM, Greenwood BM (1992) Isolation of maternal mononuclear cells from placentas for use in in vitro functional assays. J Immunol Methods 146:185–193

Redd SC, Wirima JJ, Steketee RW, Breman JG, Heymann DL (1996) Transplacental transmission of Plasmodium falciparum in rural Malawi. Am J Trop Med Hyg 55:57–60

Reeder JC, Cowman AF, Davern KM, Beeson JG, Thompson JK, Rogerson SJ, Brown GV (1999) The adhesion of Plasmodium falciparum-infected erythrocytes to chondroitin sulfate A is mediated by P. falciparum erythrocyte membrane protein 1. Proc Natl Acad Sci USA 96:5198–5202

Reeder JC, Hodder AN, Beeson JG, Brown GV (2000) Identification of glycosaminogly-can binding domains in *Plasmodium falciparum* erythrocyte membrane protein 1 of a chondroitin sulfate A-adherent parasite. Infect Immun 68:3923–3926

Ricke CH, Staalsoe T, Koram K, Akanmori BD, Riley EM, Theander TG, Hviid L (2000) Plasma antibodies from malaria-exposed pregnant women recognize variant surface antigens on *Plasmodium falciparum*-infected erythrocytes in a parity-dependent manner and block parasite adhesion to chondroitin sulfate A. J Immunol 165:3309–3316

Rogerson SJ, Beeson JG, Mhango CG, Dzinjalamala FK, Molyneux ME (2000) *Plasmodium falciparum* rosette formation is uncommon in isolates from pregnant women. Infect Immun 68:391–393

Rogerson SJ, Brown HC, Pollina E, Abrams ET, Tadesse E, Lema VM, Molyneux ME (2003) Placental tumor necrosis factor alpha but not gamma interferon is associated with placental malaria and low birth weight in Malawian women. Infect Immun 71:267–270

Rowe JA, Kyes SA, Rogerson SJ, Babiker HA, Raza A (2002) Identification of a conserved *Plasmodium falciparum var* gene implicated in malaria in pregnancy. J Infect Dis 185:1207–1211

Salanti A, Jensen AT, Zornig HD, Staalsoe T, Joergensen L, Nielsen MA, Khattab A, Arnot DE, Klinkert MQ, Hviid L, Theander TG (2002) A sub-family of common and highly conserved *Plasmodium falciparum var* genes. Mol Biochem Parasitol 122:111–115

Salanti A, Staalsoe T, Lavstsen T, Jensen AT, Sowa MP, Arnot DE, Hviid L, Theander TG (2003) Selective upregulation of a single distinctly structured var gene in chondroitin sulphate A-adhering *Plasmodium falciparum* involved in pregnancy-associated malaria. Mol Microbiol 49:179–191

Salihu HM, Naik EG, Tchuinguem G, Bosny JP, Dagne G (2002) Weekly chloroquine prophylaxis and the effect on maternal haemoglobin status at delivery. Trop Med Int Health 7:29–34

Schultz LJ, Steketee RW, Macheso A, Kazembe P, Chitsulo L, Wirima JJ (1994) The efficacy of antimalarial regimens containing sulfadoxine-pyrimethamine and/or chloroquine in preventing peripheral and placental *Plasmodium falciparum* infection among pregnant women in Malawi. Am J Trop Med Hyg 51:515–522

Sereni S (1902) Bullettino della Reale Accademia Medica di Roma 29:55–88

Shulman CE, Dorman EK, Cutts F, Kawuondo K, Bulmer JN, Peshu N, Marsh K (1999) Intermittent sulphadoxine-pyrimethamine to prevent severe anaemia secondary to malaria in pregnancy: a randomised placebo-controlled trial. Lancet 353:632–636

Shulman CE, Dorman EK, Talisuna AO, Lowe BS, Nevill C, Snow RW, Jilo H, Peshu N, Bulmer JN, Graham S, Marsh K (1998) A community randomized controlled trial of insecticide-treated bednets for the prevention of malaria and anaemia among primigravid women on the Kenyan coast. Trop Med Int Health 3:197–204

Simister NE (2003) Placental transport of immunoglobulin G. Vaccine 21:3365–3369

Singh N, Saxena A, Chand SK, Valecha N, Sharma VP (1998) Studies on malaria during pregnancy in a tribal area of central India (Madhya Pradesh). Southeast Asian J Trop Med Public Health 29:10–17

Singh N, Shukla MM, Sharma VP (1999) Epidemiology of malaria in pregnancy in central India. Bull WHO 77:567–572

Smoak BL, Writer JV, Keep LW, Cowan J, Chantelois JL (1997) The effects of inadvertent exposure of mefloquine chemoprophylaxis on pregnancy outcomes and infants of US Army servicewomen. J Infect Dis 176:831–833

Staalsoe T, Shulman CE, Bulmer JN, Kawuondo K, Marsh K, Hviid L (2004) Variant surface antigen-specific IgG and protection against clinical consequences of pregnancy-associated *Plasmodium falciparum* malaria. Lancet 363:283–289

Steketee RW, Wirima JJ, Hightower AW, Slutsker L, Heymann DL, Breman JG (1996a) The effect of malaria and malaria prevention in pregnancy on offspring birthweight, prematurity, and intrauterine growth retardation in rural Malawi. Am J Trop Med Hyg 55:33–41

Steketee RW, Wirima JJ, Slutsker L, Khoromana CO, Heymann DL, Breman JG (1996b) Malaria treatment and prevention in pregnancy: indications for use and adverse events associated with use of chloroquine or mefloquine. Am J Trop Med Hyg 55:50–56

Steketee RW, Wirima JJ, Slutsker WL, Khoromana CO, Breman JG, Heymann DL (1996c) Objectives and methodology in a study of malaria treatment and prevention in pregnancy in rural Malawi: The Mangochi Malaria Research Project. Am J Trop Med Hyg 55:8–16

Suguitan AL Jr, Leke RG, Fouda G, Zhou A, Thuita L, Metenou S, Fogako J, Megnekou R, Taylor DW (2003) Changes in the levels of chemokines and cytokines in the placentas of women with *Plasmodium falciparum* malaria. J Infect Dis 188:1074–1982

Sullivan AD, Nyirenda T, Cullinan T, Taylor T, Harlow SD, James SA, Meshnick SR (1999) Malaria infection during pregnancy: intrauterine growth retardation and preterm delivery in Malawi. J Infect Dis 179:1580–1583

Sullivan AD, Nyirenda T, Cullinan T, Taylor T, Lau A, Meshnick SR (2000) Placental haemozoin and malaria in pregnancy. Placenta 21:417–421

Taylor HM, Kyes SA, Harris D, Kriek N, Newbold CI (2000) A study of var gene transcription in vitro using universal var gene primers. Mol Biochem Parasitol 105:13–23

Taylor WR, White NJ (2004) Antimalarial drug toxicity: a review. Drug Saf 27 25–61

ter Kuile FO, Terlouw DJ, Phillips-Howard PA, Hawley WA, Friedman JF, Kariuki SK, Shi YP, Kolczak MS, Lal AA, Vulule JM, Nahlen BL (2003) Reduction of malaria during pregnancy by permethrin-treated bed nets in an area of intense perennial malaria transmission in western Kenya. Am J Trop Med Hyg 68:50–60

Tkachuk AN, Moormann AM, Poore JA, Rochford RA, Chensue SW, Mwapasa V, Meshnick SR (2001) Malaria enhances expression of CC chemokine receptor 5 on placental macrophages. J Infect Dis 183:967–972

Valiyaveettil M, Achur RN, Alkhalil A, Ockenhouse CF, Gowda DC (2001) *Plasmodium falciparum* cytoadherence to human placenta: evaluation of hyaluronic acid and chondroitin 4-sulfate for binding of infected erythrocytes. Exp Parasitol 99:57–65

van den Broek N (1996) The aetiology of anaemia in pregnancy in West Africa. Trop Doct 26:5–7

van Eijk AM, Ayisi JG, ter Kuile FO, Misore AO, Otieno JA, Rosen DH, Kager PA, Steketee RW, Nahlen BL (2003) HIV increases the risk of malaria in women of all gravidities in Kisumu, Kenya. AIDS 17:595–603

van Hung L (1951) Revue du Paludisme at de Medecine Tropicale 9:75–112

Verhoeff FH, Brabin BJ, Chimsuku L, Kazembe P, Russell WB, Broadhead RL (1998) An evaluation of the effects of intermittent sulfadoxine-pyrimethamine treatment in pregnancy on parasite clearance and risk of low birthweight in rural Malawi. Ann Trop Med Parasitol 92:141–150

Verhoeff FH, Brabin BJ, Hart CA, Chimsuku L, Kazembe P, Broadhead RL (1999) Increased prevalence of malaria in HIV-infected pregnant women and its implications for malaria control. Trop Med Int Health 4:5–12

Verhoeff FH, Brabin BJ, van Buuren S, Chimsuku L, Kazembe P, Wit JM, Broadhead RL (2001) An analysis of intra-uterine growth retardation in rural Malawi. Eur J Clin Nutr 55:682–689

Wegmann TG, Lin H, Guilbert L, Mosmann TR (1993) Bidirectional cytokine interactions in the maternal-fetal relationship: is successful pregnancy a TH2 phenomenon? Immunol Today 14:353–536

Wickramasuriya GAW (1935) J Obstet Gynaecol BE 42:816–834

Wickramasuriya GAW (1936) Malaria and ankylostomiasis in the pregnant woman. Oxford University Press, London

Winstanley P (2003) The contribution of clinical pharmacology to antimalarial drug discovery and development. Br J Clin Pharmacol 55:464–468

Winter G, Chen Q, Flick K, Kremsner P, Fernandez V, Wahlgren M (2003) The 3D7var5.2 (var COMMON) type var gene family is commonly expressed in non-placental *Plasmodium falciparum* malaria. Mol Biochem Parasitol 127:179–191

Winter G, Chen Q, Wahlgren M (2004) Meeting report: the molecular background of severe and complicated malaria. Mol Biochem Parasitol 134:37–41

Part III
Biology

CTMI (2005) 295:203–232

Host Receptors in Malaria Merozoite Invasion

S. S. Oh[1] · A. H. Chishti[2] (✉)

[1]Division of Cell Biology, Caritas St. Elizabeth's Medical Center, Tufts University School of Medicine, 736 Cambridge Street, Boston, MA 02135, USA

[2]Department of Pharmacology, UIC Cancer Center, University of Illinois College of Medicine, Molecular Biology Research Building, 900 South Ashland Avenue, Chicago, IL 60607, USA
chishti@uic.edu

Abstract The clinical manifestations of *Plasmodium falciparum* malaria are directly linked to the blood stage of the parasite life cycle. At the blood stage, the circulating merozoites invade erythrocytes via a specific invasion pathway often identified with its dependence or independence on sialic acid residues of the host receptor. The invasion process involves multiple receptor–ligand interactions that mediate a complex series of events in a period of approximately 1 min. Although the mechanism by which merozoites invade erythrocytes is not fully understood, recent advances have put a new perspective on the importance of developing a multivalent blood stage-malaria vaccine. In this review, we highlight the role of currently identified host invasion receptors in blood-stage malaria.

Abbreviations

CSP	Circumsporozoite protein
TRAP	Thrombospondin-related anonymous protein
MTIP	Myosin A tail domain interacting protein
PfEMP1	*P. falciparum*-encoded proteins, erythrocyte membrane protein 1

RBP   Reticulocyte-binding protein
DARC   Duffy antigen/receptor for chemokines
SAD   Sialic acid-dependent
SAID   Sialic acid-independent
GP(A,B,C,D)   Glycophorin A,B,C,D
DBL   Duffy binding-like
EBL   Erythrocyte-binding ligand
SAO   Southeast Asian/Melanesian ovalocytosis
MSP1   Merozoite surface protein 1

1
Introduction

Malaria is a public health problem today in more than 90 countries inhabited by 40% of the world population (WHO 2004). In any given year, there are 300 million to 500 million clinical cases of malaria, and mortality due to the disease is estimated to be over one million deaths. The majority of deaths are in young children living in sub-Saharan Africa, typically under the age of 5 years. Statistically, at least one African child is killed by malaria every 30 s, a death toll that far exceeds the mortality rate of other childhood diseases in the region (WHO 2001). Many children who survive an episode of severe malaria suffer from brain damage and cognitive disability (Holding and Snow 2001). Pregnant women living in endemic areas and nonimmune travelers entering the region are also at high risk. Moreover, the economic and social burden of malaria is enormously high for the endemic countries, which in turn has devastating consequence on the global economy (Gallup and Sachs 2001; Sachs and Malaney 2002).

The causative agent for malaria is single-celled protozoan parasites of the genus *Plasmodium*. Although malaria parasites undergo a complex life cycle, the clinical manifestation of the disease is only associated with the blood stage where haploid parasites are asexually propagated within host erythrocytes. An indispensable step of this bloodstream cycle is the invasion of erythrocytes by a free-living extracellular merozoite stage. There has been a great interest to uncover ligand–receptor interactions between the parasite and host that occur during merozoite invasion. These interactions are thought to mediate the attachment and reorientation of merozoites on the erythrocyte surface and participate in transmembrane signal transduction that could up- or downregulate specific molecular events in the invasion process. Despite much effort in the past two decades, the molecular mechanism by which merozoites interact with erythrocytes during invasion is not well understood. Nonetheless, significant progress has been made towards the functional identification of

Plasmodium proteins localized to the merozoite surface, rhoptries, and micronemes that bind to erythrocytes. Many of these parasite ligands have been the subject of recent reviews (Cowman et al. 2000; Preiser et al. 2000; Chitnis 2001). This article highlights the role of currently identified host receptors in merozoite invasion of human erythrocytes.

2
Life Cycle

The malaria parasite is commonly transmitted to a human host from an infected female mosquito of the genus *Anopheles* upon taking a blood meal. There are four malaria species that typically infect humans: *P. vivax*, *P. falciparum*, *P. malariae*, and *P. ovale*. *P. knowlesi*, primarily a species of monkeys, has recently been implicated in a large human outbreak in Malaysia (Singh et al. 2004). Infection by *P. vivax* and *P. falciparum* are most common and the falciparum malaria is most deadly. Malaria parasites undergo a series of morphological and biochemical changes throughout their complex life cycle in both human and mosquito hosts (Wilson 1990; Gratzer and Dluzewski 1993; Ghosh et al. 2000; Sinden and Billingsley 2001; Bannister and Mitchell 2003; Cooke et al. 2004). Inoculation of malaria sporozoites into a human host occurs with an infectious mosquito bite which allows the sporozoites to travel from the mosquito salivary glands to the human bloodstream to invade hepatocytes (see illustration in Bannister and Mitchell 2003).

Sporozoite surface proteins such as circumsporozoite protein (CSP) and thrombospondin-related anonymous protein (TRAP) are thought to be involved in the specific recognition and entry into hepatocytes by sporozoites (Mota and Rodriguez 2002; Kappe et al. 2003). Sporozoite activation induced by migration through host cells is an essential step for hepatocyte infection (Mota et al. 2002). Evidence suggests the myosin A tail domain interacting protein (MTIP)/myosin A complex interacts with the actin/TRAP complex in the sporozoite plasma membrane to provide the crucial gliding motility of sporozoites during host cell invasion (Bergman et al. 2003; Kappe et al. 2004). A similar mechanism has been suggested for motility and invasion of other apicomplexans such as *Toxoplasma* and *Crytosporidium* (Meissner et al. 2002; Sibley 2004).

The malaria parasites multiply in hepatocytes for a certain period (1–2 weeks depending on the parasite species) undergoing asexual exoerythrocytic schizogony. During the liver stage, an exoerythrocytic schizont develops into more than 10,000 merozoites, which are released into the bloodstream to invade erythrocytes. The merozoites then develop into young intraerythrocytic parasites (ring stage) that mature to trophozoites and schizonts within the erythrocyte in the next 48–72 h depending on the species (asexual erythrocytic schizogony). Interestingly, in the case of *P. falciparum*, mature parasites (trophozoites and schizonts) are sequestered in the post-capillary venules by attaching to host endothelial receptors, a process known as cytoadherence (Pasloske and Howard 1994; Oh et al. 1997; Sherman et al. 2003; Cooke et al. 2004). Sequestration is thought to be an important survival strategy for blood-stage *P. falciparum* as it prevents parasite circulation to the spleen. Two *P. falciparum*-encoded proteins, erythrocyte membrane protein 1 (PfEMP1) and knob-associated histidine-rich protein embedded into the host erythrocyte membrane during parasite maturation, play an essential role in cytoadherence and thereby in the pathogenesis of blood-stage *P. falciparum* (Crabb et al. 1997; Waller et al. 1999; Oh et al. 2000; Voigt et al. 2000). Each erythrocytic schizont produces approximately 10–30 merozoites, which are then released into circulation upon erythrocyte membrane rupture. The released merozoites invade other erythrocytes to continue with asexual amplification.

Less than 1% of the merozoites differentiate in erythrocytes into sexual forms, macrogametocytes (female) and microgametocytes (male). These undergo gametogenesis in the *Anopheles* mosquito midgut following ingestion, in which a macrogamete is fertilized by an exflagellating microgamete to form a zygote. Within 24 h, the zygote differentiates into an ookinete, which penetrates the midgut wall and develops into an oocyst. Thousands of new sporozoites produced from the oocyst, migrate to the salivary glands to invade gland epithelium, and ultimately inoculate into another human host during the next blood meal, thereby repeating the life cycle.

3
Merozoite Invasion

The landmark video recording (Dvorak et al. 1975) and two ultrastructural studies (Bannister et al. 1975; Aikawa et al. 1978) of simian malarial parasite *P. knowlesi* invading erythrocytes have shown that invasion is a stepwise process requiring an attachment of the apical end of the merozoite to the erythrocyte followed by a deformation of the erythrocyte (Fig. 1). The actual penetration of the merozoite into the erythrocyte only takes about 30–60 s to complete. The internalized merozoites are transformed into the young ring stage in the next 10–15 min, which is accompanied by a second wave of erythrocyte deformation. Based on the account provided by these studies, merozoite invasion is thought to proceed through the following major events: (a) initial weak attachment of the merozoite to the erythrocyte surface following random collision; (b) reorientation of the merozoite to bring the apical end of the parasite in contact with the erythrocyte membrane forming

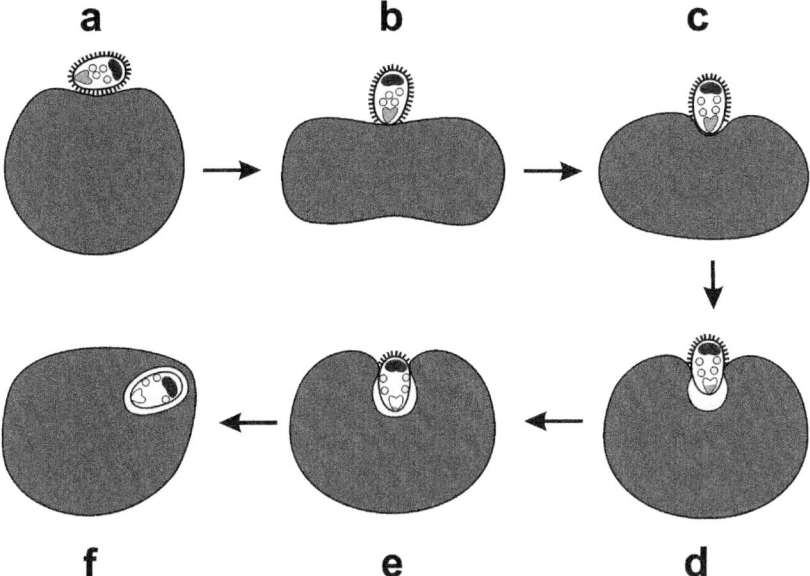

Fig. 1a–f Merozoite invasion of erythrocytes. The diagram depicts major morphological changes occur during the invasion of erythrocytes by malaria merozoites: (**a**) initial weak attachment; (**b**) reorientation; (**c**) tight junction formation; (**d, e**) secretion of apical organelle contents and invagination; (**f**) closing of the invasion pit and isolation within parasitophorous vacuole (*green*)

a distinctive tight junction between the two cells; (c) secretion of the contents of apical organelles such as the rhoptries and micronemes; (d) invagination of the erythrocyte forming an invasive pit with the tight junction moving along the surface of merozoite as it penetrates into the erythrocyte; (e) closing of the invasion pit on the erythrocyte membrane and isolation of the parasite within parasitophorous vacuole in the erythrocyte cytoplasm.

Similar lines of evidence obtained with human malaria parasite *P. falciparum*, rodent malaria parasite *P. berghei*, and avian malaria parasite *P. gallinaceum* have suggested that erythrocyte invasion involves an analogous course of events in all *Plasmodium* species (Ladda et al. 1969; Langreth et al. 1978; Bannister and Dluzewski 1990; Gratzer and Dluzewski 1993). However, the molecular requirements that warrant a successful entry of merozoites into erythrocytes are distinct amongst *Plasmodium* species (Table 1). For example, *P. vivax* exclusively invades reticulocytes, the young subpopulation that is typically present in less than 3% or 4% of total erythrocytes (Mons 1990; Galinski et al. 1992), whereas *P. falciparum* is able to invade reticulocytes as well as normocytes, the mature erythrocytes (Pasvol et al. 1980). A similar

Table 1 Host–parasite interactions during erythrocyte invasion by malaria parasites

Parasite Species	Ligand	Vertebrate Host Species	Erythrocyte receptor	Remarks
P. knowlesi	DBP	Human, simian	DARC DARC	Essential interaction for erythrocyte invasion
P. vivax	DBP	Human	DARC	Essential interaction for erythrocyte invasion
	RBP-1		Unknown,	RBP-1 and RBP-2 are involved in selective invasion
	RBP-2		unknown	into reticulocytes
P. falciparum	NBP1 (RH1)	Human	Hypothetical receptor Y	Homologous to PvRBPs, but exact function unknown in *P. falciparum*. Y is a trypsin resistant
	RBP2-Ha (NBP2a)		Unknown	and neuraminidase sensitive receptor.
	RBP2-Hb (NBP2b)		Hypothetical receptor Z	Z is a chymotrypsin sensitive, trypsin resistant, and neuraminidase resistant receptor
	RH3 (NBP3)		Unknown	
	RH4 (NBP4)		Unknown	
	EBA-175		GPA	The DBL domain of EBA-175 interacts predominantly with the sialic acid residues of GPA. Evidence also suggests that GPA peptide backbone participates in the interaction
	EBA-140 (EBP2 or BAEBL)		GPC	BAEBL (presumably through the DBL domain) interacts with GPC by a sialic acid-dependent mechanism
	BBA-181 (JESEBL)		Hypothetical receptor E	E is a chymotrypsin sensitive, trypsin resistant, and neuraminidase sensitive receptor
	Unknown		Hypothetical receptor X	X is a trypsin sensitive and neuraminidase resistant receptor

Table 1 (continued)

Parasite Species	Ligand	Vertebrate Host Species	Erythrocyte receptor	Remarks
	MSP1$_{42}$ MSP9 (ABRA)		Band 3 Band 3	The interaction is sialic acid-independent MSP1$_{42}$ and MSP9 form a co-ligand complex binding to band 3
P. yoelii	Unknown	Murine	DARC	DARC is essential for mature erythrocyte invasion, but nonessential for reticulocyte invasion by the non-lethal strain of *P. yoelii* 17X
	Py235		Unknown	Py235 is homologous to PvRBP, but its function is unclear

ABRA, Acidic basic repeat antigen; DARC, Duffy antigen/receptor chemokines; DBP, Duffy-binding protein; EBA, erythrocyte binding antigen; MSP1$_{42}$, 42-kDa processing product of merozoite surface protein 1 (MSP1); MSP9, merozoite surface protein 9; NBP, normocyte-binding protein; RBP, reticulocyte-binding protein; RH, reticulocyte-binding protein (RBP) homolog.

type of host cell selection during invasion has been observed among rodent malaria parasites. In laboratory mice, *P. berghei* (lethal) and *P. yoelii* 17X (nonlethal) preferentially invade reticulocytes, whereas, some *P. yoelii* lines such as 17XL (lethal) and YM (lethal) are capable of invading both normocytes and reticulocytes (Jarra and Brown 1989; Shear 1993; Hood et al. 1996). *P. vinckei* (lethal) has no absolute preference (Zuckerman 1958). *P. chabaudi* (nonlethal) invades predominantly normocytes, but will also invade reticulocytes when the reticulocyte level is elevated in mice under stress conditions (Carter and Walliker 1975). In fact, a reticulocyte-binding protein (RBP) complex (PvRBP-1 and Pv-RBP-2) localized to the rhoptries in *P. vivax* has been suggested to be involved in reticulocyte selection during invasion (Galinski et al. 1992; Cantor et al. 2001). Currently, a number of RBP homologs have been identified in *P. yoelii* (Py235) (Ogun and Holder 1996; Ogun et al. 2000; Khan et al. 2001) and *P. falciparum* (PfNBP1, PfRBP2-Ha, PfRBP2-Hb, PfRH3, PfRH4) (Rayner et al. 2000, 2001; Taylor et al. 2001; Triglia et al. 2001; Kaneko et al. 2002), although ambiguity remains about their in vivo function. The erythrocyte receptor for any of these proteins is not yet known. A recent study has suggested that Py235 is also involved in the sporozoite invasion of hepatocytes in *P. yoelii* (Preiser et al. 2002). In *P. falciparum*, mounting evidence now suggests that there is strain variability in the field (Okoyeh et al. 1999; Baum et al. 2003) as well as in the culture (Hadley et al. 1987; Dolan et al. 1990; Binks and Conway 1999; Mayer et al. 2002; Duraisingh et al. 2003; Mayer et al. 2004) that distinguishes the parasite invasion phenotype on the basis of specific ligand–receptor requirements for invading erythrocytes.

Persistent interactions between the malaria parasite and human host over the course of human history have selected erythrocyte polymorphisms that have provided some protection against malaria infection (Nagel and Roth 1989; Fortin et al. 2002; Zimmerman et al. 2003). In particular, a number of unique genetic mutations of erythrocyte surface proteins such as Duffy antigen, glycophorins, and band 3 are known to resist erythrocyte invasion by malaria parasites. Indeed, the Duffy antigen, glycophorins, and band 3 along with a number of other hypothetical antigens have been implicated as erythrocyte receptors for merozoite invasion. Interactions of these host receptors with the parasite counterparts (ligands) may mediate the attachment of merozoites on the erythrocyte surface or activate a transmembrane signaling mechanism that could regulate the secretion of apical organelle contents, the movement of merozoite into the erythrocyte invagination, and the development of the intracellular parasite upon internalization.

4
Duffy Blood Group Antigen

Over 95% of West Africans (Bray 1958) and about 70% of African–Americans (Young et al. 1955) are resistant to erythrocytic infection by *P. vivax*, whereas they are completely vulnerable to blood-stage infection by other species of human malaria parasites. Seminal work by Miller and coworkers (Miller et al. 1975, 1976; Mason et al. 1977) has led to the discovery of Duffy blood group antigen or Duffy antigen/receptor for chemokines (DARC) as the first host receptor identified for the invasion of malaria parasites into erythrocytes. *P. vivax* and *P. knowlesi*, the latter a simian malaria species that can infect human cells, require DARC for invading human erythrocytes, as the Duffy-negative Fy(a−b−) erythrocyte phenotype is refractory to invasion by either of the two *Plasmodium* species. An ultrastructural study using *P. knowlesi* merozoites treated with cytochalasin B, which allows merozoites to attach to erythrocytes to form the tight junction but prohibits merozoites from penetrating into erythrocytes, has shown that the Duffy-negative erythrocytes and the attached merozoites are unable to form the tight junction necessary for invasion to proceed to completion (Miller et al. 1979).

Human DARC is a 36-kDa acidic glycoprotein (336 amino acids) predicted to have seven α-helical transmembrane domains with an extracellular N-terminal domain of 60 amino acids having two N-glycosylation sites at residues 17 and 28 (Chaudhuri et al. 1989, 1993). The function of DARC in erythrocytes and other cells is not well understood, although it has been reported as a scavenger for interleukin-8 (Darbonne et al. 1991; Horuk et al. 1993). The DARC protein appears to be nonessential for erythrocyte survival, as individuals with the Duffy-negative Fy(a−b−) phenotype are healthy (Buchanan et al. 1976; Mallinson et al. 1995). Similarly, Duffy gene knockout in a murine model showed no apparent phenotype suggesting DARC is functionally a redundant protein (Luo et al. 2000).

The malaria parasites' (both *P. vivax* and *P. knowlesi*) binding domain has been identified to 35 residues in the N terminus between Ala8 and Asp43 of DARC (Chitnis et al. 1996). In both *P. vivax* and *P. knowlesi*, the Duffy-binding protein was identified as the parasite ligand interacting with DARC (Haynes et al. 1988; Miller et al. 1988; Wertheimer and Barnwell 1989). The erythrocyte-binding domain is located within the N-terminal cysteine-rich region (region II) of the Duffy-binding protein (Chitnis and Miller 1994; Ranjan and Chitnis 1999).

The Duffy-negative Fy(a−b−) phenotype, defined by a homozygous *FYB* allele, is prevalent among individuals living in West Africa where *P. vivax* has disappeared (Sanger et al. 1955), but is uncommon in Southeast Asia where *P. vivax* is endemic (Chandanayingyong et al. 1979; Breguet et al. 1982).

Most Duffy-negative Fy(a−b−) individuals in West Africa carry a silent *FY*B* allele as a result of a single nucleotide polymorphism (T46C) in the GATA box of the DARC promoter region (Tournamille et al. 1995). This mutation impairs the promoter activity specifically in erythroid cells, which supports the hypothesis that Duffy-negative phenotype prevalent in those of West African descent is restricted to erythrocytes as an adaptive response to resist malaria. Interestingly, the same point mutation identified in a population from Papua New Guinea has been linked to a new *FY*A* allele (Zimmerman et al. 1999). Individuals with this new allele are heterozygous ($FY*A/FY*A^{null}$) and have 50% less DARC on their erythrocyte surface. Although it is uncertain whether the heterozygous individuals are less susceptible to *P. vivax* infection, a significant reduction in the binding of *P. vivax* Duffy-binding protein to erythrocytes has been noted (Michon et al. 2001). It has been reported that the Duffy-negative Fy(a−b−) phenotype found in Southeast Asia individuals is linked to *FYA/FYA* and *FYA/FYB* alleles based on gene typing by polymerase chain reaction–restriction fragment length polymorphism analysis (Shimizu et al. 2000). Although the sample size was small, the results show a discrepancy between the genotype predicted by phenotypes and that deduced by DNA analysis, suggesting a possible existence of cryptic *FYA* and *FYB* alleles in the Fy(a−b−) individuals studied (Shimizu et al. 2000). It is not known whether this particular Fy(a−b−) phenotype among southeast Asians is resistant to *P. vivax* infection.

A recent study using the DARC knockout mouse model has shown that DARC is an essential host receptor for invading normocytes by a murine malaria parasite line *P. yoelii* 17X (Swardson-Olver et al. 2002). The same study, however, has shown that DARC does not play any role during infection of *P. yoelii* 17X in reticulocytes (Swardson-Olver et al. 2002). In *P. vivax*, there is apparently another interaction that is required to select reticulocytes as the primary invasion site in the blood-stage infection. This interaction presumably involves the RBP complex identified in the parasite (Galinski et al. 1992) and an unidentified receptor(s) on the reticulocytes.

5
Glycophorins

Earlier investigations on the mechanism of *P. falciparum* infection suggested that there are two distinct erythrocyte invasion pathways involving either a sialic acid-dependent (SAD) or sialic acid-independent (SAID) mechanism (Mitchell et al. 1986; Dolan et al. 1990). Evidence for the existence of the SAD invasion pathway comes from the studies where neuraminidase-treated erythrocytes lacking sialic acid (*N*-acetylneuraminic acid) residues on the cell surface were found to be significantly resistant to invasion by certain *P. falciparum* strains such as FCR-3, Dd2, W2-mef, Camp, and FVO (Mitchell et al. 1986; Dolan et al. 1990; Soubes et al. 1997). In contrast, there are *P. falciparum* strains such as 3D7, 7G8, FCB-1, HB3, and Thai-2 that show no significant difference in their capacity to invade sialic acid-depleted and wild-type erythrocytes, thus implying the existence of SAID invasion pathway (Mitchell et al. 1986; Dolan et al. 1990; Binks and Conway 1999).

Erythrocyte glycophorins have long been considered as potential host receptors in *P. falciparum* merozoite invasion of erythrocytes (Pasvol 1984; Perkins 1984; Pasvol et al. 1989). Indeed, in the SAD invasion pathway, the sialic acid moiety on erythrocyte glycophorin A (GPA) was first identified as the invasion receptor binding the *P. falciparum* merozoite ligand EBA-175 (175-kDa erythrocyte-binding antigen) (Camus and Hadley 1985; Orlandi et al. 1992; Sim et al. 1994). O-linked tetrasaccharides in the extracellular region of GPA serve as the binding site for EBA-175 (Orlandi et al. 1992; Sim et al. 1994). The GPA peptide backbone, presumably providing a unique conformation of the sialic acid residues, also appears to be important for the binding of EBA-175 to GPA (Sim et al. 1994). EBA-175, a microneme protein, belongs to a family of *Plasmodium ebl* genes that encode erythrocyte-binding ligands characterized by a 5' cysteine-rich motif called Duffy binding-like (DBL) domain or region II and a second cysteine motif at the 3' end (Adams et al. 2001). Like the *P. vivax* Duffy-binding protein, the C terminus (F2 domain) of the N-terminal cysteine-rich region (region II) of *P. falciparum* EBA-175 was identified as the GPA-binding site (Sim et al. 1994).

P. falciparum BAEBL (also known as EBP-2 and EBA-140), another member of the *ebl* family, is expressed in micronemes and has been shown to interact with erythrocyte GPC by a SAD mechanism (Mayer et al. 2001; Narum et al. 2002; Lobo et al. 2003; Maier et al. 2003). The BAEBL-binding site in GPC appears to be restricted to amino acid residues 14–22 within exon 2, at least in the case of the 3D7 strain of *P. falciparum* (Lobo et al. 2003). However, evidence suggests that host receptor specificity for BAEBL could change in different *P. falciparum* lines due to polymorphism in the DBL domain (region II) of BAEBL that interacts with erythrocytes (Mayer et al. 2002). Antibodies against the region II/F2 domain of BAEBL inhibited *P. falciparum* invasion into erythrocytes (Maier et al. 2003), underscoring the importance of the DBL domain in the SAD invasion mechanism.

Nonetheless, human En (a-) erythrocytes lacking GPA and M^kM^k erythrocytes lacking both GPA and GPB confer only partial resistance to *P. falciparum* invasion (Miller et al. 1977; Pasvol et al. 1982; Hadley et al. 1987). Several recent studies have suggested that the SAD invasion pathway principally relying on host receptors GPA, GPC, and perhaps GPB is dispensable in *P. falciparum*. When the Dd2 line exhibiting a SAD invasion phenotype was propagated in neuraminidase-treated erythrocytes where the sialic acid residues were removed from the cell surface, the parasite switched to a new Dd2/Nm line having a SAID invasion phenotype (Dolan et al. 1990). Disrupting the C-terminal region of EBA-175 in the Dd2/Nm clone by gene targeting had no effect on erythrocyte invasion (Kaneko et al. 2000). A similar truncation of EBA-175 switched the SAD W2-mef line to a SAID invasion phenotype (Reed et al. 2000). Moreover, lacking the expression of either EBA-175 (Duraisingh et al. 2003) or BAEBL (Maier et al. 2003) had little effect on erythrocyte invasion in both W2-mef (SAD) and 3D7 (SAID) strains.

Interestingly, in the *P. falciparum* EBA-175 knockout study (Duraisingh et al. 2003), the wild-type W2-mef line invaded trypsin-treated erythro-cytes (which have no intact GPA on the cell surface) as efficiently as EBA-175 knockout lines generated in both W2-mef and 3D7 strains. However, both EBA-175 knockout/W2-mef and EBA-175 knockout/3D7 lines invaded chymotrypsin-treated erythrocytes inefficiently as compared to their parent wild-type clones. The authors have subscribed to the theory that the inefficiency of invasion in the EBA-175 knockout lines is due to the loss of the GPA-EBA-175 interaction, as they believe GPA is chymotrypsin resistant. In contrast, several experiments have shown that treating intact human ery-throcytes with chymotrypsin results in the fragmentation of GPA (Dzandu et al. 1985; Roggwiller et al. 1996; Goel et al. 2003). Therefore, the erythro-cyte invasion data reported in the EBA-175 knockout study (Duraisingh et al. 2003) could be interpreted somewhat differently in that there is an alternate host receptor interacting with EBA-175, most likely by a SAID mechanism. In fact, the conclusion of the study (Duraisingh et al. 2003) that EBA-175 is functional in both SAD and SAID invasion pathways has similar implica-tions. A majority of *P. falciparum* field isolates (12 out of 15 isolates) from different regions of India were shown to use the SAID invasion pathway (Okoyeh et al. 1999). There was also a considerable use of the SAID invasion pathway (up to 40%) by *P. falciparum* isolated from patients in a Gambian study (Baum et al. 2003). Despite its importance in *P. falciparum* pathogen-esis, the SAID invasion pathway has remained elusive at the molecular level (Soubes et al. 1999).

In fact, there is mounting evidence suggesting that segments of regions II and III–V of EBA-175 interact with erythrocytes by a SAID mechanism (Sim et al. 1990, 1994; Kain et al. 1993; Jakobsen et al. 1998; Narum et al. 2000; Ro-driguez et al. 2000). Although it has been reported that a biotinylated peptide (residues 1085–1096) derived from EBA-175 bound to desialylated GPA and GPB (Jakobsen et al. 1998), in light of the EBA-175 knockout study (Durais-ingh et al. 2003), it is unlikely that the peptide backbone of GPA is involved in the SAID interaction of EBA-175 with erythrocytes. The involvement of GPB in binding EBA-175 to erythrocytes was ruled out in an earlier study

(Dolan et al. 1994), because native EBA-175 from a SAD strain (FCR3/A2) of *P. falciparum* bound to trypsin-treated GPB-deficient erythrocytes (which lack both GPB and GPA) at a level comparable to trypsin-treated wild-type erythrocytes (which lack only GPA). The *P. falciparum* merozoite protein binding to GPB, a trypsin resistant, chymotrypsin sensitive, and neuraminidase sensitive (Furthmayr 1978; Reid and Lomas-Francis 1997) sialoglycoprotein, is currently unknown (Dolan et al. 1994). Furthermore, the alternate host receptor for EBA-175 remains to be identified.

6
Band 3

The naturally occurring mutations in human genes encoding erythrocyte membrane proteins have been valuable in understanding host receptor requirements for malaria parasite invasion. For example, the successful invasion of human En (a-) and M^kM^k erythrocytes that are completely deficient in either GPA or both GPA and GPB, by a SAD *P. falciparum* strain (Camp) argues that GPA alone is not sufficient as the host receptor for parasite invasion into erythrocytes (Miller et al. 1977; Hadley et al. 1987). Further findings that erythrocytes lacking GPC and GPD (Pasvol et al. 1984; Chishti et al. 1996) are partially resistant to *P. falciparum* invasion support the notion that one or more key receptors exist on the erythrocyte membrane in addition to glycophorins. On the other hand, largely due to the paucity of homozygous mutations in the *AE1* gene that encodes erythroid band 3, the role of band 3 in the invasion process has remained ambiguous. Perhaps, an exception is southeast Asian/Melanesian ovalocytosis (SAO) where a deletion of 27 base pairs (*AE1Δ27*) in the band 3 gene that encode nine amino acids at the boundary of cytoplasmic and membrane domains has been linked to *P. falciparum* resistance (Jarolim et al. 1991). The prevalence of the *AE1Δ27* mutation has been reported in Malaysia, Philippines, Papua New Guinea, Indonesia, Mauritius, and South Africa (Jarolim et al. 1991; Tanner et al. 1991; Ravindranath et al. 1994; Takeshima et al. 1994; Mgone et al. 1996). Although the *AE1Δ27* trait is thought to be lethal when homozygous, the heterozygous mutation causes significant structural changes in both normal and mutant band 3 in SAO membranes, altering functional and hematological properties of heterozygous SAO mutant erythrocytes (Chambers et al. 1999; Kuma et al. 2002). An added complexity in the heterozygous SAO mutation is that these ovalocytes have increased membrane rigidity, leaving open the possibility that altered membrane properties of the SAO erythrocytes, rather than the band 3 receptor per se, could be the cause of resistance to malaria infection (Mohandas et al. 1992; Liu et al. 1995).

Nevertheless, there has been circumstantial evidence suggesting a possible involvement of erythroid band 3 in the malaria parasite invasion of erythrocytes. Notably: (a) a monoclonal antibody binding to rhesus erythrocytes blocked *P. knowlesi* invasion and immunoprecipitated band 3 from rhesus erythrocyte ghosts solubilized in Triton X-100 (Miller et al. 1983); (b) human erythrocyte membrane fraction enriched in band 3 and incorporated into liposomes inhibited *P. falciparum* invasion of human erythrocytes (Okoye and Bennett 1985); (c) a monoclonal antibody against the extracellular epitopes of band 3 inhibited *P. falciparum* invasion of human erythrocytes (Clough et al. 1995); (d) metabolically radiolabeled *P. falciparum* proteins associated with band 3-enriched erythrocyte membrane components (Jones and Edmundson 1991); (e) proteolytic degradation of band 3 by a serine protease during the invasion of erythrocytes by *P. falciparum* as well as rodent *P. chabaudi* has been demonstrated (McPherson et al. 1993; Roggwiller et al. 1996).

Earlier, we observed (unpublished data) that band 3 (−/−) mouse erythrocytes were completely resistant to *P. falciparum* (3D7 strain) invasion in culture when band 3 (+/−) and wild-type mouse erythrocytes obtained showed a typical invasion profile known for mouse erythrocytes (Klotz et al. 1987). Similarly, band 3 (−/−) mice were refractory to blood-stage infection by rodent malaria *P. yoelii* 17XL whereas band 3 (+/−) and wild-type mice became infected at a comparable rate (our unpublished data). These band 3 (−/−) mouse erythrocytes display a secondary loss of GPA and protein 4.2 in the erythrocyte membrane as a consequence of the targeted disruption of the erythroid band 3 (*AE1*) gene (Southgate et al. 1996; Hassoun et al. 1998). Protein 4.2 (−/−) mice (Peters et al. 1999) showed a normal course of *P. yoelii* 17XL infection similar to the wild-type (our unpublished data). These observations argue that band 3 could be a crucial host receptor either independently or in conjunction with GPA in erythrocyte invasion by *P. falciparum*. However, possible indirect effects such as apparent fragility of band 3 (−/−) erythrocytes arising from the loss of band 3 and GPA complex in the erythrocyte membrane could not be ruled out as an alternate cause and made the interpretation of our data complicated.

Using more direct molecular and biochemical approaches, we have recently shown that two nonglycosylated regions of human erythroid band 3, presumably together, function as a host invasion receptor for *P. falciparum* merozoite (3D7 strain) interacting with the parasite ligand, merozoite surface protein 1 (MSP1) (Goel et al. 2003). The band 3 peptides termed 5ABC and 6A representing two nonglycosylated regions of human erythroid band 3 (amino acids 720–761 and 807–826, respectively) blocked *P. falciparum* (3D7 strain) invasion into erythrocytes in vitro (Goel et al. 2003). A significant part of each peptide represents a putative ectoplasmic region based on recent band 3 topology models (Crandall and Sherman 1994; Popov et al. 1997; Fujinaga et al. 1999). Other parts of the band 3 ectodomains had no effect on blocking *P. falciparum* invasion in culture (Goel et al. 2003). Subsequently, it was shown that both native and recombinant forms of the 42-kDa processing product of MSP1 (MSP1$_{42}$) derived from the 3D7 strain of *P. falciparum* interacted specifically with these two regions (5ABC and 6A) of band 3 (Goel et al. 2003). Further evidence showing direct binding of MSP1$_{42}$ and band 3 is as follows (Goel et al. 2003): (a) native MSP1$_{42}$ bound to trypsin-treated erythrocytes (having unmodified band 3, but no intact GPA on the cell surface) and neuraminidase-treated erythrocytes (no sialic acid residues), but not to chymotrypsin-treated erythrocytes (having cleaved and/or truncated band 3 and GPA fragments in the erythrocyte membrane); (b) MSP1$_{42}$ bound to wild-type human and mouse erythrocytes, but not to band 3 ($-/-$) mouse erythrocytes lacking band 3 and GPA (Southgate et al. 1996; Hassoun et al. 1998) as assessed by immunofluorescence assays; (c) the binding of native MSP1$_{42}$ to erythrocytes was completely abrogated when soluble 5ABC segment of band 3 was present in the assay mixture. In fact, the conserved C-terminal secondary processing product of MSP1$_{42}$ known as MSP1$_{19}$ was the domain responsible for binding to the nonglycosylated segment of band 3 receptor on the erythrocyte surface (Goel et al. 2003). Taken together, these findings establish that band 3 is the first identified host receptor for *P. falciparum* merozoites in the SAID invasion pathway.

It was reported that the same region (amino acids 720–761 and 807–826) of band 3 also interacted specifically with a number of unidentified *P. falciparum* merozoite proteins in addition to MSP1$_{42}$ (Goel et al. 2003). Subsequent screening of a *P. falciparum* (3D7 strain) cDNA library in a yeast two-hybrid system showed that the 5ABC domain of band 3 used as bait interacts with two regions of merozoite surface protein 9 (MSP9) termed MSP9/Δ1a and MSP9/Δ2 (Li et al. 2004). Moreover, native MSP9 as well as recombinant MSP9/Δ1a and MSP9/Δ2 interacted with 5ABC as well as intact erythrocytes in solution assays. When soluble 5ABC was added to the assay mixture, the binding of MSP9 to erythrocytes was significantly decreased. Recombinant MSP9/Δ1a and MSP9/Δ2 present in the culture medium blocked erythrocyte invasion by *P. falciparum* (3D7) presumably by binding to the 5ABC domain of band 3 on the erythrocyte surface (Li et al. 2004). Earlier, it was suggested that *P. falciparum* MSP9 may interact with erythrocyte band 3 (Kushwaha et al. 2002), although the study used a mixture of erythrocyte membrane proteins solubilized with a mild non-ionic detergent as the source of band 3, which leaves open the possibility of indirect binding via proteins present in the detergent-solubilized fraction.

Furthermore, native MSP9 and MSP1$_{42}$, which interact with the band 3 receptor 5ABC, were found to exist as a stable complex in both the culture supernatant and the solubilized parasite lysate (Li et al. 2004). This interesting

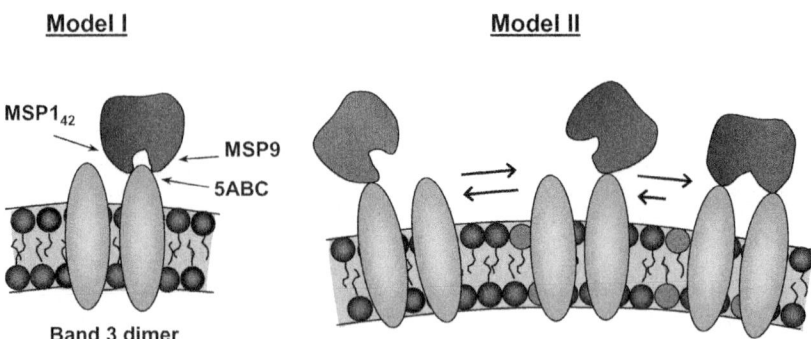

Fig. 2 The interaction of a malaria parasite co-ligand complex with host erythrocyte band 3. *Model I*: $MSP1_{42}$ and MSP9 in a single co-ligand complex interact simultaneously with the 5ABC domain of the same band 3 molecule. The co-ligand complex could also contain other unidentified parasite proteins. *Model II*: only one of the two co-ligands interacts with a single 5ABC domain of band 3. Upon equilibrium, each $MSP1_{42}$ and MSP9 in a single co-ligand complex binds to a different 5ABC domain within the band 3 dimer or tetramer in the erythrocyte membrane

finding illustrates a new concept that merozoites exploit a complex of co-ligands on their surface to interact specifically with a single erythrocyte receptor as part of the complex erythrocyte invasion mechanism. Native band 3 exists mainly as homo-dimers and tetramers in the normal erythrocyte membrane (Yu and Steck 1975; Casey and Reithmeier 1991; Vince et al. 1997; Colfen et al. 1998; Zhang et al. 2000). The existence of parasite co-ligands is postulated either to provide redundancy in the merozoite-band 3 binding process or to allow merozoites to engage into a more stable host-parasite interaction by anchoring co-ligands to adjacent 5ABC domains in the band 3 homo-dimer or tetramer on the erythrocyte surface (Li et al. 2004) (Fig. 2).

Considering well-established evidence that band 3 and GPA are in close proximity forming a complex in the erythrocyte membrane (Nigg et al. 1980; Bruce et al. 1995; Hassoun et al. 1998; Poole 2000; Auffray et al. 2001), it is interesting to speculate here that the receptor function of band 3 and GPA is mediated by a macro-molecular protein complex in the host cell membrane. This speculation is particularly interesting in view of reported cases where the *P. falciparum* invasion phenotype is switched from SAD to SAID due to clonal selection, known alterations in the host membrane proteins, and the adaptive responses due to targeted gene disruption interventions (Mitchell et al. 1986; Hadley et al. 1987; Dolan et al. 1990; Soubes et al. 1997; Kaneko et al. 2000; Reed et al. 2000; Duraisingh et al. 2003; Gilberger et al. 2003a, 2003b; Maier et al. 2003). Consistent with this view are several studies suggesting that regions II and III–V of EBA-175 might be participating in the SAID invasion pathway (Sim et al. 1990, 1994; Kain et al. 1993; Jakobsen et al. 1998; Narum et al. 2000; Rodriguez et al. 2000).

Whether the two host receptors band 3 and GPA complement each other to accommodate both SAID and SAD invasion pathways in a cooperative manner or function independently to provide two unique invasion pathways is not clear at present. It is pertinent to note that a key limitation for resolving the individual contributions of band 3 and GPA is the lack of naturally occurring and/or genetically altered erythrocytes that retain a normal complement of GPA without any detectable band 3 in the plasma membrane. Nonetheless, the available experimental data guide us to propose an invasion pathway model (Fig. 3) to entertain the following argument. In both SAID and SAD invasion pathways, band 3 could function as a crucial erythrocyte receptor for *P. falciparum* invasion. In the SAID pathway, band 3 might be an independent receptor (*yellow arrows*) or complements with GPA to take part in the SAD pathway (*orange arrow*). In the latter invasion pathway, the dominant GPA-EBA-175 interaction could be coupled with the band 3-MSP1/MSP9 interaction (*gray arrows*) at a certain stage of the invasion process or remain independent (*black arrow*). The formation of a macromolecular complex that includes band 3, MSP1/MSP9, GPA, and EBA-175 by direct or indirect interactions is an attractive theory, simply because such a protein complex could provide a mechanism allowing firm attachment between merozoites and erythrocytes, regardless of the invasion phenotype. In this cooperative invasion model, inclusion of a third component regulating the preference of first attachment to either band 3 or GPA that in turn determines the parasite invasion phenotype would need to be invoked. Hypothetically, the cytoplasmic tail of EBA-175, as reported in a recent study (Gilberger et al. 2003a), could play the role of such a molecular switch together with another component (called a putative adaptor protein in the study) interacting with the switch. In the same study (Gilberger et al. 2003a), truncation of the cytoplasmic domain of EBA-175 switched the *P. falciparum* W2-mef line from the SAD to SAID invasion phenotype. The same truncation in the 3D7 (SAID) parent line did not have an appreciable effect. More importantly, when the cytoplasmic domain of EBA-175 was substituted by the cytoplasmic domain of the sporozoite protein TRAP that had no obvious homology to EBA-175, there was no change of the invasion phenotype in the W2-mef line and the mutant parasite invaded erythrocytes at a rate comparable to the control (Gilberger et al. 2003a). Apparently, the tertiary structure of the two cytoplasmic domains provided an equivalent function in erythrocyte invasion. Based on these results, we speculate that the cytoplasmic tail of EBA-175 could be functioning as the molecular switch in the cooperative receptor invasion model (Fig. 3).

On the other hand, it remains possible that the band 3-MSP1/MSP9 interaction is responsible for the initial attachment of merozoites to erythrocytes.

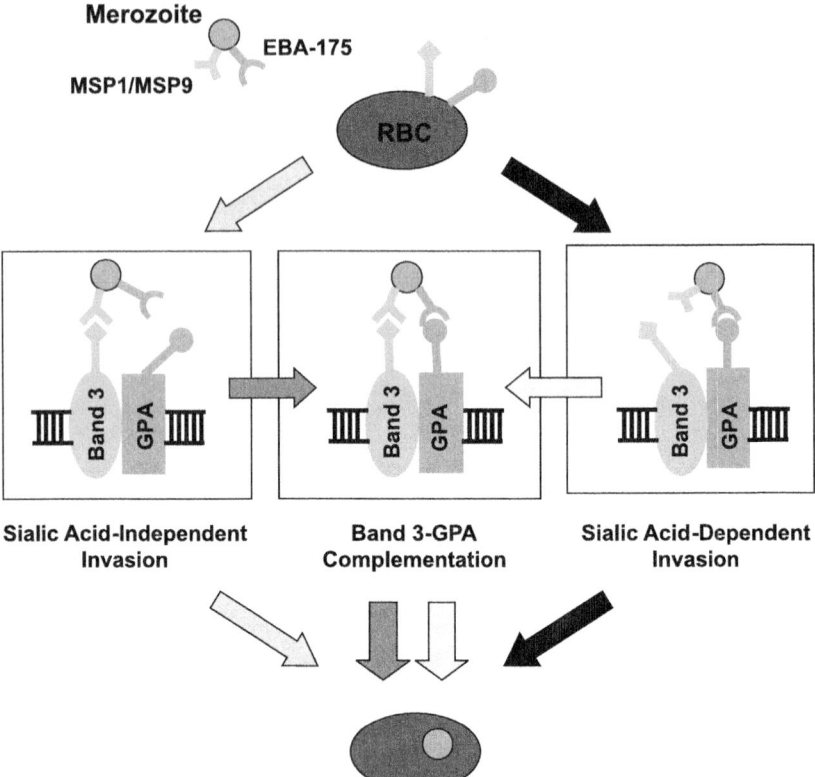

Fig. 3 Cooperative invasion receptor model. The proposed model illustrates that band 3 and GPA present as a complex in the erythrocyte membrane could be complimentary to each other in their function as a host invasion receptor for *P. falciparum* merozoites favoring either sialic acid-independent or -dependent invasion pathway

Presumably, the initial attachment is a critical event requiring both the selectivity of the interaction and efficiency of the invasion process at the cellular level. Band 3 and MSP1 especially appear to satisfy the intricate requirements of the initial attachment process at the molecular level, since they are respectively the most abundant protein on the erythrocyte and merozoite surface, highly conserved throughout the respective species especially at the presumed binding interface, and yet interacting with a remarkable specificity (Goel et al. 2003).

Since EBA-175 is localized to the micronemes, it is plausible that the GPA–EBA-175 interaction might occur about the time of formation of the irreversible tight junction between the apical end of merozoite and erythrocyte followed by secretion of the contents from apical organelles. However, the mechanism by which EBA-175 is secreted and the precise timing of the translocation of the protein during invasion process are unclear. Sialic acid moieties conjugated to GPA are the main determinants of the negative surface charge of erythrocytes (Eylar et al. 1962), and presumably prevent erythrocyte aggregation (Rogers et al. 1992). In considering the physical configuration of the erythrocyte surface, it can be argued that GPA might be involved in an earlier stage of merozoite attachment rather than in the formation of tight junction where many of erythrocyte membrane components are thought to have been extensively rearranged, altered, or removed during the course of junction formation (Bannister and Dluzewski 1990). It is noteworthy here that the apical end of merozoite is positively charged, and therefore this positive charge at the apical end might be playing an important role in the invasion of negatively charged erythrocytes by merozoites (Akaki et al. 2002). Similarly, the timing of the GPC–BAEBL interaction during invasion is unknown, and therefore its potential role in the attachment process remains uncertain.

7
Hypothetical Receptors

Several hypothetical host receptors have been proposed primarily based on their sensitivity to a set of enzymes including neuraminidase, trypsin, and chymotrypsin. The classical case is receptor X which has been defined as a host receptor resistant to neuraminidase and sensitive to trypsin treatment on the intact erythrocyte surface (Mitchell et al. 1986; Hadley et al. 1987; Dolan et al. 1994). At the molecular level, the putative receptor X may be a single molecule or a group of different host proteins since this assignment was proposed on the basis of different receptor requirements observed among several *P. falciparum* lines. The *P. falciparum* merozoite ligand interacting with the putative receptor X is also unknown. More recently, it has been suggested that the *P. falciparum* RBP homolog 1 (PfRH1; also known as PfNBP-1) binds to erythrocytes via a putative receptor Y which is a neuraminidase-sensitive and trypsin-resistant host protein (Rayner et al. 2001). Another hypothetical receptor Z, which is resistant to both neuraminidase and trypsin and sensitive to chymotrypsin, has been proposed as the erythrocyte protein binding protein for *P. falciparum* PfRH2b (also known as PfNBP-2b) (Duraisingh et al. 2003). Similarly, a hypothetical erythrocyte sialoglycoprotein termed receptor E has been proposed to interact with *P. falciparum* JESEBL (also known as EBA-181) (Gilberger et al. 2003b). Receptor E has been characterized as resistant to trypsin but sensitive to chymotrypsin and neuraminidase. However, the elimination of JESEBL expression by targeted gene disruption in the *P. falci-*

parum W2-mef strain (SAD invasion phenotype) did not affect erythrocyte invasion efficiency by this knockout line; on the other hand, the same gene knockout in the *P. falciparum* 3D7 line (SAID invasion phenotype) appeared to be deleterious (Gilberger et al. 2003a). These results appear to be somewhat inconsistent with the biochemically determined requirements of receptor E for JESEBL, making the interpretation of the data complicated.

8
Conclusions

It is now widely believed that malaria parasites, particularly *P. falciparum*, have evolved to possess a repertoire of proteins that may be used to interact with host erythrocytes during invasion. Not surprisingly, these parasite proteins are anchored by a variety of erythrocyte components including membrane proteins and oligosaccharides. To date, only a few specific host receptor–parasite ligand interactions have been definitively identified in a number of *Plasmodium* species (Table 1). Some interactions such as band 3-MSP1 and band 3-MSP9 in *P. falciparum* seem to be redundant, and others might be playing distinct roles in the invasion process. Recently, a new paradigm has emerged that merozoites exploit a specific complex of co-ligands on their surface to target a single erythrocyte receptor. Furthermore, mounting evidence now suggests that not all *P. falciparum* strains rely on the same set of receptor–ligand interactions and that many are able to switch their invasion phenotype by changing the repertoire of interactions depending upon the selective pressure of the environment. This complexity of interactions is thought to promote redundancy in the mechanism that would allow malaria parasites to invade erythrocytes successfully at any given time regardless of the host environment. Such a fail-proof strategy may be rationalized by considering several host factors that could otherwise profoundly alter the invasion process: (a) host cell polymorphisms—redundancy in the invasion mechanism would allow parasites to overcome host resistance due to erythrocyte polymorphisms; (b) host immunity—similarly, redundant invasion mechanism would allow parasites to utilize different ligands in different strains during their interaction with a host receptor; this would help to evade the host immune system by diffusing previously acquired immunity to blood-stage malaria; (c) stress of circulating conditions—redundant interactions provide efficiency required for the merozoite attachment and penetration process which is normally completed in less then 1 min under the stress of circulating conditions.

Recent advances in the identification and characterization of host invasion receptors and parasite ligands have provided new concepts in host–parasite interactions, and have undoubtedly underscored the importance and urgency of developing a multivalent malaria vaccine with a new perspective. The availability of the genome sequence of several *Plasmodium* species clearly offers new opportunities and challenges. Notwithstanding the potential impact of these studies in malaria vaccine development, future studies that would decipher *Plasmodium* invasion phenotypes at the molecular level and uncover the mechanism regulating the timing of invasion events would also be of significant general interest.

Acknowledgements This work was supported by National Institutes of Health Grants AI054532 (to S.S.O.) and HL60961 (to A.H.C.) and by Tufts University Earl P. Charlton Award (to S.S.O.). We are grateful to former and current members of our labs who have contributed to studies described herein. We thank Ms. Donna-Marie Mironchuk for the artwork.

References

Adams JH, Blair PL, Kaneko O, Peterson DS (2001) An expanding ebl family of *Plasmodium falciparum*. Trends Parasitol 17:297–299

Aikawa M, Miller LH, Johnson J, Rabbege J (1978) Erythrocyte entry by malarial parasites. A moving junction between erythrocyte and parasite. J Cell Biol 77:72–82

Akaki M, Nagayasu E, Nakano Y, Aikawa M (2002) Surface charge of *Plasmodium falciparum* merozoites as revealed by atomic force microscopy with surface potential spectroscopy. Parasitol Res 88:16–20

Auffray I, Marfatia S, de Jong K, Lee G, Huang CH, Paszty C, Tanner MJ, Mohandas N, Chasis JA (2001) Glycophorin A dimerization and band 3 interaction during erythroid membrane biogenesis: in vivo studies in human glycophorin A transgenic mice. Blood 97:2872–2878

Bannister L, Mitchell G (2003) The ins, outs and roundabouts of malaria. Trends Parasitol 19:209–213

Bannister LH, Butcher GA, Dennis ED, Mitchell GH (1975) Structure and invasive behaviour of *Plasmodium knowlesi* merozoites in vitro. Parasitology 71:483–491

Bannister LH, Dluzewski AR (1990) The ultrastructure of red cell invasion in malaria infections: a review. Blood Cells 16:257–292

Baum J, Pinder M, Conway DJ (2003) Erythrocyte invasion phenotypes of *Plasmodium falciparum* in The Gambia. Infect Immun 71:1856–1863

Bergman LW, Kaiser K, Fujioka H, Coppens I, Daly TM, Fox S, Matuschewski K, Nussenzweig V, Kappe SH (2003) Myosin A tail domain interacting protein (MTIP) localizes to the inner membrane complex of *Plasmodium* sporozoites. J Cell Sci 116:39–49

Binks RH, Conway DJ (1999) The major allelic dimorphisms in four *Plasmodium falciparum* merozoite proteins are not associated with alternative pathways of erythrocyte invasion. Mol Biochem Parasitol 103:123–127.

Bray RS (1958) The susceptibility of Liberians to the Madagascar strain of *Plasmodium vivax*. J Parasitol 44:371–373

Breguet G, Ney R, Grimm W, Hope SL, Kirk RL, Blake NM, Narendra IB, Toha A (1982) Genetic survey of an isolated community in Bali, Indonesia. I. Blood groups, serum proteins and hepatitis B serology. Hum Hered 32:52–61

Bruce LJ, Ring SM, Anstee DJ, Reid ME, Wilkinson S, Tanner MJ (1995) Changes in the blood group Wright antigens are associated with a mutation at amino acid 658 in human erythrocyte band 3: a site of interaction between band 3 and glycophorin A under certain conditions. Blood 85:541–547.

Buchanan DI, Sinclair M, Sanger R, Gavin J, Teesdale P (1976) An Alberta Cree Indian with a rare Duffy antibody, anti-Fy 3. Vox Sang 30:114–121

Camus D, Hadley TJ (1985) A *Plasmodium falciparum* antigen that binds to host erythrocytes and merozoites. Science 230:553–556

Cantor EM, Lombo TB, Cepeda A, Espinosa AM, Barrero CA, Guzman F, Gunturiz ML, Urquiza M, Ocampo M, Patarroyo ME, Patarroyo MA (2001) *Plasmodium vivax*: functional analysis of a highly conserved PvRBP-1 protein region. Mol Biochem Parasitol 117:229–234.

Carter R, Walliker D (1975) New observations on the malaria parasites of rodents of the Central African Republic - *Plasmodium vinckei petteri subsp. nov.* and *Plasmodium chabaudi Landau, 1965.* Ann Trop Med Parasitol 69:187–196

Casey JR, Reithmeier RA (1991) Analysis of the oligomeric state of Band 3, the anion transport protein of the human erythrocyte membrane, by size exclusion high performance liquid chromatography. Oligomeric stability and origin of heterogeneity. J Biol Chem 266:15726–15737

Chambers EJ, Bloomberg GB, Ring SM, Tanner MJ (1999) Structural studies on the effects of the deletion in the red cell anion exchanger (band 3, AE1) associated with South East Asian ovalocytosis. J Mol Biol 285:1289–1307

Chandanayingyong D, Bejrachandra S, Metaseta P, Pongsataporn S (1979) Further study of Rh, Kell, Duffy, P, MN, Lewis and Gerbiech blood groups of the Thais. Southeast Asian J Trop Med Public Health 10:209–211

Chaudhuri A, Polyakova J, Zbrzezna V, Williams K, Gulati S, Pogo AO (1993) Cloning of glycoprotein D cDNA, which encodes the major subunit of the Duffy blood group system and the receptor for the *Plasmodium vivax* malaria parasite. Proc Natl Acad Sci U S A 90:10793–10797

Chaudhuri A, Zbrzezna V, Johnson C, Nichols M, Rubinstein P, Marsh WL, Pogo AO (1989) Purification and characterization of an erythrocyte membrane protein complex carrying Duffy blood group antigenicity. Possible receptor for *Plasmodium vivax* and *Plasmodium knowlesi* malaria parasite. J Biol Chem 264:13770–13774

Chishti AH, Palek J, Fisher D, Maalouf GJ, Liu SC (1996) Reduced invasion and growth of *Plasmodium falciparum* into elliptocytic red blood cells with a combined deficiency of protein 4.1, glycophorin C, and p55. Blood 87:3462–3469

Chitnis CE (2001) Molecular insights into receptors used by malaria parasites for erythrocyte invasion. Curr Opin Hematol 8:85–91

Chitnis CE, Chaudhuri A, Horuk R, Pogo AO, Miller LH (1996) The domain on the Duffy blood group antigen for binding *Plasmodium vivax* and *P. knowlesi* malarial parasites to erythrocytes. J Exp Med 184:1531–1536

Chitnis CE, Miller LH (1994) Identification of the erythrocyte binding domains of *Plasmodium vivax* and *Plasmodium knowlesi* proteins involved in erythrocyte invasion. J Exp Med 180:497–506

Clough B, Paulitschke M, Nash GB, Bayley PM, Anstee DJ, Wilson RJ, Pasvol G, Gratzer WB (1995) Mechanism of regulation of malarial invasion by extraerythrocytic ligands. Mol Biochem Parasitol 69:19–27

Colfen H, Boulter JM, Harding SE, Watts A (1998) Ultracentrifugation studies on the transmembrane domain of the human erythrocyte anion transporter band 3 in the detergent C12E8. Eur Biophys J 27:651–655

Cooke BM, Mohandas N, Coppel RL (2004) Malaria and the red blood cell membrane. Semin Hematol 41:173–188

Cowman AF, Baldi DL, Healer J, Mills KE, O'Donnell RA, Reed MB, Triglia T, Wickham ME, Crabb BS (2000) Functional analysis of proteins involved in *Plasmodium falciparum* merozoite invasion of red blood cells. FEBS Lett 476:84–88

Crabb BS, Cooke BM, Reeder JC, Waller RF, Caruana SR, Davern KM, Wickham ME, Brown GV, Coppel RL, Cowman AF (1997) Targeted gene disruption shows that knobs enable malaria-infected red cells to cytoadhere under physiological shear stress. Cell 89:287–296

Crandall I, Sherman IW (1994) Cytoadherence-related neoantigens on *Plasmodium falciparum* (human malaria)-infected human erythrocytes result from the exposure of normally cryptic regions of the band 3 protein. Parasitology 108:257–267

Darbonne WC, Rice GC, Mohler MA, Apple T, Hebert CA, Valente AJ, Baker JB (1991) Red blood cells are a sink for interleukin 8, a leukocyte chemotaxin. J Clin Invest 88:1362–1369

Dolan SA, Miller LH, Wellems TE (1990) Evidence for a switching mechanism in the invasion of erythrocytes by *Plasmodium falciparum*. J Clin Invest 86:618–624

Dolan SA, Proctor JL, Alling DW, Okubo Y, Wellems TE, Miller LH (1994) Glycophorin B as an EBA-175 independent *Plasmodium falciparum* receptor of human erythrocytes. Mol Biochem Parasitol 64:55–63

Duraisingh MT, Maier AG, Triglia T, Cowman AF (2003) Erythrocyte-binding antigen 175 mediates invasion in *Plasmodium falciparum* utilizing sialic acid-dependent and -independent pathways. Proc Natl Acad Sci U S A 100:4796–4801

Duraisingh MT, Triglia T, Ralph SA, Rayner JC, Barnwell JW, McFadden GI, Cowman AF (2003) Phenotypic variation of *Plasmodium falciparum* merozoite proteins directs receptor targeting for invasion of human erythrocytes. EMBO J 22:1047–1057

Dvorak JA, Miller LH, Whitehouse WC, Shiroishi T (1975) Invasion of erythrocytes by malaria merozoites. Science 187:748–750

Dzandu JK, Deh ME, Wise GE (1985) A re-examination of the effects of chymotrypsin and trypsin on the erythrocyte membrane surface topology. Biochem Biophys Res Commun 126:50–58

Eylar EH, Madoff MA, Brody OV, Oncley JL (1962) The contribution of sialic acid to the surface charge of the erythrocyte. J Biol Chem 237:1992–2000

Fortin A, Stevenson MM, Gros P (2002) Susceptibility to malaria as a complex trait: big pressure from a tiny creature. Hum Mol Genet 11:2469–2478

Fujinaga J, Tang XB, Casey JR (1999) Topology of the membrane domain of human erythrocyte anion exchange protein, AE1. J Biol Chem 274:6626–6633

Furthmayr H (1978) Glycophorins A, B, and C: a family of sialoglycoproteins. Isolation and preliminary characterization of trypsin derived peptides. J Supramol Struct 9:79–95

Galinski MR, Medina CC, Ingravallo P, Barnwell JW (1992) A reticulocyte-binding protein complex of *Plasmodium vivax* merozoites. Cell 69:1213–1226

Gallup JL, Sachs JD (2001) The economic burden of malaria. Am J Trop Med Hyg 64:85–96

Ghosh A, Edwards MJ, Jacobs-Lorena M (2000) The journey of the malaria parasite in the mosquito: hopes for the new century. Parasitol Today 16:196–201

Gilberger TW, Thompson JK, Reed MB, Good RT, Cowman AF (2003a) The cytoplasmic domain of the *Plasmodium falciparum* ligand EBA-175 is essential for invasion but not protein trafficking. J Cell Biol 162:317–327

Gilberger TW, Thompson JK, Triglia T, Good RT, Duraisingh MT, Cowman AF (2003b) A novel erythrocyte binding antigen-175 paralogue from *Plasmodium falciparum* defines a new trypsin-resistant receptor on human erythrocytes. J Biol Chem 278:14480–14486

Goel VK, Li X, Chen H, Liu SC, Chishti AH, Oh SS (2003) Band 3 is a host receptor binding merozoite surface protein 1 during the *Plasmodium falciparum* invasion of erythrocytes. Proc Natl Acad Sci U S A 100:5164–5169

Gratzer WB, Dluzewski AR (1993) The red blood cell and malaria parasite invasion. Semin Hematol 30:232–247

Hadley TJ, Klotz FW, Pasvol G, Haynes JD, McGinniss MH, Okubo Y, Miller LH (1987) Falciparum malaria parasites invade erythrocytes that lack glycophorin A and B (M^kM^k). Strain differences indicate receptor heterogeneity and two pathways for invasion. J Clin Invest 80:1190–1193

Hassoun H, Hanada T, Lutchman M, Sahr KE, Palek J, Hanspal M, Chishti AH (1998) Complete deficiency of glycophorin A in red blood cells from mice with targeted inactivation of the band 3 (AE1) gene. Blood 91:2146–2151

Haynes JD, Dalton JP, Klotz FW, McGinniss MH, Hadley TJ, Hudson DE, Miller LH (1988) Receptor-like specificity of a *Plasmodium knowlesi* malarial protein that binds to Duffy antigen ligands on erythrocytes. J Exp Med 167:1873–1881

Holding PA, Snow RW (2001) Impact of *Plasmodium falciparum* malaria on performance and learning: review of the evidence. Am J Trop Med Hyg 64:68–75

Hood AT, Fabry ME, Costantini F, Nagel RL, Shear HL (1996) Protection from lethal malaria in transgenic mice expressing sickle hemoglobin. Blood 87:1600–1603

Horuk R, Chitnis CE, Darbonne WC, Colby TJ, Rybicki A, Hadley TJ, Miller LH (1993) A receptor for the malarial parasite *Plasmodium vivax*: the erythrocyte chemokine receptor. Science 261:1182–1184

Jakobsen PH, Heegaard PM, Koch C, Wasniowska K, Lemnge MM, Jensen JB, Sim BK (1998) Identification of an erythrocyte binding peptide from the erythrocyte binding antigen, EBA-175, which blocks parasite multiplication and induces peptide-blocking antibodies. Infect Immun 66:4203–4207

Jarolim P, Palek J, Amato D, Hassan K, Sapak P, Nurse GT, Rubin HL, Zhai S, Sahr KE, Liu SC (1991) Deletion in erythrocyte band 3 gene in malaria-resistant Southeast Asian ovalocytosis. Proc Natl Acad Sci U S A 88:11022–11026

Jarra W, Brown KN (1989) Invasion of mature and immature erythrocytes of CBA/Ca mice by a cloned line of *Plasmodium chabaudi* chabaudi. Parasitology 2:157–163

Jones GL, Edmundson HM (1991) *Plasmodium falciparum* polypeptides interacting with human red cell membranes show high affinity binding to Band-3. Biochim Biophys Acta 1097:71–76

Kain KC, Orlandi PA, Haynes JD, Sim KL, Lanar DE (1993) Evidence for two-stage binding by the 175-kD erythrocyte binding antigen of *Plasmodium falciparum* . J Exp Med 178:1497–1505

Kaneko O, Fidock DA, Schwartz OM, Miller LH (2000) Disruption of the C-terminal region of EBA-175 in the Dd2/Nm clone of *Plasmodium falciparum* does not affect erythrocyte invasion. Mol Biochem Parasitol 110:135–146

Kaneko O, Mu J, Tsuboi T, Su X, Torii M (2002) Gene structure and expression of a *Plasmodium falciparum* 220-kDa protein homologous to the *Plasmodium vivax* reticulocyte binding proteins. Mol Biochem Parasitol 121:275–278

Kappe SH, Buscaglia CA, Bergman LW, Coppens I, Nussenzweig V (2004) Apicomplexan gliding motility and host cell invasion: overhauling the motor model. Trends Parasitol 20:13–16

Kappe SH, Kaiser K, Matuschewski K (2003) The *Plasmodium* sporozoite journey: a rite of passage. Trends Parasitol 19:135–143

Khan SM, Jarra W, Preiser PR (2001) The 235 kDa rhoptry protein of Plasmodium (*yoelii*) *yoelii*: function at the junction. Mol Biochem Parasitol 117:1–10

Klotz FW, Chulay JD, Daniel W, Miller LH (1987) Invasion of mouse erythrocytes by the human malaria parasite, *Plasmodium falciparum* . J Exp Med 165:1713–1718

Kuma H, Abe Y, Askin D, Bruce LJ, Hamasaki T, Tanner MJ, Hamasaki N (2002) Molecular basis and functional consequences of the dominant effects of the mutant band 3 on the structure of normal band 3 in Southeast Asian ovalocytosis. Biochemistry 41:3311–3320

Kushwaha A, Perween A, Mukund S, Majumdar S, Bhardwaj D, Chowdhury NR, Chauhan VS (2002) Amino terminus of *Plasmodium falciparum* acidic basic repeat antigen interacts with the erythrocyte membrane through band 3 protein. Mol Biochem Parasitol 122:45–54.

Ladda R, Aikawa M, Sprinz H (1969) Penetration of erythrocytes by merozoites of mammalian and avian malarial parasites. J Parasitol 55:633–644

Langreth SG, Jensen JB, Reese RT, Trager W (1978) Fine structure of human malaria in vitro. J Protozool 25:443–452

Li X, Chen H, Oo TH, Daly TM, Bergman LW, Liu SC, Chishti AH, Oh SS (2004) A co-ligand complex anchors *Plasmodium falciparum* merozoites to the erythrocyte invasion receptor band 3. J Biol Chem 279:5765–5771

Liu SC, Palek J, Yi SJ, Nichols PE, Derick LH, Chiou SS, Amato D, Corbett JD, Cho MR, Golan DE (1995) Molecular basis of altered red blood cell membrane properties in Southeast Asian ovalocytosis: role of the mutant band 3 protein in band 3 oligomerization and retention by the membrane skeleton. Blood 86:349–358

Lobo CA, Rodriguez M, Reid M, Lustigman S (2003) Glycophorin C is the receptor for the *Plasmodium falciparum* erythrocyte binding ligand PfEBP-2 (baebl). Blood 101:4628–4631

Luo H, Chaudhuri A, Zbrzezna V, He Y, Pogo AO (2000) Deletion of the murine Duffy gene (Dfy) reveals that the Duffy receptor is functionally redundant. Mol Cell Biol 20:3097–3101

Maier AG, Duraisingh MT, Reeder JC, Patel SS, Kazura JW, Zimmerman PA, Cowman AF (2003) *Plasmodium falciparum* erythrocyte invasion through glycophorin C and selection for Gerbich negativity in human populations. Nat Med 9:87–92

Mallinson G, Soo KS, Schall TJ, Pisacka M, Anstee DJ (1995) Mutations in the erythrocyte chemokine receptor (Duffy) gene: the molecular basis of the Fya/Fyb antigens and identification of a deletion in the Duffy gene of an apparently healthy individual with the Fy(a-b-) phenotype. Br J Haematol 90:823–829

Mason SJ, Miller LH, Shiroishi T, Dvorak JA, McGinniss MH (1977) The Duffy blood group determinants: their role in the susceptibility of human and animal erythrocytes to *Plasmodium knowlesi* malaria. Br J Haematol 36:327–335

Mayer DC, Kaneko O, Hudson-Taylor DE, Reid ME, Miller LH (2001) Characterization of a *Plasmodium falciparum* erythrocyte-binding protein paralogous to EBA-175. Proc Natl Acad Sci U S A 98:5222–5227

Mayer DC, Mu JB, Feng X, Su XZ, Miller LH (2002) Polymorphism in a *Plasmodium falciparum* erythrocyte-binding ligand changes its receptor specificity. J Exp Med 196:1523–1528

Mayer DC, Mu JB, Kaneko O, Duan J, Su XZ, Miller LH (2004) Polymorphism in the *Plasmodium falciparum* erythrocyte-binding ligand JESEBL/EBA-181 alters its receptor specificity. Proc Natl Acad Sci U S A 101:2518–2523

McPherson RA, Donald DR, Sawyer WH, Tilley L (1993) Proteolytic digestion of band 3 at an external site alters the erythrocyte membrane organisation and may facilitate malarial invasion. Mol Biochem Parasitol 62:233–242

Meissner M, Schluter D, Soldati D (2002) Role of Toxoplasma gondii myosin A in powering parasite gliding and host cell invasion. Science 298:837–840

Mgone CS, Koki G, Paniu MM, Kono J, Bhatia KK, Genton B, Alexander ND, Alpers MP (1996) Occurrence of the erythrocyte band 3 (AE1) gene deletion in relation to malaria endemicity in Papua New Guinea. Trans R Soc Trop Med Hyg 90:228–231

Michon P, Woolley I, Wood EM, Kastens W, Zimmerman PA, Adams JH (2001) Duffy-null promoter heterozygosity reduces DARC expression and abrogates adhesion of the *P. vivax* ligand required for blood-stage infection. FEBS Lett 495:111–114

Miller LH, Aikawa M, Johnson JG, Shiroishi T (1979) Interaction between cytochalasin B-treated malarial parasites and erythrocytes. Attachment and junction formation. J Exp Med 149:172–184

Miller LH, Haynes JD, McAuliffe FM, Shiroishi T, Durocher JR, McGinniss MH (1977) Evidence for differences in erythrocyte surface receptors for the malarial parasites, *Plasmodium falciparum* and *Plasmodium knowlesi*. J Exp Med 146:277–281

Miller LH, Hudson D, Haynes JD (1988) Identification of *Plasmodium knowlesi* erythrocyte binding proteins. Mol Biochem Parasitol 31:217–222

Miller LH, Hudson D, Rener J, Taylor D, Hadley TJ, Zilberstein D (1983) A monoclonal antibody to rhesus erythrocyte band 3 inhibits invasion by malaria (*Plasmodium knowlesi*) merozoites. J Clin Invest 72:1357–1364

Miller LH, Mason SJ, Clyde DF, McGinniss MH (1976) The resistance factor to *Plasmodium vivax* in blacks. The Duffy-blood-group genotype, FyFy. N Engl J Med 295:302–304

Miller LH, Mason SJ, Dvorak JA, McGinniss MH, Rothman IK (1975) Erythrocyte receptors for (*Plasmodium knowlesi*) malaria: Duffy blood group determinants. Science 189:561–563

Mitchell GH, Hadley TJ, McGinniss MH, Klotz FW, Miller LH (1986) Invasion of erythrocytes by *Plasmodium falciparum* malaria parasites: evidence for receptor heterogeneity and two receptors. Blood 67:1519–1521

Mohandas N, Winardi R, Knowles D, Leung A, Parra M, George E, Conboy J, Chasis J (1992) Molecular basis for membrane rigidity of hereditary ovalocytosis. A novel mechanism involving the cytoplasmic domain of band 3. J Clin Invest 89:686–692

Mons B (1990) Preferential invasion of malarial merozoites into young red blood cells. Blood Cells 16:299–312

Mota MM, Hafalla JC, Rodriguez A (2002) Migration through host cells activates *Plasmodium* sporozoites for infection. Nat Med 8:1318–1322

Mota MM, Rodriguez A (2002) Invasion of mammalian host cells by *Plasmodium* sporozoites. Bioessays 24:149–156

Nagel RL, Roth EF, Jr. (1989) Malaria and red cell genetic defects. Blood 74:1213–1221

Narum DL, Fuhrmann SR, Luu T, Sim BK (2002) A novel *Plasmodium falciparum* erythrocyte binding protein-2 (EBP2/BAEBL) involved in erythrocyte receptor binding. Mol Biochem Parasitol 119:159–168

Narum DL, Haynes JD, Fuhrmann S, Moch K, Liang H, Hoffman SL, Sim BK (2000) Antibodies against the *Plasmodium falciparum* receptor binding domain of EBA-175 block invasion pathways that do not involve sialic acids. Infect Immun 68:1964–1966.

Nigg EA, Bron C, Girardet M, Cherry RJ (1980) Band 3-glycophorin A association in erythrocyte membrane demonstrated by combining protein diffusion measurements with antibody-induced cross-linking. Biochemistry 19:1887–1893

Ogun SA, Holder AA (1996) A high molecular mass *Plasmodium yoelii* rhoptry protein binds to erythrocytes. Mol Biochem Parasitol 76:321–324

Ogun SA, Scott-Finnigan TJ, Narum DL, Holder AA (2000) *Plasmodium yoelii*: effects of red blood cell modification and antibodies on the binding characteristics of the 235-kDa rhoptry protein. Exp Parasitol 95:187–195

Oh SS, Chishti AH, Palek J, Liu SC (1997) Erythrocyte membrane alterations in *Plasmodium falciparum* malaria sequestration. Curr Opin Hematol 4:148–154

Oh SS, Voigt S, Fisher D, Yi SJ, LeRoy PJ, Derick LH, Liu SC, Chishti AH (2000) *Plasmodium falciparum* erythrocyte membrane protein 1 is anchored to the spectrin-actin junction and knob associated histidine-rich protein in the erythrocyte skeleton. Mol Biochem Parasitol 108:237–247

Okoye VC, Bennett V (1985) *Plasmodium falciparum* malaria: band 3 as a possible receptor during invasion of human erythrocytes. Science 227:169–171

Okoyeh JN, Pillai CR, Chitnis CE (1999) *Plasmodium falciparum* field isolates commonly use erythrocyte invasion pathways that are independent of sialic acid residues of glycophorin A. Infect Immun 67:5784–5791

Orlandi PA, Klotz FW, Haynes JD (1992) A malaria invasion receptor, the 175-kilodalton erythrocyte binding antigen of *Plasmodium falciparum* recognizes the terminal Neu5Ac(alpha 2–3)Gal- sequences of glycophorin A. J Cell Biol 116:901–909

Pasloske BL, Howard RJ (1994) Malaria, the red cell, and the endothelium. Annu Rev Med 45:283–295

Pasvol G (1984) Receptors on red cells for *Plasmodium falciparum* and their interaction with merozoites. Philos Trans R Soc Lond B Biol Sci 307:189–200

Pasvol G, Anstee D, Tanner MJ (1984) Glycophorin C and the invasion of red cells by *Plasmodium falciparum* . Lancet 1:907–908

Pasvol G, Chasis JA, Mohandas N, Anstee DJ, Tanner MJ, Merry AH (1989) Inhibition of malarial parasite invasion by monoclonal antibodies against glycophorin A correlates with reduction in red cell membrane deformability. Blood 74:1836–1843

Pasvol G, Wainscoat JS, Weatherall DJ (1982) Erythrocytes deficiency in glycophorin resist invasion by the malarial parasite *Plasmodium falciparum* . Nature 297:64–66

Pasvol G, Weatherall DJ, Wilson RJ (1980) The increased susceptibility of young red cells to invasion by the malarial parasite *Plasmodium falciparum* . Br J Haematol 45:285–295

Perkins ME (1984) Binding of glycophorins to *Plasmodium falciparum* merozoites. Mol Biochem Parasitol 10:67–78

Peters LL, Jindel HK, Gwynn B, Korsgren C, John KM, Lux SE, Mohandas N, Cohen CM, Cho MR, Golan DE, Brugnara C (1999) Mild spherocytosis and altered red cell ion transport in protein 4. 2-null mice. J Clin Invest 103:1527–1537

Poole J (2000) Red cell antigens on band 3 and glycophorin A. Blood Rev 14:31–43.

Popov M, Tam LY, Li J, Reithmeier RA (1997) Mapping the ends of transmembrane segments in a polytopic membrane protein. Scanning N-glycosylation mutagenesis of extracytosolic loops in the anion exchanger, band 3. J Biol Chem 272:18325–18332

Preiser P, Kaviratne M, Khan S, Bannister L, Jarra W (2000) The apical organelles of malaria merozoites: host cell selection, invasion, host immunity and immune evasion. Microbes Infect 2:1461–1477

Preiser PR, Khan S, Costa FT, Jarra W, Belnoue E, Ogun S, Holder AA, Voza T, Landau I, Snounou G, Renia L (2002) Stage-specific transcription of distinct repertoires of a multigene family during *Plasmodium* life cycle. Science 295:342–345

Ranjan A, Chitnis CE (1999) Mapping regions containing binding residues within functional domains of *Plasmodium vivax* and *Plasmodium knowlesi* erythrocyte-binding proteins. Proc Natl Acad Sci U S A 96:14067–14072

Ravindranath Y, Goyette G, Jr., Johnson RM (1994) Southeast Asian ovalocytosis in an African-American family. Blood 84:2823–2824

Rayner JC, Galinski MR, Ingravallo P, Barnwell JW (2000) Two *Plasmodium falciparum* genes express merozoite proteins that are related to *Plasmodium vivax* and *Plasmodium yoelii* adhesive proteins involved in host cell selection and invasion. Proc Natl Acad Sci U S A 97:9648–9653

Rayner JC, Vargas-Serrato E, Huber CS, Galinski MR, Barnwell JW (2001) A *Plasmodium falciparum* homologue of *Plasmodium vivax* reticulocyte binding protein (PvRBP1) defines a trypsin-resistant erythrocyte invasion pathway. J Exp Med 194:1571–1581

Reed MB, Caruana SR, Batchelor AH, Thompson JK, Crabb BS, Cowman AF (2000) Targeted disruption of an erythrocyte binding antigen in *Plasmodium falciparum* is associated with a switch toward a sialic acid-independent pathway of invasion. Proc Natl Acad Sci U S A 97:7509–7514

Reid ME, Lomas-Francis C (1997) The blood group antigen facts book. Academic Press, New York, NY.

Rodriguez LE, Urquiza M, Ocampo M, Suarez J, Curtidor H, Guzman F, Vargas LE, Trivinos M, Rosas M, Patarroyo ME (2000) *Plasmodium falciparum* EBA-175 kDa protein peptides which bind to human red blood cells. Parasitology 120:225–235.

Rogers ME, Williams DT, Niththyananthan R, Rampling MW, Heslop KE, Johnston DG (1992) Decrease in erythrocyte glycophorin sialic acid content is associated with increased erythrocyte aggregation in human diabetes. Clin Sci (Lond) 82:309–313

Roggwiller E, Betoulle ME, Blisnick T, Braun Breton C (1996) A role for erythrocyte band 3 degradation by the parasite gp76 serine protease in the formation of the parasitophorous vacuole during invasion of erythrocytes by *Plasmodium falciparum* . Mol Biochem Parasitol 82:13–24

Sachs J, Malaney P (2002) The economic and social burden of malaria. Nature 415:680–685

Sanger R, Race RR, Jack J (1955) The Duffy blood groups of New York negroes: the phenotype Fy (a-b-). Br J Haematol 1:370–374

Shear HL (1993) Transgenic and mutant animal models to study mechanisms of protection of red cell genetic defects against malaria. Experientia 49:37–42

Sherman IW, Eda S, Winograd E (2003) Cytoadherence and sequestration in *Plasmodium falciparum*: defining the ties that bind. Microbes Infect 5:897–909

Shimizu Y, Ao H, Soemantri A, Tiwawech D, Settheetham-Ishida W, Kayame OW, Kimura M, Nishioka T, Ishida T (2000) Sero- and molecular typing of Duffy blood group in Southeast Asians and Oceanians. Hum Biol 72:511–518

Sibley LD (2004) Intracellular parasite invasion strategies. Science 304:248–253

Sim BK, Carter JM, Deal CD, Holland C, Haynes JD, Gross M (1994) *Plasmodium falciparum*: further characterization of a functionally active region of the merozoite invasion ligand EBA-175. Exp Parasitol 78:259–268

Sim BK, Chitnis CE, Wasniowska K, Hadley TJ, Miller LH (1994) Receptor and ligand domains for invasion of erythrocytes by *Plasmodium falciparum*. Science 264:1941–1944

Sim BK, Orlandi PA, Haynes JD, Klotz FW, Carter JM, Camus D, Zegans ME, Chulay JD (1990) Primary structure of the 175 K *Plasmodium falciparum* erythrocyte binding antigen and identification of a peptide which elicits antibodies that inhibit malaria merozoite invasion. J Cell Biol 111:1877–1884

Sinden RE, Billingsley PF (2001) *Plasmodium* invasion of mosquito cells: hawk or dove? Trends Parasitol 17:209–212

Singh B, Kim Sung L, Matusop A, Radhakrishnan A, Shamsul SS, Cox-Singh J, Thomas A, Conway DJ (2004) A large focus of naturally acquired *Plasmodium knowlesi* infections in human beings. Lancet 363:1017–1024

Soubes SC, Reid ME, Kaneko O, Miller LH (1999) Search for the sialic acid-independent receptor on red blood cells for invasion by *Plasmodium falciparum*. Vox Sang 76:107–114

Soubes SC, Wellems TE, Miller LH (1997) *Plasmodium falciparum*: A high proportion of parasites from a population of the Dd2 strain are able to invade erythrocytes by an alternative pathway. Exp Parasitol 86:79–83

Southgate CD, Chishti AH, Mitchell B, Yi SJ, Palek J (1996) Targeted disruption of the murine erythroid band 3 gene results in spherocytosis and severe haemolytic anaemia despite a normal membrane skeleton. Nat Genet 14:227–230

Swardson-Olver CJ, Dawson TC, Burnett RC, Peiper SC, Maeda N, Avery AC (2002) *Plasmodium yoelii* uses the murine Duffy antigen receptor for chemokines as a receptor for normocyte invasion and an alternative receptor for reticulocyte invasion. Blood 99:2677–2684

Takeshima Y, Sofro AS, Suryantoro P, Narita N, Matsuo M (1994) Twenty seven nucleotide deletion within exon 11 of the erythrocyte band 3 gene in Indonesian ovalocytosis. Jpn J Hum Genet 39:181–185

Tanner MJ, Bruce L, Martin PG, Rearden DM, Jones GL (1991) Melanesian hereditary ovalocytes have a deletion in red cell band 3. Blood 78:2785–2786

Taylor HM, Triglia T, Thompson J, Sajid M, Fowler R, Wickham ME, Cowman AF, Holder AA (2001) *Plasmodium falciparum* homologue of the genes for *Plasmodium vivax* and *Plasmodium yoelii* adhesive proteins, which is transcribed but not translated. Infect Immun 69:3635–3645

Tournamille C, Colin Y, Cartron JP, Le Van Kim C (1995) Disruption of a GATA motif in the Duffy gene promoter abolishes erythroid gene expression in Duffy-negative individuals. Nat Genet 10:224–228

Triglia T, Thompson J, Caruana SR, Delorenzi M, Speed T, Cowman AF (2001) Identification of proteins from *Plasmodium falciparum* that are homologous to reticulocyte binding proteins in *Plasmodium vivax*. Infect Immun 69:1084–1092

Vince JW, Sarabia VE, Reithmeier RA (1997) Self-association of Band 3, the human erythrocyte anion exchanger, in detergent solution. Biochim Biophys Acta 1326:295–306

Voigt S, Hanspal M, LeRoy PJ, Zhao PS, Oh SS, Chishti AH, Liu SC (2000) The cytoadherence ligand *Plasmodium falciparum* erythrocyte membrane protein 1 (PfEMP1) binds to the *P. falciparum* knob-associated histidine-rich protein (KAHRP) by electrostatic interactions. Mol Biochem Parasitol 110:423–428

Waller KL, Cooke BM, Nunomura W, Mohandas N, Coppel RL (1999) Mapping the binding domains involved in the interaction between the *Plasmodium falciparum* knob-associated histidine-rich protein (KAHRP) and the cytoadherence ligand *P. falciparum* erythrocyte membrane protein 1 (PfEMP1). Journal of Biological Chemistry 274:23808–23813

Wertheimer SP, Barnwell JW (1989) *Plasmodium vivax* interaction with the human Duffy blood group glycoprotein: identification of a parasite receptor-like protein. Exp Parasitol 69:340–350

WHO (2001) The world health report 2001, Annex Table 3: Burden of disease in disability-adjusted life years (DALYs) by cause, sex and mortality stratum in WHO Regions, estimates for 2000. WHO, Geneva Switzerland

WHO (2004) Fact Sheet No. 94: Malaria. WHO, Geneva Switzerland

Wilson RJ (1990) Biochemistry of red cell invasion. Blood Cells 16:237–252; discussion 253–236

Young MD, Eyles DE, Burgess RW, Jeffery GM (1955) Experimental testing of the immunity of Negroes to *Plasmodium vivax*. J Parasitol 41:315–318

Yu J, Steck TL (1975) Isolation and characterization of band 3, the predominant polypeptide of the human erythrocyte membrane. J Biol Chem 250:9170–9175

Zhang D, Kiyatkin A, Bolin JT, Low PS (2000) Crystallographic structure and functional interpretation of the cytoplasmic domain of erythrocyte membrane band 3. Blood 96:2925–2933

Zimmerman PA, Patel SS, Maier AG, Bockarie MJ, Kazura JW (2003) Erythrocyte polymorphisms and malaria parasite invasion in Papua New Guinea. Trends Parasitol 19:250–252

Zimmerman PA, Woolley I, Masinde GL, Miller SM, McNamara DT, Hazlett F, Mgone CS, Alpers MP, Genton B, Boatin BA, Kazura JW (1999) Emergence of FY*A(null) in a *Plasmodium vivax*-endemic region of Papua New Guinea. Proc Natl Acad Sci U S A 96:13973–13977

Zuckerman A (1958) Blood loss and replacement in plasmodial infections. II. *Plasmodium vinckei* in untreated weanling and mature rats. J Infect Dis 103:205–224

CTMI (2005) 295:233–250

A Post-genomic View of the Mitochondrion in Malaria Parasites

A. B. Vaidya (✉) · M. W. Mather

Center for Molecular Parasitology, Department of Microbiology and Immunology,
Drexel University College of Medicine, 2900 Queen Lane,
Philadelphia, PA 19129, USA
av27@drexel.edu

Abstract Mitochondria in *Plasmodium* parasites have many characteristics that distinguish them from mammalian mitochondria. Selective targeting of malaria parasite mitochondrial physiology has been exploited in successful antimalarial chemotherapy. At present, our understanding of the functions served by the parasite mitochondrion is somewhat limited, but the availability of the genomic sequences makes it possible to develop a framework of possible mitochondrial functions by providing information on genes encoding mitochondrially targeted proteins. This review aims to provide an overview of mitochondrial physiology in this post-genomic era. Although in many cases direct experimental proof for their mitochondrial functions may not be available at present, descriptions of these potential mitochondrial proteins can provide a basis for experimental approaches.

Abbreviations

GFP	Green fluorescent protein
ORF	Open reading frame

1
Introduction

The mitochondrion in *Plasmodium* parasites was deemed unusual on the basis of its acristate and 'empty' appearance in electron micrographs long before any significant biochemical studies were carried out (Aikawa 1971). The discovery of the *Plasmodium* mitochondrial DNA as the smallest such genome known with a bizarre ribosomal RNA gene arrangement further bolstered this view (Feagin et al. 1992; Vaidya et al. 1989, 1993a). This discovery also gave impetus to reassess the nature of a 35-kb circular DNA molecule present in malaria parasites that was believed to be the mitochondrial DNA. This reassessment has led to an important insight that malaria parasites (as well as most apicomplexan protozoa) possess both a mitochondrial as well as a plastid genome residing in separate organelles (Wilson et al. 1996). Both the mitochondrion and the plastid in malaria parasites are unconventional but essential for parasite physiology, and are targets for antimalarial drugs. Completion of the genomic sequence for *Plasmodium falciparum* has now opened up opportunities to explore proteins encoded in the nucleus and targeted to these organelles with a hope to gain better insight into their contribution to the parasite physiology (Gardner et al. 2002). In this review we aim to provide a view of the malaria parasite mitochondrion that is informed by bioinformatic examination of the parasite genome that permits an initial framework of the organellar physiology. While this framework, as schematically shown in Fig. 1, may not be yet fully supported by experimental proofs, it does aim to formulate testable hypotheses.

2
The Mitochondrial Genome of Malaria Parasites

At 6 kb in length, the *Plasmodium* mtDNA is the smallest such genome known among eukaryotes, encoding only three proteins (cytochrome *b*, and subunits I and III of cytochrome *c* oxidase) and ribosomal RNA encoded by fragmented genes distributed on both strands in scrambled arrangements (Feagin et al. 1992; Vaidya et al. 1989, 1993a). In light of the recently completed *Cryptosporidium* genome sequence revealing an absence of mtDNA (Abrahamsen et al. 2004; Xu et al. 2004), we could argue that the Phylum Apicomplexa is in a flux with reference to its mitochondrial genome: the minimalist mtDNA present in most members of the phylum is disposable in at least one genus. Interestingly, *Cryptosporidium* appears to maintain a vestigial organelle resembling a mitochondrion, presumably for some essential metabolic processes such as assembly of iron–sulfur clusters required for many redox proteins (LaGier et al. 2003; Riordan et al. 2003; Roberts et al. 2004).

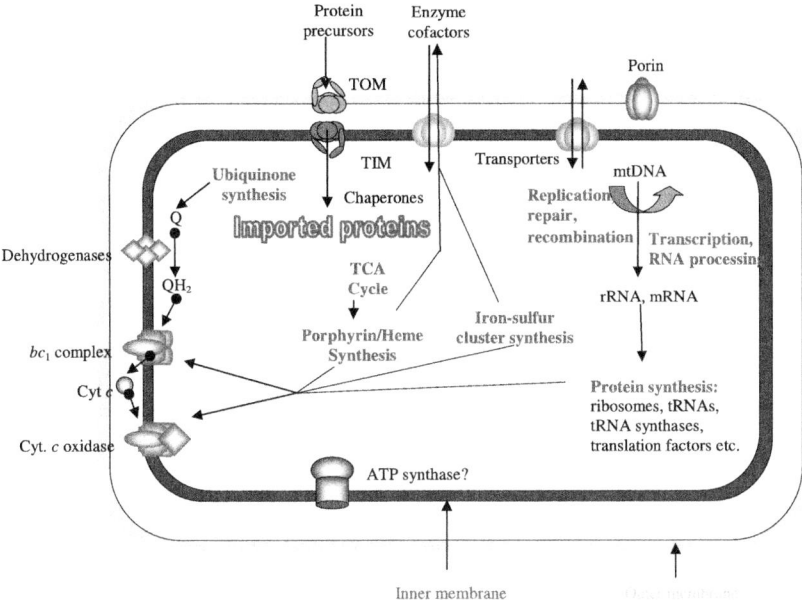

Fig. 1 A framework of mitochondrial physiology in malaria parasites. Components of the electron transport chain (described in Table 2) responsible for the generation of a proton electrochemical gradient are shown on the *left* of the figure. The mtDNA encodes only three proteins of the electron transport chain; the rest of the proteins need to be imported. Other processes relegated to the mitochondrion are shown in *red*

An unusual aspect of the malaria parasite mtDNA is its presence as multiple copies arranged in head-to-tail tandem arrays (Vaidya and Arasu 1987). This was the first example of such an arrangement for mtDNA. The number of copies appears to vary in different *Plasmodium* species, from about 30 in *P. falciparum* to about 100 in *P. yoelii*. The mtDNA does not encode any tRNA, requiring importation of the entire set of tRNA from the cytoplasm. Overall, the mtDNA is highly conserved with single nucleotide polymorphisms that can be used for establishing geographical relationships among *Plasmodium* species (Joy et al. 2003; McIntosh et al. 1998). Indeed, complete sequencing of the mtDNA from 100 *P. falciparum* isolates has provided intriguing data as to the evolutionary age of this parasite (Joy et al. 2003). The mtDNA is inherited through the female gamete during mating in the mosquito, which provides excellent markers for determining maternal lineage of progeny arising from genetic crosses between malaria parasite clones (Vaidya et al. 1993b).

3
The Challenge of Identifying Nuclearly Encoded Mitochondrial Proteins

Since the mtDNA encodes only three proteins of the mitochondrial electron transport chain, a large number of proteins encoded in the nuclear chromosomes will clearly need to be imported into parasite mitochondrion. We have made a preliminary examination of the *P. falciparum* genomic data to generate initial sets of likely and potential mitochondrial proteins (Gardner et al. 2002). Initially, we populated the list of potential mitochondrial proteins using the MitoProt II algorithm (Claros and Vincens 1996), which gives reasonably good predictions in other eukaryotes with a small number of false positives and negatives. This initial list gratifyingly included known mitochondrial proteins, such as heat shock protein 60, and many proteins that are normally mitochondrial in other organisms, such as subunits of electron transfer complexes and of the putative F-type ATP synthase. As with the *P. falciparum* genome as a whole, the number of hypothetical gene products (319) in our list outnumbered those with known or putative assignments (215). We placed the hypothetical proteins into a separate list, and culled the remainder to remove those proteins that are known or predicted to belong in other compartments. This eliminated about 48% of the remaining candidates; many of these apparent false positives were putative plastid proteins or likely cytoplasmic ribosomal proteins. To complete our working list of mitochondrial proteins, we included additional putative mitochondrial protein homologs that were in the database annotations or were reported in the literature, giving a total of 178 likely mitochondrial proteins. This list should be considered a conservative minimal catalogue of mitochondrial proteins. The list is likely to expand as proteomic data on mitochondria from other eukaryotes are used as a reference to conduct comparative genomic analysis.

Bender et al. recently developed a neural-network based program, PlasMit, specifically for predicting the mitochondrially targeted proteins of *P. falciparum* (Bender et al. 2003). They trained the neural nets using two sets of proteins that they regarded as highly likely to be targeted either to the mitochondrion or to another cellular compartment (unfortunately, there is no set of experimentally verified mitochondrial proteins in *P. falciparum* large enough to serve this purpose). Using its stringent criteria, PlasMit predicts 285 mitochondrial proteins, including hypothetical gene products. However, examination of the annotated members of this set indicates that at least 48% are probably false positives, a result similar to that given by the traditional MitoProt algorithm. From our working list of putative mitochondrial proteins generated by manual inspections informed by biochemical understanding of mitochondrial physiology in other organisms, 62% are predicted by MitoProt whereas 17% are predicted by PlasMit under stringent criteria and 65%

under nonstringent criteria. Thus, automatic predictions appear to generate unacceptable levels of false positives as well as negatives.

An experimental approach to determine mitochondrial targeting of proteins is to tag them with green fluorescent protein (GFP, or its other spectral variants) in transfection studies. The tagging is accomplished by fusing the GFP gene to the entire coding region of the protein of interest or to the putative targeting signal of the protein. Targeting of GFP fused to some of the predicted signal sequences to the malaria mitochondrion has been observed (Sato et al. 2003, 2004; Tonkin et al. 2004). The use of a strong promoter in some of these experiments, however, may compromise correct targeting as sometimes seen in other systems. Indeed, the simple act of tagging with even a small epitope has been reported to mistarget authentic mitochondrial proteins in *Saccharomyces cerevisiae* (Sickmann et al. 2003). Clearly, proteomic analysis of isolated *Plasmodium* mitochondria will be necessary to fully resolve these issues. The challenge of preparing mitochondria in acceptable purity from malaria parasites will need to be solved for this purpose.

The broad distribution of our working list of predicted mitochondrial proteins among cellular processes is shown in Fig. 2. Metabolism and translation appear to make up the bulk of the mitochondrial processes. Given the apparent lack of carbon flux through the putative tricarboxylic acid (TCA) cycle, the TCA cycle-like enzymes were assigned to anabolic, rather than catabolic, metabolism (see below). The subunits of the apparent F_1F_0 ATP synthase were placed in the ambiguous group, as the function of the ATP synthase is presently uncertain (see below). In the following sections, we will discuss several of the mitochondrial processes and pathways of interest.

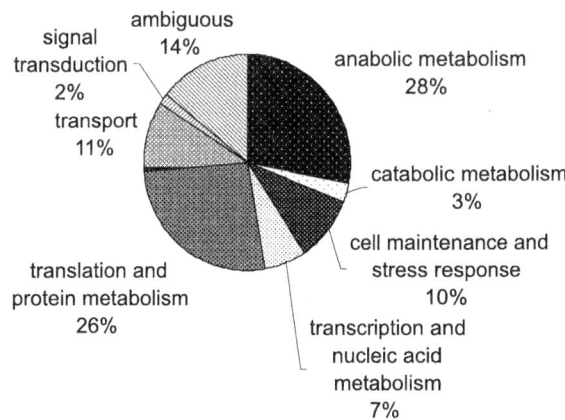

Fig. 2 Distribution among various cellular processes of 178 putative mitochondrial proteins

4
TCA Cycle Enzymes and Their Potential Roles

The TCA cycle in conjunction with oxidative phosphorylation is the central energy-yielding metabolic pathway in most aerobic organisms, and is usually localized to the mitochondrion in eukaryotes. The pathway also serves to provide intermediates to anabolic and anaplerotic pathways. Nevertheless, a number of prokaryotes and simple eukaryotes do not contain a complete TCA cycle. The existence of a conventional TCA cycle in malaria parasites has periodically been in dispute for a variety of reasons: there have been conflicting reports of the presence or absence of the activities of the various enzymes of the cycle in the blood stage forms of the parasites; ATP is derived virtually completely via glycolysis in erythrocytic stages; and the malaria genome project has revealed potential gene products corresponding to all of the enzymes required to constitute the TCA cycle (see Table 1), if expressed in the same compartment at the same time. There are also reports that some potential TCA cycle enzymes are localized in the cytoplasm rather than the mitochondrion (aconitase, Loyevsky et al. 2001; malate dehydrogenase, Lang-Unnasch 1995). The key point arguing against the presence of a canonical TCA cycle in erythrocytic stages of *P. falciparum* is the fact that essentially all of glucose consumed by the parasite is converted to lactate, depriving the TCA cycle of its key substrate, pyruvate. Furthermore, there is good evidence that the pyruvate dehydrogenase complex, which is the normal source of the input metabolite acetyl-coenzyme A for the TCA cycle, is localized to the apicoplast rather than the mitochondrion (Ralph et al. 2004). It should be remembered, however, that the TCA cycle may assume a more conventional role in other stages of the parasite life cycle, biochemical data for which are absent at this point.

Comparative and phylogenetic analyses indicate that some of the malarial enzymes are not closely related to the corresponding mitochondrial enzyme isoforms found in other eukaryotes. These include fumarate hydratase (class I) and malate-quinone oxidoreductase, which are otherwise known only in prokaryotes. The parasite malate dehydrogenase is also an isoform found predominantly in bacteria and archaea. The isocitrate dehydrogenase appears to be the $NADP^+$-dependent isoform (Chan and Sim 2003; Wrenger and Muller 2003), which in most other eukaryotes is not the isoform that participates in the TCA cycle, even though it is often located in mitochondria, its chief function being to assist in maintaining redox balance.

One hypothesis that is consistent with the apparent lack of carbon flux through the TCA cycle in erythrocytic stages is that the cycle, or portions thereof, serves primarily to generate essential metabolic intermediates, such as succinyl-CoA for the biosynthesis of heme, as alluded to in Fig. 1.

Table 1 Potential TCA-cycle enzymes in *P. falciparum*

Enzyme	PlasmoDB ID	Length (aa)	MitoP score[a]	PlasMit prediction[b]
Citrate synthase	PF10_0218	563	0.9	+/−
Aconitase/IRP	PF13_0229	909	0.95	+
Isocitrate dehydrogenase	PF13_0242	468	0.9	+
α-Ketoglutarate dehydrogenase				
E1 subunit	PF08_0045	1038	0.95	−
E2 subunit	PF13_0121	421	0.94	+/−
Succinyl-CoA synthase				
SCSa	PF11_0097	327	0.09	+
SCSb	PF14_0295	462	0.9	+/−
Succinate dehydrogenase				
SDHa	PF10_0334	631	0.91	+
SDHb	PFL0630w	321	0.13	+
Fumarate hydratase	PFI1340w	681	0.8	−
Malate dehydrogenase[c]	PFF0895w	313	0.07	−
Malate quinone oxidoreductase	PFF0815w	521	0.3	+

[a] Predicted probability that the protein is likely targeted to the mitochondrion.

[b] +, Indicates mitochondrial localization prediction under stringent criteria; +/− indicates mitochondrial localization prediction under relaxed criteria; −, indicates mitochondrial localization not predicted.

[c] Malate dehydrogenase is localized to the cytosol.

5
The Mitochondrial Electron Transport Chain

In most eukaryotes, the electron transport chain is a key part of the cell's energy transduction machinery, transferring the reducing equivalents from catabolic metabolism to oxygen in a series of energy-conserving steps that establish a transmembrane proton gradient across the mitochondrial inner membrane. The proton gradient is used to power the production of the majority of ATP in metabolically active cells, as well as to assist mitochondrial transport and other processes. In metazoans, the electron transport chain is composed of four integral membrane enzyme complexes in the mitochondrial inner membrane, NADH-ubiquinone oxidoreductase (complex I), succinate-ubiquinone oxidoreductase (complex II), ubiquinol-cytochrome c oxidodreductase (complex III), and cytochrome c oxidase (complex IV), plus ubiquinone (Coenzyme Q) and cytochrome c, which function as electron carriers completing the electron circuit between the complexes. Complexes I, III, and IV catalyze energy-conserving steps, coupling electron transfer to transmembrane proton translocation. While the mitochondrion is not a significant source of ATP biosynthesis in *Plasmodium*, the electron transport is still essential, probably for at least two reasons: (1) to provide an electron sink for ubiquinone-dependent dehydrogenases required for cellular metabolism, such as dihydroorotate dehydrogenase (Gutteridge et al. 1979), succinate dehydrogenase, and glycerol-3-phosphate dehydrogenase and (2) to energize the transmembrane proton gradient required for the transport of metabolites and proteins across the mitochondrial membranes. In addition to the dehydrogenases mentioned above, genomic data indicate the presence of a novel malate-quinone oxidoreductase, previously unknown in eukaryotes, which may replace NAD-dependent malate dehydrogenase in the mitochondrion, although its physiological importance has yet to be investigated. A comparison of mammalian and *Plasmodium* enzymes involved in electron transport chain is given in Table 2.

The *Plasmodium* electron transport chain differs from the classical chain in lacking a complex I; however, a single subunit, nonenergy-conserving NADH dehydrogenase is present, which is homologous to peripheral membrane NADH dehydrogenases found in yeast, plants, and many bacteria, but not in animals (Luttik et al. 1998; Yagi et al. 2001). *P. falciparum* may also lack an integral membrane complex II as homologues of the membrane anchor subunits of complex II are not evident in the genome, but it does express a peripheral membrane succinate dehydrogenase (Suraveratum et al. 2000; Takeo et al. 2000).

The 'business' subunits of *Plasmodium* complex III (cytochrome b, cytochrome c_1, and the Rieske iron sulfur protein) are homologous to their corresponding mammalian orthologs, and the complex is inhibited by classical complex III inhibitors, such as myxothiazol and antimycin A, which are ubiquinol/ubiquinone antagonists. As we have previously noted, there are differences in the ubiquinol ('Q_o') binding region of cytochrome b of the *Plasmodium* complex III versus the mammalian cytochrome b (Vaidya et al. 1993a). These unique structural features are the basis for the selective toxicity of the potent antimalarial drug, atovaquone (Srivastava et al. 1997, 1999). The development and mode of action of atovaquone have been reviewed elsewhere (Vaidya 1998, 2004). The mammalian complex III contains

Table 2 Comparison of mitochondrial electron transport chain enzymes in mammals and *Plasmodium*

Enzyme	Mammalian	*Plasmodium*	Comments
NADH dehydrogenase	42–44 subunits	Single subunit	The single subunit enzyme of malaria parasites does not have an ortholog in mammals; many plants, fungi and bacteria have orthologs
Succinate dehydrogenase	4 subunits	2 subunits	Membrane anchoring subunits not detected in the parasite genome
Dihydroorotate dehydrogenase	Single subunit	Single subunit	Differential susceptibility of the mammalian and parasite enzymes to inhibitors (Baldwin et al. 2002), suggesting a therapeutic window
Malate dehydrogenase	Single subunit	Single subunit	Cytosolic localization in the parasite; mitochondrial isozyme likely replaced by malate quinone oxidoreductase (see below)
Glycerol 3-phosphate dehydrogenase	Single subunit	Single subunit	Ubiquinone is the electron acceptor
Malate quinone oxidoreductase	Absent	Single subunit	First eukaryote to possess this otherwise prokaryotic enzyme. Not seen in mammals
Ubiquinone cytochrome *c* oxidoreductase	11 subunits	7 subunits	Subtle structural difference of cytochrome *b* is the likely reason for this enzyme being a validated antimalarial target
Cytochrome *c* oxidase	13 subunits	5 subunits	Appears to have significantly streamlined subunit composition. Subunit II is encoded by two genes
$F_1 F_o$ ATP synthase	9 subunits	7 subunits	Subunits a and b of the F_o segment cannot be detected in the genome. Functionality is questioned

eight additional subunits in addition to the three containing the cofactors and ubiquinone/ubiquinol binding sites, whereas the yeast complex and several plant complexes have seven such subunits. Putative homologs for subunit I (MPPβ/core 1; PFI1625c), subunit II (MPPα/core 2; PFE1155c), subunit VII

(QPc; PF10_0120), and subunit VIII (hinge protein; PF14_0248) have been identified in the Pf genome database. The *Plasmodium* complex III, then, probably has a subunit composition similar to or slightly simpler than the complexes from yeast and plants, with some of the smaller subunits undetected in the genome database due to their short length and/or relatively low degree of conservation. There appears to be only one set of genes coding for putative Complex III 'core' 1 and 2 subunits and for the mitochondrial processing peptidase α and β subunits. Thus, the processing peptidase probably serves dual functions both as a processing enzyme and as the 'core' subunits of complex III, as reported for plants and some other lower eukaryotes (Brumme et al. 1998).

Complex IV, or cytochrome *c* oxidase, is the terminal enzyme complex of the mitochondrial electron transport chain, in which the electrons are finally donated to dioxygen in a strongly exergonic reaction producing water. The complex uses part of the energy of this reaction to 'pump' up to two protons per pair of electrons across the plane of the inner membrane, helping to establish the electrochemical proton gradient. The mammalian complex consists of 13 subunits, while simpler eukaryotes generally have fewer subunits (e.g., nine in yeast, six in *Dictyostelium*). Subunits I and II contain the active centers for electron transfer as well as oxygen binding and reduction. In most organisms, they are encoded by the mtDNA, along with subunit III. The remaining smaller subunits, some of which have regulatory functions, are encoded in the nucleus. In *Plasmodium*, subunits I and III are encoded by the mtDNA, but the subunit II gene has been transferred to the nucleus and split into two parts in the process (PF13_0327, PF14_0288). Until now, this was observed only in the Chlamydomonad algae family (Perez-Martinez et al. 2001), which suggests the possibility that the cytochrome *c* oxidase subunit II originated in the mitochondrion of the algal endosymbiont, rather than in the ancestral protist mitochondrion. Only two additional putative oxidase subunits were detected in the genome (PFI1365w, PFI1375w), apparent orthologs of mammalian subunits Vb and VIb. Thus, *Plasmodium* cytochrome *c* oxidase appears to have the most streamlined subunit composition among the eukaryotes.

6
The Mystery of F_1F_0-ATP Synthase

F_1F_0-ATP synthase (Complex V) completes the process of oxidative phosphorylation by using the energy stored in the transmembrane proton gradient to drive the endergonic synthesis of ATP from ADP and inorganic phosphate. The ATP synthase/ATPase functions as a reversible protonic motor (Boyer 2001); if the proton gradient has been depleted and ATP is available, it will use energy from the hydrolysis of ATP to translocate protons from the mitochondrial matrix across the inner membrane, regenerating the gradient. An F_1F_0-ATP synthase is a component of the inner membrane of mitochondria in all cases that have been characterized, and is also found in chloroplasts and in bacteria. The F_1 sector of the complex is detachable from the membrane under relatively mild conditions and consists of α (three copies), β (three copies), γ, δ, ε, and oligomycin sensitivity conferring protein (OSCP) subunits. The F_0 subcomplex is integral to the membrane, where it acts as a proton channel consisting of three subunits: a, b, and ~12 copies of c.

The *P. falciparum* genome contains putative subunits corresponding to all of the expected ATP synthase subunits, except for the F_0 a and b subunits. Since a small portion of the *P. falciparum* genome was refractory to sequencing, it is possible that these subunits are encoded in the unsequenced DNA regions. This is unlikely, however, due to the similar inability to detect a or b subunit homology in the genome of any other *Plasmodium* species, including *P. yoelii*, which was sequenced at an average coverage of five times (Carlton et al. 2002); nor could they be detected in the large body of genomic/cDNA/EST data from a number of other *Plasmodium* species. Subunit a is essential for F_1F_0 ATP synthase function. Protons from the matrix channel through a to the c subunits, the latter forming a ring that rotates in the membrane. This rotational movement of the c subunit ring is linked to the central stalk of the F_1 portion of the enzyme, and ultimately to the synthesis of ATP at the active site in one of the β subunits. We are not aware of any report of the confirmed absence of these subunits in any species. Thus, the apparent absence of these subunits is puzzling. One possibility is that the functions of the missing subunits may be served by recruitment of novel polypeptides. It is also possible that the ATP synthase subunits may serve unconventional functions. Seeking the solution to this mystery will be a very worthwhile endeavor.

7
Replication, Transcription, and Translation of mtDNA

Except for the fragmented rRNA molecules, all of the components necessary for replication, transcription and translation of mtDNA will need to be encoded in the nucleus for transport to the mitochondrion. Based on morphological and electrophoretic properties of mtDNA, Preiser et al. proposed a rolling circle mode of replication with extensive gene conversion and recombination for the parasite mtDNA (Preiser et al. 1996). This would require participation of a large number of proteins such as those involved in the complex process of recombination, in addition to the usual DNA replication and repair enzymes. At present, little is known about these proteins in malaria parasites. The genome sequence failed to reveal the presence of DNA polymerase γ, the canonical mitochondrial DNA polymerase. However, there are two genes encoding proteins with homology to DNA polymerase I. PF14_0112, currently annotated as POM1, encodes a 2,016 amino acid open reading frame (ORF) with homology to DNA polymerase I at its C terminus, which is preceded by regions with homology to a $3' \rightarrow 5'$ exonuclease and a DNA helicase/primase. The predicted protein also has an apicoplast targeting signal at its N terminus, and the protein appears to be post-translationally processed (Michael Barrett et al. personal communication). PFF1225c encodes a1, 444-amino acid ORF with a DNA polymerase domain located at the C-terminal region. Further investigations on these gene products will be informative as to their functions and localization.

The mtDNA is transcribed in a complex manner, generating almost 20 discrete stable RNA molecules (Feagin et al. 1992; Ji et al. 1996; Suplick et al. 1990). Three of these are mRNAs encoding the proteins, whereas the rest are fragments of rRNA or apparent precursors from which these rRNA fragments are generated. An RNA polymerase with similarity to bacteriophage T3/T7 RNA polymerase has been identified, encoded by the PF11_0264 gene (Li et al. 2001). This protein of 1,531 amino acids is likely to be the polymerase that transcribes the parasite mtDNA. Other components of the mitochondrial transcription machinery have not yet been identified.

The mitochondrial ribosomes in malaria parasites have a highly unusual organization: multiple rRNA fragments will need to interact in trans to form the core small and large subunit rRNA structures, which will in turn interact with imported ribosomal proteins to form functional ribosomes (Feagin et al. 1992; Vaidya et al. 1993a). At present, little is known about the components and the process underlying the assembly of these bizarre ribosomes. Through a bioinformatics approach, a number of nuclearly encoded putative mitochondrial ribosomal proteins have been identified, and the N-terminal signal of one has been experimentally shown to target a fused GFP to the mitochondrion (Perrault and Vaidya, unpublished data). The parasite mtDNA does not encode any tRNAs or tRNA synthetases, all of which will need to be imported. Genomic sequences do not reveal any tRNA genes other than those for cytoplasmic protein synthesis, thus tRNAs will require dual targeting in malaria parasites. Indeed, there do not appear to be sufficient tRNA synthetase genes to independently serve the three compartments–the cytoplasm, the mitochondrion and the apicoplast–in which protein synthesis occurs in malaria parasites. Hence some of the tRNA synthetases will have dual or triple targeting. A similar situation is likely to exist for a number of different components required for protein synthesis.

8
Metabolic Functions of the Mitochondrion

Although the mitochondrion does not appear to be a source of ATP in erythro-cytic stages of malaria parasites, it does serve essential functions in several metabolic processes. A major function proposed originally by Gutteridge et al. (1979) is in de novo pyrimidine biosynthesis by serving as the electron sink for dihydroorotate dehydrogenase, which is present within the mitochon-drion and uses ubiquinone as the electron acceptor. Since malaria parasites are incapable of pyrimidine salvage and rely solely on de novo synthesis, interference with mitochondrial functions will have a deleterious effect by limiting the pyrimidine supply. Indeed, mitochondrial electron transport is a validated target for antimalarial drugs such as atovaquone (Fry and Pudney 1992; Srivastava et al. 1997).

Heme biosynthesis in many eukaryotes is initiated within the mitochon-drion, and malaria parasites encode δ-aminolevulinate synthase, the rate limiting enzyme in heme synthesis, targeted to the mitochondrion (Sato et al. 2004). Whereas in most other organisms several of the subsequent steps in heme biosynthesis occur in the cytosol, in malaria parasites these steps appear to be relegated to the apicoplast (Dhanasekaran et al. 2004; Foth and McFad-den 2003; Sato et al. 2004; Wilson 2002). Ferrochelatase and heme lyases, the last set of enzymes in hemoprotein assembly, are usually located within the mitochondrion. A recent report, however, suggests apicoplast localization of the parasite-encoded ferrochelatase (Varadharajan et al. 2004). Malaria parasites possess a-, b- and c-type cytochromes (present in the respiratory chain) and possibly also a cytochrome P450 (as inferred from the presence of a gene encoding cytochrome P450 reductase, although a conventional P450 monooxygenase cannot be detected). Thus, the mitochondrion will be critical in assembly of these proteins.

Iron-sulfur [Fe–S] clusters are components of many proteins in which they participate in a variety of different ways, such as electron transfer reactions, gene regulation and sensing of environmental signals (Beinert 2000). Malaria parasites also possess many [Fe–S] proteins. Biogenesis of [Fe–S] clusters is a complex process requiring participation of several enzymes and chap-erones, often present within the mitochondria (Lill and Kispal 2000). Indeed, many amitochondriate protozoa appear to possess vestigial mitochondrion-like organelles to which components of [Fe–S] assembly are localized (Seeber 2002). In P. falciparum, components for [Fe–S] cluster generation have been identified to be targeted to the apicoplast (Ellis et al. 2001; Seeber 2002). How-ever, the genomic sequence also reveals another set of genes that may encode [Fe–S] cluster generating proteins that may be targeted to the mitochondrion. For example, elemental sulfur needed for [Fe–S] clusters is generated by cysteine desufurase; the *P. falciparum* genome has one gene that may target the protein to the apicoplast (PF07_0068) and another that may be specific for the mitochondrion (MAL7P1.150), both present on chromosome 7.

Close collaboration between the mitochondrion, apicoplast and the cytoplasm would clearly require the activity of a number of transporters for the exchange of metabolites and precursors. Since almost all of the proteins necessary for mitochondrial functions are imported from the cytoplasm, protein transporters located within the outer and inner mitochondrial membranes (TOMs and TIMs) will clearly be required, and the genome reveals several candidate components of these complexes. In other eukaryotes, mitochondria play a significant role in calcium homeostasis. And one can envision a similar role for the organelle in malaria parasites as well. Appropriate import of other cations, such as Cu and Fe, as well as their appropriate assembly into metalloproteins, are also likely to be critical features of the mitochondrial physiology in malaria parasites. At present, the identity of proteins that carry out metal transport in mitochondria is not clear.

9
Perspectives

The mitochondrion of malaria parasites clearly has unconventional features. The genome has revealed sequences that are likely to be critical for mitochondrial contributions to the parasite physiology. That the mitochondrial physiology is a validated target for antimalarial chemotherapy has been amply proven by the action of atovaquone. The novel features of the mitochondrially targeted sequences need closer examination, which may provide leads for development of antimalarial compounds as well as a better understanding of a minimalist mitochondrion.

Acknowledgements We thank our colleagues for helpful discussions. This work was supported by a grant (AI028398) from the National Institutes of Health.

References

Abrahamsen MS, Templeton TJ, Enomoto S, Abrahante JE, Zhu G, Lancto CA, Deng M, Liu C, Widmer G, Tzipori S, Buck GA, Xu P, Bankier AT, Dear PH, Konfortov BA, Spriggs HF, Iyer L, Anantharaman V, Aravind L, Kapur V (2004) Complete genome sequence of the apicomplexan, *Cryptosporidium parvum*. Science 304:441–445
Aikawa M (1971) Parasitological review. *Plasmodium*: the fine structure of malarial parasites. Exp Parasitol 30:284–320

Baldwin J, Farajallah AM, Malmquist NA, Rathod PK, Phillips MA (2002) Malarial dihydroorotate dehydrogenase. Substrate and inhibitor specificity. J Biol Chem 277:41827–41834

Beinert H (2000) Iron-sulfur proteins: ancient structures, still full of surprises. J Biol Inorg Chem 5:2–15

Bender A, van Dooren GG, Ralph SA, McFadden GI, Schneider G (2003) Properties and prediction of mitochondrial transit peptides from *Plasmodium falciparum*. Mol Biochem Parasitol 132:59–66

Boyer PD (2001) New insights into one of nature's remarkable catalysts, the ATP synthase. Mol Cell 8:246–247

Brumme S, Kruft V, Schmitz UK, Braun HP (1998) New insights into the co-evolution of cytochrome c reductase and the mitochondrial processing peptidase. J Biol Chem 273:13143–13149

Carlton JM, Angiuoli SV, Suh BB, Kooij TW, Pertea M, Silva JC, Ermolaeva MD, Allen JE, Selengut JD, Koo HL, Peterson JD, Pop M, Kosack DS, Shumway MF, Bidwell SL, Shallom SJ, van Aken SE, Riedmuller SB, Feldblyum TV, Cho JK, Quackenbush J, Sedegah M, Shoaibi A, Cummings LM, Florens L, Yates JR, Raine JD, Sinden RE, Harris MA, Cunningham DA, Preiser PR, Bergman LW, Vaidya AB, van Lin LH, Janse CJ, Waters AP, Smith HO, White OR, Salzberg SL, Venter JC, Fraser CM, Hoffman SL, Gardner MJ, Carucci DJ (2002) Genome sequence and comparative analysis of the model rodent malaria parasite Plasmodium *yoelii yoelii*. Nature 419:512–519

Chan M, Sim TS (2003) Recombinant *Plasmodium falciparum* NADP-dependent isocitrate dehydrogenase is active and harbours a unique 26 amino acid tail. Exp Parasitol 103:120–126

Claros MG, Vincens P (1996) Computational method to predict mitochondrially imported proteins and their targeting sequences. Eur J Biochem 241:779–786

Dhanasekaran S, Chandra, NR, Chandrasekhar Sagar, BK, Rangarajan PN, Padmanaban G (2004) Delta-aminolevulinic acid dehydratase from *Plasmodium falciparum*: indigenous versus imported. J Biol Chem 279:6934–6942

Ellis KE, Clough B, Saldanha JW, Wilson RJ (2001) Nifs and Sufs in malaria. Mol Microbiol 41:973–981

Feagin JE, Werner E, Gardner MJ, Williamson DH, Wilson RJ (1992) Homologies between the contiguous and fragmented rRNAs of the two *Plasmodium falciparum* extrachromosomal DNAs are limited to core sequences. Nucleic Acids Res 20:879–887

Foth BJ, McFadden GI (2003) The apicoplast: a plastid in *Plasmodium falciparum* and other Apicomplexan parasites. Int Rev Cytol 224:57–110

Fry M, Pudney M (1992) Site of action of the antimalarial hydroxynaphthoquinone, 2-[trans-4-(4'-chlorophenyl) cyclohexyl]-3-hydroxy-1,4-naphthoquinone (566C80). Biochem Pharmacol 43:1545–1553

Gardner MJ, Hall N, Fung E, White O, Berriman M, Hyman RW, Carlton JM, Pain A, Nelson KE, Bowman S, Paulsen IT, James K, Eisen JA, Rutherford K, Salzberg SL, Craig A, Kyes S, Chan MS, Nene V, Shallom SJ, Suh B, Peterson J, Angiuoli S, Pertea M, Allen J, Selengut J, Haft D, Mather MW, Vaidya AB, Martin DM, Fairlamb AH, Fraunholz MJ, Roos DS, Ralph SA, McFadden GI, Cummings LM, Subramanian GM, Mungall C, Venter JC, Carucci DJ, Hoffman SL, Newbold C, Davis RW, Fraser CM, Barrell B (2002) Genome sequence of the human malaria parasite *Plasmodium falciparum*. Nature 419:498–511

Gutteridge WE, Dave D, Richards WH (1979) Conversion of dihydroorotate to orotate in parasitic protozoa. Biochim Biophys Acta 582:390–401

Ji YE, Mericle BL, Rehkopf DH, Anderson JD, Feagin JE (1996) The *Plasmodium falciparum* 6 kb element is polycistronically transcribed. Mol Biochem Parasitol 81:211–223

Joy DA, Feng X, Mu J, Furuya T, Chotivanich K, Krettli AU, Ho M, Wang A, White NJ, Suh E, Beerli P, Su XZ (2003) Early origin and recent expansion of *Plasmodium falciparum*. Science 300:318–321

LaGier MJ, Tachezy J, Stejskal F, Kutisova K, Keithly JS (2003) Mitochondrial-type iron-sulfur cluster biosynthesis genes (IscS and IscU) in the apicomplexan *Cryptosporidium parvum*. Microbiology 149: 3519–3530

Lang-Unnasch N (1995) *Plasmodium falciparum*: antiserum to malate dehydrogenase. Exp Parasitol 80:357–359

Li J, Maga JA, Cermakian N, Cedergren R, Feagin JE (2001) Identification and characterization of a *Plasmodium falciparum* RNA polymerase gene with similarity to mitochondrial RNA polymerases. Mol Biochem Parasitol 113:261–269

Lill R, Kispal G (2000) Maturation of cellular Fe-S proteins: an essential function of mitochondria. Trends Biochem Sci 25:352–356

Loyevsky M, LaVaute T, Allerson CR, Stearman R, Kassim OO, Cooperman S, Gordeuk VR, Rouault TA (2001) An IRP-like protein from *Plasmodium falciparum* binds to a mammalian iron-responsive element. Blood 98:2555–2562

Luttik MA, Overkamp KM, Kotter P, de Vries S, van Dijken JP, Pronk JT (1998) The *Saccharomyces cerevisiae* NDE1 and NDE2 genes encode separate mitochondrial NADH dehydrogenases catalyzing the oxidation of cytosolic NADH. J Biol Chem 273:24529–24534

McIntosh MT, Srivastava R, Vaidya AB (1998) Divergent evolutionary constraints on mitochondrial and nuclear genomes of malaria parasites. Mol Biochem Parasitol 95:69–80

Perez-Martinez X, Antaramian A, Vazquez-Acevedo M, Funes S, Tolkunova E, d'Alayer J, Claros MG, Davidson E, King MP, Gonzalez-Halphen D (2001) Subunit II of cytochrome c oxidase in Chlamydomonad algae is a heterodimer encoded by two independent nuclear genes. J Biol Chem 276:11302–11309

Preiser PR, Wilson RJ, Moore PW, McCready S, Hajibagheri MA, Blight KJ, Strath M, Williamson DH (1996) Recombination associated with replication of malarial mitochondrial DNA. EMBO J 15: 684–693

Ralph SA, Van Dooren GG, Waller RF, Crawford MJ, Fraunholz MJ, Foth BJ, Tonkin CJ, Roos DS, McFadden GI (2004) Tropical infectious diseases: Metabolic maps and functions of the *Plasmodium falciparum* apicoplast. Nat Rev Microbiol 2:203–216

Riordan CE, Ault JG, Langreth SG, Keithly JS (2003) *Cryptosporidium parvum* Cpn60 targets a relict organelle. Curr Genet 44:138–147

Roberts CW, Roberts F, Henriquez FL, Akiyoshi D, Samuel BU, Richards TA, Milhous W, Kyle D, McIntosh L, Hill GC, Chaudhuri M, Tzipori S, McLeod R (2004) Evidence for mitochondrial-derived alternative oxidase in the apicomplexan parasite *Cryptosporidium parvum*: a potential anti-microbial agent target. Int J Parasitol 34:297–308

Sato S, Clough B, Coates L, Wilson RJ (2004) Enzymes for heme biosynthesis are found in both the mitochondrion and plastid of the malaria parasite *Plasmodium falciparum*. Protist 155:117–125

Sato S, Rangachari K, Wilson RJ (2003) Targeting GFP to the malarial mitochondrion. Mol Biochem Parasitol 130:155–158

Seeber F (2002) Biogenesis of iron-sulphur clusters in amitochondriate and apicomplexan protists. Int J Parasitol 32:1207–1217

Sickmann A, Reinders J, Wagner Y, Joppich C, Zahedi R, Meyer HE, Schonfisch B, Perschil I, Chacinska A, Guiard B, Rehling P, Pfanner N, Meisinger C (2003) The proteome of *Saccharomyces cerevisiae* mitochondria. Proc Natl Acad Sci USA 100:13207–13212

Srivastava IK, Morrisey JM, Darrouzet E, Daldal F, Vaidya AB (1999) Resistance mutations reveal the atovaquone-binding domain of cytochrome b in malaria parasites. Mol Microbiol 33:704–711

Srivastava IK, Rottenberg H, Vaidya AB (1997) Atovaquone, a broad spectrum antiparasitic drug, collapses mitochondrial membrane potential in a malarial parasite. J Biol Chem 272:3961–3966

Suplick K, Morrisey J, Vaidya AB (1990) Complex transcription from the extrachromosomal DNA encoding mitochondrial functions of *Plasmodium yoelii*. Mol Cell Biol 10:6381–6388

Suraveratum N, Krungkrai SR, Leangaramgul P, Prapunwattana P, Krungkrai J (2000) Purification and characterization of *Plasmodium falciparum* succinate dehydrogenase. Mol Biochem Parasitol 105:215–222

Takeo S, Kokaze A, Ng CS, Mizuchi D, Watanabe JI, Tanabe K, Kojima S, Kita K (2000) Succinate dehydrogenase in *Plasmodium falciparum* mitochondria: molecular characterization of the SDHA and SDHB genes for the catalytic subunits, the flavoprotein (Fp) and iron-sulfur (Ip) subunits. Mol Biochem Parasitol 107:191–205

Tonkin CJ, van Dooren GG, Spurck TP, Struck NS, Good RT, Handman E, Cowman AF, McFadden GI (2004) Localization of organellar proteins in *Plasmodium falciparum* using a novel set of transfection vectors and a new immunofluorescence fixation method. Mol Biochem Parasitol 137:13–21

Vaidya AB (1998) Mitochondrial physiology as a target for atovaquone and other antimalarials. In: Sherman IW (ed.) Malaria: Parasite biology, pathogenesis, and protection. ASM Press, Washington DC, pp 355–368

Vaidya AB (2004) Mitochondrial and plastid functions as antimalarial drug targets. Curr Drug Targets Infect Disord 4:11–23

Vaidya AB, Akella R, Suplick K (1989) Sequences similar to genes for two mitochondrial proteins and portions of ribosomal RNA in tandemly arrayed 6-kilobase-pair DNA of a malarial parasite. Mol Biochem Parasitol 35:97–107

Vaidya AB, Arasu P (1987) Tandemly arranged gene clusters of malarial parasites that are highly conserved and transcribed. Mol Biochem Parasitol 22:249–257

Vaidya AB, Lashgari MS, Pologe LG, Morrisey J (1993a) Structural features of *Plasmodium* cytochrome b that may underlie susceptibility to 8-aminoquinolines and hydroxynaphthoquinones. Mol Biochem Parasitol 58:33–42

Vaidya AB, Morrisey J, Plowe CV, Kaslow DC, Wellems TE (1993b) Unidirectional dominance of cytoplasmic inheritance in two genetic crosses of *Plasmodium falciparum*. Mol Cell Biol 13:7349–7357

Varadharajan S, Sagar BK, Rangarajan PN, Padmanaban G (2004) Localization of ferrochelatase in *Plasmodium falciparum*. Biochem J 384:429–436

Wilson RJ (2002) Progress with parasite plastids. J Mol Biol 319:257–274

Wilson RJ, Denny PW, Preiser PR, Rangachari K, Roberts K, Roy A, Whyte A, Strath M, Moore DJ, Moore PW, Williamson DH (1996) Complete gene map of the plastid-like DNA of the malaria parasite *Plasmodium falciparum*. J Mol Biol 261:155–172

Wrenger C, Muller S (2003) Isocitrate dehydrogenase of *Plasmodium falciparum*. Eur J Biochem 270:1775–1783

Xu P, Widmer G, Wang Y, Ozaki LS, Alves JM, Serrano MG, Puiu D, Manque P, Akiyoshi D, Mackey AJ, Pearson WR, Dear PH, Bankier AT, Peterson DL, Abrahamsen MS, Kapur V, Tzipori S, Buck GA (2004) The genome of *Cryptosporidium hominis*. Nature 431:1107–1112

Yagi T, Seo BB, Di Bernardo S, Nakamaru-Ogiso E, Kao MC, Matsuno-Yagi A (2001) NADH dehydrogenases: from basic science to biomedicine. J Bioenerg Biomembr 33:233–242

CTMI (2005) 295:251–273

The Plastid of *Plasmodium* spp.: A Target for Inhibitors

S. Sato (✉) · R. J. M. Wilson

National Institute for Medical Research, Mill Hill, London NW7 1AA, UK
ssato@nimr.mrc.ac.uk

Abstract Determined efforts are being made to explore the non-photosynthetic plastid organelle of *Plasmodium falciparum* as a target for drug development. Certain antibiotics that block organellar protein synthesis are already in clinical use as antimalarials. However, all the indications are that these should be used only in combination with conventional antimalarials. The use of antibiotics such as doxycycline and clindamycin may reduce the development of drug resistant parasites and such means to avoid drug resistance should be explored hand-in-hand with drug development. Genomic information predicts that fatty acid type II (FAS II) and isoprenoid biosynthetic pathways are localized to the plastid. However, clinical trials with fosmidomycin (a specific inhibitor of DOXP reductase in the non-mevalonate pathway for isoprenoids) suggest it too should only be used in drug combinations. Prospects for more potent antimalarial compounds have emerged from studies of several of the enzymes involved in the FAS II pathway. Lead antibiotics such as thiolactomycin (an inhibitor of β-ketoacyl-ACP synthase) and triclosan (a specific inhibitor of enoyl-ACP reductase) have led to structurally similar, active compounds that rapidly kill ring- and trophozoite-stage parasites. The FAS II pathway is of particular interest to the pharma-industry.

Abbreviations

ACC	Acetyl-CoA carboxylase
FA	Fatty acid
FAS	Fatty acid synthesis
KAS	β-Ketoacyl-ACP synthase
KAR	β-Ketoacyl-ACP reductase
HAD	β-Hydroxyacyl-ACP reductase
ENR	Enoyl-ACP reductase
PDF	Peptide deformylase
TLM	Thiolactomycin
TC	Triclosan
PBGS or ALAD	Porphobilinogen synthase

1
Introduction

The plastid organelle of *Plasmodium falciparum* offers a promising target for inhibitors. The organelle is essential, its biochemistry is prokaryotic, and it is evidently accessible to antimicrobial agents. These few introductory words sum up a decade or more of research—for extensive reviews see Ralph et al. 2001; Roos et al. 2002; Wilson 2002; Foth and McFadden 2003; Seeber 2003; Ralph et al. 2004. Yet we are still far from understanding the function of the organelle and a long way from finding ideal, practically useful inhibitory drugs. This article appraises the present situation and indicates some of the problems the plastid presents as a therapeutic target.

The plastid of *Plasmodium* is a semi-autonomous organelle carrying a small 35-kb genome. The genome's highly restricted coding capacity (Wilson et al. 1996) clearly indicates that imported proteins encoded in the nucleus are essential for the organelle's primary functions. A large number of these proteins have now been predicted from their genomic sequence. As with other secondary plastids, nucleus-encoded plastid proteins of *Plasmodium* begin with an N-terminal leader sequence encoding a bipartite targeting signal (Waller et al. 1998). For many proteins of unknown function, this N-terminal extension is the only means of predicting a plastidic association. The number of proteins imported from the cytosol into the *Plasmodium* plastid remains uncertain, but the predictive algorithm known as PlasmoAP (Foth et al. 2003) gave a figure of ~500 (corresponding to ~10% of the coding capacity of the genome). This number is likely to be modified as predicted leader sequences are tested, and still other proteins whose properties lie outside the parameters set for the algorithm, are discovered.

A recent global transcriptional analysis pointed to some 150 genes whose products are predicted to be imported into the organelle. The amount of mRNA from these genes reached maximal levels as the parasite developed from the late trophozoite to early schizont stage in the erythrocytic cycle (Bozdech et al. 2003). By this time the plastid has grown from a small ovoid body to a large tubular (ultimately branched) structure, resembling the mitochondrion in gross appearance but distinct from it (Fig. 1). Up-regulation of transcription of the plastid's 35-kb genome is co-ordinated with that of the plastid-associated nuclear genes, though how this is mediated is not clear.

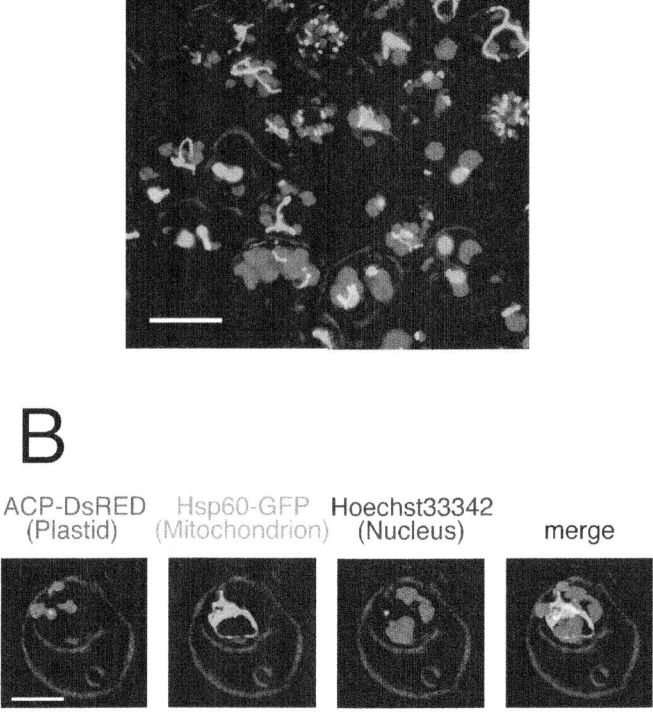

Fig. 1A,B Plastids (*green*) in living *P. falciparum* parasites transfected with the plasmid pSSPF2*Pf*ACP-GFP (Sato and Wilson 2003). Nucleus (*blue*) stained with Hoechst 33342. The parasites are in various stages of development and show plastids that grow from a small ovoid body to a large elongated tubular structure. Fluorescent image captured using a DeltaVision microscope system (Applied Precision). *Scale bar*, 10 μm. **B** The plastid (*red*) is distinct from the mitochondrion (*green*). A living *P. falciparum* schizont doubly transfected with plasmids pSSPF2*Pf*ACP-DsRed and pSSPF2*Pf*Hsp60-GFP (Sato and Wilson 2003). Nucleus (*blue*) stained with Hoechst 33342. *Scale bar*, 5 μm

Proteins predicted to be imported into the plastid range from subunits for DNA gyrase, to ribosomal proteins and enzymes for anabolic processes (see Appendix). Up to now, the import of very few of these has been verified in *P. falciparum* by transfection technology with fluorescent reporters (Waller et al. 1998), or detected by their enzymatic activity in extracts of parasites (for IspC see Wiesner et al. 2000; for ENR see Surolia and Surolia 2001). The Appendix shows that a wide range of proteins can be considered as potential targets for inhibitors. Selecting the best options is the issue that has to be addressed in the development of novel antimalarial compounds.

Although the plastid organelle has changed radically from its ancient free-living cyanobacterial state, one approach is to look for *Plasmodium* genes whose bacterial orthologues have already been identified as essential by microbiologists. This approach for the discovery of new antibacterials was recently reviewed (Miesel et al. 2003). Many of the categories of essential genes found for *Bacillus subtilis* by Kobayashi et al. (2003) are represented in the Appendix of predicted plastid proteins of *Plasmodium*. However, the number of *Plasmodium* proteins identified in most of the categories remains meagre. Only four pathways have substantial representation: protein synthesis, fatty acid (FA) biosynthesis (FAS II), isoprenoid biosynthesis (non-mevalonate pathway), and haem biosynthesis. Whilst DNA replication and RNA synthesis also undoubtedly occur, Roos et al. (2002) expressed the opinion that other major pathways are not likely to have been overlooked.

An important ancillary question is to find ways to minimize the selection of resistant parasites should inhibitors of the plastid be used extensively in the future. One strategy (though a costly one) is to introduce a cocktail of inhibitors simultaneously (White 1999). This requires that independent targets are found lying at different points in the same pathway, or that synergistic inhibitors are found that block coupled parts of the organelle's biosynthetic machinery. It is encouraging that plastidic candidates for both of these criteria have been found. Nevertheless, very little attention has been paid to the problem of drug resistance for compounds that have emerged as potential lead inhibitors of the plastid. As most of these compounds are based on antimicrobials with known activities, e.g., clindamycin, the issue is an important one; the introduction of new antimalarials that might have wide spectrum antibiotic activity has obvious dangers. In some cases, earlier experience with the use of these lead compounds as antimicrobials has already indicated how drug resistance can develop.

2
Plastid Inhibitors

Potential inhibitors of the *Plasmodium* plastid can be divided conveniently into two groups—(a) those suppressing the expression of proteins encoded on the plastid genome by affecting its replication, transcription and translation; and (b) those blocking the functions of imported proteins involved with plastidic anabolism.

2.1
Group A

2.1.1
DNA Replication

Inhibitors of DNA gyrase, such as ciprofloxacin, can specifically block DNA replication in the plastid of *P. falciparum* by selective linearization of the 35-kb circular genome (Weissig et al. 1997). However, clinical trials with norfloxacin and ciprofloxacin have shown no substantial benefit to patients with malaria in Thailand, Africa or India (Watt et al. 1991; McClean et al. 1992; Tripathi et al. 1993) and novobiocin is no longer used to treat microbial infections in humans because of side effects. The more potent cyclothialidines have problems with cellular uptake or efflux. Grepafloxacin and norfloxacin were the most effective of a range of fluoroquinolones cytotoxic for the parasite in vitro and future developments with new antibiotics of this type against hepatic stages might be of interest (Mahmoudi et al. 2003).

2.1.2
Transcription

Rifampicin, the inhibitor of multi-subunit prokaryotic RNA polymerases (RNAP) acted slowly and did not produce radical cures of patients infected with *P. vivax* (Pukrittayakamee et al. 1994). However, this antibiotic was well tolerated and has been used successfully in drug combinations (Goerg et al. 1999). Rifampicin is not recommended for use with quinine because it induces more rapid turnover to the less active 3-hydroxyquinine (Pukrittayakamee et al. 2003). We are not aware of any antimalarial work with the remarkable new small molecule inhibitors of RNAP that have become available in recent years (see Darst 2004).

2.1.3
Protein Synthesis

Minocycline reduced the level of transcripts of two mitochondrial genes as well those of the plastidic *rpoB/C* genes but not for two nucleus-encoded genes (Lin et al. 2002). This result suggests that at least some tetracycline related antibiotics can target both the mitochondrion and plastid in *P. falciparum*.

Discovery of a gene specifying the peptide elongation factor EF1A (previously known as EF-Tu) on the plastid genome of *P. falciparum* initiated early trials with a range of antibiotics that block the elongation cycle (Fig. 2). Several of these antibiotics had antimalarial activity to varying degrees (Clough and Wilson 2001), but whether they inhibited EF1A in the mitochondrion or plastid, or both, is often difficult to determine. Usually an opinion has been inferred from rather circumstantial evidence. These interpretational difficulties stem in part from the unknown structure and incomplete sequence information available for mitochondrial rRNAs of *P. falciparum* (Feagin 1997), as well as from the lack of purified organelles, and the absence of assays to detect translated organellar products.

Several crystal structures of bacterial EF1A in complex with antibiotic inhibitors have been obtained in recent years (Anderson et al. 2003). These

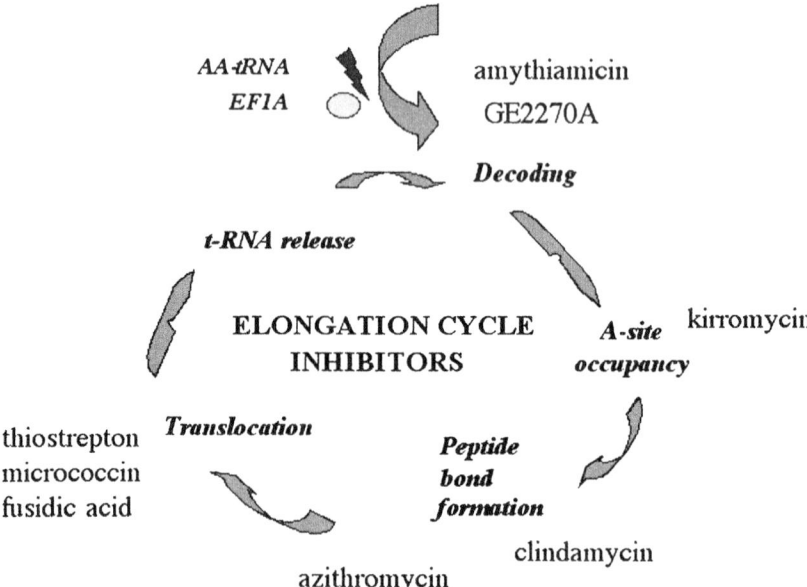

Fig. 2 Representation of the plastidic elongation cycle showing points targeted by antibiotics with known antimalarial activity

structures are being used to study details of the effects of inhibitors on factor–ribosome interactions. As indicated in Fig. 2, cyclic thiopeptides such as amythiamicin and GE2270 prevent aminoacylated tRNA forming a ternary complex with EF1A–GTP. A computer-modelled 3D structure of EF1A from *P. falciparum* suggested that the plastidic translational machinery might be affected by these same two inhibitors (Sato et al. 2000). In cultures of *P. falciparum*, the IC_{50} for amythiamicin was 10 nM (Clough et al. 1999). Unpublished studies with smaller synthetic derivatives of promothiocin A, a related thiopeptide antibiotic (Bagley et al. 2000), found that inhibition could be achieved, albeit at a lower level, without cyclization of the inhibitor (C.J. Moody and B. Clough, personal communication).

Kirromycin is a narrow spectrum, non-toxic antibiotic that prevents release of EF1A from the ribosome after GTP hydrolysis (Vogeley et al. 2003). Aurodox, the methyl derivative of kirromycin, had an IC_{50} of 50 μM for *P. falciparum* in vitro (Strath et al. 1993). Binding of kirromycin to a recombinant version of *Plasmodium* EF1A was demonstrated in vitro (Clough et al. 1999) despite the fact that the protein has the A387S substitution correlated with drug resistance in *B. subtilis* (see Mesters et al. 1994). Kirromycin acted relatively rapidly, reducing the incorporation of radioactive tracers within 5–10 h of treating *P. falciparum* in vitro. This swift effect brought the specificity of its inhibitory action into question.

By contrast with the above the slow acting antibiotic clindamycin, that blocks formation of the peptide bond, has been used successfully in combination with other antimalarials e.g., quinine or atovaquone, to treat uncomplicated malaria in the clinic (Adehossi et al. 2003; Lell and Kremsner 2002). The recovery from laboratory cultures of spontaneous mutants of *Toxoplasma gondii* highly resistant to clindamycin (Camps et al. 2002) was pointed out as a contrary sign (Foth and McFadden 2003). However, the isolation of these mutants required stringent conditions and repeated attempts. The mutants carried a G1857U transversion in domain V of the plastidic 23S rRNA, the known binding site for clindamycin. At lower concentrations of the antibiotic, inhibition of parasite growth only took place in division cycles after invasion of fresh cells—the so-called delayed death phenotype (Fichera and Roos 1997). In *T. gondii*, it was noted that parasites undergoing delayed death following exposure to clindamycin form multi-nucleated cells blocked in late endodyogeny (the stage when daughter cells are separated from the mother cell by the enveloping growth of its plasma membrane). One possible explanation for the antimalarial activity of clindamycin is that interruption of protein synthesis in the plastid blocks production of the Clp protein encoded on the 35-kb genome. This chaperone is thought to be involved in the import of cytosolic proteins that in this case would include the enzymes required for FA biosynthesis (see below). The products of the FAS II pathway might be required for parasite membrane formation following replication or invasion of new cells. However, this explanation has to be set against the observation that *T. gondii* can grow and divide several times without a plastid and only requires one upon invasion of a fresh cell (He et al. 2001a, 2001b).

The clindamycin-resistant *T. gondii* referred to above showed some cross-resistance to doxycycline and azithromycin, antibiotics that have been used successfully either alone as prophylactics (Anderson et al. 1998) or in combination with standard antimalarial drugs (Yeo et al. 1997). 'Cures' were effected in a small group of patients with acute infections of *P. vivax* when treated with 1.2 g azithromycin for 7 days. However, this did not prevent relapses from hypnozoites (Ranque et al. 2002). Ohrt et al. (2002) reported that phase II dose-ranging studies with azithromycin in combination with other antimalarials are in progress.

We found no effect on the growth of *Plasmodium* cultures with viomycin. However, another classical small molecule inhibitor of EF-G, namely fusidic acid, was reported at an early date to have antimalarial activity (Black et al. 1985). No further work seems to have built upon this observation. More complex thiopeptide inhibitors of translocation such as thiostrepton and micrococcin prevent oscillating conformations of the ribosome that are triggered by binding GTP. The specificity of these antibiotics for the plastid of *P. falciparum* has been associated with the presence of preferred nucleotides in the sequence of the plastidic 23S rRNA (Clough et al. 1997; Rogers et al. 1997; Conn et al. 1999). Inhibition tests in vitro showed micrococcin was the more potent with an IC_{50} of 35 nM (Rogers et al. 1998). Unfortunately this antibiotic seems virtually unobtainable for further experimental work.

The plastidic ribosomal proteins and rRNAs of *P. falciparum* are highly divergent in sequence from those of bacteria, algae and higher plants, and few details are known about their fine structure. Hence the multiple interactive sites of the ribosomal proteins (mostly imported from the cytosol) might provide novel targets additional to the conformational and active sites of the ribosomal complex that all the antibiotics referred to above are directed.

2.1.4
Peptide Deformylase

Like their ancient prokaryotic ancestors, the plastid and mitochondrial organelles of eukaryotic cells modify the first methionine of newly synthesized proteins with an *N*-formyl group prior to incorporation into a polypeptide. In plastids, but apparently not in the mitochondrial organelles of mammalian cells (Nguyen et al. 2003) deformylation catalysed by peptide deformylase (PDF) is necessary before the N-terminus of the nascent protein is processed with methionine aminopeptidase (Giglione and Meinnel 2001). The genome of *P. falciparum* encodes one gene specifying a PDF (class 1B) with a predicted plastid-targeting sequence at its N terminus. Like the *Escherichia coli* enzyme, native peptide deformylase of *P. falciparum* has a highly unstable Fe^{2+} in its active site, but recombinant versions stabilized by replacing Fe^{2+} with Co^{2+} have been made and scrutinized for inhibitors (Bracchi-Richard et al. 2001). Differences found in and around the enzyme's active site explain its decreased affinity (~100-fold) for substrates and inhibitors, compared with PDF from *E. coli* (Kumar et al. 2002). A recent crystal structure shows the PDF of *P. falciparum* complexed with a synthetic inhibitor (Robien et al. 2004). Such structural studies could provide co-ordinates for designing *Plasmodium*-specific inhibitors; PDF is a metallo-protease, a group of enzymes for which there is an excellent track record in the design of mechanism-based inhibitors. The need for such an approach is evident from the finding that cultures of erythrocytes infected with *P. falciparum* are rather insensitive to Actinonin, a typical inhibitor of PDF (IC_{50}=3 μM) (Wiesner et al. 2001). Other new potent derivatives of *N*-formyl hydroxamic acids have been developed (Hackbarth et al. 2002; Waller and Clements 2002) and should be tested against the PDF of *P. falciparum* or used for structural studies.

2.2
Group B

2.2.1
FA Synthesis

A set of enzymes that form a complete biosynthetic pathway for FAs from acetyl-CoA are encoded in the nuclear genome of *P. falciparum*. Acetyl-CoA carboxylase (ACC) that catalyses carboxylation of acetyl-CoA to produce malonyl-CoA is the first enzyme in this pathway (see http://sites.huji.ac.il/malaria/maps/facidsynthesispath.html). Like other eukaryotes, apicomplexan ACC has both acetyl-CoA carboxyltransferase and biotin carboxylase activities and contains a biotin-carboxyl-carrier protein domain (Zuther et al. 1999). Two genes specifying different ACCs have been cloned from *T. gondii*. The polypeptide encoded by one of them (ACC1) has a functional plastid-targeting sequence at its N-terminus (Jelenska et al. 2001). A DNA fragment encoding an orthologue of *Tg*ACC1 has been cloned from two *Plasmodium* species, *P. falciparum* and *P. knowlesi*, whereas *Cryptosporidium parvum*, the exceptional apicomplexan parasite that is supposed to have lost its plastid, seems to lack the gene for ACC1. Instead it only has a gene specifying an orthologue of the cytosolic *Tg*ACC2. Although the plastidic ACC of most plant species consists of four different subunits like those of bacteria, grasses are an exception. They have a multi-domain

plastidic ACC that resembles apicomplexan ACC1. Aryloxyphenoxypropionate derivatives are strong inhibitors of the multi-domain plastidic ACC
of grasses, though these compounds have no effect on other multi-domain
ACCs, including those of humans. These herbicides effectively suppressed
growth of *T. gondii*, despite the fact that the cytosolic *Tg*ACC2 is resistant to
these inhibitors (Jelenska et al. 2002). Clodinafop was the most effective of the
herbicides tested against *T. gondii* in vitro ($IC_{70}=10$ µM). These herbicides
might control the growth of other apicomplexan parasites with an ACC1
orthologue in the plastid, although the danger of mutations has been pointed
out (Seeber 2003).

Using acetyl-CoA and malonyl-CoA as substrates, *P. falciparum* synthe-
sizes FAs with a type II 'fatty acid synthase' system (FAS II). FAS II consists
of six separate enzymes: acetyltransferase, malonyltransferase, β-ketoacyl-
ACP synthase (KAS), β-ketoacyl-ACP reductase (KAR), β-hydroxyacyl-ACP
dehydratase (HAD) and enoyl-ACP reductase (ENR). By contrast, animals including humans have a type I fatty acid synthase (FAS I). This is a homodimer
of a multi-functional polypeptide carrying all six activities plus thioesterase
activity. The plastid-lacking *C. parvum* has a FAS I that is probably located in
the cytosol (Zhu et al. 2000). Other coccidians (*T. gondii* and *Eimeria tenella*)
also have been reported to possess a FAS I of extraordinarily large size (over
10,000 amino acid residues) besides the plastid-localizing FAS II (Crawford
et al. 2003; Molecular Parasitology Meeting, Woodshole, Abstract). However,
the gene specifying FAS I apparently is missing from *P. falciparum*. As there
seems to be only one *Plasmodium* gene specifying (FAS II) KAR, and it has
a putative plastid-targeting sequence at the N-terminus (Pillai et al. 2003), FA
synthesis in *P. falciparum* is inferred to take place solely through FAS II in the
plastid.

The KAS enzymes of FAS II are susceptible to thiolactomycin (TLM), an
antibiotic that binds to the active site, mimicking the substrate malonyl-
ACP. TLM targets KAS I and KAS II enzymes (encoded by *fabB* and *fabF*,
respectively in *E. coli*) being specific for mid to longer FA chains and is far less
effective against the short chain-specific KAS III enzyme (encoded by *fabH*).
These observations suggested that KAS I/II of *P. falciparum* rather than KAS
III would be the likely target of TLM.

TLM inhibits *P. falciparum* in blood cultures (Waller et al. 1998), parasites
in the early ring stages being most susceptible, thereafter steadily becoming
less so. Analogues of TLM showed increasing activity as the length of the side
chain increased (Waller et al. 2003). This is probably because TLM does not
fully occupy a hydrophobic side pocket at the active site of *Pf*KAS I/II (Price
et al. 2001). As TLM inhibits KAS enzymes in FAS II with minimal effects
on FAS I, this inhibitor seems a suitable lead compound for further drug
development. Moreover, TLM binds to multiple essential residues that form

the active site, so resistant mutants are unlikely to arise. Attempts to make mycobacteria resistant to TLM largely failed, though it could be achieved either by over-expression of KAS I or through the activation of a non-specific efflux pump (Campbell and Cronan 2001). Prigge et al. (2003) reported that two 1,2-dithiole-3-one compounds that have a structural similarity to TLM, inhibit *Pf* KASIII ($IC_{50}<10$ μM) as much as 100 times more effectively than TLM.

HAD (FabZ) of *P. falciparum* also has been expressed in active recombinant form (Sharma et al. 2003) and used for the design of inhibitors. Two small molecule inhibitors (NAS91 and NAS21)–the first for HAD of any species– killed the parasites within a single growth cycle in cultures of erythrocytes ($IC_{50}=7.4$ and 100 μM, respectively). Both inhibitors were less effective in vivo (5- and 50-fold, respectively) and further development is indicated, including the use of high throughput screens.

ENR is the FAS II component that is specifically inhibited by Triclosan (TC). Although this compound has been in widespread use for some two decades as an antimicrobial and *E. coli* can acquire resistance by over-expressing ENR or carrying missense mutations (Heath et al. 1998), field isolates of resistant bacteria have never been recorded (Surolia and Surolia 2002). TC is a potent inhibitor of *Plasmodium* ENR ($IC_{50}=0.7-7.0$ μM) (McLeod et al. 2001; Surolia and Surilia 2002; Perozzo et al. 2002). Like TLM, TC acts on the ring and young trophozoite stages and not on schizonts (Waller et al. 2003). The action of TC is rapid and it can cure mice in experimental infections. The ENR of *P. falciparum* has an extensive loop at the active site by comparison with other known orthologues. However, a crystal structure of the ternary complex with NAD+ and TC revealed that the mode of TC binding is very similar to other ENRs (Perozzo et al. 2002). Comparing the structure of 20 synthesized analogues (of which two were active), critical groups in both rings of TC were identified. Two pyrazole compounds, found in a high-throughput screen with the ENR of *Mycobacterium tuberculosis*, inhibited the activity of purified *Pf* ENR as well as the growth of *P. falciparum* in blood cultures (Kuo et al. 2003). Potency ($IC_{50}=\sim20$ μM) was similar in both situations suggesting good cellular uptake of the inhibitors.

FAS II seems the most attractive of the *Plasmodium* spp. plastidic pathways for drug discovery at the present time. It has yielded several sequential targets as well as lead compounds with promising activity. The pharma-industry is well acquainted with this pathway and is actively interested in it for the development of new antimicrobials.

2.2.2
Isoprenoid Biosynthesis

Unlike humans, *Plasmodium* spp. synthesize isoprenoids via 1-deoxy-D-xylulose 5-phosphate (DOXP) (Jomaa et al. 1999). This non-mevalonate pathway is absent from mammals but is found in several species of bacteria and plant plastids. It has attracted interest in the search for selective plastidic inhibitors of *P. falciparum*. Localization experiments with the N-terminus of *Plasmodium* DOXP reductoisomerase indicated that it is targeted to the plastid (Jomaa et al. 1999). The activity of the second enzyme in the pathway—DOXP reductoisomerase, is detectable in extracts of the parasite by radiometric assay (Wiesner et al. 2000). Another enzyme that operates later in the pathway, namely 2C-methyl-D-erythritol-2,4-cyclodiphosphate synthase, has been expressed as a recombinant protein (Rohdich et al. 2001).

It is known that DOXP reductoisomerase is inhibited specifically by fosmidomycin. This antibiotic has been evaluated pharmacologically and used in phase II clinical trials as an antimicrobial agent (Kuemmerle et al. 1985). The serum half-life of fosmidomycin in humans is about 2 h and it is excreted without being metabolized. Malaria parasites exposed to the antibiotic show arrested development in the late schizont stage. This resembles the rapid effect of TC but is unlike the slow acting effect of antibiotics that interfere with intrinsic plastid processes such as DNA replication (ciprofloxacin) or protein synthesis (clindamycin).

The short half-life of fosmidomycin indicates that multiple doses (seven daily doses in non-immunes) would be required to prevent the survival and recrudescence of very young ring stages present at the first administration of the inhibitor. Clinical trials have been carried out on malarial patients in Gabon (Missinou et al. 2002) and Thailand (Lell et al. 2003). After treatment for 5–7 days, parasites were cleared rapidly and symptoms such as fever were resolved. But the antibiotic was only partially effective as it did not prevent recrudescence (50% of cases) or gametocytaemia. The viability of such gametocytes is a contra-indication as they could enable the spread of drug resistance. Studies of the effect of derivatives of fosmidomycin on murine malaria showed that a *bis*-(4-methoxyphenyl) ester derivative was the most active compound for oral administration. But it was not clear whether the increased efficacy was due to improved absorption, a longer half-life, or some other factor (Reichenberg et al. 2001).

Like other antibiotics that have been catalogued, fosmidomycin and its derivatives should be used only as part of a combination therapy, preferably with another compound with similar pharmacokinetics (White 1999). Studies with cultures of erythrocytes infected with *P. falciparum* revealed synergy between fosmidomycin and clindamycin, or its natural precursor lincomycin, but not with other commonly used antimalarials (atovaquone, quinine, artemisinin and proguanil) (Wiesner et al. 2002). Both fosmidomycin and clindamycin are available in water-soluble form and clinical studies have shown that the combination is well tolerated and clears parasitaemias in African children (Wiesner et al. 2003; Borrmann et al. 2004).

2.2.3
Haem Biosynthesis

Plasmodium species that have been investigated have a complete set of genes specifying enzymes required for haem biosynthesis. The intracellular distribution of these intrinsic enzymes is peculiar and unique and implies that both the mitochondrion and plastid take part in haem biosynthesis (Sato et al. 2004). Studies with the inhibitor succinylacetone have shown that the parasite depends on haem synthesized de novo (Surolia and Padmanaban 1992), hence disruption of this metabolism is worthy of consideration.

We are currently examining this possibility for the second enzyme in the pathway, namely porphobilinogen synthase (PBGS or ALAD). The N-terminal targeting sequence of *Plasmodium* PBGS localizes it to the plastid (Sato et al. 2004) and the enzyme has plant-like features that clearly distinguish it from the host's enzyme (Sato et al. 2000; Sato and Wilson 2002). Recombinant versions of PBGS from *P. falciparum* have been characterized enzymatically (Dhanasekeran et al. 2003; Sato et al. 2004) and the active site loop is being examined in detail to ascertain whether differences from the host enzyme might be used to advantage in the development of inhibitors.

From an extensive set of experiments, Padmanaban and his group proposed that the malaria parasite imports and utilizes for its own purposes erythrocytic enzymes for haem biosynthesis (Padmanaban 2003). They claim that the intrinsic PBGS is utilized only for a minor 'plastidic pathway', whereas the bulk of haem is synthesized in the mitochondrion largely utilizing extrinsic enzymes imported from the host cell (Dhanasekeran et al. 2003). By contrast, we proposed that as the parasite has all the genes required for a complete pathway and some of the intrinsic enzymes are localized in the plastid, haem production in the mitochondrion depends on substrates generated in the plastid. Hopefully, these discrepancies will be resolved with further work.

3
Conclusion

The formidable problems of developing and marketing new antimalarials that block the functions of the *Plasmodium* plastid were spelled out by Beeson et al. (2001). However, progress has been made in finding important pathways and selecting targets. Whilst there is a world of difference between a lead compound and a drug, high-resolution structural information is presently being gathered to tailor the construction of specific inhibitory ligands for plastidic enzymes of *P. falciparum*. Both synthetic and combinatorial chemistry is being used to direct the search from initial leads to smaller libraries of compounds. Screens, such as the use of other microorganisms and the development of high throughput assays, should not be forgotten, especially in the search for natural products with activities in the nmol range For example, an approach taken with *T. gondii* to produce a stable transfectant expressing a bright yellow fluorescent tandem protein offers a simple system for monitoring growth in microwell culture plates without manipulation (Gubbels et al. 2003). A similar system might be devised with a suitably chosen fluorophore for intra-erythrocytic *P. falciparum*. Transcription profiling with microarrays (Bozdech et al. 2003) might offer a new approach to monitoring the mode of action and effects of plastidic inhibitors.

In an earlier review, McFadden and Roos (1999) ended by posing eight questions for future research. On nearly every point we have come a long way towards an answer. However, at the biological level it remains unresolved why antiplastidic compounds result in either slow or rapid parasite death. Some options are discussed by Surolia et al. (2002). Although we still know almost nothing about how the life cycle of *Plasmodium* is maintained by plastidic processes nor how these are regulated, steady progress is being made in the evaluation and the development of novel plastidic inhibitors for *P. falciparum*.

Acknowledgements SS and RJMW are supported by the British Medical Research Council.

Appendix

A.1
Proteins Predicted to Be Imported into the Plastid of *P. falciparum*[a]

- DNA metabolism
    - [b]DNA gyrase
    - 5′-3′ Exonuclease, N-teminal resolvase-like domain
    - DNA ligase 1
    - [b,c]POM1 (PREX—plastid replication/repair enzyme complex)
    - TatD-like deoxyribonuclease
    - [b]Single strand binding protein
    - AP endonuclease (DNA-apurinic or apyrimidinic lyase)
    - A/G-specific adenine glycosylase
    - UMP/CMP kinase

- RNA metabolism
  - [b]DNA-directed RNA polymerase, alpha chain
  - Pseudouridine synthetase
  - Pseudouridylate synthase
  - Queuine tRNA ribosyltransferase
  - N^2, N^2 dimethylguanosine tRNA methyltransferase
  - tRNA Δ^2-isopentenylpyrophosphatetransferase
  - ADP-ribosylation factor
  - Dimethyladenosine transferase
  - ATP-dependent helicase
  - DEAD-box helicase

- Protein synthesis
  - [b]tRNA synthetase (A, C, D, E, F, G, H, I, L, K, M, N, P, Q, R, T, S, V, W, Y)
  - [b]Glutamyl–tRNA amidotransferase
  - [b]Methionyl-tRNA formyltransferase
  - [b,c]30S Ribosomal subunit (S6, 9,10, 14)
  - [b,c]50S Ribosomal subunit (L1, 2, 7/12, 10, 14, 15, 18, 27–29, 35)
  - Ribosome releasing factor
  - [b]Peptide chain release factor
  - [b]Elongation factors EF1B and EF2
  - [b]Translation initiation factor IF-1
  - Peptide deformylase (PDF)
  - Met-10+ like protein

- Fatty acid synthesis
  - Phosphoenolpyruvate/phosphate translocator
  - Pyruvate kinase
  - [b]Pyruvate dehydrogenase E1 component (α and β subunits)
  - Lipoamide dehydrogenase
  - Dihydrolipamide acetyl transferase
  - Lipoate synthase
  - Lipoate protein lygase
  - Acetyl-CoA synthetase
  - Dephospho-CoA kinase
  - [b,c]Biotin carboxylase subunit of acetyl-CoA carboxylase
  - [b,c]Acyl carrier protein (ACP)
  - [b]Holo-ACP synthase
  - [b]Malonyl-CoA–ACP transacyclase (FabD or MCAT)
  - [b]β-Ketoacyl-ACP synthase III (FabH or KAS III)
  - [b]β-Ketoacyl-ACP reductase (FabG or KAR)

  – β-Hydroxyacyl-ACP dehydratase (FabZ or HAD)
  – Enoylacyl-ACP reductase (FabI or ENR)
  – [b]β-Ketoacyl-ACP synthase I/II (FabB/F or KAS I/II)
  – Aldo-keto reductase
- Glycerolipid synthesis
  – Glycerol-3-phosphate acyltransferase
  – Diacylglycerol kinase
  – Phospholipid or glycerol acyltransferase
  – Stearoyl-CoA desaturase (acyl-CoA desaturase)
  – Phosphotidylcholine-sterol acyltransferase
- Isoprenoid biosynthesis
  – Triose or hexose phosphate/phosphate translocator
  – [b]Triose phosphate isomerase
  – [b]1-Deoxy-D-xylulose-5-phosphate synthase
  – [b,c]1-Deoxy-D-xylulose-5-phosphate reductoisomerase (IspC)
  – [b]4-Diphosphocytidyl-2C-methyl-D-erythritol kinase (IspE)
  – [b]2C-Methyl-D-erythritol-2,4-cyclodiphosphate synthase (IspF)
  – GcpE (IspG)
  – LytB (IspH)
- Haem biosynthesis
  – [c]Porphobininogen synthase (PBGS)
  – [c]Hydroxymethylbilane synthase (HMBS)
  – Uroporphyrinogen decarboxylase (UroD)
- Redox-associated proteins
  – Ferredoxin
  – [c]Ferredoxin-NADP$^+$ reductase
  – Antioxidant protein
  – Glutathione peroxidase
  – [b,c]Cysteine desulfurase (SufS)
- Proteases
  – Aspartyl (acid) protease
  – Leucine aminopeptidase
  – [b]Methione aminopeptidase
  – M1 family aminopeptidase
  – [c]M16 Metalloendopeptidase
  – [b]O-Sialoglycoprotein endopeptidase
  – Peptidase
  – Serine protease (subtilisin family)

- – ATP-dependent Clp protease
- – ATP-dependent Clp protease, proteolytic subunit
- – Alpha/beta hydrolase
- Chaperones
  - – [c]Cpn60 ('mitochondrial precursor')
  - – Heat shock protein DnaJ homolog Pfj2
  - – Heat shock protein 101
  - – DnaJ
- Various
  - – [b]GTP-binding protein
  - – Guanidine nucleotide exchange factor
  - – TetQ family GTPase
  - – Phosphatase 1 regulatory subunit
  - – Pyridine nucleotide transhydrogenase
  - – Glutamate dehydrogenase
  - – Glyoxalase 1
  - – Methylase like protein
  - – Rho GTPase activating protein
  - – Glucosamine-fructose-6-phosphate aminotransferase
  - – Aspartate carbamoyl transferase
  - – UBA/THIF-type NAD/FAD binding protein
  - – PDZ-domain protein
  - – Ubiquitin protein ligase
  - – Tmp21
  - – Nucleotide binding protein
  - – Ribose-phosphate pyrophosphokinase
  - – Serine/threonine protein kinase
  - – DHHC-type zinc finger protein
  - – Glucose inhibited division protein A
  - – ABC-transporter
  - – ATP-dependent transporter subunit
- Unknown ORFs
  - – ~350

[a] Note that the localization of only a small number of the plastid ORFs predicted from PlasmoDB has been confirmed at present; others are likely to have escaped detection and yet others will have been identified by investigators currently at work. Accordingly, this list is only a provisional guide.
[b] Essential in *B. subtilis* (Kobayashi et al. 2003).
[c] Plastid localization confirmed in *P. falciparum*.

References

Adehossi E, Parola P, Foucault C, Delmont J, Brouqui P, Badiaga S, Ranque S (2003) Three day quinine-clindamycin treatment of uncomplicated falciparum malaria imported from the tropics. Antimicrob Agents Chemother 47:1173.

Andersen GR, Nissen P, Nyborg J (2003) Elongation factors in protein synthesis. Trends Biochem Sci 28:434–441

Andersen SL, Oloo AJ, Gordon DM, Ragama OB, Aleman GM, Berman JD, Tang DB, Dunne MW, Shanks GD (1998) Successful double-blinded, randomized, placebo-controlled field trial of azithromycin and doxycycline as prophylaxis for malaria in western Kenya. Clin Infect Dis 26:146–150

Bagley MC, Bashford KE, Hesketh CL, Moody CJ (2000) Total synthesis of the thiopeptide promothiocin A. J Am Chem Soc 122:3301–3313

Beeson JG, Winstanley PA, McFadden GI, Brown GV (2001) New agents to combat malaria. Nature Med 7:149–150

Black FT, Wildfang IL, Borgbjerg K (1985) Activity of fusidic acid against *Plasmodium falciparum in vitro*. Lancet 325:578–579

Borrmann S, Adegnika AA, Matsiegui P-B, Issifou S, Schindler A, Mawili-Mboumba DP, Baranek T, Wiesner J, Jomaa H, Kremsner PG (2004). Fosmidomycin-clindamycin for *Plasmodium falciparum* infections in African children. J Infect Dis 189:901–908

Bozdech Z, Llinas M, Pulliam BL, Wong ED, Zhu J, DeRisi JL (2003) The transcriptome of the intraerythrocytic developmental cycle of *Plasmodium falciparum*. PLoS Biol 1:85–100

Bracchi-Richard V, Nguyen KT, Zhou Y, Rajagopalan PTR, Chakrabarti D, Pei D (2001) Characterization of an eukaryotic peptide deformylase from *Plasmodium falciparum*. Arch Biochem Biophys 396:162–170

Campbell JW, Cronan JE (2001) Bacterial fatty acid biosynthesis: Targets for antibacterial drug discovery. Ann Rev Microbiol 55:305–332

Camps M, Arrizabalaga G, Boothroyd J (2002) An rRNA mutation identifies the apicoplast as the target for clindamycin in *Toxoplasma gondii*. Mol. Microbiol 43:1309–1318

Clough B, Rangachari K, Strath M, Preiser PR, Wilson RJM (1999) Antibiotic inhibitors of organellar protein synthesis in *Plasmodium falciparum*. Protist 150:189–195

Clough B, Strath M, Preiser P, Denny P, Wilson RJM (1997) Thiostrepton binds to malarial plastid rRNA. FEBS Lett 406:123–125

Clough B, Wilson RJM (2001) Antibiotics and the plasmodial plastid organelle. In: Rosenthal PJ (ed) Antimalarial chemotherapy: mechanisms of action, resistance, and new directions in drug discovery. Humana Press Inc., Totowa, NJ, pp 265–286

Conn GL, Draper DE, Lattman EE, Gittis AG (1999) Crystal structure of a conserved ribosomal protein–RNA complex. Science 284:1171–1174

Darst SA (2004) New inhibitors targeting bacterial RNA polymerase. Trends Biochem Sci 29:159–162

Dhanasekaran S, Chandra NR, Chandrasekhar BK, Rangajarin PN, Padmanaban G (2003) δ-aminolevulinic acid dehydratase from *Plasmodium falciparum* - indigenous vs imported. J Biol Chem 279:6934–6942

Feagin JE, Mericle BL, Werner E, Morris M (1997) Identification of additional rRNA fragments encoded by the *Plasmodium falciparum* 6 kb element. Nucl Acids Res 25:438–446

Fichera ME, Roos DS (1997) A plastid organelle as a drug target in apicomplexan parasites. Nature 390:407–409

Foth BJ, McFadden GI (2003) The apicoplast: A plastid in *Plasmodium falciparum* and other apicomplexan parasites. Int Rev Cytol 224:57–110

Foth BJ, Dissecting apicoplast targeting in the malaria parasite *Plasmodium falciparum*. Science 299:705–708

Gardner MJ, Hall N, Fung E, White O, Berriman M, Hyman RW, Carlton JM et al. (2002) Genome sequence of the human malaria parasite *Plasmodium falciparum*. Nature 419:498–511

Giglione C, Meinnel T (2001) Organellar peptide deformylases: universality of the N-terminal methionine cleavage mechanism. Trends Plant Sci 6:566–572

Goerg H, Ochola SA, Goerg R (1999) Treatment of malaria tropica with a fixed combination of rifampicin, co-trioxazole and isoniazid: a clinical study. Chemotherapy 45:68–76

Gubbels M-J, Li C, Striepen B (2003) High throughput growth assay for *Toxoplasma gondii* using yellow fluorescent protein. Antimicrob Agents Chemother 47:309–316

Hackbarth CJ, Chen DZ, Lewis JG, Clark K, Mangold JB, Cramer JA, Margolis PS, Wang W, Koehn J, Wu C, et al. (2002) N-alkyl urea hydroxamic acids as a new class of peptide deformylase inhibitors with antibacterial activity. Antimicrob Agents Chemother 46:2752–2764

He CY, Striepen B, Pletcher CH, Murray JM, Roos DS (2001a) Targeting and processing of nuclear-encoded apicoplast proteins in plastid segregation mutants of *Toxoplasma gondii*. J Biol Chem 276:28436–28442

He CY, Shaw MK, Pletcher CH, Striepen B, Tilney LG, Roos DS (2001b) A plastid segregation defect in the protozoan parasite *Toxoplasma gondii*. EMBO J 20:330–339

Heath RJ, Yu Y-T, Shapiro MA, Olson E, Rock CO (1998) Broad spectrum antimicrobial biocides target the FabI component of fatty acid synthesis. J Biol Chem 273:30316–30320

Jelenska J, Crawford MJ, Harb OS, Zuther E, Haselkorn R, Roos DS, Gornicki P (2001) Subcellular localization of acetyl-CoA carboxylase in the apicomplexan parasite *Toxoplasma gondii*. Proc Natl Acad Sci USA 98:2723–2728

Jelenska J, Sirikhachornkit A, Haselkorn R, Gornicki P (2002) The carboxyltransferase activity of the apicoplast acetyl-CoA carboxylase of *Toxoplasma gondii* is the target of aryloxyphenoxypropionate inhibitors. J Biol Chem 277:23208–23215

Jomaa H, Wiesner J, Sanderbrand S, Altincicek B, Weidemeyer C, Hintz M, Turbachova I, Eberl M, Zeidler J, Lichtenthaler HK, Sodati D, Beck E (1999) Inhibitors of the nonmevalonate pathway of isoprenoid biosynthesis as antimalarial drugs. Science 285:1573–1576

Kobayashi K, Ehrlich SD, Albertini A, Amati G, Anderson KK, Arnaud M, Asai K, Ashikaga S, Aymerich S, et al. (2003) Essential *Bacillus subtilis* genes. Proc Natl Acad Sci USA 100:4678–4683

Kuemmerle H-P, Murakawa T, Sakamoto H, Sato N, Konishi T, De Santis F (1985) Fosmidomycin, a new phosphonic acid antibiotic. PartII : Human pharmacokinetics 2. Preliminary early phase IIa clinical studies. Int J Clin Pharm Ther Tox 23:521–528

Kumar A, Nguyen KT, Srivathsan S, Ornstein B, Turley S, Hirsh I, Pei D, Hol WGJ (2002) Crystals of peptide deformylase from *Plasmodium falciparum* reveal critical characteristics of the active site for drug design. Structure 10:357–367

Kuo MR, Morbidoni HR, Alland D, Sneddon SF, Gourlie BB, Staveski MM, Leonard M, Gregory JS, Janjigian AD, et al. (2003) Targeting tuberculosis and malaria through inhibition of enoyl reductase. J Biol Chem 278:20851–20859

Lell P, Kremsner PG (2002) Clindamycin as an antimalarial drug: Review of clinical trials. Antimicrob Agents Chemother 46:2315–2320

Lell B, Ruangweerayut R, Wiesner J, Missinou AM, Schindler A, Baranek T, Hintz M, Hutchinson DB, Jomaa H, Kremsner PG (2003) Fosmidomycin, a novel chemotherapeutic agent for malaria. Antimicrob Agents Chemother 47:735–738.

Lin Q, Katakura K, Suzuki M (2002) Inhibition of mitochondrial and plastid activity of *Plasmodium falciparum* by minocycline. FEBS Lett 515:71–74

McClean KL, Hitchman D, Shafran SD (1992) Norfloxacin is inferior to chloroquine for falciparum malaria in northwestern Zambia: a comparative clinical trial. J Infect Dis 165:904–907

McConkey GA, Rogers MJ, McCutchan TF (1997) Inhibition of *Plasmodium falciparum* protein synthesis. Targeting the plastid-like organelle with thiostrepton. J Biol Chem 272:2046–2049

McFadden GI, Roos DS (1999) Apicomplexan plastids as drug targets. Trends Microbiol 7:328–333

McLeod R, Muench SP, Rafferty JB, Kyle DE, Mui EJ, Kirisits MJ, Mack DG, Roberts CW, Samuel BU, Lyons RE, Dorris M, Milhous WK, Rice DW (2001) Triclosan inhibits the growth of *Plasmodium falciparum* and *Toxoplasma gondii* by inhibition of Apicomplexan FabI. Int J Parasitol 31:109–113

Mahmoudi N, Ciceron L, Franetich J-F, Farhati K, Silvie O, Eling W, Sauerwein R, Danis M, Mazier D, Derouin F (2003) *In vitro* activities of 25 quinolones and fluoroquinolones against liver and blood stage *Plasmodium* spp. Antimicrob Agents Chemother 47:2636–2639

Mesters JR, Zeef LAH, Hilgenfeld R, de Graaf JM, Kraal B, Bosch L (1994) The structural and functional basis for the kirromycin resistance of mutant EF-Tu species in *Escherichia coli*. EMBO J 13:4877–4885

Miesel L, Greene J, Black TA (2003) Genetic strategies for antibacterial drug discovery. Nature Rev Genet 4:442–456

Missinou AM, Borrmann S, Schindler A, Issifou S, Adegnika AA, Matsiegui P-B, Binder R, Lell B, Wiesner J, Baranek T, Jomaa H, Kremsner PG (2002) Fosmidomycin for malaria. Lancet 360:1941–1942

Nguyen KT, Hu X, Colton C, Chakrabarti R, Zhu MX, Pei D (2003) Characterization of a human peptide deformylase: Implications for antibacterial drug design. Biochemistry 42:9952–9958

Ohrt C, Willingmyre GD, Lee PJ, Knirsch C, Milhous WK (2002) Assessment of azithromycin in combination with other antimalarial drugs against *Plasmodium falciparum in vitro*. Antimicrob Agents Chemother 46:2518–2524

Padmanaban G (2003) Drug targets in malaria parasites. Adv Biochem Engin/Biotechnol 84:123–141

Perozzo R, Kuo M, Sidhu AS, Valiyaveettil JT, Bittman R, Jacobs WR Jr, Fidock DA, Sacchettini JC (2002) Structural elucidation of the specificity of the antibacterial agent Triclosan for malarial enoyl acyl carrier protein reductase. J Biol Chem 277:13106–13114

Pillai S, Rajagopal C, Kapoor M, Kumar G, Gupta A, Surolia N (2003) Functional characterization of β-ketoacyl-ACP reductase (FabG) from *Plasmodium falciparum*. Biochem Biophys Res Comun 303:387–392

Price AC, Choi K-H, Heath RJ, Li Z, White SW, Rock CO (2001) Inhibition of β-ketoacyl-acyl carrier protein synthases by thiolactomycin and cerulenin. J Biol Chem 276:6551–6559

Prigge ST, He X, Gerena L, Waters NC, Reynolds KA (2003) The initiating steps of type II fatty acid synthase in *Plasmodium falciparum* are catalyzed by pfACP, pfMCAT, and pfKASIII. Biochemistry 42:1160–1169

Pukrittayakamee S, Viravan C, Charoenlarp P, Yeamput C, Wilson RJM, White NJ (1994) Antimalarial effects of rifampicin in vivax malaria. Antimicrob Agents Chemother 38:511–514

Pukrittayakamee S, Prakongpan S, Wanwimolruk S, Clemens R, Loareesuwan S, White NJ (2003) Adverse effect of rifampicin on quinine efficiency in uncomplicated falciparum malaria. Antimicrob Agents Chemother 47:15009–15013

Ralph SA, D'Ombrain MC, McFadden GI (2001) The apicoplast as a drug target. Drug Res Updates 4:145–151

Ralph SA, van Dooren GG, Waller RF, Crawford M, Fraunholz MJ, Foth BJ, Tonkin CJ, Roos DS, McFadden GI (2004) Metabolic maps and functions of the *Plasmodium falciparum* apicoplast. Nat Rev Microbiol 2:203–216

Ranque S, Badiaga S, Delmont J, Brouqui P (2002) Triangular test applied to the clinical trial of azithromycin against relapses in *Plasmodium vivax* infections. Malaria J 1:13

Reichenberg A, Wiesner J, Weidemeyer C, Dreiseidler E, Sanderbrand S, Altincicek B, Beck E, Schlitzer M, Jomaa H (2001) Diaryl ester prodrugs of FR900098 with improved *in vivo* antimalarial activity. Bioorganic Med Chem Lett 11:833–835

Robien MA, Nguyen KT, Kumar A, Hirsh I, Turley S, Pei D, Hol WGJ (2004) An improved crystal form of *Plasmodium falciparum* peptide deformylase. Prot Sci 13:1155–1163

Rohdich F, Eisenreich W, Wungsintaweekul J, Hecht S, Schuhr CA, Bacher A (2001) 2C-methyl-D-erythritol 2,4-cyclodiphosphate synthase (IspF) from *Plasmodium falciparum*. Eur J Biochem 268:3190–3197

Rogers MJ, Bukhman YV, McCutchan TF, Draper DE (1997) Interaction of thiostrepton with an RNA fragment derived from plastid-encoded ribosomal RNA of the malaria parasite. RNA 3:815–820

Rogers MJ, Cundliffe E, McCutchan TF (1998) The antibiotic micrococcin is a potent inhibitor of growth and protein synthesis in the malaria parasite. Antimicrob Agents Chemother 42:715–716

Roos DS, Crawford MJ, Donald RGK, Fraunholz M, Harb OS, He CY, Kissinger J, Shaw MK, Striepen B (2002) Mining the *Plasmodium* genome database to define organellar function: what does the apicoplast do? Phil Trans R Soc Lond B 357:35–46

Sato S, Clough B, Coates L, Wilson RJM (2004) Enzymes for heme biosynthesis are found in both the mitochondrion and plastid of the malaria parasite *Plasmodium falciparum*. Protist 155:117–125

Sato S, Tews I, Wilson RJM (2000) Impact of an endocytobiont on the genome of apicomplexans. Int. J. Parasitol 30:427–439

Sato S, Wilson RJM (2002) The genome of *Plasmodium falciparum* encodes an active δ-aminolevulinate dehydratase. Curr Genet 40:391–398

Sato S, Wilson RJM (2003) The use of DsRed in single- and dual-colour fluorescence labeling of live malarial organelles. Mol Biochem Parasitol 134:175–179

Seeber F (2003) Biosynthetic pathways of plastid-derived organelles as potential drug targets against parasitic Apicomplexa. Current Drug Targets- Immune, Endocrine Metabolic Disorders 3:99–109

Sharma SK, Kapoor M, Ramya TNC, Kumar S, Kumar G, Modak R, Sharma S, Surolia N, Surolia A (2003) Identification, characterization, and inhibition of *Plasmodium falciparum* β-hydroxyacyl-acyl carrier protein dehydratase (FabZ). J Biol Chem 278:45661–45671

Strath M, Scott-Finnigan T, Gardner M, Williamson D, Wilson I (1993). Antimalarial activity of rifampicin *in vitro* and in rodent models. Trans R Soc Trop Med Hyg 87:211–216

Surolia N, Ramachandra RSP, Surolia A (2002) Paradigm shifts in malaria parasite biochemistry and anti-malarial chemotherapy. BioEssays 24:192–196

Surolia N, Padmanaban G (1992) *De novo* biosynthesis of heme offers a new chemotherapeutic target in the human malarial parasite. Biochem Biophys Res Commun 187:744–750

Surolia N, Surolia A (2001). Triclosan offers protection against blood stages of malaria by inhibiting enoyl-ACP reductase of *Plasmodium falciparum*. Nat Med 7:167–173

Tripathi KD, Sharma AK, Valecha N, Kulpati DD (1993) Curative efficiacy of norfloxacin in falciparum malaria. Ind J Med Res 97:176–178

Vogeley L, Palm GF, Mesters JR, Higenfeld R (2001) Conformational change of elongation factor Tu (EF-Tu) induced by antibiotic binding. J Biol Chem 276:17149–17155

Waller AS, Clements JM (2002) Novel approaches to antimicrobial therapy: peptide deformylase. Curr Opin Drug Discov Dev 5:785–792

Waller RF, Keeling PJ, Donald RGK, Striepen B, Handman E, Lang-Unnasch N, Cowman AF, Besra GS, Roos D. Proc Natl Acad Sci USA 95:12352–12357

Waller RF, Ralph SA, Reed MB, Su V, Douglas JD, Minnikin DE, Cowman AF, Besra GS, McFadden GI (2003) A type II pathway for fatty acid biosynthesis presents drug targets in *Plasmodium falciparum*. Antimicrob Agents Chemother 47:297–301

Watt G, Shanks GD, Edstein MD, Pavanand K, Webster HK, Wechgritaya S (1991) Ciprofloxacin treatment of drug-resistant malaria. J Infect Dis 164:602–604

Weissig V, Vetro-Widenhouse TS, Rowe TC (1997) Topoisomerase II inhibitors induce cleavage of nuclear and 35-kb plastid DNAs in the malarial parasite *Plasmodium falciparum*. DNA Cell Biol 16:1483–1492

White NJ (1999) Delaying antimalarial drug resistance with combination chemotherapy. Parassitologia 41:301–308

Wiesner J, Hintz M, Altincicek B, Sanderbrand S, Weidemeyer C, Beck E, Jomaa H (2000) *Plasmodium falciparum*: Detection of the deoxyxylulose 5-phosphate reductoisomerase activity. Exp Parasitol 96:182–186

Wiesner J, Borrmann S, Jomaa H (2003) Fosmidomycin for the treatment of malaria. Parasitol Res 90:S71–S76

Wiesner J, Henschker D, Hutchinson DB, Beck E, Jomaa H (2002) In vitro and in vivo synergy of fosmidomycin, a novel antimalarial drug, with clindamycin. Antimicrob. Agents Chemother 46:2889–2894

Wiesner J, Sanderbrand S, Altincicek B, Beck E, Jomaa H (2001) Seeking new targets for antiparasitic agents. TRENDS Parasitol 17:7–8

Wilson RJM (2002) Progress with parasite plastids. J Mol Biol 319:257–274

Wilson RJM, Denny PW, Preiser PR, Rangachari K, Roberts K, Roy A, Whyte A, Strath M, Moore DJ, Moore PW, Williamson DH (1996) Complete gene map of the plastid-like DNA of the malaria parasite *Plasmodium falciparum*. J Mol Biol 261:155–172

Yeo EAT, Edstein MD, Shanks GD, Rieckmann KH (1997) Potentiation of the antimalarial activity of atovaquone by doxycycline against *Plasmodium falciparum in vitro*. Parasitol Res 83:489–491

Zhu G, Marchewka MJ, Woods KM, Upton SJ, Keithly JS (2000) Molecular analysis of a Type I fatty acid synthase *in Cryptosporidium parvum*. Mol Biochem Parasitol 105:253–260

Zuther E, Johnson JJ, Haselkorn R, McLeod R, Gornicki P (1999) Growth of *Toxoplasma gondii* is inhibited by aryloxyphenoxypropionate herbicides targeting acetyl-CoA carboxylase. Proc Natl Acad Sci USA 96:13387–13392

CTMI (2005) 295:275–291

Hemoglobin Degradation

D. E. Goldberg (✉)

Howard Hughes Medical Institute. Departments of Medicine and Molecular Microbiology, Washington University, 660 S. Euclid Ave., St. Louis, MO 63110, USA
goldberg@borcim.wustl.edu

Abstract Hemoglobin degradation by *Plasmodium* is a massive catabolic process within the parasite food vacuole that is important for the organism's survival in its host erythrocyte. A proteolytic pathway is responsible for generating amino acids from hemoglobin. Each of the enzymes involved has its own peculiarities to be exploited for development of antimalarial agents that will starve the parasite or result in build-up of toxic intermediates. There are a number of unanswered questions concerning the cell biology, biochemistry and metabolic roles of this crucial pathway.

Abbreviations

RBC   Red blood cell
HAP   Histo-aspartic protease
MPP   Mitochondrial processing peptidase
ER   Endoplasmic reticulum

1
Introduction

Malaria parasites in the bloodstream reside within host erythrocytes (red blood cells; RBCs). About 95% of the soluble RBC protein is hemoglobin, which is present at a concentration of 340 mg/ml. This serves as a rich nutrient source for parasite metabolism. An estimated 75% of the hemoglobin is consumed by *Plasmodium falciparum* during its brief intraerythrocytic stay [1–3]. Thus, hemoglobin degradation is a massive and rapid catabolic process. A number of the enzymes involved have been studied in detail and have some unusual features. A description of these enzymes, their roles in hemoglobin degradation, their biosynthesis and targeting, forms the basis for this review.

1.1
Purpose of Hemoglobin Degradation

Plasmodium parasites utilize hemoglobin as an amino acid source for protein synthesis. The evidence for this is that amino acids from radiolabeled hemoglobin get incorporated into parasite proteins [4, 5], and that despite limited ability for de novo amino acid synthesis, *P. falciparum* can survive in medium supplying just five amino acids that are in limited supply or absent from hemoglobin [6, 7]. Use of amino acids for growth appears to be important, because parasites grown in the five amino acid medium are more sensitive to hemoglobin degradation inhibitors than those grown in full medium [7].

Amino acids can also be used as an energy source, though the metabolic significance of this is unknown. Hemoglobin degradation appears to have nonanabolic roles as well. A significant portion of amino acids released from hemoglobin is excreted by the intraerythrocytic parasite [8,9], and it has been proposed that the parasite is making room for itself in its host cell [9,10], or that it is controlling RBC osmotic stability [11]. Even in medium containing all 20

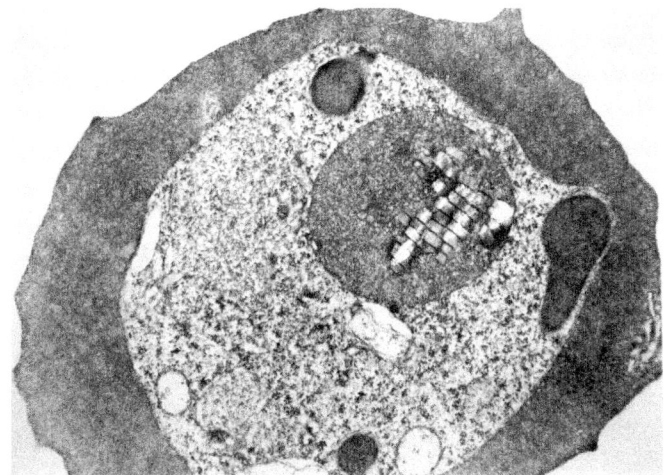

Fig. 1 Transmission electron micrograph of a *P. falciparum* trophozoite within an erythrocyte. At the *top* of the field, a cytostome is seen ingesting hemoglobin for delivery to the food vacuole, which is already filled with hemozoin crystals

amino acids, hemoglobin degradation inhibitors are effective, suggesting that exogenous amino acid supplementation cannot fully override hemoglobin proteolysis blockade.

1.2
Site of Degradation

Plasmodium ingests hemoglobin from the host cell through an opening called the cytostome, an invagination of the parasitophorous vacuolar and parasite plasma membranes. Hemoglobin (and other RBC content) is transported to the acidic food vacuole for degradation (Fig. 1). Little is known about the cell biology of hemoglobin ingestion. It has been hypothesized that hemoglobin degradation may start in transport vesicles before delivery to the vacuole [12]. There is no solid evidence to support this interesting possibility. Clearly the food vacuole is a major site of hemoglobin degradation. Its function has been reviewed in detail [13].

1.3
Degradation Pathway

A multitude of proteases have been localized to the food vacuole and proposed to play a role in hemoglobin degradation. These include a group of aspartic proteases called plasmepsins, a group of cysteine proteases called falcipains,

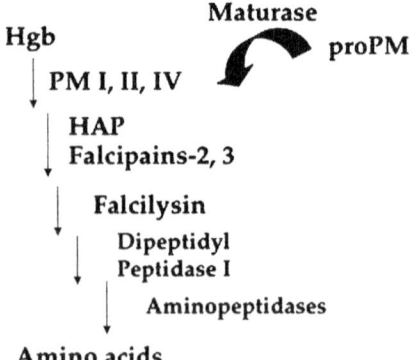

Fig. 2 Proposed hemoglobin degradation pathway. *Hgb*, hemoglobin; *PM*, plasmepsin

a metalloprotease called falcilysin, and at least one dipeptidylpeptidase I. Aminopeptidases are thought to be involved, but their site of action has not been established. There is evidence (discussed below) that the degradative enzymes function in a semi-ordered pathway (Fig. 2), with plasmepsins making the initial cleavage in intact hemoglobin, followed by secondary cleavages by plasmepsins and falcipains. Falcilysin appears to recognize only short peptides generated by upstream enzymes, while the dipeptidylpeptidases and aminopeptidases are presumed to function most efficiently in terminal degradation/amino acid release. An alternative view of the pathway has falcipains participating in the initial cleavage (see Sect. 3). It is clear that inhibitors of multiple classes of protease involved in hemoglobin degradation kill the parasites in culture and/or in animal models and therefore merit development as antimalarial drug targets.

Early in the degradative pathway, heme is released and is detoxified by assembly in a crystalline array called hemozoin. Antimalarial 4-aminoquinolines appear to function by disrupting this sequestration, leading to an accumulation of toxic heme products. This topic is discussed in the chapter by Bray et al. and Scholl et al., this volume.

2
Plasmepsins

2.1
Genomics

The plasmepsins are a group of aspartic proteases whose members were first discovered and purified by following their hemoglobinase activities [7, 14].

P. falciparum has 10 plasmepsins in its genome [15]. Other human and animal parasite species have orthologs of IV–X [16]. In *P. falciparum* or a near ancestor, IV appears to have undergone multiple gene duplications, giving rise to I, II and HAP (histo-aspartic protease, formerly plasmepsin III, see below). Plasmepsins I, II, IV and HAP genes are clustered on chromosome 14 and encode proteins that have 60%–70% amino acid identity. In contrast, these sequences are quite distant from those of the other (non-*Plasmodium*) plasmepsins (10%–20% identity).

2.2
Temporal and Spatial Location

Plasmepsins I, II, IV and HAP are all expressed in intraerythrocytic parasites and are located in the food vacuole [17]. Transcripts for plasmepsins I and IV are detected early in intraerythrocytic development (ring stage) while those for plasmepsin II and HAP are detected later (trophozoite stage) [18, 19]. Protein levels and even plasmepsin biosynthesis persist over a wider swathe of the intraerythrocytic cycle [17, 18]. How these expression patterns relate to function of the different plasmepsins has not been established. Plasmepsins II and IV have been shown to cleave spectrin in vitro and could therefore play an additional role later in intraerythrocytic development [20, 21]. The other six plasmepsins are expressed in other stages or are located in other parts of the parasite.

2.3
Specificity

All four food vacuole plasmepsins have some capacity to cleave native hemoglobin in vitro, though plasmepsin I may be the best under the conditions studied [17, 22]. It needs to be pointed out that crude gel assays do not allow real quantification or kinetic determination of hemoglobin cleavage. Specificity of native hemoglobin cleavage by plasmepsin I has been studied in detail [14, 22]. Initial cleavage occurs between 33Phe and 34Leu on the alpha chain of hemoglobin. Following this cleavage several other cleavages can occur. In 'less native' alpha globin preparations, initial cleavage at multiple sites occurs, giving rise to the model that the initial 33–34 cleavage in intact hemoglobin unravels the hemoglobin molecule so that other sites become accessible. Plasmepsin II is also capable of cleaving at the alpha 33–34 peptide bond but its secondary cleavage sites differ from those of plasmepsin I [22]. Plasmepsin IV has also been shown to make an early cleavage at 33–34 [21].

A number of techniques has been used to study the cleavage of synthetic peptides by plasmepsins. Using chromogenic substrates substituted with a series of residues one position at a time, plasmepsin II was found to prefer hydrophobic residues at P3, P2 and P3' [23] (nonprime numbering starts N terminal to the cleavage and counts upstream; prime side numbering starts C terminal to the cleavage and proceeds downstream). Proline was preferred at P4 and alanine at P2'. In these studies, basic amino acids were not tolerated in P3. This finding is inconsistent with the fact that there is an arginine at P3 of the initial cleavage site of hemoglobin. Also, fluorogenic peptides with P3 arginine are cleaved well [24]. This suggests that neighboring amino acids can influence specificity in native substrate hemoglobin, and that higher order structural features may shape specificity.

Combinatorial peptide inhibitor libraries have also been used to probe plasmepsin II specificity, and a preference for P2 branched amino acids fits nicely with the hemoglobin cleavage data [25]. Random peptide libraries have been used to probe prime side specificity [26]. The data reveal a strong preference for leucine in P1', as is found with native hemoglobin as substrate. P1' specificity could not be assessed in the chromogenic substrate assays because this residue is fixed as a nitrophenol reporter moiety.

Plasmepsins I, IV, as well as the IV orthologs from *P. vivax* and *P. malariae* have been studied using chromogenic substrates [27]. These studies emphasize differences at the P2 position. Interestingly, when recombinant plasmepsin II was made with nine amino acid substitutions to recapitulate the plasmepsin I active site surface, the specificity for hemoglobin, peptides and inhibitors remained the same as for plasmepsin II [26]. This suggested that active site geometry is more important than amino acid functionality in explaining differences in specificity between these homologous enzymes, and may be influenced by distal amino acids.

2.4
Structure and Mechanism

Plasmepsins are made in the cell as inactive proenzymes. A convertase cleaves them to generate the mature enzyme (see below). Recombinant proplasmepsins II and IV and to a limited extent proHAP are able to autoactivate by cleaving themselves fortuitously at a site near the natural cleavage site [21, 24, 28, 29]. Recombinant proplasmepsin I can autoactivate if a construct with a mutation in the propiece is made [30]. The structure of autoactivated recombinant *P. falciparum* plasmepsin II has been solved [31]. More recently a structure of *P. vivax* plasmepsin IV has been elucidated and is quite similar [32]. The corresponding proenzyme structures have also been determined [32, 33]. A structure of *P. falciparum* plasmepsin IV has been deposited in the protein database but an analysis has not been published.

The plasmepsin II structure reveals a typical two-lobe eukaryotic aspartic protease fold [31]. Indeed the plasmepsins are about 30% identical to mammalian aspartic proteases such as cathepsin D and renin. The inhibitor pepstatin sits in the active site of plasmepsin II and mammalian homologs similarly, but both ends of the molecule show conformational differences when comparing host and parasite enzymes. The thermodynamics of pepstatin binding also differ [34]. These results suggest that selective inhibitors are feasible, and indeed this has been confirmed experimentally (see Sect. 2.5).

The plasmepsins mentioned above are dimeric in the crystalline state and have extensive subunit interfaces [35]. It appears that these enzymes are dimeric in solution as well and that dimerization is important for activity and specificity [36]. This feature may be exploitable for development of selective chemotherapy.

The plasmepsins appear to function as typical aspartic proteases, using two aspartates for acid-base catalytic activation of a water molecule to promote peptide bond hydrolysis. This may also be the case for HAP, the paralog that has a histidine in place of the first catalytic aspartate [17]. Studies with peptide substrates show that HAP has kinetics that are similar to those of other plasmepsins and that it is potently inhibited by pepstatin, a transition-state mimic that forms hydrogen bonds with both aspartates in its action on aspartic proteases.

The proforms of the plasmepsins are quite unusual [32, 33]. The propiece forces open the active site and distorts it so that catalytic activity is prevented. Upon maturation (see below), extensive N-terminal refolding and rotation bring the catalytic machinery to the appropriate geometry for substrate hydrolysis to occur.

2.5
Inhibitors

A variety of inhibitors have been developed to target the plasmepsins. A comprehensive discussion of these compounds and of drug development efforts is beyond the scope of this review and is covered elsewhere [37, 38]. Some general comments will be made here. Many of the compounds generated so far are quite potent against isolated enzymes and some are quite selective for parasite over host enzymes, but they have insufficient activity against cultured parasites. The most active agents have mid to high nanomolar culture potencies[30, 37, 39, 40] and attempts to improve their potency have not yet been successful. Two reasonable explanations for this are that inhibitors have poor bioavailability or that they are potent against only a subset of nonessential plasmepsins. Most rational and combinatorial drug efforts have focused on plasmepsin II because it has been the plasmepsin for which substantial quantities of recombinant enzyme can be generated and for which a crystal structure has been determined. Unfortunately, drug studies and recent gene knockout experiments suggest that plasmepsin II is not an essential gene, nor are the other food vacuole plasmepsins [41, 42]. Indeed it is possible that the redundant function of plasmepsins is extensive enough that an inhibitor must block most or all of the food vacuole plasmepsins to kill parasites. An attempt to develop adaptive inhibitors that bind the conserved portions of the plasmepsin active site and can rotate an asymmetric functional group to interact well with the unconserved part of the substrate binding pocket, appears promising [29].

3
Falcipains

3.1
Genomics

The falcipains are papain family cysteine proteases initially identified by their role in hemoglobin degradation [22, 43, 44]. *P. falciparum* has four falcipain genes, 1, 2, 3 and 2'—a gene that is 99% identical to 2 in the mature protein but is quite divergent in the propiece. All except falcipain-1 are clustered on chromosome 11. Falcipains-2 and 3 share 53% identity, while falcipain-1 is more distantly related [45]. Rodent falcipain homologs have been characterized [46–48], though a comprehensive evolutionary analysis has not yet been carried out.

3.2
Temporal and Spatial Location

Falcipains-2 and 3 are expressed in intraerythrocytic trophozoites and schizonts [45,49]. Falcipain-3 may turn on slightly later than 2. Antisera to falcipain-2 have localized this protein to the food vacuole as well as to regions outside the food vacuole [49,50]. A role in cleavage of ankyrin during host cell exit has been proposed [50,51]. Falcipain-1 appears to be located in an apical organelle of late-stage parasites and may play a role in invasion [52].

3.3
Specificity

Falcipains-2 and 3 prefer leucine at the P2 position of synthetic peptide substrates [45,49]. Falcipain-3 catalysis is enhanced by valine at P3. It is the subject of some debate whether the falcipains are capable of cleaving native hemoglobin. Little hemoglobin cleavage is detected unless a reducing agent is added to the reaction [53]. Reducing agents denature hemoglobin [54], though under mild reducing conditions where hemoglobin denaturation cannot be detected spectrophotometrically, some hemoglobin cleavage is seen [53]. The possibility of undetected partial denaturation has not been excluded. Whether or not the falcipains have a small amount of native hemoglobin-degrading activity, they clearly work much better on denatured substrate [22, 53]. Falcipain inhibitor treatment of cultured parasites leads to hemoglobin accumulation in the food vacuole after prolonged incubation; this has been argued as being in favor of an initial role for falcipains in hemoglobin degradation [43, 55]. An alternative interpretation is that this is an indirect effect since shorter treatment does not yield hemoglobin accumulation but does allow heme release from hemoglobin, an action that is blocked by plasmepsin inhibitors [56–58]. A possible mechanism for the indirect effect of falcipain inhibitors has been proposed. Accumulation of peptide fragments from the action of upstream enzymes (plasmepsins) leads to the osmotic swelling of the food vacuole seen with cysteine protease inhibitor treatment, leading to food vacuole dysfunction and hemoglobin accumulation [59].

3.4
Structure and Mechanism

No crystal structure has been determined for the falcipains, though homology modeling based on other cysteine proteases such as papain has been performed [60]. The enzymes are blocked by standard cysteine protease inhibitors and appear to have a typical papain-family thiol protease mechanism of action.

3.5
Inhibitors

A variety of potent falcipain inhibitors in different classes have been identified and/or developed. These efforts are reviewed in detail elsewhere [61]. A few general comments will be made here. Many are potent against cultured parasites and some work in rodent malaria models. There is synergism with plasmepsin inhibitors in the test tube, in culture and in the rodent model [22, 62, 63]. Certain falcipain inhibitors show substantial promise and are under development by the Medicines for Malaria Venture [64]. It is still unclear which of the falcipains need to be inhibited to kill the parasite. A gene disruption of falcipain-2 grows normally, but is more sensitive to aspartic protease inhibitors [65].

4
Falcilysin

4.1
Genomics

Falcilysin is an M16 family metalloprotease identified in a search for a food vacuole activity that could cleave hemoglobin fragments at polar residues [66]. It is a single copy gene on falciparum chromosome 14. There are a number of other metalloproteases in the falciparum genome, all quite distantly related [67]. There are several other M16 family members; one is an apicoplast enzyme [68] and several others appear to have mitochondrial targeting sequences. Falcilysin may be the only food vacuole metalloprotease, though the possibility that others reside there has not been excluded.

4.2
Temporal and Spatial Location

Falcilysin is expressed in trophozoites and schizonts, similar to other globinases [69]. Immunolocalization studies show that the enzyme is in the food vacuole [70]. It is also located in endoplasmic reticulum (ER)-like membranes. Whether it has a separate function there has not been established (see next section).

4.3
Specificity

Falcilysin does not degrade native or denatured hemoglobin but recognizes hemoglobin peptides of 10–20 amino acids [66]. Its specificity has been studied in detail using a random peptide library. These experiments have shown that the enzyme has quite different specificities at acidic and neutral pH [70]. With some substrates the enzyme appears to be a neutral-to-alkaline protease, while with others it is clearly an acidic protease. This finding makes a second function outside the food vacuole seem entirely possible.

4.4
Structure and Mechanism

The crystal structure of falcilysin has not yet been solved. The structure of one M16 family member that has 20% identity with falcilysin, mitochondrial processing peptidase (MPP), is known but is a dimeric enzyme with its catalytic pocket at the subunit interface [71]. This is unlikely to be the case for falcilysin, which has poor homology with MPP in the dimer interface sequence and is a larger protein with its catalytic residues near the N terminus. The function of the rest of the molecule is unknown. How falcilysin achieves its dual pH-dependent specificity remains to be determined.

4.5
Inhibitors

Metal chelators block the activity of falcilysin. No selective agents have yet been identified. A preliminary attempt to disrupt the falcilysin gene by homologous recombination using positive/negative selection was unsuccessful. This raises the possibility that falcilysin is essential for parasite viability and therefore that the enzyme may be a good drug target.

5
Other Proteases

The *P. falciparum* genome contains several genes encoding proteases that may be located in the food vacuole and that may have a role in hemoglobin degradation [67]. There are three dipeptidyl peptidase-1 homologs, at least one of which is in the food vacuole. There are oligopeptidases that might function in the food vacuole—in other systems these can be degradative enzymes. There is a number of aminopeptidases [67, 72, 73]; one has been localized, in part, to a rim around the food vacuole [74].

Extracts of food vacuoles were capable of breaking down hemoglobin into small peptides [75]. No free amino acids were detected. This raised the possibility that the food vacuole generates peptides and exports them for terminal degradation by cytosolic aminopeptidases. An alternative explanation is that the food vacuole does generate amino acids in vivo but that the downstream enzymes were not active in the food vacuole extracts under the conditions used. This is not an entirely academic issue because in the first case, peptide transporters would be required at the food vacuole membrane, while in the latter case, amino acid transporters would be needed. Both classes of transporter exist in the genome and could be interesting drug targets.

6
Biosynthesis

Biosynthesis of the plasmepsins has been studied most extensively. The plasmepsins are synthesized in the ER as type II integral membrane proteins, anchored by a hydrophobic stretch in the proregion [18]. Antibody [7] and green fluorescent protein tagging [76] studies have revealed that the pro-plasmepsins go through the secretory pathway to the surface of the parasite, perhaps directly to the cytostome. From there, the proplasmepsins are internalized along with their eventual substrate hemoglobin. Targeting signals have not yet been identified. At some point in the delivery pathway, most likely upon reaching the food vacuole, the plasmepsins are cleaved from the membrane by an acid convertase, resulting in activation [18, 76, 77]. Cleavage occurs after a conserved Pro-Gly motif in the proregion and is mediated by an ALLN-sensitive enzyme with acid pH-optimal activity [18, 77]. This processing protease has not yet been isolated.

Falcipains also have substantial propieces and are activated by cleavage [45, 49]. Their biosynthesis has not been extensively studied. Recombinant falcipains-2 and 3 can be generated by activation of the proenzymes, and mature profalcipain-2 can be folded without the prodomain by inclusion in cis or in trans of a small chaperone peptide found as an N-terminal extension to the mature protease [78, 79].

Falcilysin does not have a propiece, but rather is synthesized as the mature form [69]. It is a peripheral membrane protein and might be targeted by association with another protein. Its trafficking is brefeldin A-insensitive.

7
Unanswered Questions

A number of issues remain to be answered in the field. Among them are the following:

- Why does the parasite degrade hemoglobin? For nutrients? For osmolar balance? To make room in its host cell?

- How does the cytostome form and function in hemoglobin ingestion?

- How does a protease recognize and cleave a specific peptide bond in a large, folded protein substrate? Specifically, how do the plasmepsins recognize the B helix on the alpha chain of hemoglobin and access a peptide bond that is wound up in the helix?

- Does having a substantial complement of proteases improve the efficiency of degradation compared with having fewer, less specific proteases?

- What are the proteases involved in downstream steps of proteolysis?

- Does the food vacuole generate free amino acids or does it export peptides for terminal degradation in the cytosol?

- How are the hemoglobin-degrading enzymes targeted to the food vacuole?

- What is the proplasmepsin maturase that activates the hemoglobin degradation pathway?

- How can we better exploit the eccentricities of this pathway to design potent and selective inhibitors?

It is worth continued effort to understand this important metabolic process. Biochemical, genetic and chemical studies have the potential to lead us to new antimalarial chemotherapies based on interference with the hemoglobin degradation pathway.

Acknowledgements I wish to thank Drs. Eva Istvan and Michael Klemba for critical review of this manuscript. The author is a Burroughs Wellcome Fund Scholar in Molecular Parasitology.

References

1. Morrison DB, Jeskey HA (1948) Alterations in some constituents of the monkey erythrocyte infected with *Plasmodium knowlesi* as related to pigment formation. J Nat Malar Soc 7:259–264
2. Ball EG, et al. (1948) Studies on malarial parasites: ix. chemical and metabolic changes during growth and multiplication in vivo and in vitro. J Biol Chem 175:547–571
3. Loria P, et al. (1999) Inhibition of the peroxidative degradation of haem as the basis of action of chloroquine and other quinoline antimalarials. Biochem J 339:363–370
4. McCormick GJ (1970) Amino acid transport and incorporation in red blood cells of normal and *Plasmodium knowlesi*-infected rhesus monkeys. Exp Parasitol 27:143–149
5. Sherman IW, Tanigoshi L (1970) Incorporation of 14C-amino acids by malaria. Int J Biochem 1:635–637
6. Divo AA, et al. (1985) Nutritional requirements of *Plasmodium falciparum* in culture. I. Exogenously supplied dialyzable components necessary for continuous growth. J Protozool 32:59–64
7. Francis SE, et al. (1994) Molecular characterization and inhibition of a *Plasmodium falciparum* aspartic hemoglobinase. EMBO J 13:306–317
8. Zarchin S, Krugliak M, Ginsburg H (1986) Digestion of the host erythrocyte by malaria parasites is the primary target for quinolone-containing antimalarials. Biochem Pharmacol 35:2435–2442
9. Krugliak M, Zhang J, Ginsburg H (2002) Intraerythrocytic *Plasmodium falciparum* utilizes only a fraction of the amino acids derived from the digestion of host cell cytosol for the biosynthesis of its proteins. Mol Biochem Parasitol 119:249–256
10. Ginsburg H (1990) Some reflections concerning host erythrocyte-malarial parasite interrelationships. Blood Cells 16:225–235
11. Lew VL, Tiffert T, Ginsburg H (2003) Excess hemoglobin digestion and the osmotic stability of *Plasmodium falciparum*-infected red blood cells. Blood 101:4189–4194

12. Hempelmann E, et al. (2003) *Plasmodium falciparum*: sacrificing membrane to grow crystals? Trends Parasitol 19:23–26

13. Banerjee R, Sullivan DJ Jr, Goldberg DE (2001) The *Plasmodium* food vacuole. In: Rosenthal PJ (ed.) ntimalarial chemotherapy: mechanisms of action, resistance and new directions in drug discovery.Humana Press: Totowa, NJ, ch 4

14. Goldberg DE, et al. (1991) Hemoglobin degradation in the human malaria pathogen *Plasmodium falciparum*: A catabolic pathway initiated by a specific aspartic protease. J Exp Med 173:961–969

15. Coombs GH, et al. (2001) Aspartic proteases of *Plasmodium falciparum* and other parasitic protozoa as drug targets. Trends Parasitol 17:532–537

16. Dame JB, et al. (2003) Plasmepsin 4, the food vacuole aspartic proteinase found in all *Plasmodium* spp. infecting man. Mol Biochem Parasitol 130:1–12

17. Banerjee R, et al. (2002) Four plasmepsins are active in the *Plasmodium falciparum* food vacuole, including a protease with an active-site histidine. Proc Natl Acad Sci USA 99:990–995

18. Francis SE, Banerjee R, Goldberg DE (1997) Biosynthesis and maturation of the malaria aspartic hemoglobinases plasmepsins I and II. J Biol Chem 272:14961–14968

19. Bozdech Z, et al. (2003) The transcriptome of the intraerythrocytic developmental cycle of *Plasmodium falciparum*. PLoS Biol 1:E5

20. Le Bonniec S, et al. (1999) Plasmepsin II, an acidic hemoglobinase from the *Plasmodium falciparum* food vacuole, is active at neutral pH on the host erythrocyte membrane skeleton. J Biol Chem 274:14218–14223

21. Wyatt DM, Berry C (2002) Activity and inhibition of plasmepsin IV, a new aspartic proteinase from the malaria parasite, *Plasmodium falciparum*. FEBS Lett 513:159–162

22. Gluzman IY, et al. (1994) Order and specificity of the *Plasmodium falciparum* hemoglobin degradation pathway. J Clin Invest 93:1602–1608

23. Westling J, et al. (1999) Active site specificity of plasmepsin II. Protein Sci 8:2001–2009

24. Luker KE, et al. (1996) Kinetic analysis of plasmepsins I and II, aspartic proteases of the *Plasmodium falciparum* digestive vacuole. Mol Biochem Parasitol 79:71–78

25. DiIanni Carroll C, et al. (1998) Identification of potent inhibitors of *Plasmodium falciparum* plasmepsin II from an encoded statine combinatorial library. Bioorg Med Chem Lett 8:2315–2320

26. Siripurkpong P, et al. (2002) Active site contribution to specificity of the aspartic proteases plasmepsins I and II. J Biol Chem 277:41009–41013

27. Westling J, et al. (1997) *Plasmodium falciparum*, *P. vivax*, and *P. malariae*: a comparison of the active site properties of plasmepsins cloned and expressed from three different species of the malaria parasite. Exp Parasitol 87:185–193

28. Dame JB, et al. (1994) Sequence, expression and modeled structure of an aspartic proteinase from the human malaria parasite *Plasmodium falciparum*. Mol Biol Parasitol 64:177–190

29. Nezami A, et al. (2003) High-affinity inhibition of a family of *Plasmodium falciparum* proteases by a designed adaptive inhibitor. Biochemistry 42:8459–8464

30. Moon RP, et al. (1997) Expression and characterization of plasmepsin I from *Plasmodium falciparum*. Eur J Biochem 244:552–560

31. Silva AM, et al. (1996) Structure and inhibition of plasmepsin II, a hemoglobin-degrading enzyme from *Plasmodium falciparum*. Proc Natl Acad Sci USA 93:10034–10039

32. Bernstein NK, et al. (2003) Structural insights into the activation of *P. vivax* plasmepsin. J Mol Biol 329:505–524

33. Bernstein NK, et al. (1999) Crystal structure of the novel aspartic proteinase zymogen proplasmepsin II from plasmodium falciparum. Nat Struct Biol 6:32–37

34. Xie D, et al. (1997) Dissection of the pH dependence of inhibitor binding energetics for an aspartic protease: direct measurement of the protonation states of the catalytic aspartic acid residues. Biochemistry 36:16166–16172

35. Asojo OA, et al. (2003) Novel uncomplexed and complexed structures of plasmepsin II, an aspartic protease from *Plasmodium falciparum*. J Mol Biol 327:173–181

36. Istvan ES, Goldberg DE (2003) Dimerization of *P. falciparum* plasmepsins: implications for catalysis and drug design. Molecular Parasitology Meeting XIV, Woods Hole, MA, p.16E

37. Boss C, et al. (2003) Inhibitors of the *Plasmodium falciparum* parasite aspartic protease plasmepsin II as potential antimalarial agents. Curr Med Chem 10:883–907

38. Klemba M, Goldberg DE (2002) Biological roles of proteases in parasitic protozoa. Annu Rev Biochem 71:275–305

39. Haque TS, et al. (1999) Potent, low-molecular-weight non-peptide inhibitors of malarial aspartyl protease plasmepsin II. J Med Chem 42:1428–1440

40. Jiang S, et al. (2001) New class of small nonpeptidyl compounds blocks *Plasmodium falciparum* development in vitro by inhibiting plasmepsins. Antimicrob Agents Chemother 45:2577–2584

41. Dame JB, et al. (2003) Molecular and phenotypic characterization of gene knockouts of each of the four food vacuole plasmepsins of *Plasmodium falciparum*. Molecular Parasitology Meeting XIV, Woods Hole, MA, p.277C

42. Liu J, Drew M, Goldberg DE (in preparation, 2004)

43. Rosenthal PJ, et al. (1988) A malarial cysteine proteinase is necessary for hemoglobin degradation by *Plasmodium falciparum*. J Clin Invest 82:1560–1566

44. Rosenthal PJ, et al. (1989) *Plasmodium falciparum*: inhibitors of lysosomal cysteine proteinases inhibit a trophozoite proteinase and block parasite development. Mol Biochem Parasitol 35:177–184

45. Sijwali PS, et al. (2001) Expression and characterization of the *Plasmodium falciparum* haemoglobinase falcipain-3. Biochem J 360:481–489

46. Rosenthal PJ, Lee GK (1993) Inhibition of a *Plasmodium vinkei* cysteine proteinase cures murine malaria. J Clin Invest 91:1052–1056

47. Rosenthal PJ (1993) A *Plasmodium vinckei* cysteine proteinase shares unique features with its *Plasmodium falciparum* analogue. Biochem Biophys Acta 1173:91–93

48. Rosenthal PJ (1996) Conservation of key amino acids among the cysteine proteinases of multiple malarial species. Mol Biochem Parasitol 75:255–260

49. Shenai BR, et al. (2000) Characterization of native and recombinant falcipain-2, a principal trophozoite cysteine protease and essential hemoglobinase of *Plasmodium falciparum*. J Biol Chem 275:29000–29010

50. Dhawan S, et al. (2003) Ankyrin peptide blocks falcipain-2-mediated malaria parasite release from red blood cells. J Biol Chem 278:30180–30186
51. Dua M, et al. (2001) Recombinant falcipain-2 cleaves erythrocyte membrane ankyrin and protein 4.1. Mol Biochem Parasitol 116:95–99
52. Greenbaum DC, et al. (2002) A role for the protease falcipain 1 in host cell invasion by the human malaria parasite. Science 298:2002–2006
53. Shenai BR, Rosenthal PJ (2002) Reducing requirements for hemoglobin hydrolysis by *Plasmodium falciparum* cysteine proteases. Mol Biochem Parasitol 122:99–104
54. Atamna H, Ginsburg H (195) Heme degradation in the presence of glutathione: A proposed mechanism to account for the high levels of non-heme iron found in the membranes of hemoglobinopathic red blood cells. J Biol Chem 42:24876–24883
55. Gamboa de Dominguez ND, Rosenthal PJ (1996) Cysteine proteinase inhibitors block early steps in hemoglobin degradation by cultured malaria parasites. Blood 87:4448–4454
56. Bray PG, et al. (1998) Access to hematin: the basis of chloroquine resistance. Mol Parmacol 54:170–179
57. Mungthin M, et al. (1998) Central role of hemoglobin degradation in mechanisms of action of 4-aminoquinolines, quinoline methanols, and phenanthrene methanols. Antimicrob Agents Chemother 42:2973–2977
58. Bray PG, et al. (1999) Cellular uptake of chloroquine is dependent on binding to ferriprotoporphyrin IX and is independent of NHE activity in *Plasmodium falciparum.*
59. Francis SE, et al. (1996) Characterization of native falcipain, an enzyme involved in *Plasmodium falciparum* hemoglobin degradation. Mol Biochem Parasitol 83:189–200
60. Ring CS, et al. (1993) Structure-based inhibitor design by using protein models for the development of antiparasitic agents. Proc Natl Acad Sci USA 90:3583–3587
61. Rosenthal PJ, et al. (2002) Cysteine proteases of malaria parasites: targets for chemotherapy. Curr Pharm Des8:1659–1672
62. Bailly E, et al. (1992) *Plasmodium falciparum*: differential sensitivity in vitro to E-64 (cysteine protease inhibitor) and pepstatin (aspartic protease inhibitor). J Protozool 39:593–599
63. Semenov A, Olson JE, Rosenthal PJ (1998) Antimalarial synergy of cysteine and aspartic protease inhibitors. Antimicrob Agents Chemother 42:2254–2258
64. Ridley RG (2002) Medical need, scientific opportunity and the drive for antimalarial drugs. Nature 415:686–693
65. Sijwali PS, Lee BJ, Rosenthal PJ (2003) Knock-down of falcipain-2 supports a cooperative role for cysteine and aspartic proteases in hemoglobin hydrolysis by *P. falciparum*. Molecular Parasitology Meeting XIV, Woods Hole, MA, p 277C
66. Eggleson KK, Duffin KL, Goldberg DE (1999) Identification and characterization of falcilysin, a metallopeptidase involved in hemoglobin catabolism within the malaria parasite *Plasmodium falciparum*. J Biol Chem 274:
67. Wu Y, et al. (2003) Data-mining approaches reveal hidden families of proteases in the genome of malaria parasite. Genome Res 13:601–616
68. van Dooren GG, et al. (2002) Processing of an apicoplast leader sequence in *Plasmodium falciparum* and the identification of a putative leader cleavage enzyme. J Biol Chem 277:23612–23619

69. Murata CE, Goldberg DE (2003) *Plasmodium falciparum* falcilysin: an unprocessed food vacuole enzyme. Mol Biochem Parasitol 129:123–126
70. Murata CE, Goldberg DE (2003) *Plasmodium falciparum* falcilysin: a metalloprotease with dual specificity. J Biol Chem 278:38022–38028
71. Taylor AB, et al. (2001) Crystal structures of mitochondrial processing peptidase reveal the mode for specific cleavage of import signal sequences. Structure (Camb) 9:615–625
72. Gavigan CS, Dalton JP, Bell A (2001) The role of aminopeptidases in haemoglobin degradation in *Plasmodium falciparum*-infected erythrocytes. Mol Biochem Parasitol 117:37–48
73. Florent I, et al. (1998) A *Plasmodium falciparum* aminopeptidase gene belonging to the M1 family of zinc-metallopeptidases is expressed in erythrocytic stages. Mol Biochem Parasitol 97:149–160
74. Allary M, Schrevel J, Florent I (2002) Properties, stage-dependent expression and localization of *Plasmodium falciparum* M1 family zinc-aminopeptidase. Parasitology 125:1–10
75. Kolakovich KA, et al. (1997) Generation of hemoglobin peptides in the acidic digestive vacuole of *Plasmodium falciparum* implicates peptide transport in amino acid production. Mol Biochem Parasitol 87:123–135
76. Klemba M, et al. (2004) Trafficking of plasmepsin II to the food vacuole of the malaria parasite *Plasmodium falciparum*. J Cell Biol (in press)
77. Banerjee R, Francis SE, Goldberg DE (2003) Food vacuole plasmepsins are processed at a conserved site by an acidic convertase activity in *Plasmodium falciparum*. Mol Biochem Parasitol 129:157–165
78. Sijwali PS, Shenai BR, Rosenthal PJ (2002) Folding of the *Plasmodium falciparum* cysteine protease falcipain-2 is mediated by a chaperone-like peptide and not the prodomain. J Biol Chem 277:14910–14915
79. Pandey KC, et al. (2003) Independent intramolecular mediators of folding, activity, and inhibition for the *Plasmodium falciparum* cysteine protease falcipain-2. J Biol Chem

CTMI (2005) 295:293–324

Bioavailable Iron and Heme Metabolism in *Plasmodium falciparum*

P. F. Scholl[1] · A. K. Tripathi[2] · D. J. Sullivan[2] (✉)

[1]Department of Environmental Health Sciences, Baltimore, MD 21205, USA

[2]W. Harry Feinstone Department of Molecular Microbiology and Immunology, Bloomberg School of Public Health, Johns Hopkins University, E5628, 615 North Wolfe Street, Baltimore, MD 21205, USA
dsulliva@jhsph.edu

Abstract Iron metabolism is essential for cell function and potentially toxic because iron can catalyze oxygen radical production. Malaria-attributable anemia and iron deficiency anemia coincide as being treatable diseases in the developing world. In absolute amounts, more than 95% of *Plasmodium* metal biochemistry occurs in the acidic digestive vacuole where heme released from hemoglobin catabolism forms heme crystals. The antimalarial quinolines interfere with crystallization. Despite the completion of the *Plasmodium* genome, many 'gene gaps' exist in components of the metal pathways described in mammalian or yeast cells. Present evidence suggests that parasite bioavailable iron originates from a labile erythrocyte cytosolic pool rather than from abundant heme iron. Indeed the parasite has to make its own heme within two separate organelles, the mitochondrion and the apicomplast. Paradoxically, despite the abundance of iron within the erythrocyte, iron chelators are cytocidal to the *Plasmodium* parasite. Hemozoin has become a sensitive biomarker for laser desorption mass spectrometry detection of *Plasmodium* infection in both mice and humans.

Abbreviations

DMT	Divalent metal-ion transporters
ZnPPIX	Zinc protoporphyrin IX
HRP II	Histidine-rich protien II
VEPL	Vacuole enriched parasite lysates
DFO	Desferrioxamine
LDMS	Laser desorption mass spectrometry
MALDI	Matrix-assisted laser desorption ionization

1
Human Host Iron Metabolism

Iron is an essential participant in metabolic reactions in all living cells because as a transition metal it can undergo changes in electron oxidation states and therefore can both oxidize and reduce molecules. Inorganic iron is the second most abundant metal in the Earth's crust, yet from a biologic perspective is difficult to extract from its insoluble oxides. Likewise in humans 95% percent of the 3 g of iron is highly concentrated in erythrocytes or macrophages, with minute amounts bioavailable in tissues. Daily absorption and excretion is one-thousandth of the total pool (Fairbanks and Beutler 1995b; Hentze et al. 2004). Similarly, the intraerythrocytic *Plasmodium* parasite exists in the erythrocyte with 100 fg (20 mM) of iron per cell, but utilizes less than one-one-thousandth of this iron pool (Egan et al. 2002).

Iron exists only transiently as a free cation because it is carefully chaperoned to proteins that bind or incorporate it. Iron proteins comprise the heme proteins such as hemoglobin, myoglobin, catalase and cytochromes or iron flavoproteins such as xanthine oxidase, succinate dehydrogenase and NAD dehydrogenase (Hentze et al. 2004). Iron is essential for mitochondrial and chloroplast function as the electron transport chain, iron sulfur clusters and nearly half of the enzymes of the tricarboxylic acid cycle require iron. Because iron can also damage membranes and proteins with the production of oxygen radicals, its intracellular concentration is closely regulated with an array of storage and transport proteins that are transcriptionally regulated in eukaryotes. An iron responsive element and iron regulatory protein interact in the absence or presence of bioavailable iron to tightly regulate genes like ferritin, transferrin, aconitases, heme synthesis enzymes and the iron transporters such as divalent metal-ion transporters (DMT) or ferroportins (Hentze et al. 2004).

The human host of the malaria parasite has approximately 2,500 mg (female) to 3,500 mg (male) of total iron as depicted in Fig. 1. Hemoglobin contains approximately 1,700 mg (female)–2,400 mg (male) or about 65%

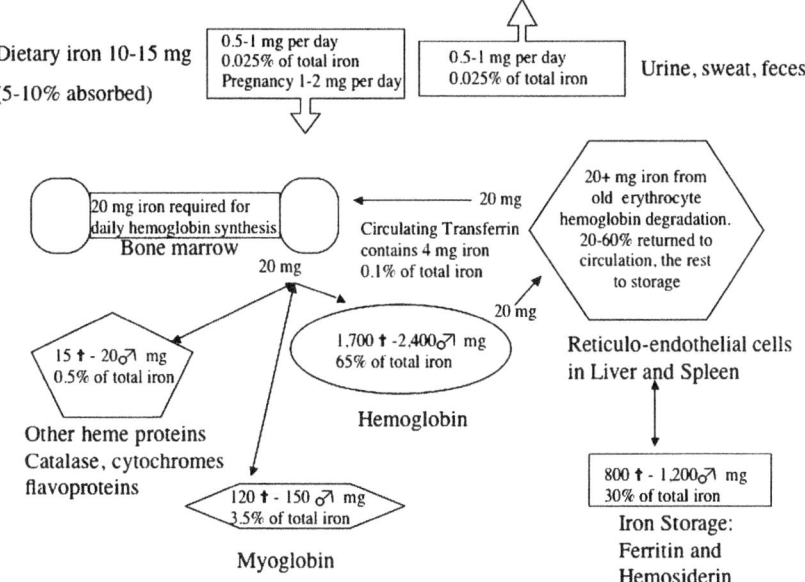

Fig. 1 Human iron compartments

of the iron total. The reticuloendothelial macrophages, predominately in the liver and spleen, contain approximately 600 mg in ferritin and hemosiderin; the bone marrow approximately 300 mg, the liver parenchyma approximately 1,000 mg and muscle and other cells approximately 400 mg. Only 3–4 mg circulate in a plasma labile pool bound to transferrin. This labile pool, also subject to diurnal variation, turns over seven to eight times a day resulting in approximately 20 mg that is loaded and delivered principally to the bone marrow for hemoglobin synthesis. While total dietary iron intake is about 10–15 mg, only 1 mg (10%) is absorbed through the intestine. One milligram is also excreted in sweat, urine or feces each day. Thus the majority of the daily hemopoietic iron requirements originates mainly from macrophage recycling of heme iron and also from ferritin store transfer. The spleen and liver reticuloendothelial cells normally digest phagocytosed erythrocytes at a rate sufficient to release 20% of the heme iron in a few hours with 20%–70% recycled back to hemoglobin in a few days (Fairbanks and Beutler 1995b; Hentze et al. 2004).

In the past 15 years, many aspects of the molecular iron metabolism in mammals have been elucidated. The majority of intestinal iron absorption is by heme endocytosis or transport into intestinal epithelial cells, called enterocytes where heme oxygenase degrades heme to release iron. The mechanism of heme transport across membranes is a large unanswered question. An alternative enterocyte pathway, that is upregulated in iron deficiency, is direct iron cation transport by DMT-1 (Gushin et al. 1997). The enterocyte exports iron to the plasma via the transporter, ferroportin (McKie et al. 2000). Cellular iron uptake from plasma is by receptor-mediated uptake of transferrin. Acidification of the endocytotic vacuole releases iron which is transported into the cell cytosol again by DMT-1 (Hentze et al. 2004). Only specialized cells, including duodenal enterocytes, macrophages, hepatocytes, astrocytes and syncytiotrophoblasts, are capable of iron export via ferroportin. Iron export from nonintestinal cells requires ceruloplasmin or hephaestin (Garrick et al. 2003; Hentze et al. 2004; Kaplan and O'Halloran 1996; Miyajima 2002). Most of the cellular requirement for iron is for mitochondrial function and DNA synthesis.

Iron deficiency anemia affects over 500 million people including more than half of all children and pregnant women (Yip 1998). Low total iron body content results most commonly from insufficient dietary intake, but can also result, amongst diverse etiologies, from chronic blood loss, diversion of iron to the fetus in pregnancy or intravascular hemolysis with hemoglobinuria. Iron depletion is first evidenced with low to absent tissue ferritin storage, with intact plasma iron and blood hemoglobin levels. Clinical iron deficiency gradually progresses from low (20% of normal) plasma iron and transferrin saturation without anemia before progressing to low hemoglobin levels manifested by a hypochromic, microcytic anemia. The normal 120-day erythrocyte life span is decreased by 10%–20%. Cellular and erythrocyte ferritin decreases, while soluble and surface transferrin receptors increase along with erythrocyte zinc protoporphyrin IX (ZnPPIX). Serum erythropoietin is elevated in pure iron deficiency anemia (Fairbanks and Beutler 1995a).

Anemia of chronic disease caused by infection, inflammation or cancer is a systemic response that is hypothesized to limit bioavailable iron from pathogenic microbes or tumors in general (Gera and Sachdev 2002). Total body iron is normal, with high ferritin stores, but plasma iron is 20% of normal and dietary absorption decreases. Erythrocyte life span is also reduced by 20%. Most studies show ineffective erythropoiesis with a blunted production and response to erythropoietin. However, exogenous recombinant erythropoietin increases erythrocyte production in many chronic inflammatory diseases (Spivak 2002).

2
Perturbations of Human Iron Metabolism by Malaria

Malaria infection and disease impacts iron metabolism in many diverse ways (see the chapter by D. Roberts et al., this volume). Malaria infection geographically coincides with iron deficiency anemia. As a chronic infectious disease, malaria further decreases iron uptake from the intestine, contributes to ineffective erythropoeisis and sequesters bioavailable iron in ferritin stores (Spivak 2002). Lysis of infected erythrocytes may increase urinary losses of hemoglobin iron. The increased destruction of both infected and uninfected erythrocytes increases demand for bone marrow production. Asymptomatic malaria in semi-immune individuals is represented with parasitemias of approximately 1,000 parasites/µl or 0.025% infected erythrocytes. This fraction would represent 2.5% of daily erythrocytes destroyed. Symptomatic infections with 400,000 parasites/µl or 10% of erythrocytes infected represents a stress on the reticuloendothelial macrophages of 10 times the normal rate of destruction of erythrocytes. During maturation in the erythrocyte, the *P. falciparum* parasite sequesters 50%–65% of erythrocyte heme into an insoluble hemozoin crystal (Egan et al. 2002; Francis et al. 1997). Hemozoin is resistant to heme oxygenase degradation and accumulates in macrophages, monocytes and polymorphonuclear neutrophils (Schwarzer et al. 1999b). Essentially the *P. falciparum* parasite removes half of bioavailable heme presented to reticuloendothelial cells because macrophages are unable to degrade hemozoin heme to release iron. From a different perspective, a single erythrocyte has approximately 90–100 fg of iron with 50 fg present in hemozoin crystals at mature schizont stages, therefore 100 billion parasites contain 5 mg of iron in hemozoin (Egan et al. 2002). A single liter of whole blood contains 3–4 trillion erythrocytes. Therefore, a 2%–3% parasitemia represents 100 billion parasites in a liter (100,000/µl) or 5 mg of iron in hemozoin. Total blood volume for children is approximately 65–75 ml/kg lean body weight (Haddad et al. 2001). A 14-kg child then would have a liter of blood from which a 2%–3% parasitemia would remove 5 mg of iron in the form of hemozoin from the human iron metabolism cycle every 2 days. A 70-kg adult with 5 l of blood and a 2%–3% parasitemia, then has 30 mg of iron removed from bioavailable pools every 2 days. Hemozoin does not stimulate heme oxygenase production in macrophages/monocytes in vitro (Schwarzer et al. 1999a) and can persist for months (Levesque et al. 1999; Taliaferro and Mulligan 1937). Hemozoin has been shown to have specific inhibitory effects on macrophages and monocytes and also may directly inhibit hemoprogenitor cells in the marrow (Arese and Schwarzer 1997; Schwarzer et al. 1992). In summary, the malaria parasite causes anemia acutely by a destruction of both infected and uninfected erythrocytes, by a negative impact on erythropoiesis, by inducing an anemia of chronic diseases state and by sequestering away tens of milligrams of iron into nonbioavailable hemozoin which can persist for months before degradation to release iron.

The proportions of anemic individuals with malaria attributable anemia (Menendez et al. 1997) versus low dietary intake (Tatala et al. 1998) or losses from intestinal helminths are often difficult to establish on a public health community level (Brooker et al. 1999; Leenstra et al. 2004; Nyakeriga et al. 2004; Stoltzfus et al. 2000). Both iron replenishment alone or malaria chemoprophylaxis or treatment increase hemoglobin levels by 1–2 g/dl (Massaga et al. 2003; Menendez et al. 2004). Often, however, with a high proportion of malaria attributable anemia in a population, iron repletion has little to no effect on hemoglobin levels (Desai et al. 2004; Stoltzfus et al. 2004). However, iron supplementation and malaria chemotherapy both increase hemoglobin values in a synergistic manner (Desai et al. 2003; Ekvall et al. 2000; Verhoef et al. 2002) although some studies still show minimal effect of combination therapy (Menendez et al. 2004).

A lingering debate in the malaria and nutrition community concerns community dietary iron supplementation, which may increase malaria disease in malaria endemic areas. Iron replenishment which improves anemia has both positive growth, development and cognitive benefits. However, early studies noted an increase in malaria disease especially with injection iron, but also with oral supplementation (Murray et al. 1975; Oppenheimer 2001). Other more recent studies mainly with oral replacement have not shown exacerbation of malaria disease (Berger et al. 2000; Mebrahtu et al. 2004; Menendez et al. 1997, 2004). A meta-analysis study did not show an increase in malaria disease from iron supplementation (Gera and Sachdev 2002). A hypothesis that seeks to balance the conflicting studies is that only the most severely anemic children with hemoglobin concentrations below 6–7 g/dl (less than 10%–20% of children in most malaria anemia study populations) are at risk for exacerbation of malaria disease with iron supplementation. The more rapid injection supplementation rather than the slower absorption by oral dosing increases the risk.

An hypothesized molecular mechanism that accounts for possible exacerbation of malaria disease with rapid iron supplementation involves high levels of ZnPPIX (Iyer et al. 2003). When bioavailable iron is low the human mitochondrial ferrochelatase inserts zinc instead of iron into protoporphyrin IX. Normal erythrocytes have concentration of ZnPPIX of 0.5 μM (25μmol ZnPPIX:1 mol heme). This can elevate to 5 μM (250 μmol ZnPPIX:1 mol heme) in severe iron deficiency anemia. Moderate iron deficiency anemia or anemia of chronic disease do not elevate ZnPPIX to these levels (Hastka et al. 1993; Lamola and Yamane 1974). ZnPPIX resides in the heme pocket of hemoglobin and can inhibit hemozoin extension in vitro with an IC_{50} of 5 μM (Iyer et al. 2003). Individuals with severe anemia have subpopulations of erythrocytes with greater elevations of ZnPPIX which are inhospitable to malaria infection alongside erythrocytes with slightly lower ZnPPIX levels that are able to support infection. Rapid iron repletion stimulates production of erythrocytes with lower levels of ZnPPIX and malaria disease is possibly exacerbated in a few weeks with the replacement by a cohort of new erythrocytes capable of supporting high parasitemias.

3
Plasmodium Iron Metabolism

The intraerythrocytic *P. falciparum* parasite encounters two important issues that relate to iron. The first is how to handle the excess heme iron produced from hemoglobin catabolism and the second is how to acquire the iron necessary for cellular metabolism.

In absolute amounts more than 95% of *Plasmodium* metal biochemistry involves the sequestration of reactive, toxic heme in an acidic, oxygen-rich vacuole into an inert heme crystal called hemozoin that removes the heme iron moiety from solution chemistry (Egan et al. 2002; Francis et al. 1997). The appearance of microscopic birefringent hemozoin marks the *Plasmodium* transition from ring stage to trophozoite stage seen in Fig 2. Microscopically, schizont stage is the appearance of more than one nucleus. Gametocytes, ookinetes and even oocysts retain the hemozoin crystals, while sporozoites, liver stages and merozoites lack hemozoin. The intraerythrocytic parasite ingests 60%–80% of the host's 5 mM hemoglobin both as a source of amino acids and possibly to provide room for the parasite (Egan et al. 2002; Lew et al. 2003; Zhang et al. 1999). Interestingly, the *P. falciparum* parasite ingests 10 times the amount of hemoglobin protein than it retains for amino acid building blocks (Krugliak et al. 2002). *Plasmodium* aspartic and cysteine proteases efficiently degrade ingested hemoglobin releasing approximately 1 fmol of heme in a 2 fl volume of the acidic digestive vacuole. Conceptually, this approximates to almost 0.4 M heme in an oxygen-rich acidic environment capable of generating oxygen radicals by the Fenton reaction (Francis et al. 1997). The parasite lacks heme oxygenase activity or iron storage proteins

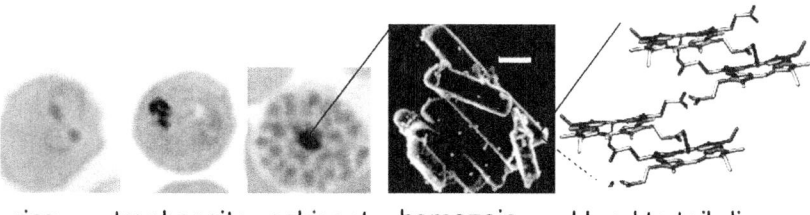

ring   trophozoite   schizont   hemozoin   Head to tail dimer

Fig. 2 Progressive magnification of hemozoin. Giemsa stain (×100) of ring, trophozoite and schizont stage intraerythrocytic *P. falciparum* with electron microscopy of crystals (*scale bar*, 200 nm) and model of hydrogen bonding of two heme crystal dimers

like ferritin. Indeed, the parasite has to synthesize its own heme rather than scavenge from the erythrocyte (Surolia and Padmanaban 1992).

Within the intraerythrocytic cycle, a single parasite undergoes rapid proliferation, multiplying 8–32-fold within 48 h. In a few weeks time, billions of infected erythrocytes result. This rapid replication of DNA requires iron for the ribonucleotide reductase for nucleotide synthesis (Mabeza et al. 1999). Despite thriving in a host cell with 20 mM concentrations of heme iron, the parasite is killed by low concentrations of iron chelators indicating that the amount of bioavailable iron is limited and crucial for parasite growth and survival (Mabeza et al. 1999; Raventos-Suarez et al. 1982).

3.1
Hemozoin Structure and Function

Even though the black malaria pigment or hemozoin was known to contain heme as described by Carbone in 1891 (Carbone 1891), 100 years passed before this molecule was demonstrated to be β-hematin, which spontaneously forms in vitro if heme is incubated at high (>50°C) temperatures in acidic, aqueous solutions (Fitch and Kanjananggulpan 1987; Slater et al. 1991). In 1991, Slater and Cerami reported, by different spectroscopic techniques, the unique iron–oxygen bond coordinating the propionate carboxylate oxygen of one heme to the central ferric molecule of another (Slater et al. 1991). They later demonstrated that parasite lysates were capable of catalyzing hemozoin formation in vitro (Slater and Cerami 1992). The initial model was that of heme polymers (Slater 1992). Later powder diffraction data by Bohle's group demonstrated instead that the biocrystal consists of two hemes linked by a reciprocal iron (Fe-1) head to carboxylate (O-31) tail (Pagola et al. 2000). The two coordinate hemes made a triclinic unit cell, $a=12.196$ (2) A°, $b=14.684$ (2) A°, $c=8.040$ (1) A°, a=90.22 (1)8, b=96.80 (1)8, γ=97.92 (1)8; $V=1416.0$ (3) A° 3; $\varrho_{exp}=1.45$ (1) g/cm^3; $Z=2$ (Pagola et al. 2000). As shown in Fig. 2 the head-to-tail dimers crystallize by hydrogen bonding of the remaining carboxylate groups into a lattice with a rectangular crystal morphology of size dimension 100 nm×100 nm×500 nm in the case of *P. falciparum* (Noland et al. 2003). β-hematin formation is a chemical process governed by time and temperature that absolutely requires an acidic pH in aqueous solutions (Egan 2002; Egan et al. 1994b). Blauer has described intermediate transition states by FTIR called B-hematin (Blauer and Akkawi 1997, 2000). Preformed heme crystals allow for rapid extension at 37°C (Chong and Sullivan 2003). Kinetics of formation is sigmoidal, characterized by a nucleation and then an extension process (Chong and Sullivan 2003; Egan 2002; Egan et al. 2001). Single head-to-tail dimers have not been isolated as they incorporate rapidly into the larger aggregate structure.

Presently no consensus exists about how hemozoin is made within the parasite. Diverse hypotheses with some supporting in vitro data include spontaneous initiation (Egan et al. 1994b), seeded from preformed crystals (Ridley 1996), enzymatic by a heme 'polymerase or crystallase' (Slater and Cerami 1992), initiation by lipids (Bendrat et al. 1995; Fitch et al. 1999; Hempelmann and Egan 2002; Tripathi et al. 2002) or nucleation by parasite proteins (Choi et al. 1999; Sullivan et al. 1996a). The Pf histidine-rich protein II (HRP II)–that has 35% of its amino acids as histidine–is present in the digestive vacuole; it can bind heme at acidic pH and is able to initiate in vitro heme crystal formation inhibited by the quinolines (Sullivan et al. 1996a). However, the laboratory progeny, 3B-D7 from parents HB3 and Dd2, lacks both HRP II and HRP III and still makes hemozoin (Sullivan 2002b). While both *P. reichnowi* and *P. lophurae* have a protein with HRP II primary structure homology, the three mouse malaria's sequenced and *P. vivax* and *P. knowlesi*, all lack a close homolog of PfHRP II or III (Bahl et al. 2002). This indicates that PfHRP II is sufficient in vitro, but not absolutely necessary in vivo, to initiate hemozoin formation (Sullivan 2002a, 2002b). Tilley has also demonstrated that only a small fraction (but still several μM) of PfHRP II reaches the digestive vacuole (Papalexis et al. 2001).

Attention has recently focused on lipid initiation of heme crystal formation first described by Bendrat and Cerami, with further evidence of in vitro formation by Fitch et al. 1999, and also by Hempelmann et al. 2003, Pandey et al. 2003, and Tekwani (Tripathi et al. 2002). Unsaturated fatty acids and phospholipids accelerate hemozoin formation possibly by increasing solubility of heme in micelles. Likewise increasing dimethyl sulfoxide concentrations also accelerate hemozoin formation. However, some lipid formulations such as erythrocyte ghosts or the cholesterol-rich lipoproteins do not initiate hemozoin formation as efficiently (Fitch et al. 2003). While the lipid bilayers may contain a few unit head-to-tail crystal dimers, how the dimers are able to stack into the larger regular array of hemozoin crystals outside of a membrane remains unanswered.

3.2
Biochemical Formation In Vivo and In Vitro

Many different assays have been developed to compare the formation of hemozoin and inhibition by drugs. The assays can be divided by: (1) use of radioactivity (Dorn et al. 1995; Hawley et al. 1998; Slater and Cerami 1992) versus direct absorbance measurement of purified heme crystal (Basilico et al. 1998); (2) initiation by parasite lysates (Fitch and Chou 1996; Slater and Cerami 1992), preformed crystals (Dorn et al. 1998a, 1998b; Sullivan et al. 1996b), proteins (Sullivan et al. 1996a), or lipids (Bendrat et al. 1995;

Kurosawa et al. 2000; Tripathi et al. 2004) or nothing (Basilico et al. 1998); (3) purification by centrifugation or filter washes (Egan et al. 1994a) in mild basic solutions based on heme crystal being insoluble and unincorporated heme being soluble. The final product is quantified by dissolution of crystal in strong basic solutions of pH 14. Table 1 compares the elements of the published crystallization assays and the purification methods.

The assays range essentially from precipitation assays with millimolar heme at an acidic pH, to substrate binding assays with excess preformed heme crystal template. Only 3–5 μM heme is soluble at the pH less than 6 that is absolutely required for crystallization. Measurable product does not form at this low concentration, so 50 μM heme is the lowest concentration for routine assays even though most of the heme is precipitated. A nonradioactive, crystal extension assay with no centrifugation or filter washes was developed (Chong and Sullivan 2003). The basis of this assay lies in the difference in solubility and difference in visible absorption of free versus crystalline heme. Free heme is soluble in weak base, and has a very high molar extinction coefficient of 100,000 M/cm at 400 nm, whereas under these conditions heme crystal is insoluble and is spectroscopically invisible at the low concentrations used. In this assay, first, the absorption at 405 nm is measured of free heme mixed with β-hematin dissolved in a weak base and then the absorption is measured again after heme is decrystallized by concentrated NaOH. The difference in the two absorptions is then used to calculate the amount of heme crystal. The heme crystal extension reaction is kinetically and morphologically identical whether initiated with parasite-derived hemozoin or chemically synthesized β-hematin (Chong and Sullivan 2003). By using this more rapid assay the effects of pH, substrate and preformed crystal on the reaction kinetics were studied.

Presently all in vitro synthesis reactions have approximated but not duplicated the morphology of heme crystals isolated from the parasites in Fig. 3A. Duplication of the anhydrous anaerobic synthesis by Bohle to obtain larger crystals of dimension 200 nm×200 nm×500 nm (Bohle and Helms 1993) shows that these crystals in Fig. 3B most closely resemble *P. falciparum* hemozoin. An investigation into the effect of pH and the resultant morphology of initiation of heme crystallization by mono-oleoylglycerol, oleic acid, palmitoleic acid, low density liproprotein and high density lipoprotein, phosphatidylcholine and purified erythrocyte ghosts showed initiation of heme crystallization for all except the erythrocyte ghosts and high density lipoprotein. The lipids that initiated heme crystallization performed near equivalent to a hemozoin seeded reaction at pH 4 and 4.8. At pH 5.6, the lipids did not initiate heme crys- tallization, whereas the hemozoin seeded reaction was still at 100%. In Fig. 3C and D, field emission in-lens scanning electron microscopy examination of the in vitro incorporation on lipid formulations had long tapered edges consis-

Table 1 Published crystallization assays and purification methods.

Reference	Inititiator	Heme concentration (nmol)	Buffer; pH; temperature; time; volume	Washes	Comments and quinoline IC_{50}
Slater and Cerami 1992a	Trophozoite lysates 500 μg (approximately 125 nmol hemozoin)	400 μM and 140 μM ^{14}C hemin (400 and 140)	500 mM acetate; pH 5.0; 37°C; overnight; (?1 ml)	2× in 2% SDS/NaHCO$_3$ pH=9.1	Radioactive assay. First demonstration of hemozoin formation. IC_{50}: chloroquine, 120 μM; quinine, 300 μM; quinidine, 90 μM
Egan et al. 1994	Nothing	2.2 mM (11,000)	2.2 M acetate; pH 4.8; 60°C; 30 min; 5.04 ml	Water on filter	An acid precipitation with hemozoin formed. IC_{99}: 6 mM or three equivalents for chloroquine, quinine, quinidine
Dorn et al. 1995	Trophozoite lysates 25 μg, purified hemozoin or β-hematin – (?nmol)	140 μM ^{14}C hemin (140 ?)	500 mM acetate; pH 4.8; 37°C; overnight; (?1 ml)	2× in 2% SDS/NaHCO$_3$ pH=9.1	Lipid, preformed hemozoin promote formation. IC_{50}: chloroquine, 80–90 μM
Bendrat et al. 1995	Acetonitrile extract	50 μM (50)	165 mM acetate; pH 5.0; 37°C; overnight; 1 ml	2× in NaHCO$_3$ pH=9.1	Lipid component promotes formation
Sullivan et al. 1996a, 1996b	10 pmol HRP II or III; 5 nmol preformed hemozoin	50 μM (25)	500 mM acetate; pH 4.8; 37°C; overnight; 500μl	2% SDS/NaHCO$_3$ pH=9.1; 2%SDS	Nonradioactive initiation (HRP II) or extension of hemozoin formation. IC_{50}: chloroquine, 5 μM; quinine and quinidine, 1 μM

Table 1 (continued)

Reference	Initititator	Heme concentration (nmol)	Buffer; pH; temperature; time; volume	Washes	Comments and quinoline IC_{50}
Basilico et al. 1998	Nothing	2 mM (400)	4 M; pH 3.0; 37°C; overnight; 200 µl	1× in DMSO	Acid precipitation with hemozoin formed. IC_{50}: chloroquine, 4 mM; quinidine, 8 mM (2–4×heme)
Hawley et al. 1998	50 µg β-hematin, 76 µmol	140 µM ^{14}C hemin (14)	500 mM acetate; pH 4.8; 37°C; overnight; 100µl	2% SDS/NaHCO$_3$, pH=9.1; NaHCO$_3$ pH=9.1; then 50 mM Tris pH 7.5	A binding assay as more than 1,000 times more preformed product than substrate added. IC_{50}: chloroquine, 24 µM; quinine, 64 µM; quinidine, 24 µM
Dorn et al. 1998a, 1998b	10–20 µmol β-hematin, Trophozoites lysates 25 µg=6 nmol hemozoin	140 µM ^{14}C hemin (14)	500 mM acetate; pH 4.8; 37°C; overnight; 100 µl	Filters. 2% SDS/NaHCO$_3$ pH 9.1; NaHCO$_3$ pH 9.1; 50 mM Tris pH 7.5	A binding assay as above. Trophozoite lysates. IC_{50}: chloroquine, 400 µM; quinine, 430 µM. Hemozoin template IC_{50}: chloroquine, 45 µM; quinine, 160 µM
Kurosawa et al. 2000	Acetonitrile extract of trophozoite lysates	100 µM cold and 0.56 nCi ^{14}C hemin (10)	500 mM acetate; pH 4.8; 37°C; overnight; 100 µl	Filters-2× 0.2% SDS/NaHCO$_3$, 2× 50 mM Tris pH7.5	No preformed polymer, just lipids to initiate. High throughput. IC_{50}: chloroquine, 80 µM

Table 1 (continued)

Reference	Intititator	Heme concentration (nmol)	Buffer; pH; temperature; time; volume	Washes	Comments and quinoline IC_{50}
Chong and Sullivan 2003	1–5 nmol β-hematin	50 μM (10–50)	100 mM acetate; pH 5; 37°C; overnight; 1,000 or 200 μl	None	No centrifugation or filter washes based on low solubility and absorbance of β-hematin vs heme. IC_{50}: chloroquine, 4 μM
Tripathi et al. 2004	Lipid extracts	100 μM (100 or 20)	100 mM acetate; pH 5; 37°C overnight; 1,000 or 200μl	3× Tris/SDS; SDS-NaHCO₃;DW	Assay involves transfer to filters with three washes, then quantification

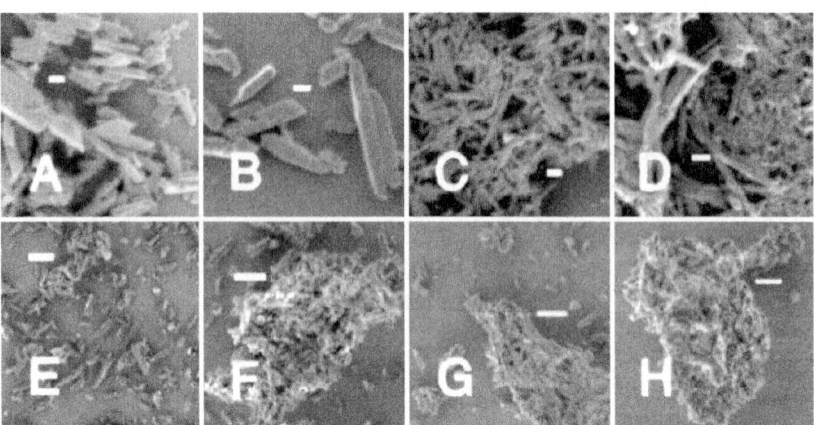

Fig. 3A–H   Hemozoin from *P. falciparum* is similar to anhydrous base abstraction prod-uct, but contrasts with lipid initiated or HRP initiated heme crystals. **A** *P. falciparum* hemozoin. **B** β-Hematin produced from anhydrous base abstraction by the Bohle syn-thesis. Purified heme crystal product initiated by: **C** 50 μM mono-oleoylglycerol; **D** 50 μM palmitoleic acid; **E** VEPL + hemoglobin; **F** VEPL + 50 μM heme; **G** VEPL + HRP II + 50 μM heme; **H** HRP II and 50 μM heme. *Scale bar*, 200 nm for **A–D** and 1 μM for **E–H**

tent with rapid growth distinct from the apparently more controlled growth in parasites. Buller has modeled the rapid incorporation at 'growth faces' (Buller et al. 2002). Oxidized Fe^{3+} heme is a requirement for heme crystal for-mation. Strictly reduced conditions that have been demonstrated in aqueous conditions have not resulted in heme crystal formation (Monti et al. 1999). Protease treatment of acidified hemoglobin does not result in heme crystal formation.

A preparation of digestive vacuole enriched parasite lysates (VEPL) also known as 'crude vacuoles' (Goldberg et al. 1990) was used to initiate heme crystal incorporation in the standard acetate buffer pH 5.0. Field emission in-lens scanning electron microscopy showed closest resemblance to *P. falci-parum* hemozoin with addition of only hemoglobin to the VEPL template in Fig. 3E, although new crystal incorporation was only 1.2 nmol. VEPL incu-bated with 50 μM heme (Fig. 3F), 1 μM recombinant HRP II and 50 μM heme (Fig. 3G), or just 1 μM recombinant HRP II and 50 μM heme and no VEPL (Fig. 3H) all showed a nonrectangular, almost amorphous addition of heme crystal dimers. The new synthesis of heme crystals was 7, 12 and 14 nmol, respectively, for the latter three conditions.

3.3
Heme Crystal Inhibition

Much of the interest in heme crystal formation derives from the extensive data showing many antimalarials like the quinolines, phenanthrenes and hydroxyxanthones inhibit β-hematin formation. Extensive work has been done on the heme binding interaction by the quinoline class of drugs (Egan et al. 1997; Fitch and Kanjananggulpan 1987; Leed et al. 2002; Moreau et al. 1985). Other drugs that bind heme have also been explored as antimalarials (Huy et al. 2002; Kalkanidis et al. 2002; Kristiansen and Jepsen 1985; Vennerstrom et al. 2000). Importantly, not all drugs that bind heme also inhibit hemozoin formation (Slater 1993). Heme binding affinity does not correlate with inhibition of heme crystallization in vitro (Egan et al. 2000). Not all drugs that bind heme and also inhibit hemozoin formation accumulate in the acidic digestive vacuole compartment to inhibit the parasite. Work has also been done that implicates the addition of basic side chains to heme binding drugs, such as the phenothiazines, to improve digestive vacuole localization and parasite inhibition (Kalkanidis et al. 2002).

Two mechanisms of β-hematin inhibition and an additional hybrid have been hypothesized. One is that the drug binds heme, sequestering away substrate from incorporation into the head-to-tail dimer or larger crystal (Fitch 1998; Leed et al. 2002; see also the chapter by Bray et al., this volume). Another is that drug binds directly to a growing face to prevent heme crystal dimer addition (Buller et al. 2002). The third, a hybrid, is that a drug–heme complex incorporates at the growing face to inhibit crystal growth (Chong and Sullivan 2003; Iyer et al. 2003; Sullivan et al. 1996b). Radiolabled chloroquine or quinidine copurify with hemozoin and in vitro bind measurably to growing heme crystals only with the addition of heme (Sullivan et al. 1996b, 1998). Examination of the kinetics of heme crystallization in the presence of quinoline antimalarials revealed that inhibition of heme crystallization by chloroquine and quinidine is reversible, and suggests previous studies that measured heme crystal formation at a fixed timepoint may underestimate the extent of inhibition. To distinguish between the inhibition mechanisms, the amount of heme substrate or heme crystal was increased to try to overcome drug inhibition. Addition of increasing amounts of heme substrate has no effect on the extent of chloroquine or quinidine inhibition, suggesting that these drugs do not act by sequestering free heme. Addition of increasing amount of heme crystal, however, reverses inhibition by these drugs, suggesting that the quinoline–heme complex somehow caps finite extension sites on the crystal (Chong and Sullivan 2003). This evidence suggests that a drug–heme complex binds at a growing face to inhibit crystal extension. Despite these and other findings, a consensus has not yet been reached on the mechanism of inhibition of heme crystallization.

A rational design strategy is hindered at present because of lack of knowledge concerning true inhibitory interaction, whether with heme, drug binding to the growing face of a crystal, drug–heme complex binding at the growing face, or competition with protein and/or lipid complex for initiation (Egan 2004). Molecular docking has been attempted to model interactions with hematin, and this had good agreement with NMR studies; this technique should be able to generate novel structures (Leed et al. 2002). Essential to this approach, but lacking at present, is the prediction of both heme binding and β-hematin inhibition. A manual docking strategy has been used to attempt to model drug alone to the well-defined structure of the β-hematin growing face (Buller et al. 2002). Egan has tried molecular mechanics and dynamics calculations to model quinoline interaction with a simplified porphyrin lacking substituents (Marques et al. 1996). The initial attempt did not correlate with structure–activity relationships or NMR evidence (Leed et al. 2002; O'Neill et al. 1997).

3.4
Plasmodium Iron Sources and Pathways

The *Plasmodium* parasite requires iron for DNA synthesis, glycolysis, pyrimidine synthesis, heme synthesis and electron transport. Debate continues on the critical source of iron for the intraerythrocytic parasite. An important paradox regarding iron is that millimolar desferrioxamine (DFO), an iron chelator, is cytostatic for mammalian cells and bacteria, while a 60-fold lower concentration of 15 μM DFO is cytocidal for *P. falciparum* despite the availability of 20 mM heme iron in hemoglobin (Mabeza et al. 1999). Postulated available sources susceptible to chelation include: (1) extracellular iron from transferrin or free iron in the media; (2) intracellular iron bound to low molecular weight proteins, heme iron or erythrocyte ferritin iron.

Despite early reports of transferrin mediated iron uptake (Haldar et al. 1986; Rodriguez and Jungery 1986), the bulk of experimental evidence excludes a requirement by parasites for extracellular iron. An important physiological study showed *Plasmodium* culture medium, which was depleted of transferrin by 500–1000-fold, supported *Plasmodium* growth well, while inhibiting fibroblasts dependent on transferrin uptake (Sanchez-Lopez and Haldar 1992). Free iron in *Plasmodium* culture medium has been determined to range from 1 to 10 μM. Dialysis removal of free iron to below 1 μM had no effect on parasite growth (Peto and Thompson 1986). Extracellular impermeable dextran-DFO also did not inhibit growth in culture (Scott et al. 1990). Intracellular low molecular weight proteins bind iron, but when DFO conjugates were loaded into resealed erythrocyte ghosts replication was not inhibited (Loyevsky et al. 1993; Scott et al. 1990). Egan has measured equal amounts of total iron in uninfected and infected erythrocytes (Egan et al. 2002). In another study iron levels in the infected erythrocyte as measured by X-ray fluorimetry remained constant while zinc levels increased two- to three-fold (Ginsburg et al. 1986). These constant iron levels suggest minimum import of extracellular iron. However with 20 mM iron levels the import of a few hundred micromolar equivalents or 1%–2% of total iron (more than enough to sustain the protozoan) may be within the standard error of measurement.

Lovesky has identified an intraerythrocytic labile pool of bioavailable iron with the fluorophore calcein (Loyevsky et al. 1999a). In the infected erythrocyte most of the labile pool of iron is in the parasite. In addition, uninfected erythrocytes had a greater total amount of labile iron than infected erythrocytes. Pollack has noted that a significant proportion of labile iron is bound to ATP in normal erythrocytes (Weaver et al. 1993; Zhan et al. 1990). This iron pool could be released with acidification, iron chelation and consumption of ATP by oxidative stress (Atamna and Ginsburg 1995; Gabay and Ginsburg 1993; Hershko and Peto 1988).

If even less than 1% of the 20 mM heme iron in hemoglobin were available, the parasite's iron requirements would be satisfied (Gabay and Ginsburg 1993; Gabay et al. 1994). The amount of heme that is uncoupled from crystallization, thus available for other metabolic needs is not known. Tilley has postulated hydrogen peroxide degradation of heme to release iron in the acidic digestive vacuole. Measurement of heme and iron in infected erythrocytes indicate that iron remains constant while total heme is decreased by 70% (Papalexis et al. 2001). At pH 7.0 reduced glutathione has also been demonstrated to catabolize heme, thereby releasing iron (Ginsburg et al. 1998). Ginsburg has proposed the transport of digestive vacuole heme to the parasite cytosol, where reduced glutathione releases iron by heme catabolism (Ginsburg and Krugliak 1999). Experiments demonstrated a loss of approximately half of the heme content. Chloroquine also interferes with heme degradation by glutathione (Famin and Ginsburg 2003; Famin et al. 1999). Elevated levels of reduced glutathione in mouse reticulocytes have also correlated with chloroquine resistance (Deharo et al. 2003). However, Egan in a separate set of experiments has demonstrated minimal loss of heme iron (Egan et al. 2002). Using subcellular fractionation and iron determinations, both infected and uninfected erythrocytes have the same amount of iron. Digestive vacuoles contained 90% of parasite iron in the form of heme with a minimal amount in parasite cytosol. This was confirmed by electron spectroscopic imaging of transmission electron microscopy.

Another potential source of iron is erythrocytic ferritin. Mature erythrocytes have 0.7 nM residual ferritin left over from heme synthesis by the reticulocyte (Gabay and Ginsburg 1993; Gabay et al. 1994). Ferritin accommodates up to 4,500 iron molecules (Mann et al. 1986). Erythrocytic ferritin could track to the digestive vacuole along with hemoglobin. Vacuolar proteases may de-

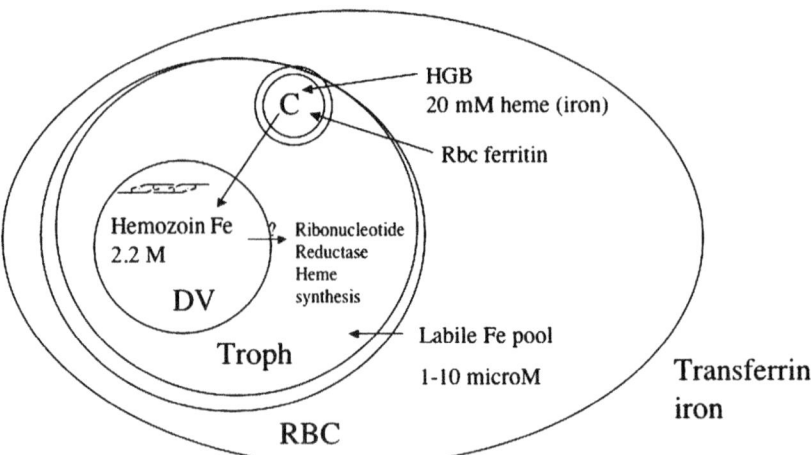

Fig. 4 *Plasmodium* heme and iron pathways. Hemoglobin (*HGB*), erythrocyte ferritin (rbc ferritin) and Pf are phagocytosed by the cytosome (*C*) and transported to the acidic digestive vacuole (*DV*) where proteases cleave hemoglobin, releasing toxic heme that is crystallized. Iron could enter the trophozoite (*TROPH*) from the erythrocyte (*RBC*) cytosol or originate from the DV from ferritin or heme. Quinolines (*Q*) enter the DV to cap crystallization

grade ferritin to release iron. Presently the data suggest that bioavailable iron is transported directly across the parasite plasma membrane rather than via the digestive vacuole or deriving from heme. Figure 4 depicts some of the postulated heme and iron pathways.

P. falciparum lacks a ferroportin, ferritin, metallothione, ferroxamine-based transport systems, or ferredoxin or bacterial iron siderophore orthologs in the sequence database (Rasoloson et al. 2004). Loyevsky has charatereized a *Plasmodium* iron regulatory protein (gb AJ012289; PF13_0229) demonstrating that it can bind mammalian IREs and also *P. falciparum* IREs which are different from the mammalian consensus sequence (Loyevsky et al. 2001, 2003). A DMT-1 ortholog (PFE1185w) localized to the plasma membrane and has not yet been expressed for functional characterization (D.J. Sullivan, unpublished results).

3.5
Heme Synthesis in Two Organelles

Plasmodium requires heme for heme dependent protein synthesis (Surolia and Padmanaban 1992) and as prosthetic group in parasitic cytochromes (Fry and Beesley 1991). Succinylacetone, a specific inhibitor of heme biosynthesis, inhibited the in vitro growth of parasite with an IC_{50} of 2 μM (Surolia and Padmanaban 1992; Wilson et al. 1996). These observations suggest that heme biosynthesis is vital for the parasite and represents an attractive target for therapeutic intervention. Whole genome sequence shows the presence of the complete heme biosynthesis pathway in *P. falciparum* (except uro-porphyrinogen III synthase whose orthologs are most disparate at primary sequence level than are other heme biosynthetic enzymes (Panek and O'Brian 2002)). Microarray analysis has shown expression of all the genes of the heme biosynthesis pathway throughout the erythrocytic stage, with the maximum level of expression during the metabolically most active trophozoite stage (Bozdech et al. 2003). Ferrochelatase has been recently cloned and reported to rescue the ferrochelatase null mutant of *Escherichia coli*, suggesting that the parasite encodes active ferrochelatase (Sato and Wilson 2003).

Protein sequence analysis of different members of the pathway divides them into two major classes: one that contains apicoplast targeting signals—ALAD, PBG deaminase and uroporphyrinogen decarboxylase;and a second group that lacks the apicoplast targeting signals—ALAS, coproporphyrinogen oxidase, protoporphyringen oxidase and ferrochelatase—which presumably are located in the mitochondria. Detailed localization studies of different enzymes of heme biosynthesis pathway is ongoing. These observations suggest that heme biosynthesis involves at least two compartments—plastids and mitochondria. The close proximity of both organelles within *Plasmodium* suggests cooperation of heme biosynthesis. The transport of intermediates to and from both the organelles would be of interest and may provide another area of therapeutic intervention.

3.6
Iron Chelation Therapy

Iron chelation therapy inhibits erythrocytic *Plasmodium* parasites in vitro and in vivo (Mabeza et al. 1999). Ultrastructural studies show a stage specific effect on a trophozoite-to-schizont stage transition with an enlarged nucleus rather than an enlarged digestive vacuole as seen with the quinolines (Atkinson et al. 1991). Iron chelation also inhibits mouse malaria development in hepatocytes at appropriate concentrations (Loyevsky et al. 1999b). The mechanism of action with diverse groups of iron chelators can be grouped to either withholding of iron or by formation of a toxic complex with iron (Mabeza et al. 1999). Iron withholding has a direct effect on the ribonucleotide reductase necessary for DNA synthesis and/or δ-aminolevulinate synthase for heme synthesis. Because of minimal effects on host cells and concurrent development of iron chelators as adjunctive cancer chemotherapy, several trials of iron chelators have been performed for both severe malaria and uncomplicated malaria. Intravenous deferioxamine B when added to quinoline therapy shortened recovery from coma and increased the rate of parasite clearance (Gordeuk et al. 1992). A later study showed an increase in mortality in cerebral malaria patients treated with deferioxamine B (Thuma et al. 1998). Other oral agents have also been tried with no therapeutic success in uncomplicated malaria. Iron chelation may potentially influence immune modulation of severe disease by interference with nitric oxide or cytokines (Mabeza et al. 1999). Iron chelation is antagonistic in vitro to artemisinin action (Eckstein-Ludwig et al. 2003). The increasing use of artemisinin combination therapy, then theoretically, precludes a role for adjunctive iron chelation therapy.

4
Hemozoin as a Biomarker of Malaria Infection

Clinical experience and light microscopic examination of blood films have endured for the past century as the primary tools for diagnosing malaria infection. However, a single low parasitemia or noninfected film can require examination for more than 30 min. Examination under polarized light detects birefringent hemozoin present in trophozoites of *P. vivax*, *P. ovale*, *P. malariae* and *P. knowlesi*, but will miss the rings of *P. falciparum* (Lawrence and Olson 1986). Theoretically, *P. falciparum* gametocytes should be detectable by hemozoin under polarized light. Immunochromatographic platforms are able to detect and speciate *Plasmodium* infections from a drop of blood in 10 min (Moody 2002). PCR based methods provide the greatest sensitivity and specificity, however they are the more labor, reagent and instrument intensive of the diagnostic techniques and are not adaptable to field work (Moody 2002). Laser desorption mass spectrometry (LDMS) was recently demonstrated to be an extremely rapid and sensitive tool for detecting hemozoin as a pan-species biomarker of infection (Demirev et al. 2002; Scholl et al. 2004).

4.1
Laser Desorption Mass Spectrometry

MS is the measurement of the relative abundance and mass-to-charge ratio (m/z) of ions generated from a sample in a vacuum. Mass spectrometric detection can be profoundly sensitive and enable the detection of femtomole amounts of heme under ideal conditions. LDMS samples (0.3–1 µL) are deposited on a metal plate and dried before introduction into the instrument. Twenty to several hundred samples can be deposited on a single plate depending on the instrument. The sample plate is held at a high (±3–20 kV) potential relative to the exit of the ion source region. Laser desorption lacks addition of denaturing matrix solutions routinely used to achieve ionization with the more common matrix assisted laser desorption ionization (MALDI) (DeHoffman et al. 1996). In LDMS, a pulse (~3 nsec) of ultraviolet or infra red wavelength laser light (~1–10 MW/cm^2) illuminates a sample area on the order of 100×50 μm resulting in its conversion to gas phase ions. These ions are accelerated by the electric field in the source region until they exit into the mass analyzer. In time of flight (TOF)-MS, the *m/z* of these ions can be calculated by measuring the time required for ions to traverse a fixed distance before striking the ion detector (Cotter 1997). A useful mass spectrum can be obtained from a single laser shot; however, 20–100 single-shot spectra are usually acquired, filtered and averaged for presentation. The sample can be interrogated by rastering the laser over its surface and thereby enable the acquisition of many spectra from fresh areas within a ~ 2-mm^2 area. Spectra can be acquired from a fixed sample position until it is depleted (Scholl et al. 2004).

By adjusting the laser intensity, ions produced by LD ionization can fragment to yield daughter ions that are structurally characteristic of the parent ion. The degree of fragmentation can be controlled by adjusting the laser intensity. Fragment ions are useful in the interpretation of mass spectra to determine chemical structure and increase specificity of analyte detection. Following the interpretation of a mass spectrum, the molecular identities of sample components can be determined, and in some circumstances, quantitatively measured. Bioinformatic approaches are being developed to automate the interpretation of mass spectra and diminish the need for technical expertise to exploit mass spectrometric detection strategies. The combined sensitivity and specificity achievable using mass spectrometric detection makes it an attractive choice for many biomedical applications including the detection of malaria parasites in blood samples.

4.2
Hemozoin Detection

The LDMS analysis of purified hemozoin crystals using a N$_2$ laser (337 nm) yields the identical mass spectrum of heme alone demonstrating that heme is desorbed from hemozoin crystal surfaces by laser illumination. Hemozoin is thus able to act as its own matrix, absorbing the laser energy to yield the intact parent ion of 616 *m/z* and characteristic structurally rich fingerprint fragment ions from consecutive cleavages of the two propionic acid side chains to yield structurally characteristic ions at *m/z* 571 (M-COOH), 557 (M-CH$_2$COOH), 512 (M-CH$_2$COOH-COOH) and 498 [M-(CH$_2$COOH)$_2$] seen in Fig. 5A (Demirev et al. 2002; Scholl et al. 2004). The head-to-tail dimer *m/z* ratio of 1,232 is not seen even if pure hemozoin is subjected to LDMS.

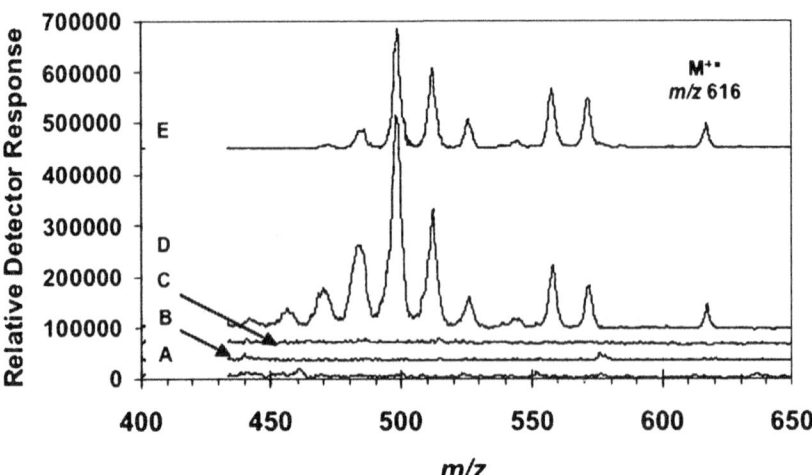

Fig. 5 LDMS spectra of tenfold diluted whole blood from the mouse presented in Fig. 1 and heme. (*A*) Day 1; (*B*) day 0, pre-infection; (*C*) day 1; (*D*) day 2 after infection; (*E*) heme standard. The parent molecular heme cation radical (*m/z* 616) fragments via consecutive cleavages of the two propionic acid side chains to yield structurally characteristic ions at *m/z* 571 (M-COOH), 557 (M-CH$_2$COOH), 512 (M-CH$_2$COOH-COOH) and 498 (M-(CH$_2$COOH)$_2$). Each spectrum represent the average of 100 individual laser shot spectra acquired from the equivalent of <0.3 µl of whole blood

Infected erythrocytes contain the bioanalyte hemozoin which has extremely high local heme concentrations of 2.2 M within the crystal (Pagola et al. 2000). A single laser energy shot is able to detect heme signal spectra consistent with hemozoin.

LDMS, surprisingly, is able to distinguish hemozoin from hemoglobin. Heme complexed with hemoglobin is also abundantly present at 20 mM concentrations in infected and noninfected blood samples. Therefore, hemoglobin was expected to preclude the specific LDMS detection of heme in hemozoin crystals. However, heme bound to hemoglobin or other serum proteins such as albumin or hemopexin is LDMS 'silent'. Denaturation of these proteins produces a heme positive signal from noninfected blood samples. Thus standard sample preparation methods that denature molecules using trifluoroacetic acid and organic solvents like acetonitrile cannot be used to specifically detect hemozoin. However, MALDI could be used to detect parasite species-specific proteins to complement LDMS hemozoin detection methods.

LDMS was initially used to detect hemozoin in 0.5-µl samples containing as few as 10 parasite equivalents/µl following centrifugal concentration and washing of hemozoin crystals from *P. falciparum* cultured in human blood. The heme spectral intensity linearly correlated with the number of parasites deposited (Demirev et al. 2002). Follow-up studies in a mouse malaria model and infected human clinical samples demonstrated that the extensive sample cleanup of blood was not necessary (Scholl et al. 2004). In a mouse time course study employing inoculating doses ranging from 10^2 to 10^6 parasites, LDMS detected *P. yoelli* infections 3–5 days earlier than light microscopy or a colorimetric hemozoin assay. Minimal sample preparation required only 10-fold dilution of whole blood samples followed by the analysis of 0.3 µl deposited on metal slides. LDMS analysis required less than approximately 1 min per sample. Although the equivalent of <0.03 µl of blood was analyzed and a four-log range of inoculating parasites used, the LDMS heme signal rapidly saturated, precluding the more interesting correlation of the mass spectrometric signal intensity with parasitemia. However, data are now emerging that demonstrate LDMS can be used to rapidly detect *P. falciparum* and *P. vivax* in human clinical blood samples and that, at low parasitemia levels (<5%), LDMS signal intensity correlates linearly with parasitemia (Nyunt et al. 2005 AJTMH in press). This has generated interest in developing LDMS for the high throughput, semi-quantitative analysis of malaria infections. The detection of *P. falciparum* infections by the LDMS detection of hemozoin crystals in blood is remarkable because the majority of erythrocytes infected with hemozoin laden trophozoites are sequestered from general circulation by attachment to tissue capillaries. Both nonmicroscopically visible hemozoin in ring stages and circulating leukocytes containing hemozoin crystals may contribute to hemozoin signal by LDMS (Scholl et al. 2004).

The unique combination of hemoglobin physical chemistry, malaria biochemistry and the operating principles of LDMS have unexpectedly enabled its use, without the need for significant sample preparation, for the potentially high throughput detection of malaria parasites. In addition to the high throughput screening of blood samples for malaria, LDMS may find additional applications in the screening of candidate anti-malaria drugs. A great deal of additional work is required to validate the use of LDMS for the detection of heme as a biomarker of malaria infection. In addition to the analytical sensitivity and specificity limits of LDMS, the persistence of hemozoin in chronically infected populations, the effect of thalassemias and sickle-cell anemia, and other viral and parasitic infections need to be evaluated. In any event, the greatest advantage in the potential use of LDMS for detecting malaria infections may ultimately be in its use to triage samples for further light microscopic examination.

5
Perspectives

Aspects of human and *Plasmodium* iron metabolism interrelate with malaria anemia and disease pathogenesis, mechanisms of malaria chemotherapy, diagnostic tests, human iron deficiency and eukaryotic iron biology. Unresolved research questions include etiology of malaria anemia both in the different physiologic setting of an iron-replete or an iron-depleted host. The exact molecular intracellular assembly of heme crystals and the inhibition mechanism of the quinoline class remains. Acquisition, intracellular shuttle and utilization of iron by *Plasmodium* in heme synthesis and other iron pathways will diversify knowledge of eukaryotic iron metabolism. A new development for *Plasmodium* hemozoin is a mass spectrometric diagnostic test.

References

Arese P, Schwarzer E (1997) Malarial pigment (haemozoin): A very active 'inert' substance. Ann Trop Med Parasitol 91:501–516

Atamna H,Ginsburg H (1995) Heme degradation in the presence of glutathione: A proposed mechanism to account for the high levels of non-heme iron found in the membranes of hemoglobinopathic red blood cells. J Biol Chem 42:24876–24883

Atkinson CT, Bayne MT, Gordeuk VR, Brittenham GM, Aikawa M (1991) Stage-specific ultrastructural effects of desferrioxamine on *Plasmodium falciparum* in vitro. Am J Trop Med Hyg 45:593–601

Bahl A, Brunk B, Coppel RL, Crabtree J, Diskin SJ, Fraunholz MJ, Grant GR, Gupta D, Huestis RL, Kissinger JC, Labo P, Li L, McWeeney SK, Milgram AJ, Roos DS, Schug J, Stoeckert CJ, Jr. (2002) Plasmodb: The *Plasmodium* genome resource. An integrated database providing tools for accessing, analyzing and mapping expression and sequence data (both finished and unfinished). Nucl Acids Res 30:87–90

Basilico N, Pagani E, Monti D, Olliaro P, Taramelli D (1998) A microtitre-based method for measuring the haem polymerization inhibitory activity (hpia) of antimalarial drugs. J Antimicrob Chemother 42:55–60

Bendrat K, Berger BJ, Cerami A (1995) Haem polymerization in malaria. Nature 378:138

Berger J, Dyck JL, Galan P, Aplogan A, Schneider D, Traissac P, Hercberg S (2000) Effect of daily iron supplementation on iron status, cell-mediated immunity, and incidence of infections in 6–36 month old Togolese children Eur J Clin Nutr 54:29–35

Blauer G, Akkawi M (1997) Investigations of b- and β-hematin. J Inorg Chem 66:145–152

Blauer G, Akkawi M (2000) On the preparation of beta-haematin. Biochem J 346:249–250

Bohle DS, Helms JB (1993) Synthesis of β-hematin by dehydrohalogenation of hemin. Biochem Biophys Res Comm 193:504–508

Bozdech Z, Llinas M, Pulliam BL, Wong ED, Zhu J, DeRisi JL (2003) The transcriptome of the intraerythrocytic developmental cycle of *Plasmodium falciparum*. PLoS Biol 1:86–100

Brooker S, Peshu N, Warn PA, Mosobo M, Guyatt HL, Marsh K, Snow RW (1999) The epidemiology of hookworm infection and its contribution to anaemia among pre-school children on the Kenyan coast. Trans R Soc Trop Med Hyg 93:240–246

Buller R, Peterson ML, Almarsson O, Leiserowitz L (2002) Quinoline binding site on malaria pigment crystal:A rational pathway for antimalaria drug design. Cryst Growth Des 2:553–562

Carbone T (1891) Sulla natura chimica del pigmento malarico G. Accad Med Torino 39:901

Choi YHC, Cerda FJ, Chu H-a, Babcock TG, Marletta AM (1999) Spectroscopic characterization of the heme-binding sites in *Plasmodium falciparum* histidine-rich protein 2. Biochemistry 38:16916–16924

Chong CR, Sullivan DJ (2003) Inhibition of heme crystal growth by antimalarials and other compounds:Implications for drug discovery. Biochem Pharm 66:2201–2212

Cotter RJ (1997) Time-of-flight mass spectrometry: Instrumentation and applications in biological research. American Chemical Society, Washington, D.C

Deharo E, Barkan D, Krugliak M, Golenser J, Ginsburg H (2003) Potentiation of the antimalarial action of chloroquine in rodent malaria by drugs known to reduce cellular glutathione levels. Biochem Pharm 66:809–817

DeHoffman E, Charette J, Stroobant V (1996) Mass spectrometry: Principles and applications. John Wiley and Sons, Chichester, UK

Demirev PA, Feldman AB, Kongkasuriyachai D, Scholl P, Sullivan D, Jr., Kumar N (2002) Detection of malaria parasites in blood by laser desorption mass spectrometry. Anal Chem 74:3262–3266

Desai MR, Dhar R, Rosen DH, Kariuki SK, Shi YP, Kager PA, Ter Kuile FO (2004) Daily iron supplementation is more efficacious than twice weekly iron supplementation for the treatment of childhood anemia in western Kenya. J Nutr 134:1167–1174

Desai MR, Mei JV, Kariuki SK, Wannemuehler KA, Phillips-Howard PA, Nahlen BL, Kager PA, Vulule JM, ter Kuile FO (2003) Randomized, controlled trial of daily iron supplementation and intermittent sulfadoxine-pyrimethamine for the treatment of mild childhood anemia in western Kenya. J Infect Dis 187:658–666

Dorn A, Stoffel R, Matile H, Bubendorf A, Ridley RG (1995) Malarial haemozoin/beta-haematin supports haem polymerization in the absence of protein. Nature 374:269–271

Dorn A, Vippagunta SR, Matile H, Bubendorf A, Vennerstrom JL, Ridley RG (1998a) A comparison and analysis of several ways to promote haematin (haem) polymerisation and an assessment of its initiation *in vitro*. Biochem Pharm 55:737–747

Dorn A, Vippagunta SR, Matile H, Jaquet C, Vennerstrom JL, Ridley RG (1998b) An assessment of drug-haematin binding as a mechanism for inhibition of haematin polymerisation by quinoline antimalarials Biochem Pharm 55:727–736

Eckstein-Ludwig U, Webb RJ, Van Goethem ID, East JM, Lee AG, Kimura M, O'Neill PM, Bray PG, Ward SA, Krishna S (2003) Artemisinins target the SERCA of *Plasmodium falciparum*. Nature 424:957–961

Egan T, Ross D, Adams P (1994a) Quinoline anti-malarial drugs inhibit spontaneous formation of beta-haematin (malaria pigment). FEBS Letters 352:54–57

Egan TJ (2002) Physico-chemical aspects of hemozoin (malaria pigment) structure and formation. J Inorg Biochem 91:19–26

Egan TJ (2004) Haemozoin formation as a target for the rational design of new anti-malarials. Drug Design Reviews Online 1:93–110

Egan TJ, Combrinck JM, Egan J, Hearne GR, Marques HM, Ntenteni S, Sewell BT, Smith PJ, Taylor D, Van Schalkwyk DA, Walden JC (2002) Fate of haem iron in the malaria parasite *Plasmodium falciparum*. Biochem J 365:343–347

Egan TJ, Hunter R, Kaschula CH, Marques HM, Misplon A, Walden J (2000) Structure-function relationships in aminoquinolines: Effect of amino and chloro groups on quinoline-hematin complex formation, inhibition of beta-hematin formation, and antiplasmodial activity. J Med Chem 43:283–291

Egan TJ, Mavuso WW, Ncokazi KK (2001) The mechanism of beta-hematin formation in acetate solution. Parallels between hemozoin formation and biomineralization processes. Biochemistry 40:204–213

Egan TJ, Mavuso WW, Ross DC, Marques HM (1997) Thermodynamic factors controlling the interaction of quinoline antimalarial drugs with ferriprotoporphyrin IX. J Inorg Biochem 68:137–145

Egan TJ, Ross DC, Adams PA (1994b) Quinoline anti-malarial drugs inhibit spontaneous formation of beta-haematin (malaria pigment). FEBS Lett 352:54–57

Ekvall H, Premji Z, Bjorkman A (2000) Micronutrient and iron supplementation and effective antimalarial treatment synergistically improve childhood anaemia. Trop Med Int Health 5:696–705

Fairbanks Vf, Beutler E. (1995a) Iron deficiency. In: Beutler E, Lichtman M, Coller B, Kipps T (eds) Williams hematology, McGraw-Hill, New York, pp490–511

Fairbanks Vf, Beutler E. (1995b). Iron metabolism. In: Beutler E, Lichtman M, Coller B, Kipps T (eds) Williams hematology, McGraw-Hill, New York, pp369–379

Famin O, Ginsburg H (2003) The treatment of plasmodium falciparum-infected erythrocytes with chloroquine leads to accumulation of ferriprotoporphyrin IX bound to particular parasite proteins and to the inhibition of the parasite's 6-phosphogluconate dehydrogenase. Parasite 10:39–50

Famin O, Krugliak M, Ginsburg H (1999) Kinetics of inhibition of glutathione-mediated degradation of ferriprotoporphyrin IX by antimalarial drugs. Biochem Pharm 58:59–68

Fitch C, Cai G, Chen Y, Shoemaker J (1999) Involvement of lipids in ferriprotoporphyrin IX polymerization in malaria. Biochim Biophys Acta 1454:31–37

Fitch CD (1998) Involvement of heme in the antimalarial action of chloroquine. Trans Am Clin Climatol Assoc 109:97–105; discussion 105–106

Fitch CD, Chen YF, Cai GZ (2003) Chloroquine-induced masking of a lipid that promotes ferriprotoporphyrin IX dimerization in malaria. J Biol Chem 278:22596–22599

Fitch CD, Chou AC (1996) Heat-labile and heat-stimulable heme polymerase activities in *Plasmodium berghei*. Mol Biochem Parasitol 82:261–264

Fitch CD, Kanjananggulpan P (1987) The state of ferriprotoporphyrin IX in malaria pigment. J Biol Chem 262:15552–15555

Francis SE, Sullivan DJ Jr, Goldberg DE (1997) Hemoglobin metabolism in the malaria parasite *Plasmodium falciparum* Annu Rev Microbiol 51:97–123

Fry M, Beesley JE (1991) Mitochondria of mammalian *Plasmodium* spp. Parasitology 102:17–26

Gabay T, Ginsburg H (1993) Hemoglobin denaturation and iron release in acidified red blood cell lysate-a possible source of iron for intraerthrocytic malaria parasites. Exp Parisitol 77:261–272

Gabay T, Krugliak M, Shalmiev G, Ginsburg H (1994) Inhibition by anti-malarial drugs of haemoglobin denaturation and iron release in acidified red blood cell lysates-a possible mechanism of their anti-malarial effect? Parasitology 108:371–381

Garrick MD, Nunez MT, Olivares M, Harris ED (2003) Parallels and contrasts between iron and copper metabolism. Biometals 16:1–8

Gera T, Sachdev HP (2002) Effect of iron supplementation on incidence of infectious illness in children: Systematic review. BMJ 325:1142

Ginsburg H, Famin O, Zhang J, Krugliak M (1998) Inhibition of glutathione-dependant degradation of heme by chloroquine and amodiaquine as a possible basis for their antimalarials mode of action. Biochem Pharmacol 56:1305–1313

Ginsburg H, Gorodetsky R, Krugliak M (1986) The status of zinc in malaria (*Plasmodium falciparum*) infected human red blood cells: Stage dependent accumulation, compartmentation and effect of dipicolinate. Biochimica et Biophysica Acta 886:337–344

Ginsburg H, Krugliak M (1999) Chloroquine—some open questions on its antimalarial mode of action and resistance. Drug Resist Update 2:180–187

Goldberg DE, Slater AFG, Cerami A, Henderson GB (1990) Hemoglobin degradation in the malaria parasite *Plasmodium falciparum*: An ordered process in a unique organelle. Proc Natl Acad Sci USA 87:2931–2935

Gordeuk V, Thuma P, Brittenham G, McLaren C, Parry D, Backenstose A, Biemba G, Msiska R, Holmes L, McKinley E, et al. (1992) Effect of iron chelation therapy on recovery from deep coma in children with cerebral malaria. N Engl J Med 327:1473–1477

Gushin H, Mackenzie B, Berger UV, Gushin Y, Romero MR, Boron WF, Nussberger S, Gollan JL, Hediger MA (1997) Cloning and characterization of a mammalian proton-coupled metal-ion transporter. Nature 388:482–487

Haddad S, Restieri C, Krishnan K (2001) Characterization of age-related changes in body weight and organ weights from birth to adolescence in humans. J Toxicol Environ Health A 64:453–464

Haldar K, Henderson CL, Cross GAM (1986) Identification of the parasite transferrin receptor of *Plasmodium falciparim*- infected erythrocytes and its acylation via 1,2-diacyl-sn-glycerol. Proc Natl Acad Sci USA 83:8565–8569

Hastka J, Lasserre J, Schwarzbeck A, Strauch M, Hehlmann R (1993) Zinc protoporphyrin in anemia of chronic disorders. Blood 81:1200–1204

Hawley SR, Bray PG, Mungthin M, Atkinson JD, O'Neill PM, Ward SA (1998) Relationship between antimalarial drug activity, accumulation, and inhibition of heme polymerization in *Plasmodium falciparum in vitro* Antimicrob Agents Chemother 42:682–686

Hempelmann E, Egan TJ (2002) Pigment biocrystallization in *Plasmodium falciparum*. Trends Parasitol 18:11

Hempelmann E, Motta C, Hughes R, Ward SA, Bray PG (2003) *Plasmodium falciparum*: Sacrificing membrane to grow crystals? Trends Parasitol 19:23–26

Hentze MW, Muckenthaler MU, Andrews NC (2004) Balancing acts: Molecular control of mammalian iron metabolism. Cell 117:285–297

Hershko C, Peto TE (1988) Deferoxamine inhibition of malaria is independent of host iron status. J Exp Med 168:375–387

Huy NT, Kamei K, Yamamoto T, Kondo Y, Kanaori K, Takano R, Tajima K, Hara S (2002) Clotrimazole binds to heme and enhances heme-dependent hemolysis: Proposed antimalarial mechanism of clotrimazole. J Biol Chem 277:4152–4158

Iyer JK, Shi L, Shankar AH, Sullivan DJ, Jr. (2003) Zinc protoporphyrin IX binds heme crystals to inhibit the process of crystallization in *Plasmodium falciparum*. Mol Med 9:175–182

Kalkanidis M, Klonis N, Tilley L, Deady LW (2002) Novel phenothiazine antimalarials: Synthesis, antimalarial activity, and inhibition of the formation of beta-haematin. Biochem Pharmacol 63:833–842

Kaplan J, O'Halloran TV (1996) Iron metabolism in eukaryotes: Mars and venus at it again. Science 271:1510–1512

Kristiansen JE, Jepsen S (1985) The susceptibility of *Plasmodium falciparum in vitro* to chlorpromazine and the stereo-isomeric compounds cis (z)- and trans (e)-clopenthixol. Acta Pathol Microbiol Immunol Scand [B] 93:249–251

Krugliak M, Zhang J, Ginsburg H (2002) Intraerythrocytic *Plasmodium falciparum* utilizes only a fraction of the amino acids derived from the digestion of host cell cytosol for the biosynthesis of its proteins. Mol Biochem Parasitol 119:249–256

Kurosawa Y, Dorn A, Kitsuji-Shirane M, Shimada H, Satoh T, Matile H, Hofheinz W, Masciadri R, Kansy M, Ridley RG (2000) Hematin polymerization assay as a high-throughput screen for identification of new antimalarial pharmacophores. Antimicrob Agents Chemother 44:2638–2644

Lamola AA, Yamane T (1974) Zinc protoporphyrin in the erythrocytes of patients with lead intoxication and iron deficiency anemia. Science 186:936–938

Lawrence C, Olson JA (1986) Birefringent hemozoin identifies malaria. Am J Clin Pathol 86:360–363

Leed A, DuBay K, Ursos LM, Sears D, De Dios AC, Roepe PD (2002) Solution structures of antimalarial drug-heme complexes Biochemistry 41:10245–10255

Leenstra T, Kariuki SK, Kurtis JD, Oloo AJ, Kager PA, ter Kuile FO (2004) Prevalence and severity of anemia and iron deficiency: Cross-sectional studies in adolescent schoolgirls in western Kenya. Eur J Clin Nutr 58:681–691

Levesque MA, Sullivan AD, Meshnick SR (1999) Splenic and hepatic hemozoin in mice after malaria parasite clearance. J Parasitol 85:570–573

Lew VL, Tiffert T, Ginsburg H (2003) Excess hemoglobin digestion and the osmotic stability of *Plasmodium falciparum*-infected red blood cells. Blood 101:4189–4194

Loyevsky M, John C, Dickens B, Hu V, Miller JH, Gordeuk VR (1999a) Chelation of iron within the erythrocytic *Plasmodium falciparum* parasite by iron chelators. Mol Biochem Parasitol 101:43–59

Loyevsky M, LaVaute T, Allerson CR, Stearman R, Kassim OO, Cooperman S, Gordeuk VR, Rouault TA (2001) An IRP-like protein from *Plasmodium falciparum* binds to a mammalian iron-responsive element. Blood 98:2555–2562

Loyevsky M, Lytton SD, Mester B, Libman J, Shanzer A, Cabantchik ZI (1993) The antimalarial action of desferal involves a direct access route to erythrocytic (*Plasmodium falciparum*) parasites. J Clin Invest 91:218–224

Loyevsky M, Mompoint F, Yikilmaz E, Altschul SF, Madden T, Wootton JC, Kurantsin-Mills J, Kassim OO, Gordeuk VR, Rouault TA (2003) Expression of a recombinant IRP-like *Plasmodium falciparum* protein that specifically binds putative plasmodial ires. Mol Biochem Parasitol 126:231–238

Loyevsky M, Sacci JB, Jr., Boehme P, Weglicki W, John C, Gordeuk VR (1999b) *Plasmodium falciparum* and *Plasmodium yoelii*: Effect of the iron chelation prodrug dexrazoxane on in vitro cultures. Exp Parasitol 91:105–114

Mabeza GF, Loyevsky M, Gordeuk VR, Weiss G (1999) Iron chelation therapy for malaria: A review. Pharmacol Ther 81:53–75

Mann S, Bannister JV, Williams RJP (1986) Structure and composition of ferritin cores isolated from human spleen, limpet (*Patella vulgata*) hemolymph and bacterial (*Pseudomonas aeruginosa*) cells. J Mol Biol 188:225–232

Marques HM, Voster K, Egan TJ (1996) The interaction of the heme-octapeptide, n-acetylmicroperoxidase-8 with antimalarial drugs: Solution studies and modeling by molecular mechanics methods. J Inorgan Biochem 64:7–23

Massaga JJ, Kitua AY, Lemnge MM, Akida JA, Malle LN, Ronn AM, Theander TG, Bygbjerg IC (2003) Effect of intermittent treatment with amodiaquine on anaemia and malarial fevers in infants in Tanzania: A randomised placebo-controlled trial Lancet 361:1853–1860

McKie AT, Marciani P, Rolfs A, Brennan K, Wehr K, Barrow D, Miret S, Bomford A, Peters TJ, Farzaneh F, Hediger MA, Hentze MW, Simpson RJ (2000) A novel duodenal iron-regulated transporter, IREG1, implicated in the basolateral transfer of iron to the circulation. Mol Cell 5:299–309

Mebrahtu T, Stoltzfus RJ, Chwaya HM, Jape JK, Savioli L, Montresor A, Albonico M, Tielsch JM (2004) Low-dose daily iron supplementation for 12 months does not increase the prevalence of malarial infection or density of parasites in young Zanzibari children J Nutr 134:3037–3041

Menendez C, Kahigwa E, Hirt R, Vounatsou P, Aponte JJ, Font F, Acosta CJ, Schellenberg DM, Galindo CM, Kimario J, Urassa H, Brabin B, Smith TA, Kitua AY, Tanner M, Alonso PL (1997) Randomised placebo-controlled trial of iron supplementation and malaria chemoprophylaxis for prevention of severe anaemia and malaria in Tanzanian infants. Lancet 350:844–850

Menendez C, Schellenberg D, Quinto L, Kahigwa E, Alvarez L, Aponte JJ, Alonso PL (2004) The effects of short-term iron supplementation on iron status in infants in malaria-endemic areas. Am J Trop Med Hyg 71:434–440

Miyajima H (2002) Genetic disorders affecting proteins of iron and copper metabolism: Clinical implications. Intern Med 41:762–769

Monti D, Vodopivec B, Basilico N, Olliaro P, Taramelli D (1999) A novel endogenous antimalarial: Fe (ii)-protoporphyrin IX alpha (heme) inhibits hematin polymerization to beta-hematin (malaria pigment) and kills malaria parasites. Biochemistry 38:8858–8863

Moody A (2002) Rapid diagnostic tests for malaria parasites. Clin Microbiol Rev 15:66–78

Moreau S, Perly B, Chachaty C, and Deleuze C (1985) A nuclear magnetic resonance study of the interactions of antimalarial drugs with porphyrins. Biochem Biophys Acta 840:107–116

Murray MJ, Murray NJ, Murray AB, Murray MB (1975) Refeeding-malaria and hyperferraemia. Lancet 1:653–654

Noland GS, Briones N, Sullivan DJ (2003) The shape and size of hemozoin crystals distinguishes diverse *Plasmodium* species. Mol Biochem Parasitol 130:91–99

Nyakeriga AM, Troye-Blomberg M, Dorfman JR, Alexander ND, Back R, Kortok M, Chemtai AK, Marsh K, Williams TN (2004) Iron deficiency and malaria among children living on the coast of Kenya. J Infect Dis 190:439–447

O'Neill PM, Willock DJ, Hawley SR, Bray PG, Storr RC, Ward SA, Park BK (1997) Synthesis, antimalarial activity, and molecular modeling of tebuquine analogues. J Med Chem 40:437–448

Oppenheimer SJ (2001) Iron and its relation to immunity and infectious disease. J Nutr 131:616S-633S; discussion 633S-635S

Pagola S, Stephens PW, Bohle DS, Kosar AD, Madsen SK (2000) The structure of malaria pigment beta-hematin. Nature 404:307–310

Pandey AV, Babbarwal VK, Okoyeh JN, Joshi RM, Puri SK, Singh RL, Chauhan VS (2003) Hemozoin formation in malaria: A two-step process involving histidine-rich proteins and lipids. Biochem Biophys Res Commun 308:736–743

Panek H, O'Brian MR (2002) A whole genome view of prokaryotic haem biosynthesis. Microbiology 148:2273–2282

Papalexis V, Siomos MA, Campanale N, Guo X, Kocak G, Foley M, Tilley L (2001) Histidine-rich protein 2 of the malaria parasite, *Plasmodium falciparum*, is involved in detoxification of the by-products of haemoglobin degradation. Mol Biochem Parasitol 115:77–86

Peto TE, Thompson JL (1986) A reappraisal of the effects of iron and desferrioxamine on the growth of *Plasmodium falciparum* 'in vitro': The unimportance of serum iron Br J Haematol 63:273–280

Rasoloson D, Shi L, Chong CR, Kafsack BF, Sullivan DJ (2004) Copper pathways in *Plasmodium falciparum* infected erythrocytes indicate an efflux role for the copper P-ATPase. Biochem J

Raventos-Suarez C, Pollack S, Nagel RL (1982) *Plasmodium falciparum*: Inhibition of *in vitro* growth by desferrioxamine. Am J Trop Med Hyg 31:919–922

Ridley RG (1996) Hemozoin formation in malaria parasites: Is there a haem polymerase? Trends Microbiol 4:253–254

Rodriguez MH, Jungery M (1986) A protein on *Plasmodium falciparum*-infected erythrocytes functions as a transferrin receptor. Nature 324:388–391

Sanchez-Lopez R, Haldar K (1992) A transferrin-independent iron uptake activity in *Plasmodium falciparum*-infected and uninfected erythrocytes. Mol Biochem Parasitol 55:9-20

Sato S, Wilson RJ (2003) Proteobacteria-like ferrochelatase in the malaria parasite Curr Genet 42:292–300

Scholl PF, Kongkasuriyachai D, Demirev PA, Feldman AB, Lin JS, Sullivan DJ, Kumar N (2004) Rapid detection of malaria infection *in vivo* by laser desorption mass spectrometry. Am J Trop Med Hyg 71:546–551

Schwarzer E, Datteis DF, Giribaldi G, Ulliers D, Valente E, Arese P (1999a) Hemozoin stability and dormant induction of heme oxygenase in hemozoin-fed human monocytes. Mol Biochem Parasitol 100:61–72

Schwarzer E, Turrini F, Ulliers D, Giribaldi G, Ginsburg H, Arese P (1992) Impairment of macrophage functions after ingestion of *Plasmodium falciparum*-infected erythrocytes or isolated malarial pigment. J Exp Med 176:1033–1041

Scott MD, Ranz A, Kuypers FA, Lubin BH, Meshnick SR (1990) Parasite uptake of desferroxamine: A prerequisite for antimalarial activit.y Br J Haematol 75:598–602

Slater AF (1993) Chloroquine: Mechanism of drug action and resistance in *Plasmodium falciparum*. Pharmacol Ther 57:203–235

Slater AF, Cerami A (1992) Inhibition by chloroquine of a novel haem polymerase enzyme activity in malaria trophozoites. Nature 355:167–169

Slater AF, Swiggard WJ, Orton BR, Flitter WD, Goldberg DE, Cerami A, Henderson GB (1991) An iron-carboxylate bond links the heme units of malaria pigment. Proc Natl Acad Sci USA 88:325–329

Slater AFG (1992) Malarial pigment. Experim Parasitol 74:362–365

Spivak JL (2002) Iron and the anemia of chronic disease. Oncology (Huntingt) 16:25–33

Stoltzfus RJ, Chway HM, Montresor A, Tielsch JM, Jape JK, Albonico M, Savioli L (2004) Low dose daily iron supplementation improves iron status and appetite but not anemia, whereas quarterly anthelminthic treatment improves growth, appetite and anemia in Zanzibari preschool children. J Nutr 134:348–356

Stoltzfus RJ, Chwaya HM, Montresor A, Albonico M, Savioli L, Tielsch JM (2000) Malaria, hookworms and recent fever are related to anemia and iron status indicators in 0- to 5-y old Zanzibari choldren and these relationships change with age. J Nutr 130:1724–1733

Sullivan D (2002a) Hemozoin: A biocrystal synthesized during the degradation of hemoglobin. In Matsumura S, Steinbüchel A (eds) Miscellaneous biopolymers, biodegradation of synthetic polymers, Wiley-VCH Verlag GmbH & Co, Weinheim, Germany, pp129–163

Sullivan DJ (2002b) Theories on malarial pigment formation and quinoline action. Int J Parasitol 32:1645–1653

Sullivan DJ Jr, Gluzman IY, Goldberg DE (1996a) *Plasmodium* hemozoin formation mediated by histidine-rich proteins. Science 271:219–222

Sullivan DJ Jr, Gluzman IY, Russell DG, Goldberg DE (1996b) On the molecular mechanism of chloroquine's antimalarial action. Proc Natl Acad Sci USA 93:11865–11870

Sullivan DJ Jr, Matile H, Ridley RG, Goldberg DE (1998) A common mechanism for blockade of heme polymerization by antimalarial quinolines. J Biol Chem 273:31103–31107

Surolia N, Padmanaban G (1992) De novo biosynthesis of heme offers a new chemotherapeutic target in the human malarial parasite. Biochem Biophys Res Commun 187:744–750

Taliaferro WH, Mulligan HW (1937) The histopathology of malaria with special reference to the function and origin of macrophages in defense. Indian Med Mem 29:1–12

Tatala S, Svanberg U, Mduma B (1998) Low dietary iron availability is a major cause of anemia: A nutrition survey in the Lindi district of Tanzania. Am J Clin Nutr 68:171–178

Thuma PE, Mabeza GF, Biemba G, Bhat GJ, McLaren CE, Moyo VM, Zulu S, Khumalo H, Mabeza P, M'Hango A, Parry D, Poltera AA, Brittenham GM, Gordeuk VR (1998) Effect of iron chelation therapy on mortality in zambian children with cerebral malaria. Trans R Soc Trop Med Hyg 92:214–218

Tripathi AK, Garg SK, Tekwani BL (2002) A physiochemical mechanism of hemozoin (beta-hematin) synthesis by malaria parasite Biochem Biophys Res Commun 290:595–601

Tripathi AK, Khan SI, Walker LA, Tekwani BL (2004) Spectrophotometric determination of de novo hemozoin/beta-hematin formation in an *in vitro* assay. Anal Biochem 325:85–91

Vennerstrom JL, Ager AL Jr, Andersen SL, Grace JM, Wongpanich V, Angerhofer CK, Hu JK, Wesche DL (2000) Assessment of the antimalarial potential of tetraoxane WR 148999. Am J Trop Med Hyg 62:573–578

Verhoef H, West CE, Nzyuko SM, de Vogel S, van der Valk R, Wanga MA, Kuijsten A, Veenemans J, Kok FJ (2002) Intermittent administration of iron and sulfadoxine-pyrimethamine to control anaemia in Kenyan children: A randomised controlled trial. Lancet 360:908–914

Weaver J, Zhan H, Pollack S (1993) Erythrocyte haemolysate interacts with atp-fe to form a complex containing iron, ATP and 13 800 MW polypeptide. Br J Haematol 83:138–144

Wilson CM, Smith AB, Baylon RV (1996) Characterization of the delta-aminolevulinate synthase gene homologue in *P. falciparum*. Mol Biochem Parasitol 79:135–140

Yip R (1998) Iron deficiency. Bull WHO 76:121–123

Zhan H, Gupta RK, Weaver J, Pollack S (1990) Iron bound to low MW ligands: Interactions with mitochondria and cytosolic proteins. Eur J Haematol 44:125–131

Zhang J, Krugliak M, Ginsburg H (1999) The fate of ferriprotoporphyrin IX in malaria infected erythrocytes in conjunction with the mode of action of antimalarial drugs. Mol Biochem Parasitol 99:129–141

CTMI (2005) 295:325–356

Plasmodium Permeomics:
Membrane Transport Proteins in the Malaria Parasite

K. Kirk[1] (✉) · R. E. Martin[1] · S. Bröer[1] · S. M. Howitt[1] · K. J. Saliba[1,2]

[1]School of Biochemistry and Molecular Biology, The Australian National University,
0200 Canberra, ACT, Australia
Kiaran.Kirk@anu.edu.au

[2]Medical School, The Australian National University, 0200 Canberra, ACT, Australia

Abstract Membrane transport proteins are integral membrane proteins that mediate the passage across the membrane bilayer of specific molecules and/or ions. Such proteins serve a diverse range of physiological roles, mediating the uptake of nutrients into cells, the removal of metabolic wastes and xenobiotics (including drugs), and the generation and maintenance of transmembrane electrochemical gradients. In this chapter we review the present state of knowledge of the membrane transport mechanisms underlying the cell physiology of the intraerythrocytic malaria parasite and its host cell, considering in particular physiological measurements on the parasite and parasitized erythrocyte, the annotation of transport proteins in the *Plasmodium* genome, and molecular methods used to analyze transport protein function.

Abbreviations

NPP	New permeability pathways
PVM	Parasitophorous vacuole membrane
TMD	Transmembrane domains
PfHT	*P. falciparum* hexose transporter
PfCRT	*P. falciparum* chloroquine resistance transporter
MC	Mitochondrial carrier
Pgh1	P-glycoprotein homologue 1
PfAQP	*P. falciparum* aquaglyceroporin
PfNT1	*P. falciparum* nucleoside transporter
RNAi	RNA interference

1
Introduction

In recent years physiological studies on *Plasmodium*-infected erythrocytes, on parasites functionally 'isolated' from their host erythrocytes and, to a limited extent, on isolated parasite organelles have provided substantial insight into the membrane transport properties of both host and parasite membranes. A number of membrane transport proteins encoded by the parasite have been expressed and, in some cases, characterized in heterologous expression systems, and an increasing number of genes encoding putative transport proteins are being identified in the *P. falciparum* genome. Nevertheless, in comparison with our understanding of other animal and plant systems, our knowledge of the membrane physiology of the malaria parasite remains at a preliminary level. A variety of membrane transport processes has been characterized in some detail in the parasite and its host cell, and a small number of transport proteins has been characterized to at least some extent in heterologous expression systems.

'The Permeome' is a term used to describe the total complement of proteins involved in membrane permeability in a given organism [73]. It encompasses the full range of channels and transporters encoded in the genome. In this chapter we provide an overview of what is currently known of the permeome of the malaria parasite. We start with an overview of those membrane transport processes that have been characterized in physiological studies of the parasitized erythrocyte and the intracellular parasite. We consider issues relating to the identification of genes encoding membrane transport proteins, and summarize the studies carried out to date on the localization of membrane transport proteins within the parasitized erythrocyte. The use of heterologous expression systems for the expression and characterization of *Plasmodium* transport proteins is discussed in some detail, focusing in particular on the *Xenopus laevis* oocyte and yeast systems. Finally, we consider the methods available for manipulation of gene expression in the parasite itself, and how these might be used in conjunction with physiological measurements on the parasite to provide insights into the function of candidate transporters and channels.

2
Physiological Measurements in Parasitized Erythrocytes and Isolated Parasites

2.1
Methods

2.1.1
Cell Preparations

The majority of investigations of membrane transport in *Plasmodium* parasites and their host cells has been carried out with *P. falciparum*, cultured in human erythrocytes in vitro. Various techniques are available for synchronizing cultures and for concentrating the infected cells to obtain a high parasitemia [69,82]. There is also a number of different approaches that have been used to permeabilize the host erythrocyte membrane, releasing the cytosol of the host cell and allowing solutes added to the extracellular solution (e.g., radiolabeled transport solutes) ready access to the surface membranes of the intracellular parasite. Streptolysin O is a bacterial pore-forming protein which has been shown to permeabilize the erythrocyte membrane while leaving the parasitophorous vacuole membrane (PVM) intact [6]. By contrast saponin, a plant-derived detergent, induces the irreversible formation of large pores in both the erythrocyte membrane and the PVM [7]. Prolonged exposure of the parasite to saponin, and/or excessively high saponin concentrations can cause significant damage to the parasite. However, there is now ample evidence that parasites functionally 'isolated' from their host erythrocytes by brief exposure to low concentrations of saponin (typically <10 min exposure to 0.05% w/v saponin) retain their biochemical and physiological integrity. For parasites isolated in this way the plasma membrane is able to maintain a substantial electrical potential [3], as well as transmembrane concentration gradients for H^+ [95] and Ca^{2+} [5]. Furthermore, the [ATP] and the rate of phosphorylation of the essential vitamin pantothenate [93] are the same as those measured in intact parasitized cells.

Treatment of saponin-isolated parasites with another plant detergent, digitonin, under certain conditions, permeabilizes the parasite plasma membrane, while leaving the membrane of at least one intracellular organelle, the parasite's digestive vacuole, intact and able to generate and sustain a substantial [H^+] gradient [66, 97]. Intact digestive vacuoles, again able to generate and maintain a [H^+] concentration gradient, can also be isolated using published protocols [92, 97].

Figure 1 shows a schematic representation of the cell preparations that have been used successfully for the characterization of membrane transport mechanisms in the parasite/parasitized erythrocyte.

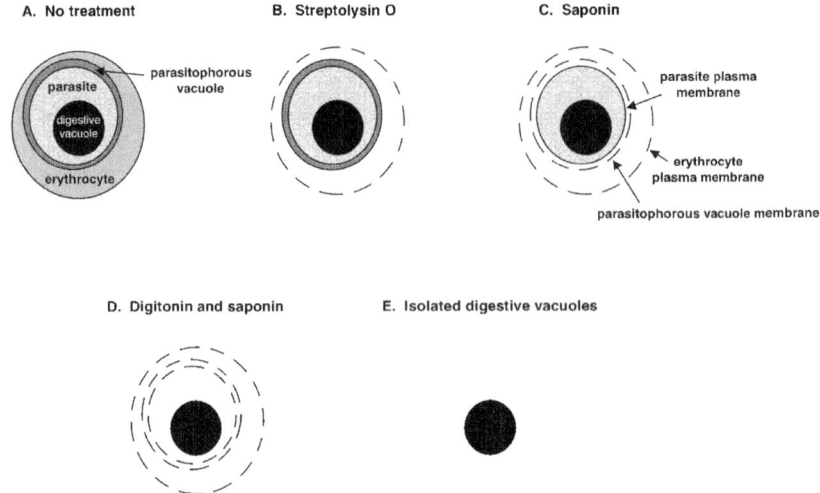

Fig. 1A–E Schematic representation of different preparations that have been used to study aspects of malaria parasite physiology. (*A*) Intact parasitized erythrocyte; (*B*) parasitized erythrocyte in which the erythrocyte membrane is permeabilized with streptolysin O; (*C*) parasitized erythrocyte in which the erythrocyte membrane and PVMs are permeabilized with saponin; (*D*) parasitized erythrocyte in which the erythrocyte membrane PVM and parasite plasma membrane are permeabilized with a combination of saponin and digitonin; (*E*) isolated digestive vacuole. The means by which the different permeabilizing agents discriminate between the different membranes of the *Plasmodium*-infected erythrocyte are not clear. (Adapted from [97])

2.1.2
Physiological Methods

A range of different physiological techniques has been applied to the different cell preparations outlined above, yielding important information about the physiology of the infected cell and the intracellular parasite.

The membrane transport properties of the infected erythrocyte membrane have been investigated using a combination of direct transport measurements and measurements of the hemolysis of parasitized cells suspended in solutions containing high concentrations (typically isosmotic with serum) of permeant solutes [41, 59, 62, 117]. Direct transport measurements have, in most cases, involved the use of radiolabeled solutes, though there are examples of the use of NMR spectroscopy [103], fluorescence detection [61], and a bioassay [94] to monitor solute uptake. In recent years several groups have applied the patch-clamp technique to the study of the electrophysiological properties of the infected erythrocyte (see below). Desai and colleagues used the same technique to study the PVM [21], and have also obtained single channel recordings from the membrane fraction of homogenized parasitized erythrocytes reconstituted into artificial membrane bilayers [22].

The transport of a range of solutes across the parasite plasma membrane has been measured directly using radiolabeled compounds in conjunction with isolated parasite preparations, as well as through the use of fluorescent ion-sensitive dyes for H^+ [46, 95], Ca^{2+} [5, 9, 39] and Na^+ [126], loaded into the parasite cytosol. Fluorescent Ca^{2+} indicators have been targeted to the parasitophorous vacuole by their inclusion in the media in which parasite invasion of the erythrocyte takes place [39]. Fluorescent H^+ [30, 97] and Ca^{2+} [9] indicators have also been loaded into the digestive vacuole, though in the case of the H^+ indicators there has been substantial controversy regarding the interpretation of the data [101].

The use of fluorescent ion-sensitive dyes has been applied both to intact parasitized erythrocytes and to isolated parasites. In most cases the former studies have involved single-cell fluorescence microscopy whereas the latter have involved the use of parasite suspensions.

2.2
The Physiology of the Parasitized Cell

With the methods outlined above we are gradually acquiring a detailed knowledge of the membrane transport mechanisms underlying the physiology of the parasitized erythrocyte. Transport pathways that have been characterized to at least some extent in the parasite/parasitized cell are represented in Fig. 2A.

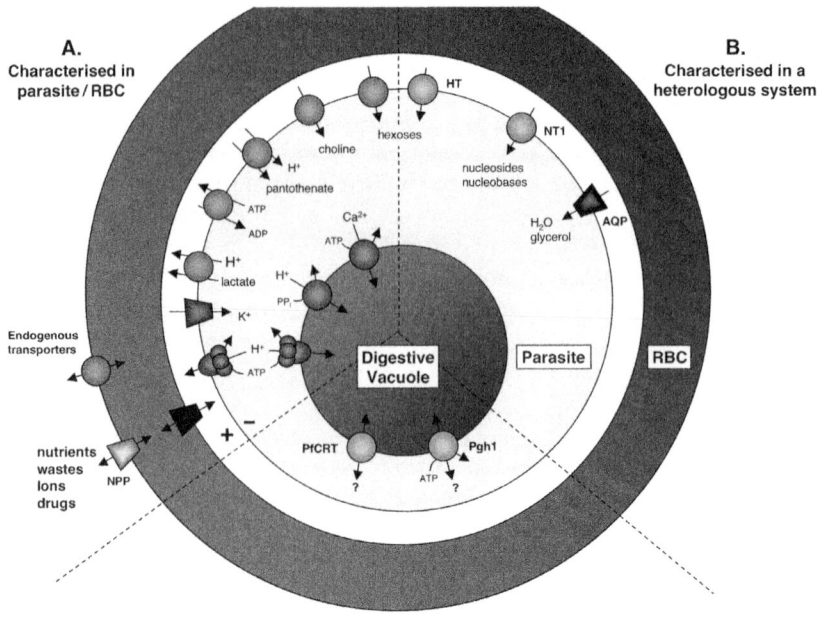

A.
Characterised in
parasite / RBC

B.
Characterised in a
heterologous system

HT

hexoses

choline

NT1

H⁺
pantothenate

nucleosides
nucleobases

ATP

Ca²⁺

ADP

ATP

H₂O
glycerol

AQP

H⁺
lactate

H⁺

PPᵢ

K⁺

Endogenous
transporters

H⁺

ATP

Digestive
Vacuole

Parasite

RBC

+   −

nutrients
wastes
Ions
drugs

NPP

PfCRT

?

Pgh1

ATP   ?

C. Unknown function

Fig. 2 Schematic representation of a *Plasmodium*-infected erythrocyte showing: (*A*) those transport pathways that have been characterized to at least some extent in physiological studies of parasitized cells/parasites; (*B*) those transport proteins that have been expressed and their transport properties characterized in a heterologous expression system; (*C*) transporters implicated in chloroquine resistance but for which a transporter function is yet to be demonstrated directly

2.2.1
The Erythrocyte Membrane

It has long been recognized that in erythrocytes infected with mature asexual stage parasites there is a profound increase in the permeability of the host cell membrane to a wide range of low molecular weight solutes, including three- to seven-carbon polyols, amino acids, nucleosides, vitamins, monocarboxylates, and inorganic monovalent anions and cations [41, 59]. The increase is attributable to what have been termed 'new permeability pathways' or NPP. The NPP are thought to serve a number of important roles in the parasitized cell. They serve as the major route (and perhaps in some cases, such as those of pantothenate [93] and glutamate [61], the only route) for the entry into the infected cell of important nutrients. They provide a pathway for the efflux from the infected cell of metabolic wastes such as lactic acid (originating from the parasite's active glycolysis) and amino acids (originating from the parasite digesting hemoglobin in quantities that far exceed its nutritional requirements; [67]). They also result in an increased leakage of Na^+ and K^+ across the erythrocyte membrane [104], gradually converting the erythrocyte compartment from a low Na^+/high K^+ environment to a high Na^+/low K^+ environment, thereby influencing the parasite's membrane potential [3] as well as generating a substantial inward Na^+ gradient across the parasite's plasma membrane.

Applying Occam's razor to the results of experiments comparing the effects of a range of different inhibitors on the parasite-induced transport of a wide range of solutes led to the hypothesis that the NPPs are a single class of channel [59], having a broad substrate specificity, but a general preference for monovalent anions over monovalent cations, with Cl^- permeating the channels several orders of magnitude faster than Na^+ and K^+. Ginsburg and Stein have argued recently that the data are better accounted for by the presence of two types of channels: one present at an estimated four copies per cell, admitting a range of organic solutes, and discriminating against cations; the other present at hundreds of copies per cell and mediating the flux of monovalent anions and nucleosides [42]. However their analysis rests on the assumption that the channels behave strictly as size-selective, water-filled pores and this is open to question.

Electrophysiological data from Sanjay Desai and colleagues are consistent with the view that the NPP take the form of a single class of anion-selective channel. In the first reported patch clamp study of *Plasmodium*-infected erythrocytes, using both whole-cell and single channel recordings, Desai et al. [23] reported the presence in infected cells of a novel, low-conductance, 'inwardly rectifying' anion-selective channel showing pharmacological characteristics and substrate-selectivity remarkably similar to those reported previously for the NPP on the basis of transport and hemolysis measurements. Subsequent studies from the same group, making quantitative comparisons between transport rates and conductances [117] and between the effects of the NPP inhibitor furosemide on transport, hemolysis and conductance [2] provided further support for the hypothesis that a single channel-type underlies the increased permeability of the infected erythrocyte to anions and low molecular weight electroneutral solutes.

Thomas and colleagues [32] have also carried out both whole-cell and single channel recordings and have again reported the observation of a single anion-selective channel type in parasitized cells, albeit with characteristics that differ somewhat from those reported by Desai and colleagues. However, Huber, Lang and colleagues [26, 27, 49], in a series of studies using the whole-cell recording configuration, have reported the presence in infected cells of three discrete conductances which they attribute to three different channel types: an 'outwardly rectifying' anion-selective channel; an 'inwardly-rectifying' anion-selective channel; and a 'nonselective cation channel'. They have subsequently presented evidence that the channel underlying the inwardly rectifying current observed in their experiments is ClC-2, a member of a well-characterized family of Cl⁻ channels, and largely impermeable to organic solutes, whereas the channel underlying the outwardly rectifying current is permeable to the organic anion lactate, and interacts with polyols [28, 50]. These data led them to postulate that it is this outwardly-rectifying channel that underlies the increased permeability of parasitized erythrocytes to low molecular weight organic solutes [28].

The question of why Desai and colleagues see a single, inwardly-rectifying anion channel, permeable to organic solutes, whereas Huber, Lang and colleagues see multiple conductances, with the outwardly-rectifying anion conductance permeable to organic solutes and the inwardly-rectifying anion conductance impermeable to organic solutes, is yet to be resolved. In a recent study, Staines et al. have reported that the nature of the conductances obtained in whole cell recordings of parasitized human erythrocytes is highly sensitive to the experimental conditions used [105]. In particular, the outwardly-rectifying anion conductance described by Huber et al. increases several-fold in the presence of trace amounts of serum in the medium. This may account for at least some of the discrepancies between the results obtained by different groups. In another recent study Verloo et al. [115] reported that: (1) whereas erythrocytes taken from normal donors and infected with *P. falciparum* show a prominent inwardly-rectifying anion current, those taken from patients with cystic fibrosis and infected with *P. falciparum* do not; and (2) infected erythrocytes from normal and cystic fibrosis donors show the same parasite-induced uptake of at least some low molecular weight organic solutes. The implication here is that the inwardly-rectifying channel observed in the patch-clamp experiments is not the pathway underlying the increased transport of solutes such as amino acids across the infected erythrocyte membrane. This is perhaps consistent with the view that it is the more elusive (serum-dependent) outwardly-rectifying conductance that is the electrophysiological manifestation of the NPP [106]. However, the findings of Verloo et al. have been disputed [2].

To summarize, while there is clear evidence for the parasite-induced NPP serving a number of important physiological roles in the parasitized erythrocyte, the number and nature of the pathways involved remains controversial. Two recent reviews address aspects of the controversy [106, 112]. There is similar uncertainty about the origin of these pathways, with some groups providing evidence that similar pathways can be activated in uninfected erythrocytes exposed to particular stresses or stimuli [32, 49] and another reporting strain-specific differences in the electrophysiological characteristics of the channel, consistent with, though not proof of, the channel being parasite-encoded [2]. The controversies are likely to continue until the protein(s) involved are identified and either expressed in heterologous systems or purified and reconstituted into artificial membrane bilayers. The recent bioinformatic analysis of the *P. falciparum* permeome [73] did identify one potential molecular candidate and this awaits experimental characterization.

There is evidence that, in addition to the NPP, there is a number of other parasite-induced transport pathways present in the parasitized erythrocyte membrane. Staines et al. have reported the activation in ATP-depleted parasitized erythrocytes of a novel Ca^{2+} transport pathway [102], and the fluorescent dye, lucifer yellow, is taken up into infected cells more rapidly than into uninfected cells, again via a pathway distinct from the NPP [61]. The increasing number of reports of parasitized erythrocytes taking up macromolecules, including proteins [33], points to the operation of additional uptake mechanisms operating at the erythrocyte surface, but these are yet to be characterized.

2.2.2
The Parasitophorous Vacuole Membrane

The PVM enclosing the intraerythrocytic parasite originates from the host erythrocyte membrane at the time of invasion. As the parasite grows within the infected cell, the PVM grows with it. Recent experiments with fluorescent Ca^{2+} dyes, present in the extracellular medium during invasion and thereby loaded into the parasitophorous vacuole, indicate that in the hours immediately following invasion the PVM retains a low Ca^{2+} permeability, and that the parasitophorous vacuole has a Ca^{2+} concentration significantly higher than that of the erythrocyte cytosol [39], perhaps through the action of Ca^{2+} pumps originating from the erythrocyte membrane [13]. However, electrophysiological experiments by Desai and colleagues on mature trophozoite-stage parasites provide evidence for the PVM enclosing the mature, metabolically active parasite being rendered freely permeable to solutes with diameters of up to 23 Å (and molecular weight 1,400 Da) by the presence of high-capacity and largely nonselective pores [21]. The finding by Nyalwidhe et al. that in parasitized cells in which the erythrocyte membrane (but not the PVM) is permeabilized by the pore-forming protein streptolysin O, a normally 'membrane impermeant' biotin derivative (600 Da) gains access to parasitophorous vacuole proteins [80] is consistent with this view.

As with the parasite-induced channels at the erythrocyte surface, the molecular identity of the broad-specificity pores underlying the high permeability of the PVM is unknown.

2.2.3
The Parasite Plasma Membrane

The parasite plasma membrane is energized through the action of a V-type H^+ ATPase which extrudes H^+ from the cytosol, into the parasitophorous vacuole [46,95]. This pump plays a key role in regulating the cytosolic pH of the parasite, maintaining it at approximately pH 7.3. It generates a significant transmembrane pH gradient; the pH in the region of the parasitophorous vacuole has been estimated as 6.9 [46]. This is lower than the pH in the bulk erythrocyte cytosol, and it is possible that the pH in the immediate vicinity of the external surface of the parasite is even lower, and the pH gradient across the membrane therefore greater than is indicated by the measurements to date.

The V-type H^+ ATPase is also the source of a substantial membrane potential, estimated to be approximately -95 mV under the ionic conditions that prevail within the infected cell [3]. The influx into the parasite of K^+, via Ba^{2+}- and Cs^+-sensitive K^+ channels [3], modulates the membrane potential and has been postulated to play a role in the volume expansion associated with parasite growth [4].

A range of nutrient transporters has been characterized in the parasite plasma membrane. Glucose has been shown to be taken up via an equilibrative (i.e., nonconcentrative) mechanism [43, 60, 98], attributed to the hexose transporter PfHT (see below). The essential vitamin pantothenate is taken up via an electroneutral H^+:pantothenate symporter [96] whereas the phospholipid precursor choline is taken up via an electrogenic transporter which is energized by the parasite's membrane potential [10,70] and which provides the major route of entry into the parasite of a number of cationic antiparasitic drugs [10]. Lactate, produced by the parasite in large quantities as a byproduct of anaerobic glycolysis, exits the parasite via an electroneutral H^+:monocarboxylate transporter [20, 35, 56].

An ATP/ADP exchanger at the parasite plasma membrane [16, 55] allows the parasite to supply ATP to the host cell compartment [55] but can also act in reverse, allowing the parasite to take up ATP from its external environment [16, 97].

2.2.4
The Digestive Vacuole

There is limited information available about the physiological properties of the parasite's acidic digestive vacuole. There is physiological evidence for the presence in the digestive vacuole membrane of two distinct H^+ pumps: a V-type H^+ ATPase and a H^+ pyrophosphatase [97]. Both pump H^+ into the vacuole, though their relative contributions to regulating the digestive vacuole pH are unclear. The actual value of the pH in the digestive vacuole has been, and remains, controversial [12, 101].

The digestive vacuole has also been shown to serve as an intracellular Ca^{2+} store and there is evidence for a thapsigargin- and cyclopiazonic acid-sensitive Ca^{2+} pump [9]. The physiological role of this store remains unclear.

3
Transport Protein Genes in the *P. falciparum* Genome

3.1
An Expanding Inventory of Transport Proteins

The original annotation of the *P. falciparum* genome identified "a very limited repertoire of membrane transporters, particularly for uptake of organic nutrients" [38]. However, it is questionable whether this reflects a genuine paucity of such proteins in this organism, or whether it simply reflects shortcomings in the annotation. Analysis of the *P. falciparum* genome using a computer program that searches a genome database on the basis of the hydropathy plots of the corresponding proteins [17] has led to the identification of a significant number of additional putative membrane transport proteins [73]. The approach is based on the observation that the polypeptides comprising transporter proteins typically possess hydrophobic TMDs and connecting hydrophilic, extra-membrane loops that are detected as 'peaks' and 'troughs', respectively, in a plot of the 'hydrophobicity index' of the polypeptide. Many transporters characterized to date have between 8 and 14 TMDs [91]. In searching for additional candidate transporters the *P. falciparum* genome was therefore scanned for proteins with seven or more TMDs.

Only 54 transport proteins were identified in the original genome annotation and many of these were designated with generic descriptions such as 'transporter, putative', from which no information can be gained about the probable mechanism of transport or substrate specificity. The hydropathy-based analysis retrieved a further 55 putative transport proteins, as well as attributing putative substrate specificities and/or transport mechanisms to those previously without. This brings the total number of putative/proven *P. falciparum*-encoded transport proteins to 109 [73]. Of these, 61 are 'porters' (i.e., uniporters, antiporters or symporters; [91]), 29 are primary transporters (i.e., transporters that utilize biochemical energy to pump solutes against an electrochemical gradient), 5 are channels and 14 are putative novel transport proteins of unknown classification. Figure 3 summarizes our current knowledge of the permeome of *P. falciparum*, including both those transporters previously identified as such (and having any number of TMDs) and the 55

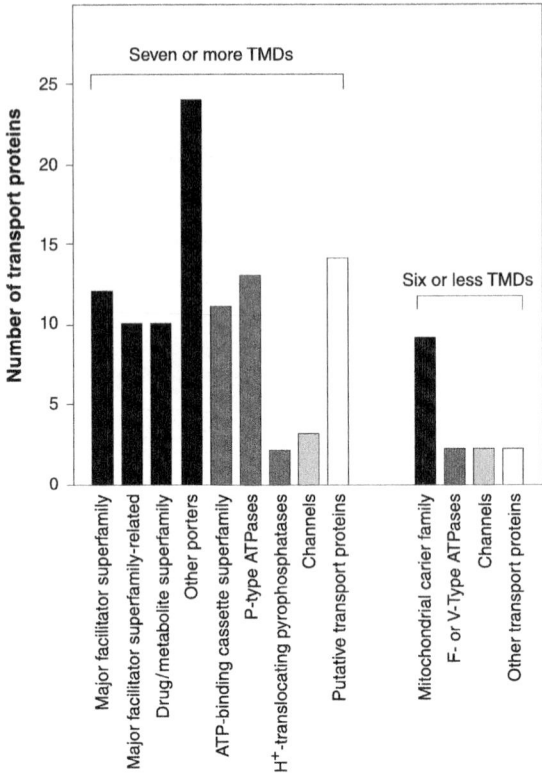

Fig. 3 Graphical overview of the permeome of *P. falciparum*. On the left are transport proteins with seven or more TMDs, identified through the analysis of the genome using a computer program that interrogates a genome database on the basis of the hydropathy plots of the corresponding proteins [17]. These include all of the putative seven or more TMD transport proteins identified in the original annotation [38], as well as 55 seven or more TMD transport proteins recently recognized as such [73]. On the *right* are transport proteins with six or fewer TMDs, sourced in the most part from the annotated genome. *Black bars* denote members of porter families (i.e., uniporters, symporters and antiporters); *dark gray bars* denote members of primary transporter families (i.e., pumps); *light gray bars* denote members of channel families; and *white bars* denote putative transporters of unknown lineage and function

additional (seven or more TMD) putative transport proteins identified in our recent analysis.

The *P. falciparum* transport proteins belong to 28 superfamilies and families; examples of these include the major facilitator superfamily, of which the PfHT [124] is a member, and several major facilitator superfamily-related families (such as the glycoside-pentoside-hexuronide:cation symporter family, of which the parasite has one member). Another large family of porter proteins, the drug/metabolite superfamily, is also well represented in the *P. falciparum* genome and includes the 'chloroquine resistance transporter' (PfCRT) [74, 113]. There are nine members of the mitochondrial carrier (MC) family and members of other, more minor porter families, including the equilibrative nucleoside transporter family, to which the *P. falciparum* nucleoside transporter (PfNT1; [14, 81]) belongs. There is a plethora of putative primary active transporters encoded in the parasite genome, including proteins of the ATP-binding cassette superfamily e.g., 'P-glycoprotein homologue 1' (Pgh1), encoded by the parasite's *PfMDR1* gene; [37, 122], the P-type ATPase superfamily, the H^+-translocating pyrophosphatase family [75] and the H^+- or Na^+-translocating F-type and V-type ATPases [57, 58]. By contrast, only a limited number of parasite-encoded channels (five) has been identified thus far; these include the aquaglyceroporin (PfAQP; [45]), a putative K^+ channel of the voltage-gated ion channel superfamily [34] and a homolog of an unusual bifunctional protein that contains an amino-terminus K^+ channel and a carboxyl-terminus adenylyl cyclase [121].

The newly designated proteins include candidate transporters for nutrients such as sugars, amino acids, nucleosides and vitamins. There are also transport proteins predicted to be involved in maintaining the ionic composition of the cell and in the extrusion of metabolic wastes such as lactate. This enrichment in the repertoire of *P. falciparum*-encoded transport proteins indicates that the parasite permeome is not as impoverished as originally thought. For instance, in the original analysis of the genome data it was proposed on the basis of the apparent absence of an obvious amino acid transporter, that the intraerythrocytic parasite must all but rely upon the ingestion and digestion of host hemoglobin for its supply of amino acids [38]. However, the discovery of several putative amino acid transporters [73], along with the previous observation that the parasite has an essential requirement for the exogenous supply of several amino acids (including isoleucine, which is absent from human hemoglobin) [24], indicates that this is unlikely to be the case. It was also reported that "no clear homologues of eukaryotic sodium, potassium or chloride ion channels could be identified" [38]; however, two putative potassium channels have since been cloned [34, 121] and there are an additional two putative ion channels identified in the genome [73; Allen RJW, unpublished].

A number of the *P. falciparum* proteins retrieved with seven or more TMDs bear no significant sequence similarity to any other proteins (transporters or otherwise) characterized previously. They do, however, have hydropathy plots that are similar to those of known transport proteins, consistent with the hypothesis that they too are transporters. Within this group is a set of related proteins that form a novel family of putative transporter proteins specific to *Plasmodium* [73] and which may therefore represent new antimalarial drug targets.

3.2
Predicting the Cellular Localization of *P. falciparum* Transport Proteins

Many transport proteins are located at the surface of the cell, where they mediate the flux of solutes across the plasma membrane. Other transport proteins are found in the membranes of intracellular compartments such as those of the apicoplast, mitochondrion and organelles of the secretory pathway. The likely destination(s) within the cell of a given transporter can often be inferred by signals present in its polypeptide sequence and/or by its close homology to a transport protein of a known cellular localization. For example, the signal peptide required for the targeting of nuclear-encoded proteins to the parasite's apicoplast has been elucidated [119, 129] and a number of *P. falciparum* transport proteins contain this type of signal. These putative apicoplast transporters include the parasite homolog of the plant chloroplast phosphoenolpyruvate:P_i antiporters. Likewise, the nine parasite MCs contain putative signals for the targeting of these transporters to the mitochondria. One of these, the putative phosphate carrier protein (MPC), has been cloned and shown experimentally to possess mitochondrial targeting signals [8]. Several transport proteins are dedicated to performing specialized tasks in the secretory pathway and specific 'retention' motifs participate in the sorting of these proteins between the membranes of the endoplasmic reticulum and the various Golgi compartments [48, 111]. The parasite's putative nucleotide sugar transporters, such as the UDP-galactose:UMP antiporter homologue (which contains a retention motif), are predicted to be residents of these organelles. In the absence of any targeting signals or sorting motifs, membrane proteins are usually destined to follow the 'default' pathway and travel through the secretory pathway to the plasma membrane [90].

3.3
Future Work

Although the analysis summarized here has increased significantly the number of transport proteins predicted to be present in *P. falciparum*, it is very likely that more parasite-encoded transport proteins remain to be uncovered. Search criteria were targeted towards identifying transporters with seven or more TMDs; however, many of the polypeptides which form ion channels possess six or fewer TMDs and several types of transporters also contain six or fewer TMDs. The fact that four putative channels (two putative K^+ channels [34, 121] and two other putative ion channels, both with seven or more TMDs) have been identified [73; Allen RJW, unpublished] indicates that these types of transport proteins are indeed present in the *P. falciparum* permeome.

The identification of genes and the prediction of intron–exon structures in the *P. falciparum* genome are still being refined (e.g., [51]). For a number of the previously unannotated transport proteins there were inappropriate predictions for the 5' end and intron–exon boundaries and it is this that led to their being overlooked in the original annotation process. Subsequent revisions of the genome data, by using the latest gene-finding tools, comparative genomics and the integration of full-length cDNA sequences, should improve the existing annotation of the genome and this, in turn, may lead to further additions to the *P. falciparium* permeome.

4
Subcellular Localization of Membrane Transport Proteins in the Parasitized Erythrocyte

The majority of *P. falciparum* genes, including most of those expressing putative transport proteins, are expressed (at least at the mRNA level) during the intraerythrocytic phase of the parasite lifecycle [11, 73]. For only a few of the putative transport proteins has their subcellular localization been investigated. In most cases this has involved low-resolution, immunofluorescence studies, though for a small number of proteins there have been studies done at higher resolution using immunoelectron microscopy. As the number of such studies increases we will gradually acquire a 'physiological atlas' of the parasitized erythrocyte.

In early work on the *P. falciparum* V-type H^+-ATPase, immunofluorescence using antibodies raised against the B subunit showed a heterogeneous distribution over most of the parasite [58]. Similarly, immunofluorescence studies of the localization of the parasite's aquaglyceroporin, PfAQP gave immunostaining covering the whole of the parasite "with some bright patches around the rim" but none in the host erythrocyte [45].

Immunofluorescence microscopy using antibodies against the parasite's putative type I pyrophosphatase (PfVP1) showed fluorescence associated both with the parasite plasma membrane and punctate intracellular inclusions, but not with the parasite's digestive vacuole [72, 75]. Similar results were obtained with a PfVP1–green fluorescent protein fusion. The localization data are at odds with the physiological and biochemical evidence for the presence of a type I H^+-pumping pyrophosphatase on the digestive vacuole membrane [97], as well as the lack of any physiological evidence for a functional H^+-pumping pyrophosphatase on the parasite plasma membrane [3, 97].

PfHT has been localized to the region of the parasite surface using immunofluorescence [86, 124], and the same is true of the putative plasma membrane Ca^{2+} ATPase, PfATP4 [29], and the parasite's nucleoside transporter, PfNT1 [86]. Using immunoelectron microscopy Rager et al. have localized PfNT1 predominantly, if not exclusively, to the parasite plasma membrane, and not to the parasitophorous vacuole or host erythrocyte membranes [86], and in the same study it was shown using immunofluorescence that PfHT co-localized with PfNT1 on the plasma membrane of parasites at the mature schizont stage. These findings lend support to the prevailing view that the mature parasite regulates the efflux and influx of solutes, and hence its internal composition, at the level of its plasma membrane, rather than at the PVM.

In a recent study a putative *P. falciparum* copper ATPase has been immunolocalized, using both immunofluorescence and electronmicroscopy, to both the parasite and the region of the erythrocyte surface [87].

The two putative transport proteins involved in chloroquine resistance, Pgh1 and PfCRT have been localized primarily (in the case of Pgh1) or exclusively (in the case of PfCRT) to the parasite's digestive vacuole, using both immunofluorescence [19, 36] and immunoelectron microscopy [18, 19].

5
Heterologous Expression of *Plasmodium*-Encoded Transport Proteins

Heterologous expression of membrane transport proteins in a functional form is often more difficult than expressing soluble proteins; not only must the proteins be synthesized and folded correctly, they must proceed through the secretory pathways where they undergo (often species-specific) translational modifications, and then need to be targeted to an appropriate membrane. In the case of *Plasmodium* membrane proteins this presents particular challenges as the parasite possesses unique trafficking pathways in addition to the secretory pathway, and it is not to clear to what extent species-specific proteins are involved in the classical secretory pathway.

Most attempts to express *Plasmodium* transporters in a heterologous system have involved either *Xenopus laevis* oocytes or yeast.

5.1
X. laevis Oocytes

Since its initial validation as an expression system for membrane transport proteins [47] the *X. laevis* oocyte has been used in numerous studies as an expression system for transporters and channels from animals, plants, a variety of lower eukaryotes, as well as prokaryotes.

The first evidence that oocytes might provide a suitable system for the expression of *Plasmodium* transport proteins came from the observations of Penny et al. [83] that 3 days after injection of oocytes with 50 ng poly(A)-RNA isolated from asexual-stage *P. falciparum*, the oocytes showed increased uptake of 2-deoxyglucose and D-adenosine. The hexose transporter PfHT, having approximately 50% similarity to some members of the GLUT family of hexose transporters, was subsequently identified from the *P. falciparum* genome data base, cloned, and expressed in *Xenopus* oocytes [124]. Orthologs of PfHT have been identified in other *Plasmodium* species [52, 54]. Like other members of the GLUT family, PfHT is a Na^+-independent hexose transporter. It has a relatively high affinity for D-glucose, with a K_m of 0.48 mM, and transport of D-glucose and 2-deoxyglucose via PfHT is inhibited by D-mannose, 3-O-methylglucose and cytochalasin B [124], as well as by a number of long-chain O-3 hexose derivatives [53].

Two groups have reported the cloning and functional expression in oocytes of a *P. falciparum* nucleoside transporter, designated PfNT1 [14] or PfENT1 [81]. The two studies differed significantly in the reported properties of the transporter. Carter et al. measured a K_m of 13 μM for adenosine, while Parker et al. reported a K_m of 320 μM. Carter et al. reported that the transporter was unable to transport nucleobases, whereas Parker et al. reported that it transported nucleobases efficiently, with K_m values similar to that for adenosine, as well as accepting the antiviral nucleoside analogues 3′-azido-3′-deoxythymidine, 2′,3′-dideoxycytidine and 2′,3′-dideoxyinosine, none of which are transported by mammalian equilibrative nucleoside transporters. Carter et al. reported the transporter to be inhibited by 10 μM dipyridamole, whereas in the study by Parker et al. it was not. Preliminary studies from our own laboratory have essentially reproduced the findings of Parker et al. with regard to substrate specificity (M.J. Downie, unpublished results). Parker et al. also showed PfNT1-mediated adenosine transport to be largely pH independent and unaffected by removal of Na^+, consistent with PfNT1 being an equilibrative transporter.

Oocytes show a limited water permeability that can be increased dramatically by the expression of aquaporin water channels [1]. The increased water permeability is usually assayed by videocapture of the increasing oocyte size after incubation in hypotonic solutions. This approach was used successfully to identify a *P. falciparum* member of the aquaporin family (PfAQP) [45]. The aquaporin family is divided into classical aquaporins, which are water-specific channels and aquaglyceroporins, which mediate the permeation of glycerol and a number of other small solutes, but have low water permeability. The sequence of the *Plasmodium* aquaporin is more closely related to that of the *Escherichia coli* glycerol facilitator (a prototypical aquaglyceroporin) than to those of mammalian aquaporins [45]. Expression of PfAQP in oocytes induces an increase of water permeability, which is comparable to that induced by rat AQP1. Surprisingly, the glycerol permeability of PfAQP is even higher than that of the rat aquaglyceroporin AQP3. Thus it appears that PfAQP allows the rapid transport of both water and glycerol, a feature that distinguishes it from its mammalian counterparts. PfAQP is also permeable to urea and to larger polyols such as erythritol (C4) and xylitol (C5). Polyols with more than five carbon atoms are impermeant. It has been speculated that a dual-function aquaglyceroporin has evolved in an effort to minimize the number of proteins required for parasite growth [45]. The apparent lack of any other aquaporin in the *P. falciparum* genome is consistent with this view.

Two Ca^{2+}-ATPases from *Plasmodium*, designated PfATP4 and PfATP6, have been expressed in oocytes [31, 65]. In both cases the activity was identified as a Ca^{2+}-dependent increase of ATPase activity, about two- to fourfold above that of noninjected oocytes. Sequence similarity and inhibition by vanadate are consistent with both PfATP4 and PfATP6 being P-type ATPases. Sequence analysis further suggests that PfATP4 represents a new subclass of Ca^{2+}-ATPases, and this is supported by its pharmacological properties. Notably, the enzyme is resistant to thapsigargin, an inhibitor of sarcoplasmic/endoplasmic reticulum Ca^{2+}-ATPases (SERCAs). This, together with the results of immunofluorescence studies [29], suggests that PfATP4 is a plasma membrane Ca^{2+}-ATPase, albeit perhaps not exclusively. By contrast, PfATP6 is more closely related to SERCAs and is also inhibited by thapsigargin as well as by artemisinins, prompting the proposal that inhibition of PfATP6 underlies the antiplasmodium action of this potent and important class of antimalarials [31].

In a very recent study Lanzer and colleagues have reported that expression of PfCRT in *Xenopus* oocytes results in a depolarized resting membrane potential and an increased cytosolic pH [79]. The membrane depolarization was attributed to activation of an endogenous nonselective cation conductance, and the increased pH to increased activity of the endogenous Na^+/H^+ exchanger. On the basis of these observations it was proposed that PfCRT serves to activate/modulate the activity of other transport proteins, though whether it has such a role in situ remains to be seen.

Xenopus oocytes clearly provide a suitable system for the expression of at least some *Plasmodium* transport proteins, and those *Plasmodium* transporters that have been successfully expressed in oocytes show transport rates of a similar magnitude to those seen for their mammalian homologs. It should be emphasized, however, that functional expression of a number of transporters has been attempted without success (unpublished results). There is a number of reasons why these attempts may have been unsuccessful [64]:

- The predicted sequences may be incomplete.

- The appropriate substrates may not have been identified.

- The mRNA may not be translated efficiently, or is degraded.

- The protein may be misfolded or degraded.

– The protein may not be targeted to the oocyte plasma membrane.

– The polypeptide may be part of a heteromeric complex.

– The transporter may be inactive due to a lack of, or incorrect, post-translational modifications, or the absence of a regulatory ligand.

Of these, only the third possibility listed has been investigated to any extent. Many laboratories now routinely use vectors for the generation of in vitro transcribed RNA that contain the 5′ and 3′ untranslated region of the *Xenopus* β-globin. Similarly, a Kozak consensus sequence [63] before the start codon appears to be beneficial for expression of transporters in oocytes. Both manipulations have been shown to enhance the translation of mammalian cRNA in numerous studies, and this has also been shown to be true for the *Plasmodium* transporter PfHT [125]. Once injected, the in vitro transcribed capped mRNA is stable for several days (unpublished observation).

One factor that is often cited as underlying difficulties in the expression of *P. falciparum* proteins in heterologous systems is the extreme AT-bias of the *P. falciparum* genome and the consequent disparity in codon usage between *Plasmodium* and the organism being used for heterologous expression. However, a comparison of the sequences of those transporters that have been expressed successfully in *Xenopus* oocytes with those that have proven more difficult does not reveal any significant difference in codon usage. More likely reasons for failures to express *Plasmodium* transporters are misfolding or mistargeting of the proteins. A systematic study of this issue is overdue.

5.2
Yeast

Yeast has many advantages as a heterologous expression system: large volumes of yeast can be grown rapidly and cheaply and can produce high yields of recombinant protein; yeast is readily amenable to genetic manipulations; and it performs many of the post-translational protein modifications typically found in secreted proteins of higher eukaryotes, such as glycosylation and disulfide bond formation. This expression system is therefore suitable for many applications including production of high levels of proteins for vaccines, expression and characterization of heterologous proteins, and drug screening. Most heterologous expression studies in yeast involve the use of vectors with strong promoters either in baker's yeast, *Saccharomyces cerevisiae* or the methylotrophic yeast, *Pichia pastoris*.

Despite these advantages, only a limited number of proteins from *Plasmodium* have been expressed in yeast and most of these are not transporters. The AT-richness of the *P. falciparum* genome may present a particular problem for expression of *P. falciparum* proteins in yeast. It has been suggested that

AT-rich sequences may cause premature termination of transcription in yeast by mimicking polyadenylation signals [77], and the presence of truncated transcripts has been demonstrated in at least one case [77]. Consistent with this, *Plasmodium* sequences have mostly been found to produce low, often undetectable, levels of protein in yeast [77, 99, 123, 127, 128] although most studies have not determined whether this is a transcriptional or translational problem. Both problems could potentially be overcome by resynthesizing the gene so that the AT content is reduced and the extended poly-A stretches replaced by alternative (GC-rich) codons. In several cases involving soluble proteins and membrane proteins, the level of protein expression in yeast has been improved by this strategy [77, 99, 123, 127, 128]. As with *Xenopus* oocytes, however, codon usage is clearly not the only consideration as the same resynthesized gene may function much better in *P. pastoris* than in *S. cerevisiae* [77,128], despite the two yeast strains having similar codon usage.

The high levels of expression obtained from most standard yeast expression systems may overload the secretory pathway, reducing plasma membrane expression [15, 44]. At least some *Plasmodium* proteins are trafficked through the secretory pathway of yeast; there have been some examples of secreted proteins being produced as potential vaccine candidates [77, 127]. However, these proteins are not membrane-localized. Interestingly, the three integral membrane proteins from *Plasmodium* that have been successfully expressed in yeast are all localized to internal membranes in the parasite. Two of these, PfCRT [128] and Pgh1 [116], are digestive vacuole membrane proteins. The third expressed membrane protein is PfGatp [99], an enzyme involved in glycerolipid synthesis and found in the endoplasmic reticulum membrane of the parasite. When expressed in yeast, Pgh1 and PfGatp were localized predominantly to internal membranes [99, 116] while in one study, PfCRT was modified at the N terminus to allow plasma membrane localization [128]. The nature of this modification was not described. In a preliminary report of another study, unmodified PfCRT was described as being found in internal membranes [110]. It is not clear whether trafficking of plasma membrane proteins from *Plasmodium* represents a serious limitation to expression studies in yeast or whether these results simply reflect the fact that expression of so few membrane proteins has been attempted.

The expression of Pgh1 in *S. cerevisiae* is one of the few examples of successful expression in yeast of a gene consisting of the native *Plasmodium* DNA sequence [116]. Expression of Pgh1 was detected as complementation of a yeast mutant deficient in Ste6, which is responsible for export of the peptide pheromone, a-factor. The yeast mutant is sterile and is unable to mate. Pgh1 was able to restore mating function while a mutant version of Pgh1 was not. Export of the a-factor was demonstrated using a quantitative pheromone assay on the culture supernatant, suggesting that Pgh1 has the ability to transport peptides. Both Pgh1 and Ste6 were suggested to function in internal membranes with the a-factor being released by endocytosis. However it cannot be concluded from these results that a peptide is the natural substrate for Pgh1 as other homologous drug resistance transporters have also been found to complement Ste6 mutants [88] and these transporters generally exhibit broad substrate specificity.

Codon-optimized PfCRT was expressed in both *S. cerevisiae* and *P. pastoris*. It was expressed at significant levels in both yeasts but at higher levels in *P. pastoris*, whereas the endogenous sequences failed to produce detectable PfCRT protein [128]. Inside-out plasma membrane vesicles were prepared from *P. pastoris* and used to analyze pH gradient formation under various conditions. It was suggested that PfCRT mediates passive Cl^- movement either directly or indirectly and therefore has a role in maintenance of the pH gradient across the digestive vacuole membrane in the parasite. A mutant form of the transporter associated with chloroquine resistance was reported to show increased proton pumping, consistent with more efficient Cl^- transport. However, there was no attempt to measure chloroquine transport.

In another study, published in abstract form, a resynthesized PfCRT gene was expressed in *P. pastoris* and it was found that membrane vesicles derived from the transfected yeast strain could accumulate chloroquine [110]. A mutation in PfCRT that confers chloroquine resistance reduced the accumulation of chloroquine in yeast membrane vesicles. Further studies will be required to determine whether PfCRT transports chloroquine directly, or has an indirect role in mediating chloroquine accumulation in the digestive vacuole.

6
Manipulation of Gene Expression in the Parasite

The difficulties inherent in the expression and characterization of *Plasmodium* transport proteins in heterologous systems raise the possibility that, for at least some such proteins, the best system in which to demonstrate function might actually be the parasite itself. As outlined in Sect. 2, methods for measuring membrane transport processes in the parasite are well developed and the combination of these with techniques for manipulating gene expression in the parasite represents an attractive way forward for 'functional permeomics' (i.e., ascribing function to genes encoding proteins involved in membrane permeability). Recent years have seen significant advances in our ability to manipulate gene expression in the parasite, though there are, as yet, few examples of the application of these methodologies to demonstrate or characterize transport protein function.

6.1
Mutation of Transporters

In situ manipulation of transporters in *P. falciparum* has so far only been carried out for Pgh1 and PfCRT. An allelic exchange approach has been used to make specific amino acid substitutions in these two proteins, and to demonstrate thereby their involvement in chloroquine resistance [89, 100]. A similar strategy could be employed to study the significance of specific mutations in other transporters. The method could be applied to essential transporters in situ, providing that the mutations intended for study do not result in a complete loss of function. It should, for example, be possible to replace the glutamine at position 169 to an asparagine (Q169N) in the *P. falciparum* hexose transporter, PfHT. This mutation has been shown, using *Xenopus* oocytes, to abolish the fructose transport capacity of the transporter, while leaving the glucose transport capacity unaltered [125, 120]. If, as has been proposed [53, 52], PfHT is the only hexose transporter in the parasite plasma membrane, parasites bearing the Q169N mutation in PfHT should grow normally in media containing glucose as the hexose source, but should differ from wild-type parasites in their ability to grow on media containing fructose as the sole hexose source.

6.2
Knockout of Transporters

Targeted gene deletion has been achieved in *P. falciparum* by double crossover recombination [25]; however this is yet to be achieved for any transport protein. The knockout of specific transporters offers significant benefits in establishing function but this strategy presents obvious problems if the transporter plays an essential role in the parasite.

Under certain circumstances knockout of a transporter which plays an essential role in nutrient uptake may be achieved if the substrate for the transporter is known (e.g., from sequence homology), and if the substrate is able to cross the membrane via an alternative route. In such instances, it may be possible to knock out an essential transporter, provided the parasites are supplemented with high concentrations of the (putative) substrate, as has been demonstrated in other systems [108, 109]. Alternatively, it may be possible to circumvent the role of the transporter by supplementing the medium with downstream metabolites of the relevant (essential) substrates, provided they have a route of entry into the parasite.

Another approach could involve replacing the gene encoding the transporter of interest with a gene from a different organism that is known to have the same function, but which has significantly different transport characteristics from those of the *Plasmodium* transporter. For example, *P. fal-*

ciparum parasites transport pantothenate (an essential nutrient) across the parasite plasma membrane via a low affinity, electroneutral, H^+/pantothenate symporter [96]. The gene predicted to be responsible for the transport of pantothenate into the parasite could, for example, be replaced with the human multivitamin transporter [120], which is a high affinity, electrogenic $2Na^+$/pantothenate symporter [84, 85, 120]. A difference in pantothenate transport characteristics should, in principle, be readily discernable under these conditions and would establish the function of the deleted gene. There are, however, a number of significant challenges that are likely to arise in using such an approach, including problems associated with post-translational modifications and trafficking of the foreign transporter to the correct parasite membrane.

6.3
RNAi-Mediated Knockdown of Transporters

There are a number of recent reports of the use of RNA interference (RNAi)–the degradation of specific gene transcripts, resulting in decreases in protein levels, using homologous double-stranded RNA–to decrease the level of expression of a number of proteins in *Plasmodium* species [68, 71, 76, 78]. Whether the data in these papers were a result of genuine RNAi or nonspecific effects, has recently been called into question, partly because of the apparent absence in the *Plasmodium* databases of genes that are known to be necessary for the RNAi pathway to be active [114]. If the initial reports suggesting that RNAi is feasible in *P. falciparum* are vindicated, the technique has the potential of bypassing a lot of the problems associated with studying transporters that are essential for parasite growth. It may be possible, for example, to knockdown an essential transporter to sufficiently low levels that a measurable transport defect is detected without causing a significant decrease in cell viability. RNAi would also greatly decrease the length of time required to achieve gene knockdown when compared to alternative methods.

6.4
Knockdown of Transporters Using Allelic Exchange

Introduction of a truncated 3′-untranslated region into the functional locus of the gene encoding PfCRT resulted in a 30%–40% decrease in the level of protein [118]. Similar strategies could potentially be used for other transporters, and may have to become the method of choice for such experiments if RNAi proves to be impractical in *Plasmodium*.

7
Conclusions

In this chapter we have tried to present an overview of the range of different approaches that are being used to investigate the permeome of the malaria parasite. Interest in this area is increasing, not least because the proteins involved are not only potential antimalarial drug targets in their own right, but provide routes by which antimalarials can be targeted to the parasitized erythrocyte and the parasite within [10, 40, 94, 107]. Transport proteins are known to play a key role in antimalarial drug resistance, either through their ability to transport drugs away from their principal site of action or by influencing the electrochemical ion gradients that energize the accumulation of drugs in the relevant compartments within the infected cell.

Working with membrane transport proteins is not straightforward, and this is particularly true for those of the malaria parasite. Transport measurements in parasitized cells are made difficult by the multi-membrane nature of the infected erythrocyte and the consequent inaccessibility of the parasite. Heterologous expression of *Plasmodium* proteins in general, and *Plasmodium* transport proteins in particular is difficult, for reasons that are not always clear. Manipulation of the level of gene expression in the parasite itself is difficult, and still relatively inefficient; the techniques involved are still in their infancy and there are as yet very few examples of their application to the study of transporters and channels. Nevertheless, with improvements in the techniques available, improvements in genome annotation, and advances in our understanding of the nature of the difficulties associated with the functional expression of *Plasmodium* transporters and channels there should be very significant progress made in this area over the coming years.

Acknowledgements Work by the authors in this area is supported by a grant from the Australian Research Council (DPO344425) and by grants from the Australian National Health and Medical Research Council (224243 and 224245).

References

1.   Agre P, Preston GM, Smith BL, Jung JS, Raina S, Moon C, Guggino WB, Nielsen S (1993) Aquaporin CHIP: the archetypal molecular water channel. Am J Physiol 265:F463–F476
2.   Alkhalil A, Cohn JV, Wagner MA, Cabrera JS, Rajapandi T, Desai SA (2004) *Plasmodium falciparum* likely encodes the principal anion channel on infected human erythrocytes. Blood 104:4279–4286
3.   Allen RJW, Kirk K (2004) The membrane potential of the intraerythrocytic malaria parasite *Plasmodium falciparum*. J Biol Chem 279:11264–11272
4.   Allen RJW, Kirk K (2004) Cell volume control in the *Plasmodium*-infected erythrocyte. Trends Parasitol 20:7–10

5. Alleva LM, Kirk K (2001) Calcium regulation in the intraerythrocytic malaria parasite *Plasmodium falciparum*. Mol Biochem Parasitol 117:121–128

6. Ansorge I, Benting J, Bhakdi S, Lingelbach K (1996) Protein sorting in *Plasmodium falciparum*-infected red blood cells permeabilized with the pore-forming protein streptolysin O. Biochem J 315:307–314

7. Ansorge I, Paprotka K, Bhakdi S, Lingelbach K (1997) Permeabilization of the erythrocyte membrane with streptolysin O allows access to the vacuolar membrane of *Plasmodium falciparum* and a molecular analysis of membrane topology. Mol Biochem Parasitol 84:259–261

8. Bhaduri-McIntosh S, Vaidya AB (1998) *Plasmodium falciparum*: import of a phosphate carrier protein into heterologous mitochondria. Exp Parasitol 88:252–254

9. Biagini GA, Bray PG, Spiller DG, White MR, Ward SA (2003) The digestive food vacuole of the malaria parasite is a dynamic intracellular Ca^{2+} store. J Biol Chem 278:27910J. Biol. Chem. 27915

10. Biagini GA, Pasini EM, Hughes RH, De Koning H, Vial HJ, O'Neill PM, Ward SA, Bray PG (2004) Characterization of the choline carrier of *Plasmodium falciparum*: a route for the selective delivery of novel antimalarial drugs. Blood 104:3372–3377

11. Bozdech Z, Llinas M, Pulliam BL, Wong ED, Zhu J, DeRisi JL (2003) The transcriptome of the intraerythrocytic developmental cycle of *Plasmodium falciparum*. PLoS Biol 1:85–100

12. Bray PG, Saliba KJ, Davies JD, Spiller DG, White MR, Kirk K, Ward SA (2002) Distribution of acridine orange fluorescence in *Plasmodium falciparum*-infected erythrocytes and its implications for the evaluation of digestive vacuole pH. Mol Biochem Parasitol 119:301–304; discussion 307–309, 311–313

13. Caldas ML, Wasserman M (2001) Cytochemical localisation of calcium ATPase activity during the erythrocytic cell cycle of *Plasmodium falciparum*. Int J Parasitol 31:776–782

14. Carter NS, Ben Mamoun C, Liu W, Silva EO, Landfear, SM, Goldberg DE, Ullman B (2000) Isolation and functional characterization of the PfNT1 nucleoside transporter gene from *Plasmodium falciparum*. J Biol Chem 275:10683–10691

15. Cereghino GP, Cregg JM (1999) Applications of yeast in biotechnology: protein production and genetic analysis. Curr Opin Biotechnol 10:422–427

16. Choi I, Mikkelsen RB (1990) *Plasmodium falciparum*: ATP/ADP transport across the parasitophorous vacuolar and plasma membranes. Exp Parasitol 71:452–462

17. Clements JD, Martin RE (2002) Identification of novel membrane proteins by searching for patterns in hydropathy profiles. Eur J Biochem 269:2101–2107

18. Cooper RA, Ferdig MT, Su XZ, Ursos LM, Mu J, Nomura T, Fujioka H, Fidock DA, Roepe PD, Wellems TE (2002) Alternative mutations at position 76 of the vacuolar transmembrane protein PfCRT are associated with chloroquine resistance and unique stereospecific quinine and quinidine responses in *Plasmodium falciparum*. Mol Pharmacol 61:35–42

19. Cowman AF, Karcz S, Galatis D, Culvenor JG (1991) A P-glycoprotein homologue of *Plasmodium falciparum* is localized on the digestive vacuole. J Cell Biol 113:1033–1042

20. Cranmer SL, Conant AR, Gutteridge WE, Halestrap AP (1995) Characterization of the enhanced transport of L- and D-lactate into human red blood cells infected with *Plasmodium falciparum* suggests the presence of a novel saturable lactate proton cotransporter. J Biol Chem 270:15045–15052

21. Desai SA, Krogstad DJ, McCleskey EW (1993) A nutrient-permeable channel on the intraerythrocytic malaria parasite. Nature 362:643–646

22. Desai SA, Rosenberg RL (1997) Pore size of the malaria parasite's nutrient channel. Proc Natl Acad Sci USA 94:2045–2049

23. Desai SA, Bezrukov SM, Zimmerberg J (2000) A voltage-dependent channel involved in nutrient uptake by red blood cells infected with the malaria parasite. Nature 406:1001–1005

24. Divo AA, Geary TG, Davis NL, Jensen JB (1985) Nutritional requirements of *Plasmodium falciparum* in culture. I. Exogenously supplied dialyzable components necessary for continuous growth. J Protozool 32:59–64

25. Duraisingh MT, Triglia T, Cowman AF (2002) Negative selection of *Plasmodium falciparum* reveals targeted gene deletion by double crossover recombination. Int J Parasitol 32:81–89

26. Duranton C, Huber SM, Lang F (2002) Oxidation induces a Cl⁻-dependent cation conductance in human red blood cells. J Physiol 539:847–855

27. Duranton C, Huber S, Tanneur V, Lang K, Brand V, Sandu C, Lang F (2003) Electrophysiological properties of the *Plasmodium falciparum*-induced cation conductance of human erythrocytes. Cell Physiol Biochem 13:189–198

28. Duranton C, Huber SM, Tanneur V, Brand VB, Akkaya C, Shumilina EV, Sandu CD, Lang F (2004) Organic osmolyte permeabilities of the malaria-induced anion conductances in human erythrocytes. J Gen Physiol 123:417–426

29. Dyer M, Jackson M, McWhinney C, Zhao G, Mikkelsen R (1996) Analysis of a cation-transporting ATPase of *Plasmodium falciparum*. Mol Biochem Parasitol 78:1–12

30. Dzekunov SM, Ursos LM, Roepe PD (2000) Digestive vacuolar pH of intact intraerythrocytic *P. falciparum* either sensitive or resistant to chloroquine. Mol Biochem Parasitol 110:107–124

31. Eckstein-Ludwig U, Webb RJ, Van Goethem ID, East JM, Lee AG, Kimura M, O'Neill PM, Bray PG, Ward SA, Krishna S (2003) Artemisinins target the SERCA of *Plasmodium falciparum*. Nature 424:957–961

32. Egée S, Lapaix F, Decherf G, Staines HM, Ellory JC, Doerig C, Thomas SL (2002) A stretch-activated anion channel is up-regulated by the malaria parasite *Plasmodium falciparum*. J Physiol 542:795–801

33. El Tahir A, Malhotra P, Chauhan VS (2003) Uptake of proteins and degradation of human serum albumin by *Plasmodium falciparum*-infected human erythrocytes. Malaria J 2:11

34. Ellekvist P, Ricke CH, Litman T, Salanti A, Colding H, Zeuthen T, Klaerke DA (2004) Molecular cloning of a K⁺ channel from the malaria parasite *Plasmodium falciparum*. Biochem Biophys Res Commun 318:477–484

35. Elliott JL, Saliba KJ, Kirk K (2001) Transport of lactate and pyruvate in the intraerythrocytic malaria parasite, *Plasmodium falciparum*. Biochem J 355:733–739

36. Fidock DA, Nomura T, Talley AK, Cooper RA, Dzekunov SM, Ferdig MT, Ursos LM, Sidhu AB, Naude B, Deitsch KW, Su XZ, Wootton JC, Roepe PD, Wellems, TE (2000) Mutations in the *P. falciparum* digestive vacuole transmembrane protein PfCRT and evidence for their role in chloroquine resistance. Mol Cell 6:861–871

37. Foote SJ, Thompson JK, Cowman AF, Kemp DJ (1989) Amplification of the multidrug resistance gene in some chloroquine-resistant isolates of *P. falciparum*. Cell 57:921–930

38. Gardner MJ, Hall N, Fung E, White O, Berriman M, Hyman RW, Carlton JM, Pain A, Nelson KE, Bowman S, Paulsen IT, James K, Eisen JA, Rutherford K, Salzberg SL, Craig A, Kyes S, Chan MS, Nene V, Shallom SJ, Suh B, Peterson J, Angiuoli S, Pertea M, Allen J, Selengut J, Haft D, Mather MW, Vaidya AB, Martin DM, Fairlamb AH, Fraunholz MJ, Roos DS, Ralph SA, McFadden GI, Cummings LM, Subramanian GM, Mungall C, Venter JC, Carucci DJ, Hoffman SL, Newbold C, Davis RW, Fraser CM, Barrell B (2002) Genome sequence of the human malaria parasite *Plasmodium falciparum*. Nature 419:498–511

39. Gazarini ML, Thomas AP, Pozzan T, Garcia CR (2003) Calcium signaling in a low calcium environment: how the intracellular malaria parasite solves the problem. J Cell Biol 161:103–110

40. Gero AM, Dunn CG, Brown DM, Pulenthiran K, Gorovits EL, Bakos T, Weis AL (2003) New malaria chemotherapy developed by utilization of a unique parasite transport system. Curr Pharm Des 9:867–877

41. Ginsburg H, Kutner S, Krugliak M, Cabantchik ZI (1985) Characterization of permeation pathways appearing in the host membrane of *Plasmodium falciparum* infected red blood cells. Mol Biochem Parasitol 14:313–322

42. Ginsburg H, Stein WD (2004) The new permeability pathways induced by the malaria parasite in the membrane of the infected erythrocyte: comparison of results using different experimental techniques. J Membr Biol 197:113–134

43. Goodyer ID, Hayes DJ, Eisenthal R (1997) Efflux of 6-deoxy-D-glucose from *Plasmodium falciparum*-infected erythrocytes via two saturable carriers. Mol Biochem Parasitol 84:229–239

44. Griffith DA, Delipala C, Leadsham J, Jarvis SM, Oesterhelt D (2003) A novel yeast expression system for the overproduction of quality-controlled membrane proteins. FEBS Lett 553:45–50

45. Hansen M, Kun JF, Schultz JE, Beitz E (2002) A single, bi-functional aquaglyceroporin in blood-stage *Plasmodium falciparum* malaria parasites. J Biol Chem 277:4874–4882

46. Hayashi M, Yamada H, Mitamura T, Horii T, Yamamoto A, Moriyama Y (2000) Vacuolar H^+-ATPase localized in plasma membranes of malaria parasite cells, *Plasmodium falciparum*, is involved in regional acidification of parasitised erythrocytes. J Biol Chem 275:34353–34358

47. Hediger MA, Coady MJ, Ikeda TS, Wright EM (1987) Expression cloning and cDNA sequencing of the Na^+/glucose co-transporter. Nature 330:379–381

48. Hong W, Tang BL (1993) Protein trafficking along the exocytotic pathway. Bioessays 15:231–238

49. Huber SM, Uhlemann AC, Gamper NL, Duranton C, Kremsner PG, Lang F (2002) *Plasmodium falciparum* activates endogenous Cl⁻ channels of human erythrocytes by membrane oxidation. EMBO J. 21:22–30

50. Huber SM, Duranton C, Henke G, Van De Sand C, HeusslerV, Shumilina E, Sandu CD, Tanneur V, Brand V, Kasinathan RS, Lang KS, Kremsner PG, Hubner CA, Rust MB, Dedek K, Jentsch TJ, Lang F (2004) *Plasmodium* induces swelling-activated ClC-2 anion channels in the host erythrocyte. J Biol Chem

51. Huestis R, Fischer K (2001) Prediction of many new exons and introns in *Plasmodium falciparum* chromosome 2. Mol Biochem Parasitol 118:187–199

52. Joet T, Holterman L, Stedman TT, Kocken CH, Van Der Wel A, Thomas AW, Krishna S (2002) Comparative characterization of hexose transporters of *Plasmodium knowlesi, Plasmodium yoelii* and *Toxoplasma gondii* highlights functional differences within the apicomplexan family. Biochem J 368:923–929

53. Joet T, Eckstein-Ludwig U, Morin C, Krishna S (2003) Validation of the hexose transporter of *Plasmodium falciparum* as a novel drug target. Proc Natl Acad Sci USA 100:7476–7479

54. Joet T, Chotivanich K, Silamut K, Patel AP, Morin C, KrishnaS (2004) Analysis of *Plasmodium vivax* hexose transporters and characterization of a parasitocidal inhibitor. Biochem J 381:905–909

55. Kanaani J, Ginsburg H (1989) Metabolic interconnection between the human malarial parasite *Plasmodium falciparum* and its host erythrocyte. Regulation of ATP levels by means of an adenylate translocator and adenylate kinase. J Biol Chem 264:3194–3199

56. Kanaani J, Ginsburg H (1991) Transport of lactate in *Plasmodium falciparum*-infected human erythrocytes. J Cell Physiol 149:469–476

57. Karcz SR, Herrmann VR, Cowman AF (1993) Cloning and characterization of a vacuolar ATPase A subunit homologue from *Plasmodium falciparum*. Mol Biochem Parasitol 58:333–344

58. Karcz SR, Herrmann VR, Trottein F, Cowman AF (1994) Cloning and characterization of the vacuolar ATPase B subunit from *Plasmodium falciparum*. Mol Biochem Parasitol 65:123–133

59. Kirk K, Horner HA, Elford BC, Ellory JC, Newbold CI (1994) Transport of diverse substrates into malaria-infected erythrocytes via a pathway showing functional characteristics of a chloride channel. J Biol Chem 269:3339–3347

60. Kirk K, Horner HA, Kirk J (1996) Glucose uptake in *Plasmodium falciparum*-infected erythrocytes is an equilibrative not an active process. Mol Biochem Parasitol 82:195–205

61. Kirk K, Staines HM, Martin RE, Saliba KJ. (1999) In: Transport and trafficking in the malaria-infected erythrocyte. Novartis Foundation Symposium 226. Novartis Foundation, Wiley, Chichester, pp 55–73

62. Kirk K (2001) Membrane transport in the malaria-infected erythrocyte. Physiol Rev 81:495–537

63. Kozak M (1986) Point mutations define a sequence flanking the AUG initiator codon that modulates translation by eukaryotic ribosomes. Cell 44:283–292

64. Krishna S, Webb R, Woodrow C (2001) Transport proteins of *Plasmodium falciparum*: defining the limits of metabolism. Int J Parasitol 31:1331–1342

65. Krishna S, Woodrow C, Webb R, Penny J, Takeyasu K, Kimura M, East JM (2001) Expression and functional characterization of a *Plasmodium falciparum* Ca^{2+}-ATPase (PfATP4) belonging to a subclass unique to apicomplexan organisms. J Biol Chem 276:10782–10787

66. Krogstad DJ, Schlesinger PH, Gluzman IY (1985) Antimalarials increase vesicle pH in *Plasmodium falciparum*. J Cell Biol 101:2302–2309

67. Krugliak M, Zhang J, Ginsburg H (2002) Intraerythrocytic *Plasmodium falciparum* utilizes only a fraction of the amino acids derived from the digestion of host cell cytosol for the biosynthesis of its proteins. Mol Biochem Parasitol 119:249–256

68. Kumar R, Adams B, Oldenburg A, Musiyenko A, Barik S (2002) Characterisation and expression of a PP1 serine/threonine protein phosphatase (PfPP1) from the malaria parasite, *Plasmodium falciparum*: demonstration of its essential role using RNA interference. Malaria J 1:5

69. Lambros C, Vanderberg JP (1979) Synchronization of *Plasmodium falciparum* erythrocytic stages in culture. J Parasitol 65:418–420

70. Lehane AM, Saliba KJ, Allen RJW, Kirk K (2004) Choline uptake into the malaria parasite is energised by the membrane potential. Biochem Biophys Res Commun 320:311–317

71. Malhotra P, Dasaradhi PV, Kumar A, Mohmmed A, Agrawal N, Bhatnagar RK, Chauhan VS (2002) Double-stranded RNA-mediated gene silencing of cysteine proteases (falcipain-1 and -2) of *Plasmodium falciparum*. Mol Microbiol 45:1245–1254

72. Marchesini N, Luo S, Rodrigues CO, Moreno SN, Docampo R (2000) Acidocalcisomes and a vacuolar H^+-pyrophosphatase in malaria parasites. Biochem J 347:243–253

73. Martin RE, Henry RI, Abbey JL, Clements JD, Kirk K (2005) The 'permeome' of the malaria parasite: an overview of the membrane transport proteins of *Plasmodium falciparum*. Genome Biol 6:R26

74. Martin RE, Kirk K (2004) The malaria parasite's chloroquine resistance transporter is a member of the drug/metabolite transporter superfamily. Mol Biol Evol 21:1938–1949

75. McIntosh MT, Drozdowicz YM, Laroiya K, Rea PA, Vaidya AB (2001) Two classes of plant-like vacuolar-type H^+-pyrophosphatases in malaria parasites. Mol Biochem Parasitol 114:183–195

76. McRobert L, McConkey GA (2002) RNA interference (RNAi) inhibits growth of *Plasmodium falciparum*. Mol Biochem Parasitol 119:273–278

77. Milek RL, Stunnenberg HG, Konings RN (2000) Assembly and expression of a synthetic gene encoding the antigen Pfs48/45 of the human malaria parasite *Plasmodium falciparum* in yeast. Vaccine 18:1402–1411

78. Mohmmed A, Dasaradhi PV, Bhatnagar RK, Chauhan VS, Malhotra P (2003) In vivo gene silencing in *Plasmodium berghei* - a mouse malaria model. Biochem Biophys Res Commun 309:506–511

79. Nessler S, Friedrich O, Bakouh N, Fink RH, Sanchez CP, Planelles G, Lanzer M (2004) Evidence for activation of endogenous transporters in *Xenopus laevis* oocytes expressing the *Plasmodium falciparum* chloroquine resistance transporter PfCRT. J Biol Chem 279:39438–39446

80. Nyalwidhe J, Baumeister S, Hibbs AR, Tawill S, Papakrivos J, Volker U, Lingelbach K (2002) A nonpermeant biotin derivative gains access to the parasitophorous vacuole in *Plasmodium falciparum*-infected erythrocytes permeabilized with streptolysin O. J Biol Chem 277:40005–40011

81. Parker MD, Hyde RJ, Yao SY, McRobert L, Cass CE, Young JD, McConkey GA, Baldwin SA (2000) Identification of a nucleoside/nucleobase transporter from *Plasmodium falciparum*, a novel target for anti-malarial chemotherapy. Biochem J 349:67–75

82. Pasvol G, Wilson RJ, Smalley ME, Brown J (1978) Separation of viable schizont-infected red cells of *Plasmodium falciparum* from human blood. Ann Trop Med Parasitol 72:87–88

83. Penny JI, Hall ST, Woodrow CJ, Cowan GM, Gero AM, Krishna S (1998) Expression of substrate-specific transporters encoded by *Plasmodium falciparum* in *Xenopus laevis* oocytes. Mol Biochem Parasitol 93:81–89

84. Prasad PD, Wang H, Huang W, Fei YJ, Leibach FH, Devoe LD, Ganapathy V (1999) Molecular and functional characterization of the intestinal Na$^+$-dependent multivitamin transporter. Arch Biochem Biophys 366:95–106

85. Prasad PD, Srinivas SR, Wang H, Leibach FH, Devoe LD, Ganapathy V (2000) Electrogenic nature of rat sodium-dependent multivitamin transport. Biochem Biophys Res Commun 270:836–840

86. Rager N, Mamoun CB, Carter NS, Goldberg DE, Ullman B (2001) Localization of the *Plasmodium falciparum* PfNT1 nucleoside transporter to the parasite plasma membrane. J Biol Chem 276:41095–41099

87. Rasoloson D, Shi L, Chong CR, Kafsack BF, Sullivan DJ (2004) Copper pathways in *Plasmodium falciparum* infected erythrocytes indicate an efflux role for the copper P-ATPase. Biochem J 381:803–811

88. Raymond M, Gros P, Whiteway M, Thomas DY (1992) Functional complementation of yeast ste6 by a mammalian multidrug resistance mdr gene. Science 256:232–234

89. Reed MB, Saliba KJ, Caruana SR, Kirk, K, Cowman AF (2000) Pgh1 modulates sensitivity and resistance to multiple antimalarials in *Plasmodium falciparum*. Nature 403:906–909

90. Rothman JE (1987) Protein sorting by selective retention in the endoplasmic reticulum and Golgi stack. Cell 50:521–522

91. Saier MH Jr (2000) A functional-phylogenetic classification system for transmembrane solute transporters. Microbiol Mol Biol Rev 64:354–411

92. Saliba KJ, Folb PI, Smith PJ (1998) Role for the *Plasmodium falciparum* digestive vacuole in chloroquine resistance. Biochem Pharmacol 56:313–320

93. Saliba KJ, Horner HA, Kirk K (1998) Transport and metabolism of the essential vitamin pantothenic acid in human erythrocytes infected with the malaria parasite *Plasmodium falciparum*. J Biol Chem 273:10190–10195

94. Saliba KJ, Kirk K (1998) Uptake of an antiplasmodial protease inhibitor into *Plasmodium falciparum*-infected human erythrocytes via a parasite-induced pathway. Mol Biochem Parasitol 94:297–301

95. Saliba KJ, Kirk K (1999) pH regulation in the intracellular malaria parasite, *Plasmodium falciparum*: H$^+$ extrusion via a V-type H$^+$-ATPase. J Biol Chem 274:33213–33219

96. Saliba KJ, Kirk K (2001) H$^+$-coupled pantothenate transport in the intracellular malaria parasite. J Biol Chem 276:18115–18121

97. Saliba KJ, Allen RJW, Zissis S, Bray PG, Ward SA, Kirk K (2003) Acidification of the malaria parasite's digestive vacuole by a H$^+$-ATPase and a H$^+$-pyrophosphatase. J Biol Chem 278:5605–5612

98. Saliba KJ, Krishna S, Kirk K (2004) Inhibition of hexose transport and abrogation of pH homeostasis in the intraerythrocytic malaria parasite by an O-3-hexose derivative. FEBS Lett 16:93–96

99. Santiago TC, Zufferey R, Mehra RS, Coleman RA, Mamoun CB (2004) The *Plasmodium falciparum* PfGatp is an endoplasmic reticulum membrane protein important for the initial step of malarial glycerolipid synthesis. J Biol Chem 279:9222–9232

100. Sidhu AB, Verdier-Pinard D, Fidock DA (2002) Chloroquine resistance in *Plasmodium falciparum* malaria parasites conferred by PfCRT mutations. Science 298:210–213

101. Spiller DG, Bray PG, Hughes RH, Ward SA, White MR (2002) The pH of the *Plasmodium falciparum* digestive vacuole: holy grail or dead-end trail? Trends Parasitol 18:441–444

102. Staines HM, Chang W, Ellory JC, Tiffert T, Kirk K, Lew VL (1999) Passive Ca^{2+} transport and Ca^{2+}-dependent K^+ transport in *Plasmodium falciparum*-infected red cells. J Membr Biol 172:13–24

103. Staines HM, Rae C, Kirk K (2000) Increased permeability of the malaria-infected erythrocyte to organic cations. Biochim Biophys Acta 1463:88–98

104. Staines HM, Ellory JC, Kirk K (2001) Perturbation of the pump-leak balance for Na^+ and K^+ in malaria-infected erythrocytes. Am J Physiol 280: C1576–C1587

105. Staines HM, Powell T, Ellory JC, Egee S, Lapaix F, Decherf G, Thomas SL, Duranton C, Lang F, Huber SM (2003) Modulation of whole-cell currents in *Plasmodium falciparum*-infected human red blood cells by holding potential and serum. J Physiol 552:177–183

106. Staines HM, Powell T, Thomas SL, Ellory JC (2004) *Plasmodium falciparum*-induced channels. Int J Parasitol 34:665–673

107. Stead AM, Bray PG, Edwards IG, DeKoning HP, Elford BC, Stocks PA, Ward SA (2001) Diamidine compounds: selective uptake and targeting in *Plasmodium falciparum*. Mol Pharmacol 59:1298–1306

108. Stolz J, Sauer N (1999) The fenpropimorph resistance gene FEN2 from *Saccharomyces cerevisiae* encodes a plasma membrane H^+-pantothenate symporter. J Biol Chem 274:18747–18752

109. Stolz J, Caspari T, Carr AM, Sauer N (2004) Cell division defects of *Schizosaccharomyces pombe* liz1- mutants are caused by defects in pantothenate uptake. Eukaryot Cell 3:406–412

110. Tan W, Tai E, Chow LMC (2003) Expression and characterisation of PfCRT in yeast. Exp Parasitol 105:34

111. Teasdale RD, Jackson MR (1996) Signal-mediated sorting of membrane proteins between the endoplasmic reticulum and the golgi apparatus. Ann Rev Cell Dev Biol. 12:27–54

112. Thomas SL, Lew VL (2004) *Plasmodium falciparum* and the permeation pathway of the host red blood cell. Trends Parasitol 20:122–125

113. Tran CV, Saier MH Jr (2004) The principal chloroquine resistance protein of *Plasmodium falciparum* is a member of the drug/metabolite transporter superfamily. Microbiology 150:1–3

114. Ullu E, Tschudi C, Chakraborty T (2004) RNA interference in protozoan parasites. Cell Microbiol 6:509–519

115. Verloo P, Kocken CH, Van der Wel A, Tilly BC, Hogema BM, Sinaasappel M, Thomas AW, De Jonge HR (2004) *Plasmodium falciparum*-activated chloride channels are defective in erythrocytes from cystic fibrosis patients. J Biol Chem 279:10316–10322

116. Volkman SK, Cowman AF, Wirth DF (1995) Functional complementation of the ste6 gene of *Saccharomyces cerevisiae* with the PfMDR1 gene of *Plasmodium falciparum*. Proc Natl Acad Sci. USA 92:8921–8925

117. Wagner MA, Andemariam B, Desai SA (2003) A two-compartment model of osmotic lysis in *Plasmodium falciparum*-infected erythrocytes. Biophys J 84:116–123

118. Waller KL, Muhle RA, Ursos LM, Horrocks P, Verdier-Pinard D, Sidhu AB, Fujioka H, Roepe PD, Fidock DA (2003) Chloroquine resistance modulated in vitro by expression levels of the *Plasmodium falciparum* chloroquine resistance transporter. J Biol Chem 278:33593–33601

119. Waller RF, Keeling PJ, Donald RG, Striepen B, Handman E, Lang-Unnasch N, Cowman AF, Besra GS, Roos DS, McFadden GI (1998) Nuclear-encoded proteins target to the plastid in *Toxoplasma gondii* and *Plasmodium falciparum*. Proc Natl Acad SciUSA 95:12352–12357

120. Wang H, Huang W, Fei YJ, Xia H, Yang-Feng TL, Leibach FH, Devoe LD, Ganapathy V, Prasad PD (1999) Human placental Na^+-dependent multivitamin transporter. Cloning, functional expression, gene structure, and chromosomal localization. J Biol Chem 274:14875–14883

121. Weber JH, Vishnyakov A, Hambach K, Schultz A, Schultz JE, Linder JU (2004) Adenylyl cyclases from *Plasmodium*, *Paramecium* and *Tetrahymena* are novel ion channel/enzyme fusion proteins. Cell Signal 16:115–125

122. Wilson CM, Serrano AE, Wasley A, Bogenschutz MP, Shankar AH, Wirth DF (1989) Amplification of a gene related to mammalian mdr genes in drug-resistant *Plasmodium falciparum*. Science 244:1184–1186

123. Withers-Martinez C, Carpenter EP, Hackett F, Ely B, Sajid M, Grainger M, Blackman MJ (1999) PCR-based gene synthesis as an efficient approach for expression of the A+T-rich malaria genome. Protein Eng 12:1113–1120

124. Woodrow CJ, Penny JI, Krishna S (1999) Intraerythrocytic *Plasmodium falciparum* expresses a high affinity facilitative hexose transporter. J Biol Chem 274:7272–7277

125. Woodrow CJ, Burchmore RJ, Krishna S (2000) Hexose permeation pathways in *Plasmodium falciparum*-infected erythrocytes. Proc Natl Acad Sci USA 97:9931–9936

126. Wunsch S, Sanchez CP, Gekle M, Grosse-Wortmann L, Wiesner J, Lanzer M (1998) Differential stimulation of the Na^+/H^+ exchanger determines chloroquine uptake in *Plasmodium falciparum*. J Cell Biol 140:335–345

127. Yadava A, Ockenhouse CF (2003) Effect of codon optimization on expression levels of a functionally folded malaria vaccine candidate in prokaryotic and eukaryotic expression systems. Infect Immun 71:4961–4969

128. Zhang H, Howard EM, Roepe PD (2002) Analysis of the antimalarial drug resistance protein PfCRT expressed in yeast. J Biol Chem 277:49767–49775

129. Zuegge J, Ralph S, Schmuker M, McFadden GI, Schneider G (2001) Deciphering apicoplast targeting signals—feature extraction from nuclear-encoded precursors of *Plasmodium falciparum* apicoplast proteins. Gene 280:19–26

CTMI (2005) 295:357–382
© Springer-Verlag Berlin Heidelberg 2005

Plasmodium Ookinete Invasion of the Mosquito Midgut

J. M. Vinetz (✉)

Division of Infectious Diseases, University of California,
San Diego School of Medicine, 9500 Gilman Drive, La Jolla, CA 92093-0640, USA
jvinetz@ucsd.edu

Abstract The *Plasmodium* ookinete is the developmental stage of the malaria parasite that invades the mosquito midgut. The ookinete faces two physical barriers in the midgut which it must traverse to become an oocyst: the chitin- and protein-containing peritrophic matrix; and the midgut epithelial cell. This chapter will consider basic aspects of ookinete biology, molecules known to be involved in midgut invasion, and cellular processes of the ookinete that facilitate parasite invasion. Detailed knowledge of these mechanisms may be exploitable in the future towards developing novel strategies of blocking malaria transmission.

Abbreviations

CTRP	Circumsporozoite- and TRAP-related protein
WARP	von Willebrand A-domain-related protein

1
Overview of Ookinete Biology

The *Plasmodium* ookinete is the stage of the malaria parasite that invades the midgut of its definitive host, the mosquito (Fig. 1). After a mosquito takes an infectious blood meal, gametocytes are stimulated to form male

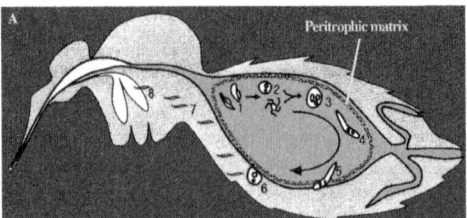

Fig. 1A,B The biology of the malaria parasite within the mosquito midgut. **A** Development stages of *Plasmodium* within the mosquito midgut. After ingestion of gametocytes (1), exflagellation (2), and fertilization to produce a zygote (3), the ookinete (4) develops over 15–24 hours, then migrates out of the blood meal, first crossing the chitin-and protein-containing peritrophic matrix (5), then penetrates midgut epithelial cells and comes to rest on the lumenal aspect of the basement membrane to form an oocyst (6), from which sporozoites (7) develop that migrate to the salivary glands (8). **B** Ookinete penetration of the peritrophic matrix. The parasite is exiting the blood meal on the left, producing chitinase to focally disrupt the peritrophic matrix, en route to the microvilli of the midgut epithelial surface seen at right. From Sieber KP, Huber M, Kaslow D, Banks SM, Torii M, Aikawa M, Miller L. The peritrophic membrane as a barrier: its penetration by *Plasmodium gallinaceum* and the effect of a monoclonal antibody to ookinetes. Exp Parasitol. 1991 Feb;72(2):145–56

and female gametes (micro- and macrogametes, respectively) induced by an obligatory drop in temperature from the vertebrate body temperature to ambient temperature, and optimized in the presence of appropriate pH, biocarbonate (Kaushal et al. 1983a), and the presence of a small molecule trigger, xanthurenic acid (Billker et al. 1998, 2000). The ookinete emerges from the zygote, differentiating over a period of 12–20 h after fertilization during a process poorly understood at the molecular level. During its time within the lumen of the mosquito midgut, the malaria parasite is in its definitive stage: two haploid genome complements are present in the zygote, and during the zygote to ookinete transformation, at some point, still not worked out in detail, genetic recombination and reductive division returning the parasite to the haploid state take place. The ookinete is motile, using what is assumed to be a process of gliding motility similar to that of the sporozoite to invade the midgut; circumsporozoite- and TRAP-related protein (CTRP) is one ookinete molecule involved in gliding motility of this parasite developmental stage (Dessens et al. 1999; Templeton et al. 2000). To arrive successfully at the luminal side of the basement membrane of the midgut where it will arrest, round up and begin its development program as an oocyst, the ookinete must penetrate at least two physical barriers: the chitin- and protein-containing peritrophic matrix; and the midgut epithelial cell surface. This process is inefficient, with only a fraction of the 100–1,000 gametocytes transforming into tens of ookinetes, then a few oocysts. Nonetheless, even one oocyst can subsequently produce thousands of salivary gland sporozoites (Alavi et al. 2003) so that this inefficient process still yields mosquitoes capable of continuing to transmit malaria. The proteolytic milieu of the mosquito midgut and soluble and intracellular mosquito innate immune defenses target zygotes and ookinetes, reducing infectivity of the parasite for the mosquito.

A substantial body of experimental evidence suggests that mosquito midgut stage parasites–from gametes to zygotes to mature ookinetes–may be immunologically targeted in transmission-blocking vaccine strategies Ever since the seminal demonstrations of Gwadz (Gwadz 1976) and Carter and Chen (Carter 1976) that parasite invasion of mosquitoes can be blocked by antibodies induced by vaccination against gametes, the study of the sexual stages of the malaria parasite has been dominated by delineating surface molecules as potential transmission-blocking vaccine candidates (Kaushal et al. 1983b; Rener et al. 1983; Vermeulen et al. 1985). With regard to human malaria, the basic strategy was to prepare hybridomas from mice vaccinated with *Plasmodium falciparum* gametes and zygotes, and to use indirect immunofluorescence to screen for hybridomas secreting monoclonal antibodies reacting with the surface of gametes and zygotes (Kaushal et al. 1983b; Rener et al. 1983; Vermeulen et al. 1985). This strategy led to the molecular identification of the first validated transmission-blocking target, a zygote/ookinete GPI-anchored surface protein, Pfs25, which has a predicted secondary structure including four epidermal growth factor-like domains of otherwise unknown function (Kaslow et al. 1988). Other transmission-blocking target proteins were identified with the same approach: Pfs28, Pfs230, Pfs48/45 (Kaslow 1993). Antibodies to these zygote proteins impaired or completely blocked *P. falciparum* invasion of the mosquito midgut as assessed by a standard membrane feeding assay. In this assay, infectious blood meals including in vitro cultivated mature *P. falciparum* gametocytes are fed to mosquitoes through water-jacketed glass feeders, which maintain the temperature of the blood meal at 37°C and allow the mosquitoes, housed in individual cartons topped by screens, to feed through a Parafilm or animal skin artificial membrane. At the end of approximately 1 week, midguts are dissected from infected mosquitoes and oocysts enumerated to compare control and experimental antisera.

This review will focus on mechanisms by which the ookinete invades the mosquito midgut, including the following considerations: barriers to invasion presented by the mosquito to the *Plasmodium* midgut stages; mechanisms of transmission-blocking immunity; processes of the parasite involved in parasite invasion and the mosquito response to these processes; and current approaches to exploiting knowledge of potential transmission-blocking targets of intervention against the parasite in the midgut, including the production of *Plasmodium*-refractory transgenic mosquitoes. A central theme in studying *Plasmodium* ookinete biology is that molecules that are targets of blocking parasite infectivity for mosquitoes will have important biological functions in parasite–mosquito interactions. Therefore, one approach to studying *Plasmodium* ookinetes is to validate the importance of ookinete secreted and surface molecules as transmission-blocking targets (whether inhibited by antibodies or chemicals). Such observations provide the basis for detailed mechanistic pursuits of how these molecules function in ookinete biology, towards the long-term goal of developing novel methods of malaria control through blocking transmission to mosquitoes.

2
Barriers Presented by the Mosquito Midgut to *Plasmodium* Invasion

Within the mosquito midgut there are two main physical barriers to parasite invasion: the chitin- and protein-containing peritrophic matrix; and the epithelial surface itself. Morphologically mature ookinetes and their invasion of midgut epithelial cells begins approximately 24 h post bloodmeal ingestion (Sieber et al. 1991; Torii et al. 1992), at least for *P. falciparum* and *P. gallinaceum*, with maximal invasion at approximately 30–35 h. Timing of invasion is substantially earlier with *P. yoelii* (Vaughan et al. 1994a) and *P. berghei* (Limviroj et al. 2002; Sinden and Canning 1972), with mature ookinetes appearing by 6–8 h post bloodmeal ingestion. This earlier appearance of ookinetes may explain quantitative differences in the effect of knocking out the chitinase genes of *P. falciparum* and *P. berghei* (Dessens et al. 2001; Tsai et al. 2001). The peritrophic matrix is incomplete or thinner when *P. yoelii* and *P. berghei* first develop into mature ookinetes, so that the peritrophic matrix might present less of a barrier to ookinete penetration by these *Plasmodium* spp. than the thicker, mature peritrophic matrix presents when *P. falciparum* and *P. gallinaceum* invade.

It remains unclear whether infection of mosquitoes by *Plasmodium* spp. reduces insect fitness, either in a natural parasite–mosquito combination such as *P. falciparum–Anopheles gambiae* under field conditions (Hogg and Hurd 1997) or unnatural combinations performed experimentally in the laboratory setting (Ferguson et al. 2003; Ferguson and Read 2002). Both oocyst and sporozoite development may affect mosquito fitness. Mosquitoes have been reported to limit sporozoite development (Luckhart et al. 1998), but not ookinete development, by induction and secretion of nitric oxide. As a leading ookinete investigator has framed this issue, "How have the natural innate defences of the insect failed?" or "What mechanisms has the parasite used to overcome these defences?"(Sinden et al. 2004). Understanding mechanisms by which the mosquito resists *Plasmodium* spp. invasion or those that the parasite uses to allow for successful invasion are key for the further development of strategies to prevent mosquito transmission of malaria.

2.1
Biochemical Barriers

Successful development of zygotes to ookinetes within the bloodmeal is inefficient for a variety of reasons that reflect the complex milieu that the parasite encounters within the midgut. Nonspecific factors inhibiting parasite development in the midgut include vertebrate-derived factors such as bicarbonate ion, lactate, complement, granulocytes and mononuclear cells that are capable of activation by cytokines and ligand-engaged pattern-recognition associated molecules, and transmission-blocking antibodies that may arise naturally during the course of chronic endemic malaria exposure. Mosquito-related factors that reduce the number and quality of invading parasites include two well-established mechanisms allowing mosquitoes to resist midgut invasion: melanization of ookinetes within the epithelial wall (Collins et al. 1986) and intra-epithelial lysis of invading ookinetes (Vernick et al. 1999). While mosquito genetic loci associated with melanotic encapsulation have been identified (Zheng et al. 1997), the molecular mechanisms underlying these phenotypes continue to be elucidated, and involve peroxidase and other biochemical pathways (Kumar et al. 2003, 2004; Niare et al. 2002), as well as innate detection of invading parasites likely mediated by pattern-associated pathogen receptors in the mosquito midgut (Dimopoulos 2003). Further, soon after bloodmeal ingestion, zygotes and immature ookinetes are more susceptible than mature ookinetes to proteolytic damage from enzymes that mosquitoes secrete into the midgut (Gass and Yeates 1979). The timing of mosquito digestive enzyme secretion is critical to this process, since peak blood meal digestion occurs at the time when ookinetes are mature and are at their peak of invading the midgut epithelium. Recent evidence indicates that two glycosyl-phosphatidylinositol-linked surface proteins that contain highly structured, cysteine-disulfide bonds within mammalian epidermal growth factor-like (EGF-like) domains, in the P25 and P28 families (known as Pfs25 and Pfs28 in *P. falciparum*), confer ookinete resistance to proteolytic activity present in the mosquito midgut (Tomas et al. 2001). Additionally both the resistance to proteolysis related to timing of enzyme secretion as well as the limited diffusion of such enzymes into the blood bolus while the proteolysis-susceptible zygote differentiates into the protease-resistant ookinete confer protection (Abraham and Jacobs-Lorena 2004). Double knockout of P25 and P28 homologs in *P. berghei* resulted in enhanced susceptibility to trypsin as well as reduced numbers of ookinetes when in vitro-cultivated ookinetes were fed to mosquitoes, while mature *P. berghei* ookinetes developed at the same rate as wild type parasites in vitro, suggesting that P25 and P28 conferred resistance to a hostile mosquito midgut milieu (Tomas et al. 2001). Ookinete surface molecules of the P25 and P28 families have been demonstrated to play an important role in ookinete survival in the mosquito midgut (Tomas et al. 2001). These molecules seem not to play a role in ookinete recognition or invasion of midgut epithelial cells as evidenced by the lack of antibody inhibition of ookinete binding to mosquito midgut epithelial sheets in vitro, despite the presumption in the past that the EGF-like domains would have some ligand–receptor interaction in the invasion process.

2.2
Physical Barrier: Peritrophic Matrix

Soon after ingesting a blood meal, mosquitoes synthesize a structure known as the peritrophic matrix that surrounds the blood bolus (Tellam et al. 1999). The peritrophic matrix, whose presumed function is to prevent physical trauma to the delicate microvillar surface of the midgut epithelial cell, is comprised of chitin (ranging from ~1% to 15% by weight) cross-linked by chitin-binding proteins known as peritrophins. The structure of the peritrophic matrix varies between mosquito species but has been demonstrated to present a physical barrier to ookinete invasion (Fig. 1; Perrone and Spielman 1988; Sieber et al. 1991), in the *P. gallinaceum–Aedes aegypti* and the *P. falciparum–An. gambiae* and *P. falciparum–An. stephensi* model systems (Li et al. 2004).

A substantial body of work has shown that the *Plasmodium* ookinete secretes chitinase(s) that facilitate the movement of the ookinete across the peritrophic matrix (Fig. 2; Huber et al. 1991; Langer and Vinetz 2001; Shahabuddin et al. 1993; Vinetz et al. 1999, 2000). Interference with the function

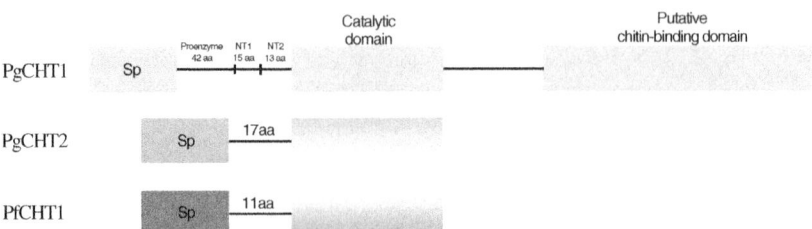

Fig. 2 Schematic representations of *P. gallinaceum* and *P. falciparum* chitinases. Note the absence of proenzyme and chitin-binding domains in PgCHT2, similar to the organization of PfCHT1, compared to PgCHT1. The epitope recognized by the anti-PfCHT1 mAb 1C3 is found within PgCHT2. Monoclonal antibody 1C3 blocks *P. gallinaceum* infectivity for mosquitoes and localizes PgCHT2 to micronemes, the apical electron dense area of the parasite, and the surface of the parasite (Langer and Vinetz 2004)

of *Plasmodium*-secreted chitinase impairs the ability of the ookinete to continue on to the sporogonic forms of the parasite, as demonstrated by chemical interference (Shahabuddin et al. 1993), antibody inhibition of chitinase activity (Langer et al. 2002a), and *Plasmodium* chitinase gene truncation/deletion studies (Dessens et al. 2001; Tsai et al. 2001). Therefore, *Plasmodium* ookinete-secreted chitinase has been validated as a target of blocking malaria transmission. More generally speaking, interruption of the ookinete's ability to cross the peritrophic matrix is an attractive candidate for development of novel methods of blocking malaria transmission, either through the potential development of transmission-blocking drugs, vaccines, or modified mosquitoes that secrete factors that inhibit chitinase function within the mosquito midgut.

Previous reports have suggested that the peritrophic matrix in *An. stephensi* and perhaps other mosquitoes lack chitin (Berner et al. 1983), potentially reducing the important of ookinete-secreted chitinase as a target of blocking mosquito midgut invasion. Recent findings that an anti-*P. falciparum* chitinase monoclonal antibody significantly impairs oocyst formation in experimentally infected mosquitoes suggest that parasite-produced chitinase is a potential malaria transmission-blocking target, despite evidence from the *P. berghei–An. stephensi* model system that knockout of the *P. berghei* chitinase PbCHT1 has a minor effect on preventing oocyst formation (Dessens et al. 2001). This latter result is likely explained by the more rapid development of *P. berghei* ookinetes in vivo and escape from the midgut lumen before the peritrophic matrix is fully developed.

Some authors have suggested that ookinete-secreted protease(s), acting synergistically with parasite-produced chitinase, are important in parasite penetration of the peritrophic matrix (Abraham and Jacobs-Lorena 2004; Langer and Vinetz 2001; Shen 1998). Recent evidence from our laboratory suggests, for the first time, that *Plasmodium* ookinetes secrete an aspartic protease, plasmepsin, the first reported presence of plasmepsins in a stage of the malaria parasite other than asexual blood stages. Aspartic protease inhibitors, peptidomimetic inhibitors (noncleavable peptides mimicking the hemoglobin chain cleavage site), and anti-plasmepsin monoclonal antibodies significantly impair ookinete invasion of the mosquito midgut, in the *P. gallinaceum–Ae. aegypti* model system (Li, Dame, J.M. Vinetz, unpublished results). The *P. gallinaceum* plasmepsin is secreted as well as localized to the surface of the apical complex of the ookinete (unpublished results), suggesting a role for this protease in 'drilling' a hole in the peritrophic matrix and allowing egress of the ookinete from the blood meal towards invasion of the epithelial surface.

The very act of ookinete egress from blood meal is a critical feature of the ability of the malaria parasite to invade the mosquito midgut. CTRP (Trottein et al. 1995) is a single-pass integral membrane protein, secreted through the micronemal pathway, present on the surface of the ookinete (Limviroj et al. 2002; Yuda et al. 1999b). Targeted disruption of this gene results in nonmotile ookinetes and eliminates oocyst production, as demonstrated in both the *P. falciparum* and *P. berghei* model systems (Templeton et al. 2000; Yuda et al. 1999a). CTRP is also an immunological target of blocking transmission, providing evidence that this protein may be a potential transmission-blocking vaccine candidate (Li et al. 2004).

Plasmodium ookinetes secrete a chitinase that is essential for *P. falciparum* and *P. gallinaceum* to penetrate the mosquito peritrophic matrix; the role of this enzyme in *P. berghei* is less clear (Dessens et al. 2001) which may be related to the observation that *P. berghei* ookinetes develop as early as 8 h after bloodmeal ingestion and likely begin to escape the bloodmeal before the

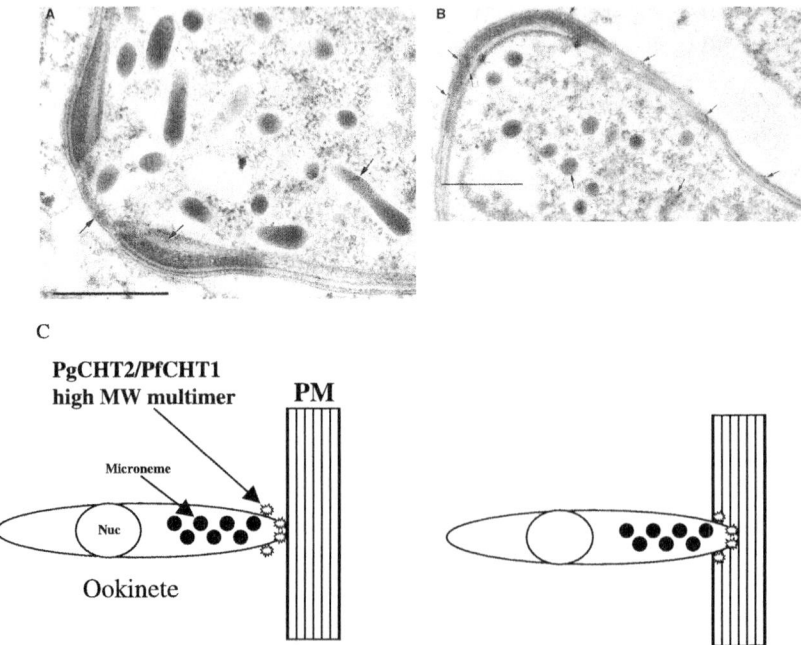

Fig. 3A–C A Immunoelectron microscopy of *P. gallinaceum* ookinete with mAb 1C3, raised against PfCHT1 that recognizes the ortholog chitinase in *P. gallinaceum*, PgCHT2. PgCHT2 is present within micronemes, the apical end of the parasite within the electron-dense region previously identified as part of a protein secretion apparatus, and on the ookinete cell surface (**B**). This electron micrograph was published as Figure 2 in Langer et al. 2002. **C** Proposed model of how a high molecular mass complex containing PgCHT2/PfCHT1 might interact with the peritrophic matrix (PM), coupling gliding motility to penetration and crossing of the PM

peritrophic matrix is fully formed (Vaughan et al. 1994a, 1994b). Chitinase is present on the *P. gallinaceum* ookinete cell surface, and probably the *P. falciparum* ookinete as well, given the similarities between the PgCHT2 chitinase of *P. gallinaceum* and PfCHT1 chitinase of *P. falciparum* (Fig. 2; Langer et al. 2002a; Vinetz et al. 2000). This short form of the chitinase, lacking pro-enzyme and putative chitin-binding domains, may be one component of a disulfide-bonded, high molecular weight complex (Langer et al. 2002a). An attractive hypothesis is that this chitinase-containing complex is lectin-like, allowing the direct physical interaction of this high molecular weight complex to be coupled to the ookinete's gliding motility, providing the parasite with physical leverage to rachet its way through the peritrophic matrix en route to the midgut epithelial cell (Fig. 3). The long form of *Plasmodium* chitinase, a pro-chitinase orthologous to PgCHT1 (Fig. 2; Tsuboi et al. 2003; Vinetz et al. 2000), is not found in *P. falciparum*, but is present in *P. vivax*, all of the rodent-infecting *Plasmodium* spp., and *P. gallinaceum*. This pro-chitinase was previously reported to be activated by a mosquito-derived trypsin-like protease (Shahabuddin et al. 1993, 1995, 1996). However, other experimental data indicate that the *P. gallinaceum* ookinete is fully capable of proteolytically activating PgCHT1 itself, in the absence of mosquito-derived proteases (Vinetz et al. 2000). Evidence suggests that the PgCHT1-activating activity is inhibited by the serine protease inhibitor 4-(2-aminoethyl)-bezenesulfonylfluoride in vitro, suggesting that a parasite-derived, trypsin-like protease is responsible for activating the pro-chitinase (JM Vinetz, unpublished results).

3
Mechanisms of Transmission-Blocking Immunity

To date, a number of *Plasmodium* sexual stage molecules have been identified that are either immunologic or enzymatic targets (i.e., chitinase, plasmepsin, as described above) of blocking parasite transmission to mosquitoes. Approaches to blocking parasite transmission to mosquitoes have traditionally focused on inducing antibodies in the vertebrate host that recognize sexual stage surface antigens that, on ingestion by mosquitoes along with *Plasmodium* gametocytes, interfere with the parasite within the mosquito midgut. More recently, alternative approaches to blocking transmission have been taken, including studies of zygote/ookinete secreted molecules (Langer et al. 2002b), novel peptides that interact with putative midgut or salivary gland receptors for parasite binding (Ghosh et al. 2001, 2002; Ito et al. 2002), or antisera binding to mosquito midgut epithelial carbohydrate and protein surface structures (Lal et al. 1994, 2001; Ramasamy et al. 1997a, 1997b). To date, however, precise mechanisms by which these molecules are responsible for interfering with parasite invasion have not been delineated. The SM1 peptide obtained from phage display library screening probably imitates the structure of a critical receptor on mosquito midgut and/or salivary gland for parasite recognition and invasion, since the indirect immunofluorescence demonstrated that the SM1 peptide bound specifically to the distal lobes of the salivary glands and the lumenal surface of the midgut, critical interfaces for parasite interaction (Ghosh et al. 2001, 2002; Ito et al. 2002).

The repertoire of *Plasmodium* zygote and ookinete molecules shown to be transmission-blocking targets includes Pfs230, Pfs48/45, Pfs25, Pfs28, chitinase, von Willebrand adhesive-related protein (WARP), and CTRP. The gametocyte surface antigen, Pfs230, is present in gametocytes. The activity of anti-Pfs230 transmission-blocking antibodies is present only prior to parasite fertilization and is also complement dependent (Graves et al. 1988; Quakyi et al. 1987; Read et al. 1994), but otherwise, the precise molecular mechanism of anti-Pfs230 antibody activity is not well defined (Williamson 2003). Antibodies to Pfs48/45, a GPI-anchored protein involved in male gamete fertility (van Dijk 2001), have transmission-blocking activity in the absence of complement, and evidence suggests inhibition of fertilization by agglutination of male gametes interfering with zygote formation (Vermeulen et al. 1985). Pfs230 and Pfs48/45 are present in blood stage gametocytes, and transmission-blocking activity to these antigens is present in some people in malaria endemic regions. Natural immunological boosting to raise titers of anti-Pfs230 and Pfs48/45 does occur, although boosting also selects for antigenic diversity of these proteins in natural *P. falciparum* populations (Foo et al. 1991; Williamson and Kaslow 1993). The concept that natural boosting of transmission-blocking titers can occur, at least for some antigens such as Pfs230 and Pfs48/45, is important in the development of transmission-blocking vaccines. *Plasmodium* antigens expressed only in the mosquito but not in the blood stages will not lead to naturally acquired transmission-blocking immunity nor will be boosted by single or recurrent infections.

The best characterized transmission-blocking targets are the EGF-like domain-containing proteins of the P25 family, in *P. falciparum* known as Pfs25. This protein is GPI-anchored on the surface of *Plasmodium* zygotes and ookinetes in the mosquito midgut but not in other stages. Naturally occurring anti-Pfs25 transmission-blocking immunity does not occur. Pfs25 was first identified as a target of transmission-blocking immunity by a monoclonal antibody recognizing a conformation- and disulfide bond-dependent epitope reacting with the surface of *P. falciparum* in in vitro obtained zygotes. The mechanism of anti-P25 transmission-blocking activity remains poorly understood, whether anti-P25 antibodies agglutinate parasites (Sieber et al. 1991), impair egress of the parasite from the blood meal (Ranawaka et al.

1994b), or induce peripheral blood mononuclear cells to kill infectious, sexual stage parasites (Ranawaka et al. 1994a) has not been determined. It has been observed in the *P. gallinaceum–Ae. aegypti* model system that anti-Pgs25 monoclonal antibody C5 was associated with a block of parasite invasion of the midgut at the level of the peritrophic matrix, although the mechanism for this phenomenon remains poorly defined (Sieber et al. 1991). The presence of C5 in the infectious bloodmeal led to fewer ookinetes being present in the ectoperitrophic space, suggesting that the monoclonal antibody against the surface protein Pgs25 could have interfered simply with the egress of the parasite across the peritrophic matrix, but not attachment of the ookinete to this structure. The alternative possibility was that the C5 monoclonal antibody could have interfered with ookinete attachment to and penetration of the epithelium, a possibility not supported by experimental data because accumulation of ookinetes occurred at the level of the peritrophic matrix, not at the epithelial cell (Sieber et al. 1991).

After the initial observation that *Plasmodium* ookinetes appear to punch a hole in the peritrophic matrix during midgut invasion, additional observations suggested, as a general principle, that at least a subset of zygote/ookinete-secreted molecules, not just surface molecules were targets of antibody-mediated transmission-blocking activity (Langer et al. 2002b). Many such molecules likely remain to be identified at the molecular level (Langer et al. 2002b). WARP, an ookinete-secreted molecule highly conserved among *Plasmodium* spp. (Li et al. 2004; Yuda et al. 2001) has been shown to be an immunological target of blocking *P. gallinaceum* and *P. berghei* infectivity for mosquitoes, despite gene knockout studies in *P. berghei* failing to demonstrate an essential role for WARP in ookinete invasion of the mosquito midgut (Yuda et al. 2001). Therefore, it cannot be concluded that negative results of zygote/ookinete-expressed gene knockout studies rule out specific molecules as transmission-blocking targets.

4
Relation of Ookinete Cell Biology to *Plasmodium* Invasion of the Mosquito Midgut

The ookinete has a number of features that are unique among the diverse developmental stages of *Plasmodium*. The ookinete does not form a parasitophorous vacuole within its target cell, the midgut epithelial cell, in contrast to invasive merozoites and sporozoites. Rather, the ookinete continues its journey through one or more epithelial cells (Han et al. 2000), exiting on the basement membrane, coming to rest on the lumenal aspect of the basement membrane, and rounding up to form the oocyst. Therefore, the

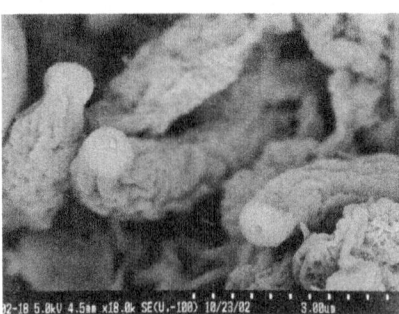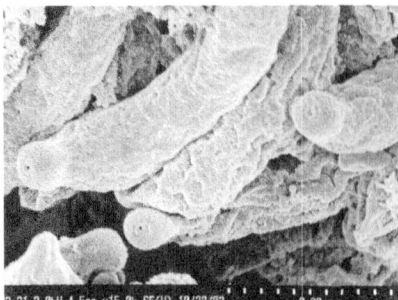

Fig. 4 High resolution scanning electron micrographs of *Plasmodium gallinaceum* ookinete. Courtesy of Dr. Vsevolod Popov and Violet Han, University of Texas Medical Branch, Galveston, Texas

final destination of the ookinete is extracellular, not intracellular, another difference between the ookinete and the other invasive stages of *Plasmodium*. Reflecting this, the ookinete does not have identifiable rhoptries or dense granules, which other *Plasmodium* stages and other apicomplexan parasites such as *Toxoplasma gondii*, contain, that function in establishment of the intracellular parasitophorous vacuole (Dubremetz et al. 1998). The apical end of the ookinete is filled with micronemes, reflecting a high level of constitutive protein secretion, in contrast to merozoites and sporozoites that secrete proteins primarily in response to cell–cell contact (Ngo et al. 2000). In recent high resolution-scanning electron micrographs of *P. gallinaceum* ookinetes, the parasite appears to have secretory pores on the apical end of the parasite (Fig. 4). Whether stimulation of microneme secretion is regulated differently in vivo vs. during in vitro culture is unknown. Indeed, the sole identifiable secretory organelle in the ookinete is the microneme (Langer et al. 2000), which appears to be the only pathway for extracellular secretion both of soluble proteins such as chitinase (Fig. 5), and integral membrane proteins such as CTRP, and putative adhesive molecules such as WARP (Li et al. 2004) (Fig. 6). It is reasonable to hypothesize that the ookinete constitutively secretes proteins to function within the unique milieu of the midgut in ways that do not involve cell–cell contact: to resist proteolytic attack, to cross the acellular peritrophic matrix, and to interact with midgut epithelial cells (Han et al. 2000), as the ookinete remains topologically outside of the mosquito until oocysts rupture to release sporozoites within the body of the mosquito.

The ookinete must be versatile insofar as it must be able to recognize and cross both cellular and acellular structures during its invasion process. Several investigators have noted that the ookinete is able to adopt a number of different, viable morphologies in vivo. First, the ookinete must squeeze

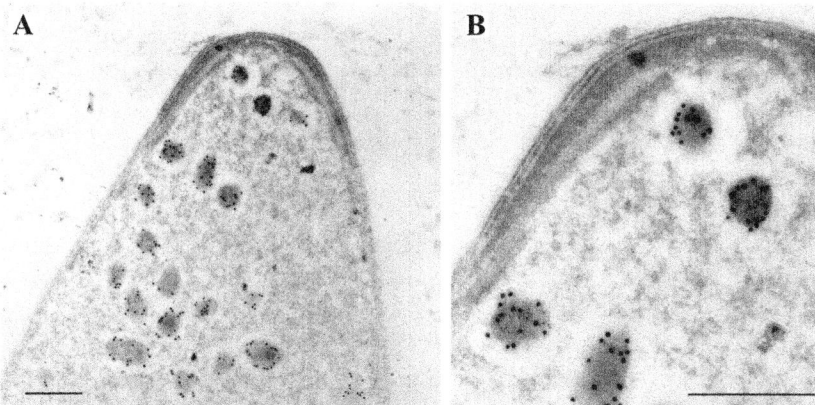

Fig. 5A,B Dual immunoelectron microscopy of a *Plasmodium gallinaceum* ookinete to determine spatial relationship of the soluble, secreted enzyme *Pg*CHT1 (5 nm gold) and the cell-membrane associated single pass integral membrane protein *Pg*CTRP (15 nm gold). Bar indicates 0.5 µM. From Li F, Templeton TJ, Popov V, Comer JE, Tsuboi T, Torii M, Vinetz JM. *Plasmodium* ookinete-secreted proteins secreted through a common micronemal pathway are targets of blocking malaria transmission. J Biol Chem. 2004 Jun 18;279(25):26635–44. Reprinted with permission from *Journal of Biological Chemistry*, American Society for Biochemistry and Molecular Biology

through an enzymatically digested hole in the peritrophic matrix (Sieber et al. 1991; Torii et al. 1992). Interacting with a structure termed the 'microvilli-associated network', and within the midgut lumen and within epithelial cells, the ookinete takes on a variety of elongated, stalked and dumbell-shaped forms (Vernick et al. 1999; Vlachou et al. 2004; Zieler et al. 1998). Associated with its differing morphologies, the ookinete displays several types of motility including stationary rotation, translocational spiraling and straight-segment motility (Vlachou et al. 2004). These morphological and functional features are present as the ookinete travels both extracellularly as well as intracellularly within one or more adjacent midgut epithelial cells.

The mechanisms by which the ookinete recognizes and traverses mosquito midgut epithelial cells are complex and incompletely understood. In vivo, the ookinete appears to interact with an acellular, network-like matrix on the midgut epithelium microvillar surface termed the microvilli-associated network (Zieler et al. 1998). The biochemical structure of this matrix remains unknown, but appears not to be a precursor of the peritrophic matrix (Zieler et al. 1998). The nature of the epithelial cell surface receptor that the ookinete recognizes appears to be a nonsialic acid carbohydrate on the epithelial cell surface but otherwise has not been identified (Zieler et al. 1999). The ookinete ligands for mosquito structures, whether the microvilli-associated network

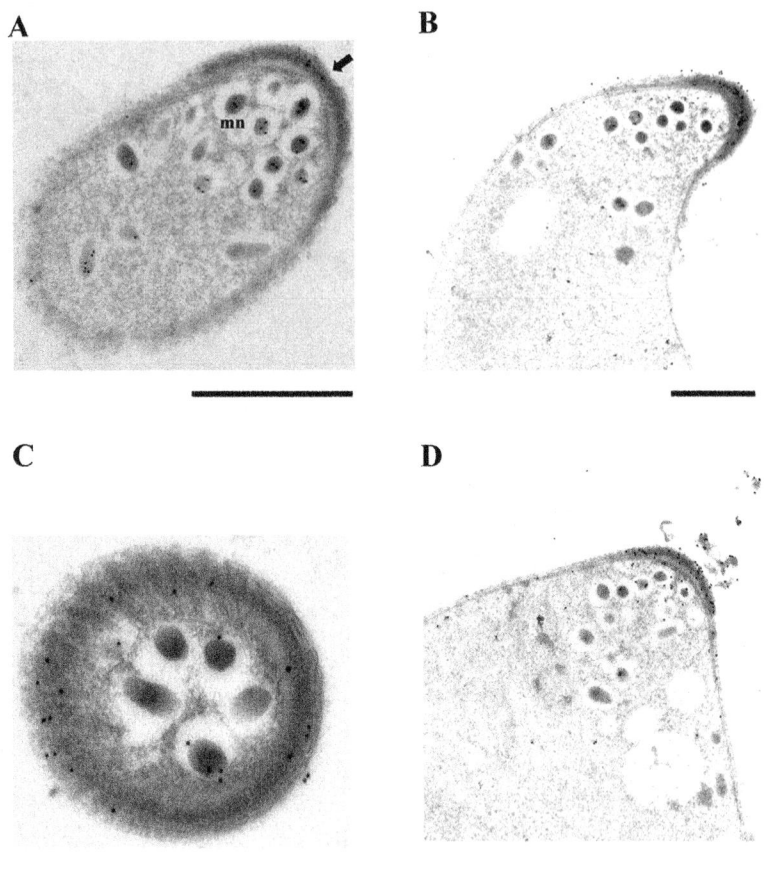

Fig. 6A–D *Plasmodium gallinaceum* ookinete secretion of the chitinase, PgCHT1, involves secretory organelle and protein traversal through the electron dense area of the apical end of the ookinete. Immunoelectron microscopy of in vitro-developed mosquito midgut stage *P. gallinaceum* parasites stained with anti-*P. gallinaceum* chitinase (PgCHT1) antibodies. (**A**) Longitudinal section of a maturing ookinete in which PgCHT1 is associated with micronemes (mn) at the apical third of the parasite (arrow). The primary antibody was mouse polyclonal antiserum raised to full-length recombinant PgCHT1. (**B**) Longitudinal section of a mature ookinete stained with anti-PgCHT1 chitin-binding domain antiserum. (**C**) Cross section of the apical end of a mature ookinete stained with anti-PgCHT1, anti-active-site serum. (**D**) Longitudinal section of a mature ookinete stained with anti-PgCHT1 chitin-binding domain serum showing extracellular PgCHT1. Bars, 1 μm. [From Langer RC, Hayward RE, Tsuboi T, Tachibana M, Torii M, *Vinetz JM*. Micronemal transport of *Plasmodium* ookinete chitinases to the electron-dense area of the apical complex for extracellular secretion. Infect Immun 68:6461–6465, 2000.] Reprinted with permission from the American Society for Microbiology

or the epithelial cell surface, remain unknown. No specific epithelial cell sub-type appears to be a target cell for the ookinete (Han et al. 2000; Sinden and Billingsley 2001; Vlachou et al. 2004; Zieler and Dvorak 2000), in contrast to previous reports in which *P. gallinaceum* was thought to invade 'Ross cells', a vesicular ATPase-expressing cell type (Shahabuddin 2002; Shahabuddin and Pimenta 1998). This finding has not been replicated. Elegant videomicroscopic studies have demonstrated that the ookinete glides smoothly over the surface of the epithelial cell, before an abrupt engagement occurs between the apical end of the ookinete and an epithelial cell surface (not distinguishable morphologically with other surrounding cells), with the ookinete plunging into the cell (Vlachou et al. 2004; Zieler and Dvorak 2000). Within minutes after invasion, the epithelial cell undergoes rapid morphological alterations: change in refractive index and visible deterioration of the cell with nuclear swelling, movement of nucleus to a more apical position in the cell, blebbing of the cell surface and Brownian motion of particles within the cell (Zieler and Dvorak 2000). Zieler first noted that these morphological changes induced by the ookinete reflect an apoptotic process, a finding subsequently confirmed by others (Han et al. 2000; Vlachou et al. 2004). Whether epithelial cell apoptosis is due simply to physical disruption of the epithelial cell plasma membrane or is specifically induced by one or more factors secreted into the epithelial cell by the ookinete is unclear (Han et al. 2000). As the ookinete exits the basolateral aspect of the epithelial cell, it carries with it a coating that appears as a thickly stained membranous structure (Vlachou et al. 2004). This coating likely arises from the mosquito cell as its does not appear in *P. berghei* ookinetes cultivated in vitro (Vlachou et al. 2004). Nitric oxide appears at a maximum rate of synthesis at about 24 h after ookinete invasion of epithelial cells, which is after the ookinete has already left the cell, as well as in the presence of enormous amounts of hemoglobin, so that role of nitric oxide in limiting parasite invasion, at least at the level of the ookinete, is unclear (Han et al. 2000; Luckhart et al. 1998).

5
Outstanding Questions in Ookinete Biology of Fundamental Biological Interest

There are a number of questions about ookinete biology that remain to be answered. Such investigations are not only of fundamental biological interest but are also relevant to understanding the fundamental biology of other development stages of the malaria parasite, may be applicable to other apicomplexan parasites, and will be useful in studying the mechanisms of mosquito responses to parasite invasion. What are the molecular bases for ookinete recognition of mosquito midgut epithelial cells and other structures such as the microvillar tubular network or peritrophic matrix? Do ookinetes require a mosquito epithelial cell-derived factor to facilitate further sporogonic development, analogous to hepatocyte growth factor receptor that has been found to be important in sporozoite establishment of infection within the vertebrate hepatocyte (Carrolo et al. 2003)? There are substantial parallels between sporozoite invasion of hepatocytes and ookinete invasion of midgut epithelial cells, particularly several observations. Under both circumstances the parasite migrates through more than one cell. Coupled to the process of gliding motility, both sporozoites and ookinetes leave a trail of GPI-anchored protein within the target cell (in the case of the sporozoite, CSP; in the case of the ookinete P25). Sporozoites and ookinetes induce actin cytoskeletal rearrangement in their target cells (Carrolo et al. 2003; Han et al. 2000). What factors, other than a subtilisin-like protease and GPI-anchored proteins do ookinetes secrete into the midgut epithelial cell, and are these specifically involved in inducing these cells to undergo apoptosis (Han et al. 2000)? Recent evidence indicates that the ookinete uses a protein in the perforin family to invade the mosquito epithelial cell (Kadota et al. 2004). A major difference between sporozoites and ookinetes, however, is the lack of a parasitophorous vacuole formed by ookinetes. What are the molecular aspects of the constitutive secretory process of the ookinete that couple gliding motility to secretion? Proteolytic activation of zymogens is likely part of the secretory process, as in *T. gondii*, but whether the process takes places within the parasite, possibly where the microneme comes into contact with the electron dense area of apical end of the parasite just prior to secretion, is unknown. The process of secretion in the ookinete must differ from those of sporozoites and *T. gondii* tachyzoites since these latter parasite stages require cell contact-stimulated secretion. What are the mechanisms by which the polar ring in the apical complex of the ookinete play a role in secretion, as shown by chitinase and WARP secretion (Fig. 7; Langer et al. 2000; Li et al. 2004).

Recently advances in systems biology are allowing us to approach the answers to these and many other questions in unprecedented ways. Systems biology can be defined as the large-scale analysis of how genomes function to produce complex biological effects, conventionally looked at in the context of patterns of gene and protein expression and their effects at the cellular level. With the recent completion of whole genome sequences of *P. falciparum* and other *Plasmodium* spp. (Carlton et al. 2002; Gardner et al. 2002), systems biology can be applied to the discovery of fundamental mechanisms of malaria parasite biology that have hitherto not been amenable to high throughout analysis.

Published results of gene expression and proteomic profiling of malaria parasites has thus far focused primarily on the asexual blood stages, and single time points from gametes, gametocytes and sporozoites (Bozdech et al. 2003; Florens et al. 2002; Le Roch et al. 2003). Notably absent from these analyses are studies of ookinetes, primarily because this developmental stage of *P. falciparum* cannot be obtained in vitro and has not been able to be obtained ex vivo from mosquito midguts in quantities sufficient for gene expression or proteomic analysis. Recent advances have been made in the proteomic analysis of *Plasmodium* zygotes, ookinetes, and proteins secreted extracellularly. Ookinetes of *P. berghei*, the genome sequence of which has recently been completed, have been an important model parasite for studies of transmission from the proteomic point of view (Trueman et al. 2004; Raine et al. 2004). Similarly, comparative proteomics of *P. gallinaceum* zygotes and ookinetes as well as secreted/released proteins of *P. gallinaceum* ookinetes is currently underway (K.P. Patra, G. Cantin, J.M. Vinetz, et al., unpublished results). These promise to accelerate the delineation of mechanisms of cell biology underlying the process of ookinete–mosquito midgut interactions.

What are the most important questions about ookinete biology, particularly mechanisms by which the ookinete invades the mosquito midgut, that these systems biology approaches can be used to address? The 'Holy Grail' is clearly to delineate the receptors used by the ookinete to recognize mosquito midgut ligands. How do the zygote and ookinete stages resist proteolytic digestion within the midgut? What is the repertoire of molecules that the ookinete uses to move directionally towards the midgut epithelium? How does the ookinete resist mosquito defensive molecules such as nitric oxide or oxidative radicals? Does the ookinete secrete molecules into the cytoplasm of mosquito epithelial cells specifically to induce apoptosis to somehow aid in the process of invasion? Does the ookinete have developmental stage-specific mechanisms of secretion? Can the extracellular ookinete be studied as a model of organelle biogenesis and secretion with applicability to other stages of the malaria parasite? Proteomic analysis of isolated *Plasmodium* ookinete micronemes, similar to the proteomic analysis of *Plasmodium* rhoptries (Sam-Yellowe et al. 2004) also will be important for defining proteins involved in ookinete biology. The ookinete must have specific mechanisms at each point along its pathway of invading the midgut to ensure its survival in the face of hostile attack and to promote its successful invasion. The recognition of specific proteins in different compartments of the ookinete–cell associated vs. secreted–will greatly facilitate answering these questions.

The proteomic approaches that study *P. berghei* and *P. gallinaceum*, respectively, are limited insofar as they do not directly study the human parasites *P. falciparum* or *P. vivax*. Using comparative approaches, we will be able to determine gene/protein expression similarities and differences between ookinetes of *P. falciparum*, and those of *P. gallinaceum* and *P. berghei*. Already, we know that *P. gallinaceum* has two chitinases, PgCHT1, the ortholog containing pro-enzyme and putative chitin-binding domains shared with all chitinases identified to date in rodent and primate malaria parasites (Tsuboi et al. 2003) with the sole exceptions being *P. falciparum* and *P. reichenowi*. *P. gallinaceum* also expresses a second chitinase that lacks pro-enzyme and putative chitin-binding domains, clearly the ortholog of the *P. falciparum* and *P. reichenowi* chitinases (Li, Patra and Vinetz, J. Infect Diseases, in press). Therefore, at least with regard to chitinase, *P. gallinaceum* has more in common with *P. falciparum* than does *P. berghei*. This observation, along with preliminary data from *P. gallinaceum* zygote/ookinete proteomic analysis, suggests that chitinase will not be the only difference between *P. falciparum* and the rodent-infecting *Plasmodium* species, suggesting that study of *P. gallinaceum* ookinetes will provide unique insights into the detailed mechanistic understanding of parasite–mosquito midgut interactions.

Recent developments in obtaining *P. vivax* (Suwanabun et al. 2001; Tsuboi et al. 2003) and *P. falciparum* ookinetes (J Sattabongkot and T Tsuboi, unpublished results) ex vivo from infected humans, along with the recent completion of the *P. falciparum* and *P. vivax* genome databases, should make these avian and mouse models of *Plasmodium* transmission extendable to the human parasites.

Detailed understandings of ookinete–mosquito interactions also have the potential to be exploited in the development of novel approaches to malaria control through the development of *Plasmodium*-refractory transgenic mosquitoes. While still in its infancy, proof of concept of being able to introduce genes into mosquitoes that interfere with known and (as yet)

--►

Fig. 7A–E Confocal immunofluorescence microscopy of immature *Plasmodium falciparum* ookinetes stained simultaneously with mouse anti-*P. falciparum* chitinase (PfCHT1) MAb 1C3 (green) and rabbit polyclonal antiserum to a recombinant *P. falciparum* zygote-ookinete surface antigen (Pfs25–Pfs28) fusion protein (8) (red). (**A** and **B**). In two different planes of focus, a maturing ookinete is shown exiting the zygote remnant of the retort toward the right. The surface of the parasite is delineated by rhodamine staining of surface proteins Pfs25 and Pfs28. Staining with MAb 1C3 demonstrates the granular appearance of PfCHT1, which is most concentrated in the zygote remnant but also appears to be present anteriorly, reaching near the apical end. (**C** to **E**) An ookinete in panel **C** shows a higher concentration of PfCHT1 at the apical end than the ookinete in panels **A** and **B**. **C**, staining with MAb 1C3 (FITC); **D**, staining with rabbit anti-Pfs25/Pfs28; **E**, colocalization of panels **A** and **B**. Staining with an isotype-matched control (not shown) showed no fluorescence. Each arrow indicates the apical end of the parasite; the arrowhead indicates a zygote remnant. Bars, 1 μm. [From Langer RC, Hayward RE, Tsuboi T, Tachibana M, Torii M, *Vinetz JM*. Micronemal transport of *Plasmodium* ookinete chitinases to the electron-dense area of the apical complex for extracellular secretion. Infect Immun 68:6461–6465, 2000.] Reprinted with permission from the American Society for Microbiology

unknown processes of parasite–mosquito interactions have been established. In addition to the SM1 peptide-transgenic *P. berghei*-resistant *An. stephesi* mosquito (Ghosh et al. 2000, 2001, 2002; Ito et al. 2002), anti-*P. berghei*

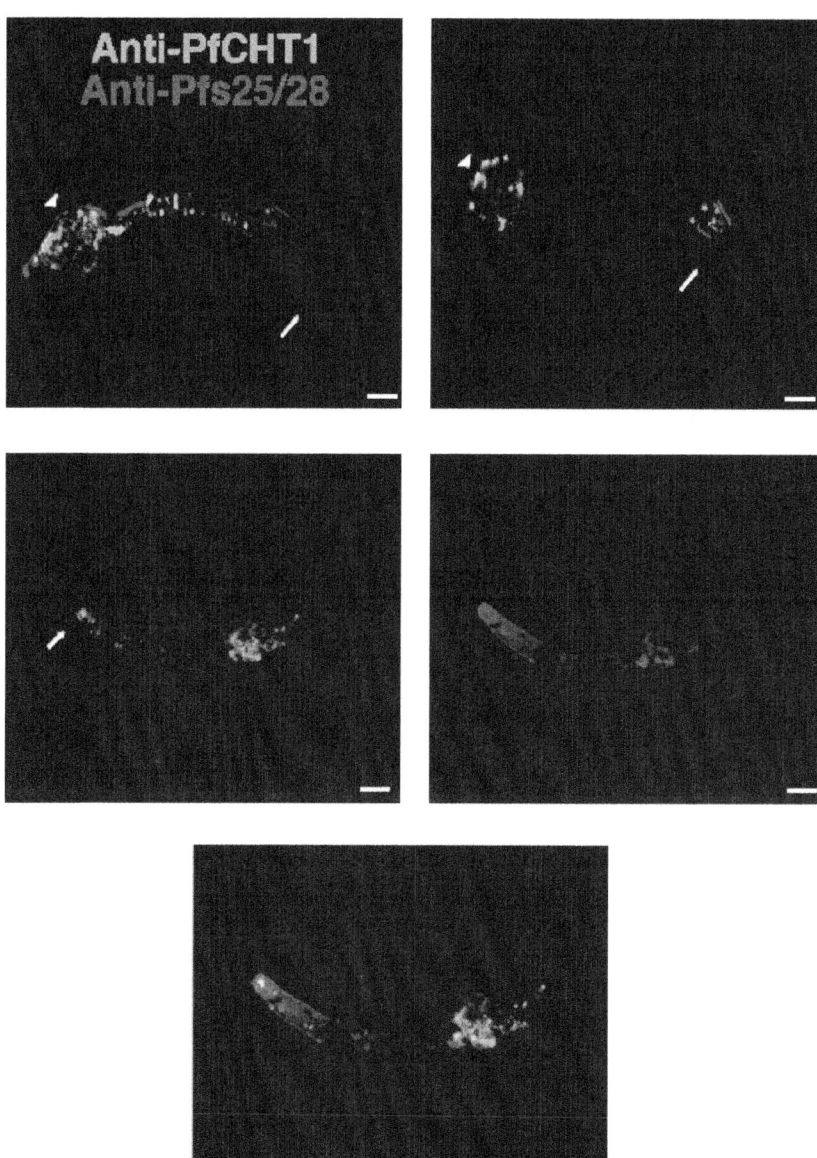

21-kDa surface protein (Yoshida et al. 1999), anti-circumsporozoite (de Lara Capurro et al. 2000), the honey bee venom phospholipase A2 gene transgenically expressed in mosquitoes (Moreira et al. 2002), and a recombinant anti-chitinase single chain antibody (Li, Patra and Vinetz, J Infect Diseases, in press) have been validated as genes with the potential of conferring refractoriness phenotypes to mosquitoes (James et al. 1999).

Acknowledgements The author's work on *Plasmodium* ookinete biology has been supported by NIH grants AI45999, AI50499, AI053781, the World Health Organization/World Bank UNDP Special Programme in Tropical Diseases, and the Culpeper Medical Sciences Scholarship of the Rockefeller Brothers Fund. The author gratefully acknowledges the contributions of Rebecca C. Langer, Fengwu Li, Kailash Patra, Karen Chin, Greg T. Cantin, Jeff Johnson, Laurence Florens, and John L. Yates III to the overall research program and unpublished data discussed within this review.

References

Abraham EG, Jacobs-Lorena M (2004) Mosquito midgut barriers to malaria parasite development. Insect Biochem Mol Biol 34:667–671

Alavi Y, Arai M, Mendoza J, Tufet-Bayona M, Sinha R, Fowler K, Billker O, Franke-Fayard B, Janse CJ, Waters A, Sinden RE (2003) The dynamics of interactions between *Plasmodium* and the mosquito: a study of the infectivity of *Plasmodium berghei* and *Plasmodium gallinaceum*, and their transmission by *Anopheles stephensi*, *Anopheles gambiae* and *Aedes aegypti*. Int J Parasitol 33:933–943

Berner R, Rudin W, Hecker H (1983) Peritrophic membranes and protease activity in the midgut of the malaria mosquito, *Anopheles stephensi* (Liston) (Insecta: Diptera) under normal and experimental conditions. J Ultrastruct Res 83:195–204

Billker O, Lindo V, Panico M, Etienne AE, Paxton T, Dell A, Rogers M, Sinden RE, Morris HR (1998) Identification of xanthurenic acid as the putative inducer of malaria development in the mosquito. Nature 392:289–292

Billker O, Miller AJ, Sinden RE (2000) Determination of mosquito bloodmeal pH in situ by ion-selective microelectrode measurement: implications for the regulation of malarial gametogenesis. Parasitology 120:547–551

Bozdech Z, Llinas M, Pulliam BL, Wong ED, Zhu J, DeRisi JL (2003) The transcriptome of the intraerythrocytic developmental cycle of *Plasmodium falciparum*. PLoS Biol 1:E5

Carlton JM, Angiuoli SV, Suh BB, Kooij TW, Pertea M, Silva JC, Ermolaeva MD, Allen JE, Selengut JD, Koo HL, Peterson JD, Pop M, Kosack DS, Shumway MF, Bidwell SL, Shallom SJ, van Aken SE, Riedmuller SB, Feldblyum TV, Cho JK, Quackenbush J, Sedegah M, Shoaibi A, Cummings LM, Florens L, Yates JR, Raine JD, Sinden RE, Harris MA, Cunningham DA, Preiser PR, Bergman LW, Vaidya AB, van Lin L H, Janse CJ, Waters AP, Smith HO, White OR, Salzberg SL, Venter JC, Fraser CM, Hoffman SL, Gardner MJ, Carucci DJ (2002) Genome sequence and comparative analysis of the model rodent malaria parasite *Plasmodium yoelii yoelii*. Nature 419:512–519

Carrolo M, Giordano S, Cabrita-Santos L, Corso S, Vigario AM, Silva S, Leiriao P, Carapau D, Armas-Portela R, Comoglio PM, Rodriguez A, Mota MM (2003) Hepatocyte growth factor and its receptor are required for malaria infection. Nat Med 9:1363–1369

Carter R, Chen DH (1976) Malaria transmission blocked by immunisation with gametes of the malaria parasite. Nature 263:57–60

Collins FH, Sakai RK, Vernick KD, Paskewitz S, Seeley DC, Miller LH, Collins WE, Campbell CC, Gwadz RW (1986) Genetic selection of a *Plasmodium*-refractory strain of the malaria vector *Anopheles gambiae*. Science 234:607–610

de Lara Capurro M, Coleman J, Beerntsen BT, Myles KM, Olson KE, Rocha E, Krettli AU, James AA (2000) Virus-expressed, recombinant single-chain antibody blocks sporozoite infection of salivary glands in *Plasmodium gallinaceum*-infected *Aedes aegypti*. Am J Trop Med Hyg 62:427–433

Dessens JT, Beetsma AL, Dimopoulos G, Wengelnik K, Crisanti A, Kafatos FC, Sinden RE (1999) CTRP is essential for mosquito infection by malaria ookinetes. EMBO J 18:6221–6227

Dessens JT, Mendoza J, Claudianos C, Vinetz JM, Khater E, Hassard S, Ranawaka GR, Sinden RE (2001) Knockout of the rodent malaria parasite chitinase pbCHT1 reduces infectivity to mosquitoes. Infect Immun 69:4041–4047

Dimopoulos G (2003) Insect immunity and its implication in mosquito-malaria interactions. Cell Microbiol 5:3–14

Dubremetz JF, Garcia-Reguet N, Conseil V, Fourmaux MN (1998) Apical organelles and host-cell invasion by Apicomplexa. Int J Parasitol 28:1007–1013

Ferguson HM, Mackinnon MJ, Chan BH, Read AF (2003) Mosquito mortality and the evolution of malaria virulence. Evolution Int J Org Evol 57:2792–2804

Ferguson HM, Read AF (2002) Why is the effect of malaria parasites on mosquito survival still unresolved? Trends Parasitol 18:256–261

Florens L, Washburn MP, Raine JD, Anthony RM, Grainger M, Haynes JD, Moch JK, Muster N, Sacci JB, Tabb DL, Witney AA, Wolters D, Wu Y, Gardner MJ, Holder AA, Sinden RE, Yates JR, Carucci DJ (2002) A proteomic view of the *Plasmodium falciparum* life cycle. Nature 419:520–526

Foo A, Carter R, Lambros C, Graves PM, Quakyi IA, Targett GA, Ponnudurai T, Lewis G (1991) Conserved and variant epitopes of target antigens of transmission-blocking antibodies among isolates of *Plasmodium falciparum* from Malaysia. Am J Trop Med Hyg 44:623–631

GardnerMJ, Hall N, Fung E, White O, Berriman M, Hyman RW, Carlton JM, Pain A, Nelson KE, Bowman S, Paulsen IT, James K, Eisen JA, Rutherford K, Salzberg SL, Craig A, Kyes S, Chan MS, Nene V, Shallom SJ, Suh B, Peterson J, Angiuoli S, Pertea M, Allen J, Selengut J, Haft D, Mather MW, Vaidya AB, Martin DM, Fairlamb AH, Fraunholz MJ, Roos DS, Ralph SA, McFadden GI, Cummings LM, Subramanian GM, Mungall C, Venter JC, Carucci DJ, Hoffman SL, Newbold C, Davis RW, Fraser CM, Barrell B (2002) Genome sequence of the human malaria parasite *Plasmodium falciparum*. Nature 419:498–511

Gass RF, Yeates RA (1979) In vitro damage of cultured ookinetes of *Plasmodium gallinaceum* by digestive proteinases from susceptible *Aedes aegypti*. Acta Tropica 36:243–252

Ghosh A, Edwards MJ, Jacobs-Lorena M (2000) The journey of the malaria parasite in the mosquito: hopes for the new century. Parasitol Today 16:196–201

Ghosh AK, Moreira LA, Jacobs-Lorena M (2002) *Plasmodium*-mosquito interactions, phage display libraries and transgenic mosquitoes impaired for malaria transmission. Insect Biochem Mol Biol 32:1325–1331

Ghosh AK, Ribolla PE, Jacobs-Lorena M (2001) Targeting *Plasmodium* ligands on mosquito salivary glands and midgut with a phage display peptide library. Proc Natl Acad Sci USA 98:13278–13281

Graves PM, Carter R, Burkot TR, Quakyi IA, Kumar N (1988) Antibodies to *Plasmodium falciparum* gamete surface antigens in Papua New Guinea sera. Parasite Immunol 10:209–218

Gwadz RW (1976) Successful immunization against the sexual stages of *Plasmodium gallinaceum*. Science 193:1150–1151

Han YS, Thompson J, Kafatos FC, Barillas-Mury C (2000) Molecular interactions between *Anopheles stephensi* midgut cells and *Plasmodium berghei*: the time bomb theory of ookinete invasion of mosquitoes. EMBO J 19:6030–6040

Hogg JC, Hurd H (1997) The effects of natural *Plasmodium falciparum* infection on the fecundity and mortality of *Anopheles gambiae* s. l. in north east Tanzania. Parasitology 114:325–331

Huber M, Cabib E, Miller LH (1991) Malaria parasite chitinase and penetration of the mosquito peritrophic membrane. Proc Natl Acad Sci USA 88:2807–2810

Ito J, Ghosh A, Moreira LA, Wimmer EA, Jacobs-Lorena M (2002) Transgenic anopheline mosquitoes impaired in transmission of a malaria parasite. Nature 417:452–455

James AA, Beerntsen BT, Capurro Mde L, Coates CJ, Coleman J, Jasinskiene N, Krettli AU (1999) Controlling malaria transmission with genetically-engineered, *Plasmodium*-resistant mosquitoes: milestones in a model system. Parassitologia 41:461–471

Kadota K, Ishino T, Matsuyama T, Chinzei Y, Yuda M (2004) Essential role of membrane-attack protein in malarial transmission to mosquito host. Proc Natl Acad Sci USA 101:16310–16315

Kaslow DC (1993) Transmission-blocking immunity against malaria and other vector-borne diseases. Curr Opin Immunol 5:557–565

Kaslow DC, Quakyi IA, Syin C, Raum MG, et al. (1988) A vaccine candidate from the sexual stage of human malaria that contains EGF-like domains. Nature 333:74–76

Kaushal DC, Carter R, Howard RJ, McAuliffe FM (1983a) Characterization of antigens on mosquito midgut stages of *Plasmodium gallinaceum*. I. Zygote surface antigens. Mol Biochem Parasitol 8:53–69

Kaushal DC, Carter R, Rener J, Grotendorst CA, Miller LH, Howard RJ (1983b) Monoclonal antibodies against surface determinants on gametes of *Plasmodium gallinaceum* block transmission of malaria parasites to mosquitoes. J Immunol 131:2557–2562

Kumar S, Christophides GK, Cantera R, Charles B, Han YS, Meister S, Dimopoulos G, Kafatos FC, Barillas-Mury C (2003) The role of reactive oxygen species on *Plasmodium melanotic* encapsulation in *Anopheles gambiae*. Proc Natl Acad Sci USA 100:14139–14144

Kumar S, Gupta L, Han YS, Barillas-Mury C (2004) Inducible peroxidases mediate nitration of anopheles midgut cells undergoing apoptosis in response to *Plasmodium* invasion. J Biol Chem 279:53475–53482

Lal AA, Patterson PS, Sacci JB, Vaughan JA, Paul C, Collins WE, Wirtz RA, Azad AF (2001) Anti-mosquito midgut antibodies block development of *Plasmodium falciparum* and *Plasmodium vivax* in multiple species of *Anopheles* mosquitoes and reduce vector fecundity and survivorship. Proc Natl Acad Sci USA 98:5228–5233

Lal AA, Schriefer ME, Sacci JB, Goldman IF, Louis-Wileman V, Collins WE, Azad AF (1994) Inhibition of malaria parasite development in mosquitoes by anti-mosquito-midgut antibodies. Infect Immun 62:316–318

Langer RC, Hayward RE, Tsuboi T, Tachibana M, Torii M, Vinetz JM (2000) Micronemal transport of *Plasmodium* ookinete chitinases to the electron-dense area of the apical complex for extracellular secretion. Infect Immun 68:6461–6465

Langer RC, Li F, Popov V, Kurosky A, Vinetz JM (2002a) Monoclonal antibody against the *Plasmodium falciparum* chitinase, PfCHT1, recognizes a malaria transmission-blocking epitope in *Plasmodium gallinaceum* ookinetes unrelated to the chitinase PgCHT1. Infect Immun 70:1581–1590

Langer RC, Li F, Vinetz JM (2002b) Identification of novel *Plasmodium gallinaceum* zygote- and ookinete-expressed proteins as targets for blocking malaria transmission. Infect Immun 70:102–106

Langer RC, Vinetz JM (2001) *Plasmodium* ookinete-secreted chitinase and parasite penetration of the mosquito peritrophic matrix. Trends Parasitol 17:269–272

Le Roch KG, Zhou Y, Blair PL, Grainger M, Moch JK, Haynes JD, De La Vega P, Holder AA, Batalov S, Carucci DJ, Winzeler EA (2003) Discovery of gene function by expression profiling of the malaria parasite life cycle. Science 301:1503–1508

Li F, Templeton T, Popov V, Comer J, Tsuboi T, Torii M, Vinetz J (2004) *Plasmodium* ookinete-secreted proteins secreted through a common micronemal pathway are targets of blocking malaria transmission. J Biol Chem 279:26635–26644

Limviroj W, Yano K, Yuda M, Ando K, Chinzei Y (2002) Immuno-electron microscopic observation of *Plasmodium berghei* CTRP localization in the midgut of the vector mosquito *Anopheles stephensi*. J Parasitol 88:664–672

Luckhart S, Vodovotz Y, Cui L, Rosenberg R (1998) The mosquito *Anopheles stephensi* limits malaria parasite development with inducible synthesis of nitric oxide. Proc Natl Acad Sci USA 95:5700–5705

Moreira LA, Ito J, Ghosh A, Devenport M, Zieler H, Abraham EG, Crisanti A, Nolan T, Catteruccia F, Jacobs-Lorena M (2002) Bee venom phospholipase inhibits malaria parasite development in transgenic mosquitoes. J Biol Chem 277:40839–40843

Ngo HM, Hoppe HC, Joiner KA (2000) Differential sorting and post-secretory targeting of proteins in parasitic invasion. Trends Cell Biol 10:67–72

Niare O, Markianos K, Volz J, Oduol F, Toure A, Bagayoko M, Sangare D, Traore SF, Wang R, Blass C, Dolo G, Bouare M, Kafatos FC, Kruglyak L, Toure YT, Vernick KD (2002) Genetic loci affecting resistance to human malaria parasites in a West African mosquito vector population. Science 298:213–216

Perrone JB, Spielman A (1988) Time and site of assembly of the peritrophic membrane of the mosquito *Aedes aegypti*. Cell Tissue Res 252:473–478

Quakyi IA, Carter R, Rener J, Kumar N, Good MF, Miller LH (1987) The 230-kDa gamete surface protein of *Plasmodium falciparum* is also a target for transmission-blocking antibodies. J Immunol 139:4213–4217

Ramasamy MS, Kulasekera R, Wanniarachchi IC, Srikrishnaraj KA, Ramasamy R (1997a) Interactions of human malaria parasites, *Plasmodium vivax* and *P. falciparum*, with the midgut of *Anopheles* mosquitoes. Med Vet Entomol 11:290–296

Ramasamy R, Wanniarachchi IC, Srikrishnaraj KA, Ramasamy MS (1997b) Mosquito midgut glycoproteins and recognition sites for malaria parasites. Biochim Biophys Acta 1361:114–122

Ranawaka GR, Alejo-Blanco AR, Sinden RE (1994a) Characterization of the effector mechanisms of a transmission-blocking antibody upon differentiation of *Plasmodium berghei* gametocytes into ookinetes in vitro. Parasitology 109:11–17

Ranawaka GRR, Fleck SL, Blanco AR, Sinden RE (1994b) Characterization of the modes of action of anti-Pbs21 malaria transmission-blocking immunity: ookinete to oocyst differentiation in vivo. Parasitology 109:403–411

Read D, Lensen AH, Begarnie S, Haley S, Raza A, Carter R (1994) Transmission-blocking antibodies against multiple, non-variant target epitopes of the *Plasmodium falciparum* gamete surface antigen Pfs230 are all complement-fixing. Parasite Immunol 16:511–519

Rener J, Graves PM, Carter R, Williams JL, Burkot TR (1983) Target antigens of transmission-blocking immunity on gametes of *Plasmodium falciparum*. J Exp Med 158:976–981

Sam-Yellowe TY, Florens L, Wang T, Raine JD, Carucci DJ, Sinden R, Yates Jr 3rd (2004) Proteome analysis of rhoptry-enriched fractions isolated from *Plasmodium* merozoites. J Proteome Res 3:995–1001

Shahabuddin M (2002) Do *Plasmodium* ookinetes invade a specific cell type in the mosquito midgut. Trends Parasitol 18:157–161

Shahabuddin M, Criscio M, Kaslow D (1995) Unique specificity of in vitro inhibition of mosquito midgut trypsin-like activity correlates with in vivo inhibition of malaria parasite infectivity. Exp Parasitol 80:212–219

Shahabuddin M, Lemos F, Kaslow D, Jacobs-Lorena M (1996) Antibody-mediated inhibition of *Aedes aegypti* midgut trypsins blocks sporogonic development of *Plasmodium gallinaceum*. Infect Immun 64:739–743

Shahabuddin M, Pimenta PF (1998) *Plasmodium gallinaceum* preferentially invades vesicular ATPase-expressing cells in *Aedes aegypti* midgut. Proc Natl Acad Sci USA 95:3385–3389

Shahabuddin M, Toyoshima T, Aikawa M, Kaslow DC (1993) Transmission-blocking activity of a chitinase inhibitor and activation of malarial parasite chitinase by mosquito protease. Proc Natl Acad Sci USA 90:4266–4270

Shen Z, Jacobs-Lorena M (1998) A type I peritrophic matrix protein from the malaria vector *Anopheles gambiae* binds to chitin. Cloning, expression, and characterization. J Biol Chem 273:17665–17670

Sieber KP, Huber M, Kaslow D, Banks SM, Torii M, Aikawa M, Miller LH (1991) The peritrophic membrane as a barrier: its penetration by *Plasmodium gallinaceum* and the effect of a monoclonal antibody to ookinetes. Exp Parasitol 72:145–156

Sinden RE, Alavi Y, Raine JD (2004) Mosquito–malaria interactions: a reappraisal of the concepts of susceptibility and refractoriness. Insect Biochem Mol Biol 34:625–629

Sinden RE, Billingsley PF (2001) *Plasmodium* invasion of mosquito cells: hawk or dove? Trends Parasitol 17:209–211

Sinden RE, Canning EU (1972) The ultrastructure of *Plasmodium berghei* ookinetes in the midgut wall of *Anopheles stephensi*. Trans R Soc Trop Med Hyg 66:6

Suwanabun N, Sattabongkot J, Tsuboi T, Torii M, Maneechai N, Rachapaew N, Yimamnuaychok N, Punkitchar V, Coleman RE (2001) Development of a method for the in vitro production of *Plasmodium vivax* ookinetes. J Parasitol 87:928–930

Tellam R, Wijffels G, Willadsen P (1999) Peritrophic matrix proteins. Insect Biochem Mol Biol 29:87–101

Templeton TJ, Kaslow DC, Fidock DA (2000) Developmental arrest of the human malaria parasite *Plasmodium falciparum* within the mosquito midgut via CTRP gene disruption. Mol Microbiol 36:1–9

Tomas AM, Margos G, Dimopoulos G, van Lin LH, de Koning-Ward TF, Sinha R, Lupetti P, Beetsma AL, Rodriguez MC, Karras M, Hager A, Mendoza J, Butcher GA, Kafatos F, Janse CJ, Waters AP, Sinden RE (2001) P25 and P28 proteins of the malaria ookinete surface have multiple and partially redundant functions. EMBO J 20:3975–3983

Torii M, Nakamura K, Sieber KP, Miller LH, Aikawa M (1992) Penetration of the mosquito (*Aedes aegypti*) midgut wall by the ookinetes of *Plasmodium gallinaceum*. J Protozool 39:449–454

Trottein F, Triglia T, Cowman AF (1995) Molecular cloning of a gene from *Plasmodium falciparum* that codes for a protein sharing motifs found in adhesive molecules from mammals and plasmodia. Mol Biochem Parasitol 74:129–141

Trueman HE, Raine JD, Florens L, Dessens JT, Mendoza J, Johnson J, Waller CC, Delrieu I, Holders AA, Langhorne J, Carucci DJ, Yates JR 3rd, Sinden RE (2004) Functional characterization of an LCCL-lectin domain containing protein family in *Plasmodium berghei*. J Parasitol 90:1062–1071

Tsai YL, Hayward RE, Langer RC, Fidock DA, Vinetz JM (2001) Disruption of *Plasmodium falciparum* chitinase markedly impairs parasite invasion of mosquito midgut. Infect Immun 69:4048–4054

Tsuboi T, Kaneko O, Eitoku C, Suwanabun N, Sattabongkot J, Vinetz JM, Torii M (2003) Gene structure and ookinete expression of the chitinase genes of *Plasmodium vivax* and *Plasmodium yoelii*. Mol Biochem Parasitol 130:51–54

van Dijk MR, Thompson J, Waters AP, Braks JAM, Dodemont HJ, Stunnenberg HG, van Gemert G-J, Sauerwein RW, Eling W (2001) A central role for P48/45 in malaria parasite male gamete fertility. Cell 104:153–164

Vaughan JA, Hensley L, Beier JC (1994a) Sporogonic development of *Plasmodium yoelii* in five anopheline species. J Parasitol 80:674–681

Vaughan JA, Noden BH, Beier JC (1994b) Sporogonic development of cultured *Plasmodium falciparum* in six species of laboratory-reared *Anopheles* mosquitoes. Am J Trop Med Hyg 51:233–243

Vermeulen AN, Ponnudurai T, Beckers PJ, Verhave JP, Smits MA, Meuwissen JH (1985) Sequential expression of antigens on sexual stages of *Plasmodium falciparum* accessible to transmission-blocking antibodies in the mosquito. J Exp Med 162:1460–1476

Vernick KD, Fujioka H, Aikawa M (1999) *Plasmodium gallinaceum*: a novel morphology of malaria ookinetes in the midgut of the mosquito vector. Exp Parasitol 91:362–366

Vinetz JM, Dave SK, Specht CA, Brameld KA, Hayward RE, Fidock DA (1999) The chitinase PfCHT1 from the human malaria parasite *Plasmodium falciparum* lacks proenzyme and chitin-binding domains and displays unique substrate preferences. Proc Natl Acad Sci USA 96:14061–14066

Vinetz JM, Valenzuela JG, Specht CA, Aravind L, Langer RC, Ribeiro JM, Kaslow DC (2000) Chitinases of the avian malaria parasite *Plasmodium gallinaceum*, a class of enzymes necessary for parasite invasion of the mosquito midgut. J Biol Chem 275:10331–10341

Vlachou D, Zimmermann T, Cantera R, Janse C, Waters A, Kafatos F (2004) Real-time, in vivo analysis of malaria ookinete locomotion and mosquito midgut invasion. Cell Microbiol 6:671–685

Williamson K, Kaslow D (1993) Strain polymorphism of *Plasmodium falciparum* transmission-blocking target antigen Pfs230. Mol Biochem Parasitol 62:125–128

Williamson KC (2003) Pfs230: from malaria transmission-blocking vaccine candidate toward function. Parasite Immunol 25:351–359

Yoshida S, Matsuoka H, Luo E, Iwai K, Arai M, Sinden RE, Ishii A (1999) A single-chain antibody fragment specific for the *Plasmodium berghei* ookinete protein Pbs21 confers transmission blockade in the mosquito midgut. Mol Biochem Parasitol 104:195–204

Yuda M, Sakaida H, Chinzei Y (1999a) Targeted disruption of the *Plasmodium berghei* CTRP gene reveals its essential role in malaria infection of the vector mosquito. J Exp Med 190:1711–1716

Yuda M, Sawai T, Chinzei Y (1999b) Structure and expression of an adhesive protein-like molecule of mosquito invasive-stage malarial parasite. J Exp Med 189:1947–1952

Yuda M, Yano K, Tsuboi T, Torii M, Chinzei Y (2001) von Willebrand Factor A domain-related protein, a novel microneme protein of the malaria ookinete highly conserved throughout *Plasmodium* parasites. Mol Biochem Parasitol 116:65–72

Zheng L, Cornel AJ, Wang R, Erfle H, Voss H, Ansorge W, Kafatos FC, Collins FH (1997) Quantitative trait loci for refractoriness of *Anopheles gambiae* to *Plasmodium cynomolgi* B. Science 276:425–428

Zieler H, Dvorak JA (2000) Invasion in vitro of mosquito midgut cells by the malaria parasite proceeds by a conserved mechanism and results in death of the invaded midgut cells. Proc Natl Acad Sci USA 97:11516–11521

Zieler H, Garon CF, Fischer ER, Shahabuddin M (1998) Adhesion of *Plasmodium gallinaceum* ookinetes to the *Aedes aegypti* midgut: sites of parasite attachment and morphological changes in the ookinete. J Eukaryot Microbiol 45:512–520

Zieler H, Nawrocki JP, Shahabuddin M (1999) *Plasmodium gallinaceum* ookinetes adhere specifically to the midgut epithelium of *Aedes aegypti* by interaction with a carbohydrate ligand. J Exp Biol 202:485–495

CTMI (2005) 295:383–415

Molecular Genetics of Mosquito Resistance to Malaria Parasites

K. D. Vernick[1] (✉) · F. Oduol[1] · B. P. Lazzaro[3] · J. Glazebrook[2] · J. Xu[1] · M. Riehle[1] · J. Li[1]

[1] Department of Microbiology, Center for Microbial and Plant Genomics, University of Minnesota, 1500 Gortner Avenue, St. Paul, MN 55108, USA
kvernick@umn.edu

[2] Department of Plant Biology, Center for Microbial and Plant Genomics, University of Minnesota, 1500 Gortner Avenue, St. Paul, MN 55108, USA

[3] Department of Entomology, Cornell University, 4138 Comstock Hall, Ithaca, NY 14853, USA

Abstract Malaria parasites are transmitted by the bite of an infected mosquito, but even efficient vector species possess multiple mechanisms that together destroy most of the parasites present in an infection. Variation between individual mosquitoes has allowed genetic analysis and mapping of loci controlling several resistance traits, and the underlying mechanisms of mosquito response to infection are being described using genomic tools such as transcriptional and proteomic analysis. Malaria infection imposes fitness costs on the vector, but various forms of resistance inflict their own costs, likely leading to an evolutionary tradeoff between infection and resistance. *Plasmodium* development can be successfully completed only in compatible mosquito-parasite species combinations, and resistance also appears to have parasite specificity.

Studies of *Drosophila*, where genetic variation in immunocompetence is pervasive in wild populations, offer a comparative context for understanding coevolution of the mosquito–malaria relationship. More broadly, plants also possess systems of pathogen resistance with features that are structurally conserved in animal innate immunity, including insects, and genomic datasets now permit useful comparisons of resistance models even between such diverse organisms.

Abbreviations

EST	Expressed sequence tag
NBS-LRR	Nucleotide binding site and leucine-rich repeat-containing protein
TLR	Toll-like receptor
ROS	Reactive oxygen species
SA	Salicylic acid

1
Introduction

The genus *Plasmodium* has more than 100 species which infect birds, reptiles and diverse mammals including humans. Different *Plasmodium* species utilize different mosquito genera for transmission, and a reptile *Plasmodium* is even transmitted by a sandfly. The vectors of human malaria parasites, however, comprise relatively few species of the *Anopheles* genus. Despite sometimes daily contact with malaria parasites in human bloodmeals, other mosquito species in malaria endemic areas never sustain infection to serve as vectors of human disease. Within the few permissive species in nature, only a proportion of females are actually involved in transmission, and even in permissive individuals, only a small proportion of parasites survive to complete successful development. Thus, malaria transmission in nature passes through a specific and narrow conduit that limits parasite numbers in many ways. Certainly the efficiency of the system should not be underestimated, because the transmission rate is sufficient to maintain malaria as one of the major public health problems of the world. However, it is reasonable to ask just how robust and stable the malaria transmission system really is, and whether there are unexploited weaknesses that could be manipulated to reduce malaria transmission below the level of population maintenance (Vernick and Waters 2004). With this goal in mind, a body of knowledge has been generated describing the molecular basis of the vector–parasite interaction and genetic variation for suppression of parasite development within permissive mosquito species.

2
Sporogonic Development

Female mosquitoes first take a bloodmeal several days after adult emergence, and continue to feed every few days thereafter. If a bloodmeal is taken from an animal carrying a compatible species of *Plasmodium*, the sporogonic stage of the malaria life cycle is initiated. But what does 'compatible' mean in this context? As with any host–pathogen interaction, compatibility requires that the parasite both finds the specific resources, such as cellular receptors, nutrients, and developmental signals, required to reproduce itself, and develops even in the presence of host immune defenses.

The human host, the vector host, and the parasite have competing evolutionary agendas, with the host seeking to eliminate infection and the parasite striving for efficient transmission. Infectious disease is generally characterized by partial host immunity, with some, but not all, individuals in a population infected at any given time. This model appears to hold for the system at hand, because mosquitoes possess mechanisms that limit, but generally do not completely prevent, parasite development (Luckhart et al. 1998; Han et al. 2000). Thus, only a small minority of malaria parasites that enter the mosquito develop completely (Gouagna et al. 1998; Vaughan et al. 1994). These baseline mechanisms already defeat most malaria parasites, so increasing their efficiencies or introducing other additive or synergistic mechanisms can result in complete resistance (Collins et al. 1986; Osta et al. 2004; Vernick et al. 1995).

This review concentrates on the second point above: vector host defenses that can suppress successful parasite development, genetic variation in those defenses, and potential molecular mechanisms underlying those defenses. Successful development requires that the virulence machinery of the parasite be a good molecular fit for the vector, allowing the parasite to acquire resources and ultimately reproduce. However, a compatible interaction does not mean a harmless one, and there is evidence that malaria parasites decrease vector reproductive fitness (discussed below). This fitness cost to the mosquito probably in part drives the mosquito to mount an active immune response. Efficient resistance to a pathogen can also be passive, and it is likely that genetic variation of critical mosquito molecules can yield parasite-resistant vector phenotypes by making the vector functionally invisible to the parasite at key developmental junctures, analogous to human genotypes with mutant CCR5 receptor for HIV invasion (Dean et al. 1996; Huang et al. 1996), or lacking the Duffy receptor for malaria merozoite invasion of erythrocytes (Miller et al. 1976; Zimmerman et al. 1999).

The sexual phase of the malaria life cycle is initiated when a proportion of parasites replicating mitotically in vertebrate host erythrocytes make a developmental switch to a terminally differentiated nonreplicating sexual form called the gametocyte (Fig. 1). Gametocytes, the only parasite stage infective for mosquitoes, are ingested by the mosquito with the bloodmeal. Within minutes in the mosquito midgut, gametocytes generate gametes that undergo fertilization to produce zygotes. Over the next 24 h, each zygote transforms into a nondividing motile form called the ookinete. This form exits the gut lumen and invades cells of the midgut epithelium by an uncharacterized mechanism. The ookinete traverses the epithelial cell, exits through the basolateral membrane and lodges between the plasma membrane and basal lamina where within hours it transforms to a rounded oocyst. Over the next 10–20 days, depending on the parasite species, about 5,000 sporozoites form within the oocyst. At the end of this period of latency, the mature oocyst releases the sporozoites into the mosquito hemocoel, and a small proportion will survive to invade the salivary glands. In subsequent bloodmeals, the mosquito injects sporozoites along with saliva into a new vertebrate host to establish infection and complete the transmission cycle.

The merits of the parasite and mosquito species and strain combinations commonly used as experimental systems have been reviewed (Sinden 1997; Vernick 1998). It is worth noting that these are mostly laboratory models that do not represent natural vector–parasite combinations, and thus much of what we see in the laboratory has not been confirmed in the natural transmis-

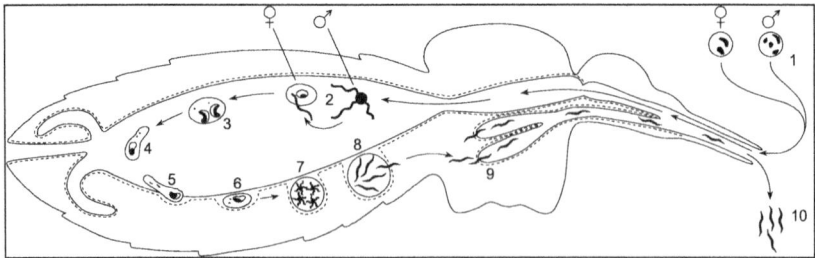

Fig. 1 Malaria parasite development in the mosquito vector. *1*, Mature gametocytes, the infective stage for the vector, are ingested in the bloodmeal. *2*, Within minutes, gametocytes produce gametes that undergo fertilization. *3*, Zygotes. *4*, Over approximately 24 h, zygotes transform into the motile ookinete stage. *5*, Ookinetes enter midgut epithelial cells from the midgut lumen and migrate to the basolateral membrane, where they exit the midgut cell. *6*, Ookinetes lodge in a position outside the plasma membrane of the epithelial cell but underneath the basal lamina (*dotted line*), where they transform into oocysts. *7*, Oocysts grow in the same location over the next 10–20 days and begin a process of differentiation to produce sporozoites. *8*, Mature oocysts rupture to release sporozoites into the hemocoel. *9*, Sporozoites invade the salivary glands, where they can remain infective for the life of the mosquito. *10*, Sporozoites are introduced along with saliva during a bloodmeal upon a vertebrate host, where they can establish a new infection

sion system. In fact, immune responses measured in non-natural laboratory models are known to differ from the natural species combinations (Tahar et al. 2002).

3
Genetically Selected Systems of Malaria Resistance

Melanotic Encapsulation in the Hemocoel   A genetic line of *An. gambiae* has been artificially selected against the simian parasite, *P. cynomolgi* (Collins et al. 1986), with resistance manifested as melanotic encapsulation of otherwise ultrastructurally normal parasites after they exit from the midgut cell into the space between the basolateral plasma membrane and the midgut basal lamina, beginning 16 h post bloodmeal (Collins et al. 1986; Paskewitz et al. 1988). Hemocytes did not appear to be directly involved in the encapsulation process (Paskewitz et al. 1988). However, insect midgut basal lamina is probably nonselectively permeable to molecules below a threshold size (Reddy and Locke 1990), so there is likely passive diffusion of melanization substrates and enzymes into the space occupied by extracellular ookinetes. This view is supported by the observation that capsules were thickest on the side of the parasite facing the hemocoel (Paskewitz et al. 1988). Only after the capsule completely surrounded the ookinete did the parasite degenerate ultrastructurally.

A model for in vivo encapsulation was developed using charged Sephadex beads. Negatively charged beads were encapsulated much more efficiently by the resistant *An. gambiae* line than by susceptible mosquitoes (Paskewitz and Riehle 1994), whereas positively charged or neutral beads were encapsulated with equally high efficiency in both genetic lines. The efficiency of bead melanization decreased with mosquito age in both genetic lines, but under the appropriate conditions the bead melanization assay allowed 80%–90% of female mosquitoes tested to be phenotypically assigned to the correct genetic strain (Chun et al. 1995). The efficiency of bead melanization was enhanced by bloodmeal in resistant, but not in susceptible, mosquitoes.

The bead assay established that the components producing the divergent encapsulation responses of resistant and susceptible mosquitoes were present in the hemocoel rather than being a feature of midgut epithelial cells. The inverse correlation between efficiency of bead melanization and mosquito age suggested that the encapsulation phenotype requires a factor that can be present in limiting amounts. This is consistent with the observation that melanization in mosquitoes of the resistant line was more efficient against low parasite numbers, also suggesting the involvement of a finite component that could be titrated by parasite number (Vernick et al. 1989).

New World and Asian strains of *P. falciparum* were efficiently encapsulated by the resistant mosquito line, but *P. falciparum* strains of African origin were not (Collins et al. 1986). Similarly, the African species *P. ovale* and *P. malariae* (along with close simian relative *P. brasilianum*) also failed to be efficiently encapsulated. The geographic range of *An. gambiae* is limited to sub-Saharan Africa. Thus, one interpretation is that parasites sympatric with *An. gambiae* have evolved local adaptations to evade recognition or effector functions of the encapsulation response. However, because these observations were made using cultured parasite strains and laboratory mosquito colonies, the genetic fidelity of either to their original natural populations could be questioned. Encapsulation of *P. falciparum* has been observed in wild *An. gambiae* in Africa (K. D. Vernick, unpublished results; Schwartz and Koella 2002), and thus it is a natural phenotype even in the sympatric combination, although at low frequency. Finally, a line of *An. dirus* was selected in which resistance to the rodent malaria parasite, *P. yoelii*, was manifested as melanotic encapsulation controlled by a polygenic mechanism (Somboon et al. 1999).

Intracellular Ookinete Lysis in the Midgut Epithelial Cell   Genetic lines of *An. gambiae* were selected to be resistant and susceptible to the avian parasite, *P. gallinaceum* (Vernick et al. 1995). Ultrastructurally, the resistant phenotype was manifested as the degeneration of ookinete cellular organelles and lysis of the parasite quickly following ookinete invasion of midgut epithelial cells. The initial number of ookinetes invading the midgut epithelium was similar in both genetic lines, suggesting that parasite killing resulted from an intracellular mechanism rather than an interaction within the bloodmeal in the midgut lumen. A genetic crossing experiment suggested that the resistance mechanism was controlled by a single main locus with dominant effect.

Whether the resistance mechanism acted against other species of *Plasmodium* was not determined, although there is no reason to believe that the response would be generalizable to human malarias as the lytic response to *P. gallinaceum* essentially represents the wild-type response of *An. gambiae* to this parasite. However, it would be interesting to know whether the lytic killing mechanism can be addressed against ookinetes of other *Plasmodium* species if they are introduced in the same bloodmeal with *P. gallinaceum*. Such an experiment would distinguish whether compatible parasite species avoid being killed by intracellular lysis because they evade recognition and induction of lysis or because they are resistant to the lytic response.

Several features appear to distinguish the lytic mechanism of parasite killing from the melanotic encapsulation response described above. Ookinetes killed by lysis did not become melanized but rather suffered rapid degeneration of cellular organelles and apparent necrosis. Encapsulated parasites appeared ultrastructurally normal until the melanotic capsule was complete, after which time they began to degenerate. Lytic killing was intracellular, with ookinetes rarely reaching the basolateral cell membrane, while encapsulation occurred after apparently healthy parasites had exited the basolateral boundary of the epithelial cell into the extracellular lymph compartment. To determine conclusively whether a relationship exists, however, it will be necessary to understand the underlying biochemistry of both systems.

Lines of *An. atroparvus* resistant and susceptible to *P. berghei* were selected in which oocyst number in resistant midguts was close to zero, and in which the infection phenotype was under polygenic control (van der Kaay and Boorsma 1977). Initial ookinete invasion of the midgut epithelium was similar between the selected lines (Sluiters et al. 1986). The subsequent failure of ookinetes to develop in the resistant line was ascribed to ookinete 'degeneration', and was not described further. Another genetic system selected in *An. gambiae* for resistance to *P. berghei* similarly was said to result from 'degenerated' sporogonic stage parasites (Al-Mashhadani et al. 1980; Al-Mashhadani and Davisdson 1976). It is not known if these two resistant systems are related in any way to the lytic response of *An. gambiae* against *P. gallinaceum*. Lines of *An. stephensi* with decreased *P. falciparum* oocyst numbers were produced by genetic selection (Feldmann et al. 1990; Feldmann and Ponnudurai 1989). The mechanism was unknown, but was based on polygenic control with influence from a possible cytoplasmic factor (Feldmann et al. 1998).

Other Resistance Mechanisms   Mosquito resistance to strains or species of malaria parasite, whether active or passive, serves as part of the ecological isolating mechanisms that define parasite niche boundaries. One example is the natural resistance of *An. gambiae* to *P. gallinaceum* described above. In that case, segregating genetic variation allowed the selection of pure susceptible and resistant lines that facilitated study of the mechanism. Another example is the natural resistance of *Culex pipiens* to infection with *P. gallinaceum*, in which parasites (in the form of gametocytes) inoculated into the hemocoel of *C. pipiens* developed ectopically until the 3-day-old oocyst stage but then degenerated and died (Weathersby and McCall 1968). Parasites inoculated into the susceptible species, *Aedes aegypti*, underwent normal (albeit ectopic) development, including sporozoite invasion of salivary glands. When *Ae. aegypti* were given an infective bloodmeal and then parabiotically joined to *C. pipiens* by a capillary to create a common hemolymph, only about 2% of the joined *Ae. aegypti* supported parasite development as compared to about 95% of unjoined *Ae. aegypti* controls, suggesting that the resistance of *C. pipiens* was caused by an diffusible toxic humoral factor (Weathersby and McCroddan 1982). Vector–parasite compatibility or its absence may be enforced at multiple steps during parasite development (Alavi et al. 2003).

An important example of resistance is the apparent failure of culicine mosquitoes to serve as vectors of mammalian malaria parasites despite frequent exposure to infective bloodmeals in natural transmission zones. Among the few publications on the topic, it was reported that a proportion of laboratory reared *C. bitaeniorhynchus* mosquitoes that fed upon infected human volunteers were susceptible to *P. falciparum*, *P. vivax* and *P. malariae* to the sporozoite stage (Williamson and Zain 1937a, 1937b). Another report described infection of the culicine mosquito *Mansonia uniformis* with *P. falciparum* by experimental feeding on an infected human volunteer (Cheong et al. 1963). In no reported case, however, did *P. falciparum* infected culicines become infective to vertebrates, although the barriers to transmission are currently unknown.

The genetic and molecular basis of barriers to malaria infection and transmission in different mosquito species remains largely unexplored. This subject could now be profitably examined with new genomic tools, and in this regard the *Ae. aegypti* genome sequence will be a useful complement to the *An. gambiae* sequence. One does not have to look far (at least from humans) to find a recent example of pathogen host range restriction. Chimpanzees can be infected with HIV, but unlike humans rarely progress to AIDS-like disease (Balla-Jhagjhoorsingh et al. 2003; Davis et al. 1998; Novembre et al. 1997). It was proposed that greatly reduced variation observed in the MHC class I gene of chimpanzees is the product of a selective sweep caused by a widespread ancient infection by an HIV relative, and that modern populations of chimpanzees are descended from survivors of that pandemic (de Groot et al. 2002). Other primates avoid disease by blocking HIV replication early after cell invasion, by action of the cellular factor, *TRIM5α* (Stremlau et al. 2004).

In mosquitoes, the species barriers to malaria could fall into a number of categories. They could be physiological, such as a thicker peritrophic matrix; cellular, if host molecules that the parasite needs to bind such as cell surface invasion receptors, developmental signaling ligands or nutritive factors are too diverged; immune, such as recognition or effector molecules that the parasite has not adapted to resist or evade. There may be important differences in gene regulatory pathways that affect any of the above. Remarkably, considering that some nonvector mosquito taxa are closely related to vector taxa, none of these possibilities seems insurmountable for the parasite, so in addition to asking how the barriers work, it is also worth asking how they have persisted as functional barriers.

4
Genetic Mapping of Resistance Loci

To date, genetic analysis of malaria resistance phenotypes has used two experimental approaches. The first approach mapped resistance in inbred laboratory lines of *An. gambiae* infected with simian or rodent malaria or charged beads. The second strategy mapped infection intensity in pedigrees of wild-caught *An. gambiae* infected with its natural parasite *P. falciparum*.

Resistance in Laboratory Strains   Resistant and susceptible strains of *An. gambiae* selected in the laboratory to encapsulate or permit development of simian malaria parasites were intercrossed for genetic mapping by linkage analysis using microsatellite markers. Mapping in five backcross families challenged with *P. cynomolgi* B malaria parasites identified one major and two minor quantitative trait loci (QTLs) associated with melanotic encapsulation of the parasite. Examination of these backcross progeny suggested a dominant effect of resistance alleles with a single locus as the major genetic determinant of encapsulation. The major QTL, *Pen1*, is located on chromosome arm 2R and accounted for approximately 54% of the variability in encapsulation response (Zheng et al. 1997). A similar study used the same inbred mosquito lines to map encapsulation of negatively charged Sephadex beads, identifying a QTL that mapped to the same region as the *Pen1* locus. In addition, encapsulation of beads and rodent malaria parasites (*P. berghei*) were shown to have similar modes of inheritance: dominant, autosomal, and controlled by a single major gene (Gorman et al. 1997). Finer scale mapping of the *Pen1* locus has identified 48 putative genes (Thomasova et al. 2002). Genetic analysis of *Pen1* remains incomplete at present.

Of the two minor QTLs, *Pen2*, was located on chromosome arm 3L and explained approximately 13% of the variation in encapsulation. The third QTL, *Pen3*, was linked to *Pen1* on chromosome 2 and affected encapsulation of *P. cynomolgi* B only slightly. Attempts to map another immune related trait, infection intensity (that is, sum total of viable plus encapsulated oocysts) in this laboratory system suggested there was no simple genetic component of inheritance, and no overlap with encapsulation QTLs (Zheng et al. 1997).

Recently, efforts to map the encapsulation trait in the same resistant and susceptible strains of *An. gambiae* challenged with a different simian parasite, *P. cynomolgi* Ceylon (it is unclear whether *P. cynomolgi* B and Ceylon differ as strains or species), suggested that different QTLs are involved in encapsulation of different parasites and that resistance to this parasite is incompletely recessive (Zheng et al. 2003). Three QTLs, *Pcen2R*, *Pcen3R* and *Pcen3L*, were identified for resistance to *P. cynomolgi* Ceylon and explained 13%, 16% and 26% of the variation in encapsulation, respectively. Two of the QTLs (*Pcen2R* and *Pcen3L*) were indistinguishable from *Pen3* and *Pen2*, the minor QTLs identified in *P. cynomolgi* B encapsulation. However, *Pcen3R* was a novel QTL not involved in resistance to *P. cynomolgi* B and *Pen1*, the major *P. cynomolgi* B locus did not contribute to the encapsulation of *P. cynomolgi* Ceylon. These results were generally consistent with previous genetic studies that also found distinct genetic mechanisms underlying encapsulation of these two parasites (Vernick and Collins 1989; Vernick et al. 1989).

Resistance in Natural Populations   Inheritance of resistance to *P. falciparum* was examined in natural populations of *An. gambiae* in Mali, west Africa (Niare et al. 2002). In an initial study, eggs were collected from single wild-caught female mosquitoes, and the resulting families were raised under environmentally controlled conditions. Three such random families were fed the same infected human blood and the difference in oocyst number among families was measured. There were significantly different infection distributions among families fed on the same blood, which demonstrated that there were frequent alleles segregating in nature with an effect on susceptibility to parasite infection. Consequently, a mapping experiment was performed.

   Genetic mapping of infection intensity (total oocyst number) in two wild pedigrees of *An. gambiae* from Mali infected with natural *P. falciparum* identified a major resistance locus in one family and a minor locus in another family. The major effect QTL (*Pfin1*), located on chromosome arm 2L, explained almost 90% of the parasite-free mosquitoes in the segregating pedigree, and was semidominant. Mosquitoes that were resistant homozygotes (by microsatellite marker) at *Pfin1* had mean 0.17 oocysts per mosquito, while susceptible homozygotes had 50.6 oocysts, indicating that natural resistance mechanisms can be quite efficient. The second QTL, *Pfin2*, was located on chromosome arm 2R near *Pen1* and showed recessive inheritance of the resistance phenotype. Despite their proximity on chromosome 2R, *Pen1* and *Pfin2* are likely to be distinct genes given the differences in their associated phenotypes (encapsulation vs. intensity, respectively) and the fact that encapsulation associated with *Pen1* was strongest against allopatric non-African malaria parasites, while *Pfin2* was discovered due to its effect on sympatric African parasites. This work on wild pedigrees represented the first concrete evidence of resistance alleles segregating in nature, and was one of the first studies to use any wild populations in identifying QTLs (Niare et al. 2002). The genetic analysis protocol established in this work is now being used to carry out a multiplexed genome-wide screen of *An. gambiae* infected with natural *P. falciparum* in Mali to determine the number and frequency of such natural refractory genotypes, and to capture alleles in extant lines for laboratory studies. A combination of single nucleotide polymorphism genotyping, finer scale microsatellite genotyping and carefully selected candidate gene analyses are being used to identify the loci underlying *Pfin1* and *Pfin2*. *Pfin1* is now restricted to an approximately 6-Mb chromosomal interval. Efforts to identify candidate genes are aided by a current expressed sequence tag (EST) project focused on immune responsive transcripts and subsequent microarray analysis of gene expression during malaria infection.

Despite identifying major QTLs for both encapsulation and infection intensities using the laboratory and field experimental approaches, no single causative locus has yet been isolated. Complete genome sequence, increasing functional characterization of genes and their protein products, finer scale genetic mapping, and continual dissection of mosquito innate immunity should allow more informed choice of candidate genes in the effort to identify malaria resistance genes.

5
Mosquito Transcriptome and Proteome

Transcriptome   The initial route taken to describe the molecular immune system of mosquitoes was the identification of individual transcriptionally responsive genes (Dimopoulos et al. 1998; Richman et al. 1997). The first transcriptome-based projects described immune ESTs from a mosquito hemocyte-like cell line (Dimopoulos et al. 2000) and from whole-mosquito subtractive hybridization (Oduol et al. 2000). These and other projects have recently yielded the resources for larger scale transcriptional profiling studies of mosquito immunity.

Microarrays fabricated from normalized cDNA libraries of an *An. gambiae* hemocyte-like cell line were hybridized with labeled cDNA produced from in vitro treated cultured cells, or from in vivo treated mosquitoes (Dimopoulos et al. 2002). Infection of mosquitoes by *P. berghei* induced the expression of 24 genes and repressed the transcription of 10. Most of the malaria-induced genes in mosquitoes were also induced in the cell line by microbial elicitors. The malaria-repressed genes in mosquitoes were mostly repressed or unchanged in response to microbial elicitors in the cell line. Among the approximately eight genes induced in mosquitoes by malaria, but not by sterile or septic injury, were two genes, *isocitrate dehydrogenase* and a dsRNA-binding RNase, that were also not induced by any treatment in the cell line.

The cell-line derived microarray was also used to analyze gene expression profiles of the above described artificially selected mosquito lines that are susceptible or resistant to malaria parasites through the melanotic encapsulation response. Some redox-related genes displayed elevated expression in the resistant line as compared to either susceptible or wild-type G3 colony mosquitoes, suggesting that the resistant line could be under constitutive oxidative stress (Kumar et al. 2003). The difference between genetic lines was manifested after an uninfected bloodmeal, with little additional effect of malaria parasite infection. This is reminiscent of the increased encapsulation of negatively charged beads after a normal bloodmeal (Chun et al. 1995). The induced gene set was enriched for genes of the mitochondrial genome, particularly those involved in mitochondrial respiration, and several nuclear genes including *thioredoxin reductase*, mitochondrial *thioredoxin*, and *xanthine dehydrogenase*. Basal expression of *catalase*, a gene involved in the clearance of oxidative free radicals, is higher in the susceptible than the resistant line. The *catalase* gene is near the *Pen3* QTL involved in parasite encapsulation, which was a minor locus for *P. cynomolgi* B but a major effect locus for *P. cynomolgi* Ceylon.

EST libraries highly enriched for repressed and induced sequences of the immune transcriptome of whole *An. gambiae* mosquitoes (Oduol et al. 2000) were spotted on microarrays and expression profiles were analyzed (J. Xu, J. Li, F. Oduol, M. Riehle and K.D. Vernick, unpublished results). Transcriptional profiles in response to the Gram-negative bacterial immune elicitor lipopolysaccharide were almost entirely distinct from the response to malaria or injury. A significant coexpressed cluster of genes was induced by injury but repressed by malaria infection, suggesting that a counter-inflammatory response may be caused by malaria parasites. The repression began soon after the infective bloodmeal, before ookinete invasion of the midgut epithelium, and expanded during midgut invasion, indicating the existence of malaria-related molecular signals that may prime the mosquito host for infection.

Whole-genome expression profiles in *Drosophila* have revealed many immune-responsive genes. In one study, microbial infection by *Escherichia coli*, *Mycobacterium luteus*, and the fungus *Beauvaria bassiana* drove the transcriptional response of 400 genes (230 induced and 170 repressed) on Affymetrix oligonucleotide array (De Gregorio et al, 2001). Of the 400, 368 were not previously characterized as immune-related genes. Among the observations were 28 new inducible small peptides, which could be new antimicrobial effectors or cytokine-like signaling molecules. The list of immune-responsive genes was arbitrarily truncated at 400, and was not claimed to be inclusive.

A related study also used an Affymetrix oligonucleotide array and the same pathogens to reveal a list of 543 genes that were upregulated at least twofold by microbial challenge (Irving et al. 2001). This number comprised 4% of the total 13,600 genes assayed. Downregulated genes were not explicitly reported in this study (although the raw data are available), but based on the ratio of up-to-down regulated genes in the previous study, there might be approximately 400 repressed genes, suggesting a conservative estimate of 7% of the *Drosophila* genome devoted to immune defense. It is striking that transcriptional expression of such a large proportion of the genome can be influenced by just three distinct pathogens. (Even more strikingly, approximately 25% of *Arabidopsis* genes can be differentially regulated during pathogen infection; Tao et al. 2003.) There were some differences but no clear expression signatures specific to the different pathogen responses in the *Drosophila* data. Most of the *Drosophila* genes with the greatest immune induction encoded hemolymph factors, including recognition proteins, antimicrobial peptides, serine protease, and protease inhibitors. Although there are hundreds of protease genes in the *Drosophila* genome, only 26 of them were induced by immune challenge. Overall, 47% of the immune-induced genes had no known function. Undoubtedly, some of these are generalized stress-responsive genes.

Proteome   Relatively few studies have been done directly on proteins involved in mosquito–parasite interactions. The hemolymph is one major site of interest, because all sporogonic malaria stages contact the hemolymph. Hemolymph is the physiological location of functions involving pathogen recognition, signaling cascades, and effector activity, and is probably the major compartment of the mosquito immune response.

Molecular cloning and two-dimensional gel electrophoresis studies of hemolymph identified hemolymph proteins, including some that were altered by inoculation of saline or Sephadex beads (Chun et al. 2000; Gorman et al. 2000; Han et al. 1999). One of the altered proteins was the serine protease AgSp14D1, which is a relative of *Drosophila* Easter and *Manduca* prophenoloxidase activating enzyme, and may be involved in immune recognition and signaling (Paskewitz et al. 1999). The *AgSp14D1* gene is located near the mapped location of the *Pen3* locus involved in melanotic encapsulation.

There is an effort underway to catalog the hemolymph proteome of *An. gambiae* by mass spectrometry and two-dimensional electrophoresis (F. Oduol, J. Li and K.D. Vernick, unpublished results). Hundreds of proteins have been identified to date, including the expected prophenoloxidases, thioester motif-containing proteins (aTEPs), and serine proteases, and also many proteins with no known function. Comparisons between resting-state and malaria-infected hemolymph have identified a number of proteomic differences, including differences in absolute and relative protein abundance, mass shifts that probably signify cleavage events, and small charge shifts that probably signify phosphorylation or other post-translational modifications involved in immune signaling.

6
Fitness Costs and Evolution of Malaria Resistance

Relatively little work has been done to examine the nature of the evolutionary relationship between malaria parasites and mosquito vectors, although an obvious expectation is that each organism places selective pressures on the other. In order for an antagonistic coevolutionary relationship between host and pathogen to arise, at least two conditions must be satisfied: (1) host genetic variants should differ in susceptibility to infection; and (2) infection should be detrimental to host fitness. Both criteria appear to be met in the mosquito–malaria system. Moreover, the intensity of selective pressure reciprocally imposed between *Anopheles* and *Plasmodium* is probably enhanced by the regularity with which mosquito hosts are exposed to the parasite and the taxonomic specificity of the host–parasite relationship. Most mosquito species are associated with only one or two closely related *Plasmodium* species in nature, and any given mosquito has a high probability of encountering the parasite over her lifetime.

Evidence for coevolutionary adaptation is also provided by the observation that the *Plasmodium* life cycle can be successfully completed only in compatible mosquito–parasite species combinations, while *Plasmodium* infections of non-natural mosquito hosts can fail at any of a number of stages (Alavi et al. 2003; Billingsley and Sinden 1997). Resistance, at least by encapsulation, also appears to include a component of parasite specificity, based on large genetic differences underlying encapsulation of *P. cynomolgi* B, *P. berghei*, and Sephadex beads, as compared to *P. cynomolgi* Ceylon. Transcriptional profiles of a panel of parasite response genes were different for *P. falciparum* and the rodent parasite *P. berghei* (Tahar et al. 2002). One could speculate that, in the case at least of encapsulation, a shared effector cassette may be controlled by distinct upstream recognition and signaling modules responsive to different malaria parasite species or strains. Caution is warranted in drawing generalizations from these studies because neither *P. cynomolgi* nor *P. berghei* are transmitted by *An. gambiae* in nature. Physiological competence for encapsulation (based on the *Pen1* bead phenotype) is further uncoupled from actual parasite encapsulation by the observation that 90% of wild-caught *An. gambiae* in Tanzania encapsulated Sephadex beads, while less than 1% of naturally infected mosquitoes carried encapsulated parasites (Schwartz and Koella 2002), again leaving room for the action of specific coevolved recognition functions.

There is a body of evidence indicating that malaria infection imposes significant fitness costs on the vector. With regard to direct mortality, laboratory studies have been equivocal, sometimes suggesting no infection-dependent mortality (Chege and Beier 1990; Robert et al. 1990) and in other cases reaching the opposite conclusion (Klein et al. 1982, 1986). However, under laboratory conditions with protection from nutritional and environmental stress, pathogens may not yield the best measure of mortality, particularly if the effects are subtle. Indeed, natural infections typically display lower oocyst numbers than experimental infections in the laboratory (Medley et al. 1993), indicating the probable existence of tradeoffs or infection-limiting factors in nature that are eliminated in the laboratory. In a study of natural populations, increased mortality was attributed to higher oocyst burdens, possibly due to the metabolic cost of infection (Lyimo and Koella 1992).

Sporozoite-infected mosquitoes probe a bloodmeal host more often than uninfected ones, and also spend more time probing (Rossignol et al. 1984, 1986; Wekesa et al. 1992). A proposed explanation was that sporozoite-infected salivary glands produce less apyrase (a platelet aggregation inhibitor) than do uninfected glands (Rossignol et al. 1984). Moreover, mosquitoes in nature positive for *P. falciparum* circumsporozoite protein (i.e., probably bearing sporozoites) bite more people per night than uninfected mosquitoes (Koella et al. 1998). Mosquitoes with parasitized salivary glands had a significantly higher feeding-associated mortality in nature (Anderson et al. 2000), which may result from greater exposure to human host defensive measures (e.g., swatting) due to the altered mosquito feeding behavior. The long latency of the oocyst before sporozoite release may defer some of the fitness costs of infection, and associated selection pressure, until late in the vector's life, after most eggs have already been laid (Koella 1999).

The presence of *Plasmodium* oocysts in the midgut has been correlated with reduced mosquito fecundity in laboratory and natural systems (Hogg and Hurd 1995a, 1995b, 1997). Decreased fecundity was also associated with alterations in utilization of the yolk protein vitellogenin by ovaries in infected versus uninfected mosquitoes (Ahmed et al. 2001; Hogg et al. 1997). It was observed that soon after development of the ovarian terminal follicles in *An. stephensi*, a significantly greater proportion of the follicles underwent resorption in infected as compared to uninfected mosquitoes, thus reducing the size of the resulting egg batch (Carwardine and Hurd 1997). The mechanism underlying the resorption appears to be apoptosis of the follicles (Hopwood et al. 2001). Thus, destruction of the follicles probably explains the reduced vitellogenin uptake by ovaries of infected mosquitoes, and the consequent reduced fecundity. These observations add malaria to the list of pathogens that subvert host reproduction to conserve critical host resources (Koella 1999).

There are probably functional tradeoffs between the costs of infection and resistance, and one category of fitness cost is the cost of resistance itself. There is likely a metabolic cost to mounting a defense response, and a parasitized host may be energetically compromised and have fewer resources available for defense against other pathogens (Ahmed et al. 2002; Brey 1994). There could also be a fitness cost from collateral damage to self caused by immune effectors, which has precedent in host defense systems such as vertebrate inflammation in sepsis (Cohen 2002; Ohta and Sitkovsky 2001) and probably the insect melanotic encapsulation response, in which reactive oxygen species in resistant mosquitoes may be harmful to both parasite and host (Armitage et al. 2003; Kumar et al. 2003; Moret and Schmid-Hempel 2000; Nappi et al. 1995). The melanotic encapsulation response might also compete with other critical functions such as eggshell and cuticle tanning (Ferdig et al. 2000; Johnson et al. 2001).

A related consideration is that the lack of robust natural resistance may at least in part result from indirect or direct parasite modulation of host defenses. Indirect effects could result from bloodmeal-related factors. For example, some malaria-responsive *An. gambiae* genes are transcriptionally regulated by an infected bloodmeal beginning before actual ookinete invasion of the midgut (J. Xu, J. Li, F. Oduol, M. Riehle, and K.D. Vernick, unpublished results; Bonnet et al. 2001; Tahar et al. 2002). This early expression could represent a response to soluble parasite-produced immune elicitors, to the quality of the bloodmeal derived from an infected vertebrate host, or to immune signaling molecules from the infected vertebrate host that might influence mosquito immune signaling pathways (Luckhart et al. 2003). Finally, the parasite may actively and directly manipulate components of host defenses. This mechanism has numerous precedents, for example insect polydnavirus proteins that inhibit host immune hemocyte function (Li and Webb 1994), the Yop virulence factors of *Yersinia pestis* that block host cell phagocytosis (Andersson et al. 1996; Rosqvist et al. 1991), plant fungus subversion of an antimicrobial compound to disrupt immune signaling (Bouarab et al. 2002), bacterial suppression of the production of antibiotic peptides in *Drosophila* (Fauvarque et al. 2002; Lindmark et al. 2001), malaria inhibition of dendritic cell functions to leave the host vulnerable to repeated reinfection (Ocana-Morgner et al. 2003), and the downregulation by herpesviruses, cytomegalovirus and HIV-1 of surface class I MHC protein to block cytotoxic T lymphocyte activity (Cohen et al. 1999; Ploegh 1998).

7
Evolution of Immune Genes in *Drosophila* Populations

Evolutionary forces acting on insect immune systems have been most thoroughly studied in *Drosophila*, where molecular and phenotypic analyses suggest that genetic variation in immunocompetence is pervasive in wild populations. Comparable data have not yet been obtained from the *Anopheles–Plasmodium* system. It has been recognized for some time that natural populations of *D. melanogaster* harbor genetic variation in the ability to resist parasitization by endoparasitic wasps (Carton and Bouletreau 1985; Hughes and Sokolowski 1996; Kraaijeveld and Godfray 1997). Although resistance to parasitization in wild flies is very low, artificial selection for true-breeding fly lines exhibiting high rates of wasp egg encapsulation has been success-ful (Carton et al. 1992; Kraaijeveld and Godfray 1997) and attributable to a small number of genes with large effects (Benassi et al. 1992; Orr and Irving 1997). In at least one case, the resistance phenotype seemed to result from in-creased hemocyte production (Kraaijeveld et al. 2001). More recently, genetic variability in antibacterial immunity has been documented among distinct genetic lines derived from a wild *D. melanogaster* population (Lazzaro et al. 2004). Extreme lines in this study differed by 10 phenotypic standard errors in bacterial load sustained following infection, although no single candidate immune-response gene explained more than 16% of the observed variance. The apparent difference in the genetic architecture of resistance to parasitoids and bacteria is suggestive, but it may be exaggerated by differences in the ex-perimental approach. The parasitoid work relied on lines artificially selected for resistance, such that mutations conferring large phenotypic effects could rapidly dominate the selected population and be readily detected by recom-bination mapping. In contrast, the candidate gene-based approach taken in the antibacterial work tends to have more power to detect small allelic effects, although the genetic basis for much of the observed phenotypic variance re-mains undetermined. A rigorous comparison of the structure of variation in defense against parasitoids and microbes will have to wait until comparable experiments have been executed with both pathogens.

Given the importance of immunocompetence to organismal fitness, it is not apparent why variation in quality of the immune response is allowed to persist in natural populations. One possibility is that functional varia-tion in *Drosophila* immunity genes exists as a result of functional tradeoffs. *D. melanogaster* larvae selected for enhanced encapsulation of parasitoid wasp eggs have been shown to be poor competitors under resource-limited conditions (Fellowes et al. 1998; Kraaijeveld and Godfray 1997), perhaps due to a twofold greater investment in generating encapsulation-competent hemocytes in resistant relative to susceptible lines (Kraaijeveld et al. 2001). Increased male courtship activity has been suggested to decrease the rate with which avirulent *E. coli* are cleared from the hemocoel of *D. melanogaster* adults (McKean and Nunney 2001). In the latter case, no mechanistic connec-tion between the resistance phenotype and the fitness cost has been defined, although hormonal differences induced by mating have been suggested to compromise immune capacity in *Tenebrio* (Rolff and Siva-Jothy 2002).

Another class of tradeoff models posits that a given host allele may confer increased resistance to one pathogen but decreased resistance to another. Innate immune systems may be especially prone to this type of tradeoff because relatively few proteins may be responsible for recognizing and combating a large diversity of potential pathogens. This model has not been well tested in *Drosophila*, but limited empirical supporting evidence has been generated with the crustacean *Daphnia magna* and its bacterial pathogen *Pasteuria ramosa*. Genetically distinct *Daphnia* clones derived from wild-collected females were most susceptible to infection by *Pasteuria ramosa* isolated along with the particular mother that founded the *Daphnia* line, suggesting that varying *Pasteuria* strains were differentially able to infect specific host genotypes (Carius et al. 2001). A similar experiment examining pathogenesis of pea aphids, however, did not find any evidence for tradeoffs due to pathogen specificity (Ferrari et al. 2001). Provided there is no universally 'best' defense allele at a locus, allelic differences in efficacy against various pathogens could, in principle, allow the maintenance of host variation as a function of pathogen diversity. If relevant pathogen diversity decreases, however, this model collapses into a more traditional host–pathogen coevolutionary 'arms race' in which evolved host resistance to a pathogen drives the pathogen to evolve means of overcoming host defenses, in turn selecting for enhanced resistance in the host, perpetuating the cycle ad infinitum (Dawkins and Krebs 1979).

While allelic differences in response to varying pathogens have not been rigorously tested in *Drosophila*, molecular sequence data from a subset of *Drosophila* immune response genes reflect rapid evolution characteristic of host–pathogen coevolutionary interactions. As a functional class, immune-related genes diverge between species more quickly than nonimmune genes in *Drosophila* (Schlenke and Begun 2003). A similar observation has been made with respect to vertebrate immune response genes (Murphy 1993) and plant *R* genes (Lehmann 2002), suggesting that frequent fixation of adaptive amino acid variants is a common property of innate defense molecules. One striking incidence of apparent *Drosophila*–pathogen coevolution occurs with Relish, a transcription factor fundamentally important to the induction of antibacterial responses (Hedengren et al. 1999). Relish shows a highly significantly accelerated rate of amino acid substitution, with most substitutions concentrated in the autoinhibitory domain of the protein (Begun and Whitley 2000). Dredd, the caspase that physically interacts with Relish to cleave the autoinhibitory domain (Stöven et al. 2003) also shows a highly accelerated rate of amino acid substitution (Schlenke and Begun 2003). Thus, one hypothesis is that bacterial molecules injected into the host cell via Type III secretion systems (Cornelis and Van Gijsegem 2000) might interfere with Relish activation, forcing the host proteins into a coevolutionary arms race to retain functionality.

Rapid amino acid divergence is not characteristic of all innate immune proteins, however. Pioneering studies of sequence variation in *D. melanogaster*

Cecropin antibacterial peptide genes revealed little evidence of adaptive evolution (Clark and Wang 1997; Date et al. 1998; Ramos-Onsins and Aguadè 1998). A later study suggested that directional selection may operate on variation generated by gene conversion between the tandemly repeated *Attacin A* and *B* antibacterial peptide genes in *D. melanogaster* (Lazzaro and Clark 2001), and more comprehensive analysis of sequence polymorphism and divergence in multiple *D. melanogaster* antibacterial peptides showed consistent indications of positive selection on antibacterial peptide genes as a functional class (Lazzaro and Clark 2003). But the rate of amino acid divergence between *D. melanogaster* and *D. simulans* in antibacterial peptides is actually lower than the substitution rate between the two species in nonimmunity genes. Instead, *D. melanogaster* peptides harbor a slight excess of nonconservative amino acid polymorphism in proteolytically processed, but not in mature antibiotic, peptide domains (Lazzaro and Clark 2003). The data from the peptide genes are not consistent with co-evolutionary arms races or selectively maintained hypervariability, but may be consistent with frequency-dependent or fluctuating selection. Finally, peptidoglycan recognition proteins appear to evolve primarily under purifying selection (Jiggins and Hurst 2003).

Drosophila immune response has been proposed as a model for description of anti-*Plasmodium* reactions in anopheline mosquitoes. Much of the existing data support the comparison, although there may also be important differences between the biological contexts of *Drosophila* immunity to parasitoids and bacteria compared to antimalarial responses in mosquitoes. *Drosophila* are plagued by several species of parasitoid wasps and an undetermined number of distinct bacteria, while *Anopheles–Plasmodium* relationships are highly specific. The likelihood of any individual *Drosophila* encountering a given pathogen species in nature is probably small, whereas anophelines in many parts of the world are virtually assured of ingesting a bloodmeal containing *P. falciparum* or *P. vivax*. These differences imply that *Anopheles–Plasmodium* relationships may specifically coevolve even where *Drosophila* immune systems are forced to maintain a strong component of generality. This hypothesis does not predict that mosquito immune response genes will be invariant, only that the variation will be tailored to variation in *Plasmodium* genotypes.

8
Models and Mechanisms of Plant Disease Resistance

Like insects, plants lack a circulating immune system and rely exclusively on a form of innate immunity to defend themselves from pathogen attack. The best-understood form of plant disease resistance is known as gene-for-gene resistance. It is so named because of early observations that particular loci in plant hosts (called *R* genes for Resistance) conferred resistance to pathogens carrying particular genes called avirulence genes (Flor 1955). Typically, a single *R* gene confers resistance to only a single pathogen strain that carries the cognate avirulence gene. This seems rather odd, as one would expect that the pathogens would quickly lose the avirulence genes. Avirulence genes actually encode virulence factors that promote pathogenicity on hosts lacking the appropriate *R* genes, providing a selection pressure for pathogen populations to retain them. Gene-for-gene resistance is effective against a wide range of pathogens, including viruses, bacteria, oomycetes, fungi, nematodes, and aphids (Dangl and Jones 2001). Resistance is associated with an enormous number of inducible defense responses, including the hypersensitive response, a form of programmed cell death that occurs in cells in direct contact with the pathogen.

At one time, the ligand–receptor model of gene-for-gene resistance was popular. This model proposes that *R* genes encode receptors for proteins or metabolites produced by pathogen avirulence genes. Binding of these molecules by the *R* proteins then triggers activation of defense responses. Isolation of *R* genes from many plant species over the last 10 years has revealed that they comprise a few classes (Ellis et al. 2000). So far, the largest class is composed of cytoplasmic proteins with nucleotide binding sites and leucine-rich repeats (NBS-LRR proteins). Other classes consist of membrane anchored proteins with LRR domains outside the membrane, or receptor kinases containing LRRs. A few *R* genes that do not fit any of these types have also been isolated. The cytoplasmic location of NBS-LRR proteins makes sense since many plant pathogens transport virulence factors into the plant cytoplasm. For example, plant pathogenic bacteria use Type III secretion systems similar to those of *Yersinia* and *Salmonella* to transport proteins into the host cytoplasm (Casper-Lindley et al. 2002; Collmer et al. 2000).

LRR-containing proteins are also components of several aspects of host defense in animals. Toll-like receptors (TLRs) are animal LRR proteins involved in sensing molecular patterns characteristic of broad groups of pathogens, such as lipopolysaccharide of Gram-negative bacteria. Distinct from most plant NBS-LRR proteins, the ligand-binding LRR region of TLRs is extracellular, and thus TLRs recognize extracellular pathogen molecules or signals derived from them, and transduce the immune signal into the cell. However, there are less well-described plant *R* genes that are also membrane anchored, with extracellular LRRs. TLRs are structurally conserved in invertebrate and vertebrate animals, and include mosquito representatives (Christophides et al. 2002; Imler and Zheng 2004), although it is not yet clear if they play a role in mosquito response to malaria parasites. More recently, it has been found that there are also intracellular NBS-LRR proteins in animals that, similarly to plant NBS-LRR proteins, bind cytosolic pathogen-derived molecules and transduce an immune signal. Despite some semantic confusion in their names, animal NBS-LRR proteins all possess an LRR region that is probably the site of ligand-binding and an NBS that mediates oligomerization and probably activation, and finally a variable domain that confers protein interaction with distinct downstream effectors (Chamaillard et al. 2003; Inohara and Nunez 2003).

Once the plant *R* genes were in hand, many groups tried to detect binding of *R* proteins to their cognate pathogen avirulence proteins. In virtually all cases, these efforts failed dismally, casting doubt on the validity of the ligand–receptor model. An additional problem with the ligand–receptor model is that it requires plants to carry a large number of *R* genes, one for each pathogen virulence factor that it could encounter in its environment. Analysis of the complete genome sequence of the model plant *Arabidopsis thaliana* revealed only 149 NBS-LRR proteins (Meyers et al. 2003), a number that seems inadequate for the task. These difficulties led to the formulation of a new model, called the Guard Hypothesis (Dangl and Jones 2001; Van der Biezen and Jones 1998). In this model, *R* proteins guard important host proteins that may be targeted by pathogen virulence factors. Interference with the guarded protein by a pathogen factor is detected by the *R* protein, and this triggers activation of defenses. This explains why direct interactions between *R* proteins and avirulence proteins have not been easy to find. In the Guard Hypothesis, the number of *R* genes required is not much larger than the number of plant proteins that are possible targets of pathogen virulence factors.

The Guard Hypothesis predicts that there should be plant proteins that interact with R proteins and with the cognate pathogen avirulence proteins. Recently, two proteins that may be guarded by *R* proteins have been identified in *Arabidopsis*. *PBS1* encodes a serine-threonine protein kinase that is degraded by the bacterial protein AvrPphB, which is a cysteine protease (Shao et al. 2003). PBS1 kinase activity is required for recognition of AvrPphB by the *R* gene *RPS5*, but not for the activity of other *R* genes (Swiderski and Innes 2001; Warren et al. 1999). *RIN4* is degraded in the presence of AvrRpt2, the cognate avirulence protein of the *R* gene *RPS2*. In the other example, RIN4 forms complexes with either of AvrRpt2 or RPS2 in vivo. Removal of RIN4 is sufficient to trigger a defense response, and this depends on RPS2 (Axtell and Staskawicz 2003; Mackey et al. 2002, 2003). Evidence that pathogens benefit from interfering with putative guarded proteins such as RIN4 and PBS1 in host backgrounds lacking appropriate *R* genes would provide additional support to the Guard Hypothesis.

There is evidence that many pathogen avirulence genes contribute to virulence in hosts lacking cognate *R* genes. For example, AvrRpt2 enhances virulence of *Pseudomonas syringae* pv. *tomato* DC3000 in hosts lacking *RPS2*, and inhibits activation of host defense responses (Chen et al. 2000). However, the molecular mechanisms by which such bacterial virulence factors contribute to virulence are not understood.

Gene-for-gene recognition of pathogen attack sets off a cascade of responses with profound effects on the responding cells. Nitric oxide and reactive oxygen species (ROS) are produced within a few hours. They may be toxic to pathogens and have been implicated in signaling. The source of most of the ROS is an NADPH oxidase similar to the enzyme responsible for ROS production in mammalian phagocytes (Torres et al. 2002). Nitric oxide is required for the hypersensitive response triggered by R genes in *Arabidopsis* (Delledonne et al. 1998), and is produced by the P-protein of glycine decarboxylase (Chandok et al. 2003). It is not yet known if loss of this activity compromises resistance.

Gene-for-gene resistance in plants is associated with production of small signal molecules such as salicylic acid (SA), which is required for expression of many defense-related genes. This signaling pathway contributes to resistance to various pathogens, but the extent to which gene-for-gene resistance depends on its function is not yet clear. Some of the molecular machinery involved in responding to the SA signal has been elucidated, and constitutes an interesting paradigm for signal transduction. Most SA-dependent responses require the activity of NPR1, an ankyrin-repeat protein (Cao et al. 1997). In the absence of SA, NPR1 is located in the cytoplasm in an oligomeric form due to disulfide bridges between monomers. Increased SA concentrations cause a change in the redox balance, resulting in de-oligomerization of NPR1, which allows it to move into the nucleus (Mou et al. 2003). Once there, it interacts with specific transcription factors (Despres et al. 2000; Zhang et al. 1999), which are required for activation of expression of certain SA-regulated defense genes (Zhang et al. 2003). This process is reminiscent of the mammalian IκB/NFκB system and a similar mechanism in insects in which binding of IκB to NFκB holds it in the cytoplasm until an appropriate signal occurs, at which time the released NFκB translocates to the nucleus (Baeuerle and Baltimore 1996; Hoffmann 2003).

The fact that there is a fitness cost to R gene mediated resistance has long been known to crop breeders who deploy R genes in the field. A recent study has documented that the presence of the R gene RPM1 in an otherwise isogenic background results in a seed yield reduction of 9% (Tian et al. 2003), although the resistance allele is strongly favored when pathogenesis is prevalent. Thus, the frequencies of resistance/susceptible alleles can cycle with the frequency of pathogenesis (Stahl et al. 1999). Deleterious effects of R genes could be a result of low-level R protein activity in the absence of pathogens, leading to low-level expression of plant defense responses. Some R proteins such as RPS2 have been shown to have activity in the absence of pathogens (Tao et al. 2000).

The fitness costs of R genes likely contribute to the enormous polymorphism in R gene repertoire in wild populations, as fitness costs are balanced against benefits of resistance.

There are far more NBS-LRR proteins in plants (149 in *Arabidopsis* to hundreds in rice) than in humans (~25), where they have only recently been identified. This class of intracellular defense molecule has not yet been described in invertebrates and may be absent in them, although human and invertebrate genomes each encode approximately 10 membrane anchored TLRs. Thus, although it is too soon to generalize, it appears that plant defense responses may be biased towards detection of intracellular attack as compared to animal defenses, perhaps consistent with differences in body plans of the organisms (Girardin et al. 2002).

9
Conclusions

Animals and plants all possess innate, genetically encoded mechanisms of host defense against pathogens. These mechanisms comprise the first line of host defense in vertebrates, and the only known line of defense in other organisms. Innate host defense systems have largely been described as responses to bacteria, viruses and fungi, pathogens that are radically different from the host. In these cases, it may be relatively straightforward for the host to recognize pathogen-associated molecular patterns that distinguish pathogen from self, and to develop effectors with specific activity against, for instance, bacterial but not eukaryotic cell membranes. Eukaryotic hosts infected with protozoan pathogens, such as mosquitoes infected by malaria parasites, lose some of these advantages. It is probably more difficult to design a defensin-like effector that forms pores in parasite but not host plasma membrane when the biophysical properties of both membranes are much more similar.

How do mosquito vectors of malaria respond to infection with malaria parasites? Because of the specificity of the association, and the likely fitness costs to mosquitoes, this probably represents an antagonistic coevolutionary relationship. Consequently, mosquito responses to malaria parasites are more likely to include pathogen-specific components, compared to the characteristically general antimicrobial responses. We do not know whether other protozoan pathogens that may infect mosquitoes in nature, such as microsporidia, are seen by the mosquito as immunologically similar to malaria parasites. Thus, it is not yet possible to say whether mosquito responses to malaria are also effective against other pathogens, or if the antimalaria response has been shaped by selection imposed by other pathogen infections and may be consequently constrained by a requirement for generality.

In the mosquito–malaria system, we do not yet know the immune elicitors, mechanisms of immune recognition and signaling, or effectors. In the laboratory model system of melanotic encapsulation, genetic loci have been mapped and there is a preliminary picture of some of the physiological features. The genetic evidence suggests that even this model response appears to be tailored to specific parasite species and possibly strains. In any case, encapsulation as described in the laboratory model does not appear to be representative of actual mosquito resistance to malaria in nature. Perhaps this is not surprising since, at least in the genetically selected line under current study, efficient encapsulation appears to be costly to the host. It is possible that individual components of the multigenic encapsulation machinery might be utilized in natural populations in different ways, which may be consistent with finding natural resistance traits (*Pfin2*) that map to the same apparent locus as one of the encapsulation loci (*Pen1*). The recent completion of the *An. gambiae* genome sequence and current work on the *Ae. aegypti* sequence offer new tools to dissect the mosquito–malaria interaction. Genomic tools are transforming the ability to analyze transcriptional and protein responses to infection, and to genetically identify and analyze the evolved mechanisms that determine mosquito resistance to malaria in nature.

References

Ahmed AM, Baggott SL, Maingon R, Hurd H (2002) The costs of mounting an immune response are reflected in the reproductive fitness of the mosquito *Anopheles gambiae*. Oikos 97:371–377

Ahmed AM, Maingon R, Romans P, Hurd H (2001) Effects of malaria infection on vitellogenesis in *Anopheles gambiae* during two gonotrophic cycles. Insect Mol Biol 10:347–356

Alavi Y, Arai M, Mendoza J, Tufet-Bayona M, Sinha R, Fowler K, Billker O, Franke-Fayard B, Janse CJ, Waters A, Sinden RE (2003) The dynamics of interactions between *Plasmodium* and the mosquito: a study of the infectivity of *Plasmodium berghei* and *Plasmodium gallinaceum*, and their transmission by *Anopheles stephensi, Anopheles gambiae* and *Aedes aegypti*. Int J Parasitol 33:933–943

Al-Mashhadani HM, Davidson G, Curtis CF (1980) A genetic study of the susceptibility of *Anopheles gambiae* to *Plasmodium berghei*. Trans R Soc Trop Med Hyg 74:585–594

Al-Mashhadani HM, Davisdson G (1976) A study of the susceptibility of *A. gambiae* species to malaria infection. Heredity 37:457

Anderson RA, Knols BG, Koella JC (2000) *Plasmodium falciparum* sporozoites increase feeding-associated mortality of their mosquito hosts *Anopheles gambiae* s.l. Parasitology 120:329–333

Andersson K, Carballeira N, Magnusson KE, Persson C, Stendahl O, Wolf-Watz H, Fallman M (1996) YopH of *Yersinia pseudotuberculosis* interrupts early phosphotyrosine signalling associated with phagocytosis. Mol Microbiol 20:1057–1069

Armitage SA, Thompson JJ, Rolff J, Siva-Jothy MT (2003) Examining costs of induced and constitutive immune investment in Tenebrio molitor. J Evol Biol 16:1038–1044

Axtell MJ, Staskawicz BJ (2003) Initiation of RPS2-specified disease resistance in Arabidopsis is coupled to the AvrRpt2-directed elimination of RIN4. Cell 112:369–377

Baeuerle PA, Baltimore D (1996) NF-kappa B: ten years after. Cell 87:13–20

Balla-Jhagjhoorsingh SS, Verschoor EJ, de Groot N, Teeuwsen VJ, Bontrop RE, Heeney JL (2003) Specific nature of cellular immune responses elicited by chimpanzees against HIV-1. Hum Immunol 64:681–688

Begun DJ, Whitley P (2000) Adaptive evolution of relish, a Drosophila NF-kappaB/IkappaB protein. Genetics 154:1231–1238

Benassi V, Frey F, Carton Y (1992) A new specific gene for wasp cellular immune resistance in Drosophila. Heredity 80: 347–352

Billingsley PF, Sinden RE (1997) Determinants of malaria–mosquito specificity. Parasitol Today 13:297

Bonnet S, Prevot G, Jacques JC, Boudin C, Bourgouin C (2001) Transcripts of the malaria vector Anopheles gambiae that are differentially regulated in the midgut upon exposure to invasive stages of Plasmodium falciparum. Cell Microbiol 3:449–458

Bouarab K, Melton R, Peart J, Baulcombe D, Osbourn A (2002) A saponin-detoxifying enzyme mediates suppression of plant defences. Nature 418:889–892

Brey P (1994) The impact of stress on immunity. Bull de l'Institut Pasteur 92:101–118

Cao H, Glazebrook J, Clarke JD, Volko S, Dong X (1997) The Arabidopsis NPR1 gene that controls systemic acquired resistance encodes a novel protein containing ankyrin repeats. Cell 88:57–63

Carius HJ, Little TJ, Ebert D (2001) Genetic variation in a host-parasite association: potential for coevolution and frequency-dependent selection. Evolution 55:1136–1145

Carton Y, Bouletreau M (1985) Encapsulation ability of Drosophila melanogaster: A genetic analysis. Dev Comp Immunol 9:211–219

Carton Y, Frey F, Nappi A (1992) Genetic determinism of the cellular immune reaction in Drosophila melanogaster. Heredity 69:393–399

Carwardine SL, Hurd H (1997) Effects of Plasmodium yoelii nigeriensis infection on Anopheles stephensi egg development and resorption. Med Vet Entomol 11:265–269

Casper-Lindley C, Dahlbeck D, Clark ET, Staskawicz BJ (2002) Direct biochemical evidence for type III secretion-dependent translocation of the AvrBs2 effector protein into plant cells. Proc Natl Acad Sci USA 99:8336–8341

Chamaillard M, Girardin SE, Viala J, Philpott DJ (2003) Nods, Nalps and Naip: intracellular regulators of bacterial-induced inflammation. Cell Microbiol 5:581–592

Chandok MR, Ytterberg AJ, van Wijk KJ, Klessig DF (2003) The pathogen-inducible nitric oxide synthase (iNOS) in plants is a variant of the P protein of the glycine decarboxylase complex. Cell 113:469–482

Chege GM, Beier JC (1990) Effect of Plasmodium falciparum on the survival of naturally infected afrotropical Anopheles (Diptera: Culicidae). J Med Entomol 27, 454–458

Chen Z, Kloek AP, Boch J, Katagiri F, Kunkel BN (2000) The Pseudomonas syringae avrRpt2 gene product promotes pathogen virulence from inside plant cells. Mol Plant Microbe Interact 13:1312–1321

Cheong WH, Eyles DE, Warren M, Wharton RH (1963) The experimental infection of *Plasmodium falciparum* in a culicine mosquito *Mansonia uniformis.* Singapore Med J 4:183–184

Christophides GK, Zdobnov E, Barillas-Mury C, Birney E, Blandin S, Blass C, Brey PT, Collins FH, Danielli A, Dimopoulos G, et al. (2002) Immunity-related genes and gene families in *Anopheles gambiae.* Science 298:159–165

Chun J, McMaster J, Han Y, Schwartz A, Paskewitz SM (2000) Two-dimensional gel analysis of haemolymph proteins from *Plasmodium*-melanizing and -non-melanizing strains of *Anopheles gambiae.* Insect Mol Biol 9, 39–45

Chun J, Riehle M, Paskewitz SM (1995) Effect of mosquito age and reproductive status on melanization of sephadex beads in *Plasmodium*-refractory and -susceptible strains of *Anopheles gambiae.* J Invertebr Pathol 66:11–17

Clark AG, Wang L (1997) Molecular population genetics of *Drosophila* immune system genes. Genetics 147:713–724

Cohen GB, Gandhi RT, Davis DM, Mandelboim O, Chen BK, Strominger JL, Baltimore D (1999) The selective downregulation of class I major histocompatibility complex proteins by HIV-1 protects HIV-infected cells from NK cells. Immunity 10:661–671

Cohen J (2002) The immunopathogenesis of sepsis. Nature 420:885–891

Collins FH, Sakai RK, Vernick KD, Paskewitz S, Seeley DC, Miller LH, Collins WE, Campbell CC, Gwadz RW (1986) Genetic selection of a *Plasmodium*-refractory strain of the malaria vector *Anopheles gambiae.* Science 234: 607–610

Collmer A, Badel JL, Charkowski AO, Deng WL, Fouts DE, Ramos AR, Rehm AH, Anderson DM, Schneewind O, van Dijk K, Alfano JR (2000) *Pseudomonas syringae* Hrp type III secretion system and effector proteins. Proc Natl Acad Sci USA 97:8770–8777

Cornelis GR, Van Gijsegem F (2000) Assembly and function of type III secretory systems. Annu Rev Microbiol 54:735–774

Dangl JL, Jones JD (2001) Plant pathogens and integrated defence responses to infection. Nature 411:826–833

Date A, Satta Y, Takahata N, Chigusa SI (1998) Evolutionary history and mechanism of the *Drosophila* cecropin gene family. Immunogenetics 47:417–429

Davis IC, Girard M, Fultz PN (1998) Loss of CD4+ T cells in human immunodeficiency virus type 1-infected chimpanzees is associated with increased lymphocyte apoptosis. J Virol 72:4623–4632

Dawkins R, Krebs JR (1979) Arms races between and within species. Proc R Soc Lond B Biol Sci 205:489–511

De Gregorio E, Spellman PT, Rubin GM, Lemaitre B (2001) Genome-wide analysis of the *Drosophila* immune response by using oligonucleotide microarrays. Proc Natl Acad Sci USA 98:12590–12595

de Groot NG, Otting N, Doxiadis GG, Balla-Jhagjhoorsingh SS, Heeney JL, van Rood JJ, Gagneux P, Bontrop RE (2002) Evidence for an ancient selective sweep in the MHC class I gene repertoire of chimpanzees. Proc Natl Acad Sci USA 99:11748–11753

Dean M, Carrington M, Winkler C, Huttley GA, Smith MW, Allikmets R, Goedert JJ, Buchbinder SP, Vittinghoff E, Gomperts E, et al. (1996) Genetic restriction of HIV-1 infection and progression to AIDS by a deletion allele of the CKR5 structural gene. Hemophilia Growth and Development Study, Multicenter AIDS Cohort Study, Multicenter Hemophilia Cohort Study, San Francisco City Cohort, ALIVE Study. Science 273:1856–1862

Delledonne M, Xia Y, Dixon RA, Lamb C (1998) Nitric oxide functions as a signal in plant disease resistance. Nature 394:585–588

Despres C, DeLong C, Glaze S, Liu E, Fobert PR (2000) The *Arabidopsis* NPR1/NIM1 protein enhances the DNA binding activity of a subgroup of the TGA family of bZIP transcription factors. Plant Cell 12:279–290

Dimopoulos G, Casavant TL, Chang S, Scheetz T, Roberts C, Donohue M, Schultz J, Benes V, Bork P, Ansorge W, et al. (2000) *Anopheles gambiae* pilot gene discovery project: identification of mosquito innate immunity genes from expressed sequence tags generated from immune-competent cell lines. Proc Natl Acad Sci USA 97:6619–6624

Dimopoulos G, Christophides GK, Meister S, Schultz J, White KP, Barillas-Mury C, Kafatos FC (2002) Genome expression analysis of *Anopheles gambiae*: responses to injury, bacterial challenge, and malaria infection. Proc Natl Acad Sci USA 99:8814–8819

Dimopoulos G, Seeley D, Wolf A, Kafatos FC (1998) Malaria infection of the mosquito *Anopheles gambiae* activates immune-responsive genes during critical transition stages of the parasite life cycle. EMBO J 17:6115–6123

Ellis J, Dodds P, Pryor T (2000) Structure, function and evolution of plant disease resistance genes. Curr Opin Plant Biol 3:278–284

Fauvarque MO, Bergeret E, Chabert J, Dacheux D, Satre M, Attree I (2002) Role and activation of type III secretion system genes in *Pseudomonas aeruginosa*-induced *Drosophila* killing. Microb Pathog 32:287–295

Feldmann AM, Billingsley PF, Savelkoul E (1990) Bloodmeal digestion by strains of *Anopheles stephensi liston* (Diptera: Culicidae) of differing susceptibility to *Plasmodium falciparum* . Parasitology 101:193–200

Feldmann AM, Ponnudurai T (1989) Selection of *Anopheles stephensi* for refractoriness and susceptibility to *Plasmodium falciparum*. Med Vet Entomol 3:41–52

Feldmann AM, van Gemert GJ, van de Vegte-Bolmer MG, Jansen RC (1998) Genetics of refractoriness to *Plasmodium falciparum* in the mosquito *Anopheles stephensi*. Med Vet Entomol 12:302–312

Fellowes MDE, Kraaijeveld AR, Godfray HCJ (1998) Trade-off associated with selection for increased ability to resist parasitoid attack in *Drosophila melanogaster*. Proc R Soc Lond B Biol Sci 265:1553–1558

Ferdig MT, Taft AS, Smartt CT, Lowenberger CA, Li J, Zhang J, Christensen BM (2000) *Aedes aegypti* dopa decarboxylase: gene structure and regulation. Insect Mol Biol 9:231–239

Ferrari J, Muller CB, Kraaijeveld AR, Godfray HC (2001) Clonal variation and co-variation in aphid resistance to parasitoids and a pathogen. Evolution Int J Org Evolution 55:1805–1814

Flor HH (1955) Host–parasite interactions in flax-its genetics and other implications. Phytopathology 45:680–685

Girardin SE, Sansonetti PJ, Philpott DJ (2002) Intracellular vs extracellular recognition of pathogens–common concepts in mammals and flies. Trends Microbiol 10:193–199

Gorman MJ, Andreeva OV, Paskewitz SM (2000) Molecular characterization of five serine protease genes cloned from *Anopheles gambiae* hemolymph. Insect Biochem Mol Biol 30:35–46

Gorman MJ, Severson DW, Cornel AJ, Collins FH, Paskewitz SM (1997) Mapping a quantitative trait locus involved in melanotic encapsulation of foreign bodies in the malaria vector, *Anopheles gambiae*. Genetics 146:965–971

Gouagna LC, Mulder B, Noubissi E, Tchuinkam T, Verhave JP, Boudin C (1998) The early sporogonic cycle of *Plasmodium falciparum* in laboratory-infected *Anopheles gambiae*: an estimation of parasite efficacy. Trop Med Int Health 3:21–28

Han YS, Chun J, Schwartz A, Nelson S, Paskewitz SM (1999) Induction of mosquito hemolymph proteins in response to immune challenge and wounding. Dev Comp Immunol 23:553–562

Han YS, Thompson J, Kafatos FC, Barillas-Mury C (2000) Molecular interactions between *Anopheles stephensi* midgut cells and *Plasmodium berghei*: the time bomb theory of ookinete invasion of mosquitoes. EMBO J 19:6030–6040

Hedengren M, Åsling B, Dushay MS, Ando I, Ekengren S, Wihlborg M, Hultmark D (1999) Relish, a central factor in the control of humoral but not cellular immunity in *Drosophila*. Mol Cell 4:827–837

Hoffmann JA (2003) The immune response of *Drosophila*. Nature 426:33–38

Hogg JC, Carwardine S, Hurd H (1997) The effect of *Plasmodium yoelii nigeriensis* infection on ovarian protein accumulation by *Anopheles stephensi*. Parasitol Res 83:374–379

Hogg JC, Hurd H (1995a) Malaria-induced reduction of fecundity during the first gonotrophic cycle of *Anopheles stephensi* mosquitoes. Med Vet Entomol 9:176–180

Hogg JC, Hurd H (1995b) *Plasmodium yoelii nigeriensis*: the effect of high and low intensity of infection upon the egg production and bloodmeal size of *Anopheles stephensi* during three gonotrophic cycles. Parasitology 111:555–562

Hogg JC, Hurd H (1997) The effects of natural *Plasmodium falciparum* infection on the fecundity and mortality of *Anopheles gambiae* s. l. in north east Tanzania. Parasitology 114:325–331

Hopwood JA, Ahmed AM, Polwart A, Williams GT, Hurd H (2001) Malaria-induced apoptosis in mosquito ovaries: a mechanism to control vector egg production. J Exp Biol 204:2773–2780

Huang Y, Paxton WA, Wolinsky SM, Neumann AU, Zhang L, He T, Kang S, Ceradini D, Jin Z, Yazdanbakhsh K, et al. (1996) The role of a mutant CCR5 allele in HIV-1 transmission and disease progression. Nat Med 2:1240–1243

Hughes K, Sokolowski MB (1996) Natural selection in the laboratory for a change in resistance by *Drosophila melanogaster* to the parasitoid wasp *Asobara tabida*. J Insect Behav 9:477–491

Imler JL, Zheng L (2004) Biology of Toll receptors: lessons from insects and mammals. J Leukoc Biol 75:18–26

Inohara N, Nunez G (2003) NODs: intracellular proteins involved in inflammation and apoptosis. Nat Rev Immunol 3:371–382

Irving P, Troxler L, Heuer TS, Belvin M, Kopczynski C, Reichhart JM, Hoffmann JA, Hetru C (2001) A genome-wide analysis of immune responses in *Drosophila*. Proc Natl Acad Sci USA 98:15119–15124

Jiggins FM, Hurst GD (2003) The evolution of parasite recognition genes in the innate immune system: purifying selection on *Drosophila melanogaster* peptidoglycan recognition proteins. J Mol Evol 57:598–605

Johnson JK, Li J, Christensen BM (2001) Cloning and characterization of a dopachrome conversion enzyme from the yellow fever mosquito, *Aedes aegypti*. Insect Biochem Mol Biol 31:1125–1135

Klein TA, Harrison BA, Andre RG, Whitmire RE, Inlao I (1982) Detrimental effects of *Plasmodium cynomolgi* infections on the longevity of *Anopheles dirus*. Mosquito News 42:265–271

Klein TA, Harrison BA, Grove JS, Dixon SV, Andre RG (1986) Correlation of survival rates of *Anopheles dirus* A (Diptera: Culicidae) with different infection densities of *Plasmodium* cynomolgi. Bull World Health Org 64:901–907

Koella JC (1999) An evolutionary view of the interactions between anopheline mosquitoes and malaria parasites. Microbes Infect 1:303–308

Koella JC, Sorensen FL, Anderson RA (1998) The malaria parasite, *Plasmodium falciparum* , increases the frequency of multiple feeding of its mosquito vector, *Anopheles gambiae*. Proc R Soc Lond B Biol Sci 265:763–768

Kraaijeveld AR, Godfray HCJ (1997) Trade-off between parasitoid resistance and larval competitive ability in *Drosophila melanogaster*. Nature 389:278–280

Kraaijeveld AR, Limentani EC, Godfray HCJ (2001) Basis of the trade-off between parasitoid resistance and larval competetive ability in *Drosophila melanogaster*. Proc R Soc Lond B Biol Sci 268:259–261

Kumar S, Christophides GK, Cantera R, Charles B, Han YS, Meister S, Dimopoulos G, Kafatos FC, Barillas-Mury C (2003) The role of reactive oxygen species on *Plasmodium* melanotic encapsulation in *Anopheles gambiae*. Proc Natl Acad Sci USA 100:14139–14144

Lazzaro BP, Clark AG (2001) Evidence for recurrent paralogous gene conversion and exceptional allelic divergence in the Attacin genes of *Drosophila melanogaster*. Genetics 159:659–671

Lazzaro BP, Clark AG (2003) Molecular population genetics of genes encoding inducible antibacterial peptides in *Drosophila melanogaster*. Mol Biol Evol 20:914–923

Lazzaro BP, Sceurman BK, Clark AG (2004) Genetic basis of natural variation in *D. melanogaster* antibacterial immunity. Science 303:1873–1876

Lehmann P (2002) Structure and evolution of plant disease resistance genes. J Appl Genet 43:403–414

Li X, Webb BA (1994) Apparent functional role for a cysteine-rich polydnavirus protein in suppression of the insect cellular immune response. J Virol 68:7482–7489

Lindmark H, Johansson KC, Stoven S, Hultmark D, Engstrom Y, KS (2001) Enteric bacteria counteract lipopolysaccharide induction of antimicrobial peptide genes. J Immunol 167:6920–6923

Luckhart S, Vodovotz Y, Cui L, Rosenberg R (1998) The mosquito *Anopheles stephensi* limits malaria parasite development with inducible synthesis of nitric oxide. Proc Natl Acad Sci USA 95:5700–5705

Luckhart S, Crampton AL, Zamora R, Lieber MJ, Dos Santos PC, Peterson TM, Emmith N, Lim J, Wink DA, Vodovotz Y (2003) Mammalian transforming growth factor beta1 activated after ingestion by *Anopheles stephensi* modulates mosquito immunity. Infect Immun 71:3000–3009

Lyimo EO, Koella JC (1992) Relationship between body size of adult *Anopheles gambiae* s.l. and infection with the malaria parasite *Plasmodium falciparum*. Parasitology 104:233–237

Mackey D, Belkhadir Y, Alonso JM, Ecker JR, Dangl JL (2003) *Arabidopsis* RIN4 is a target of the type III virulence effector AvrRpt2 and modulates RPS2-mediated resistance. Cell 112:379–389

Mackey D, Holt BF, Wiig A, Dangl JL (2002) RIN4 interacts with *Pseudomonas syringae* type III effector molecules and is required for RPM1-mediated resistance in *Arabidopsis*. Cell 108:743–754

McKean KA, Nunney L (2001) Increased sexual activity reduces male immune function in *Drosophila melanogaster*. Proc Natl Acad Sci USA 98:7904–7909

Medley GF, Sinden RE, Fleck S, Billingsley PF, Tirawanchai N, Rodriguez MH (1993) Heterogeneity in patterns of malarial oocyst infections in the mosquito vector. Parasitology 106:441–449

Meyers BC, Kozik A, Griego A, Kuang H, Michelmore RW (2003) Genome-wide analysis of NBS-LRR-encoding genes in *Arabidopsis*. Plant Cell 15:809–834

Miller LH, Mason SJ, Clyde DF, McGinniss MH (1976) The resistance factor to *Plasmodium vivax* in blacks. The Duffy-blood-group genotype, FyFy. N Engl J Med 295:302–304

Moret Y, Schmid-Hempel P (2000) Survival for immunity: the price of immune system activation for bumblebee workers. Science 290:1166–1168

Mou Z, Fan W, Dong X (2003) Inducers of plant systemic acquired resistance regulate NPR1 function through redox changes. Cell 113:935–944

Murphy PM (1993) Molecular mimicry and the generation of host defense protein diversity. Cell 72:823–826

Nappi AJ, Vass E, Frey F, Carton Y (1995) Superoxide anion generation in *Drosophila* during melanotic encapsulation of parasites. Eur J Cell Biol 68:450–456

Niare O, Markianos K, Volz J, Oduol F, Toure A, Bagayoko M, Sangare D, Traore SF, Wang R, Blass C, Dolo G, Bouare M, Kafatos FC, Kruglyak L, Toure YT, Vernick KD (2002) Genetic loci affecting resistance to human malaria parasites in a West African mosquito vector population. Science 298:213–216

Novembre FJ, Saucier M, Anderson DC, Klumpp SA, O'Neil SP, Brown CR 2nd, Hart CE, Guenthner PC, Swenson RB, McClure HM (1997) Development of AIDS in a chimpanzee infected with human immunodeficiency virus type 1. J Virol 71:4086–4091

Ocana-Morgner C, Mota MM, Rodriguez A (2003) Malaria blood stage suppression of liver stage immunity by dendritic cells. J Exp Med 197:143–151

Oduol F, Xu J, Niare O, Natarajan R, Vernick KD (2000) Genes identified by an expression screen of the vector mosquito *Anopheles gambiae* display differential molecular immune response to malaria parasites and bacteria. Proc Natl Acad Sci USA 97:11397–11402

Ohta A, Sitkovsky M (2001) Role of G-protein-coupled adenosine receptors in downregulation of inflammation and protection from tissue damage. Nature 414:916–920

Orr HA, Irving S (1997) The genetics of adaptation: The genetic basis of resistance to wasp parasitism in *Drosophila melanogaster*. Evolution 51:1877–1885

Osta MA, Christophides GK, Kafatos FC (2004) Effects of mosquito genes on *Plasmodium* development. Science 303:2030–2032

Paskewitz S, Riehle MA (1994) Response of *Plasmodium* refractory and susceptible strains of *Anopheles gambiae* to inoculated Sephadex beads. Dev Comp Immunol 18:369–375

Paskewitz SM, Brown MR, Lea AO, Collins FH (1988) Ultrastructure of the encapsulation of *Plasmodium* cynomolgi (B strain) on the midgut of a refractory strain of *Anopheles gambiae*. J Parasitol 74:432–439

Paskewitz SM, Reese-Stardy S, Gorman MJ (1999) An easter-like serine protease from *Anopheles gambiae* exhibits changes in transcript abundance following immune challenge. Insect Mol Biol 8:329–337

Ploegh HL (1998) Viral strategies of immune evasion. Science 280:248–253

Ramos-Onsins S, Aguadè M (1998) Molecular evolution of the Cecropin multigene family in *Drosophila*: functional genes vs. pseudogenes. Genetics 150:157–171

Reddy JT, Locke M (1990) The size limited penetration of gold particles through insect basal laminae. J Insect Physiol 36:397–407

Richman AM, Dimopoulos G, Seeley D, Kafatos FC (1997) *Plasmodium* activates the innate immune response of *Anopheles gambiae* mosquitoes. EMBO J 16:6114–6119

Robert V, Verhave JP, Carnevale P (1990) *Plasmodium falciparum* infection does not increase the precocious mortality rate of *Anopheles gambiae*. Trans R Soc Trop Med Hyg 84:346–347

Rolff J, Siva-Jothy MT (2002) Copulation corrupts immunity: a mechanism for a cost of mating in insects. Proc Natl Acad Sci USA 99:9916–9918

Rosqvist R, Forsberg A, Wolf-Watz H (1991) Intracellular targeting of the Yersinia YopE cytotoxin in mammalian cells induces actin microfilament disruption. Infect Immun 59:4562–4569

Rossignol PA, Ribeiro JM, Spielman A (1984) Increased intradermal probing time in sporozoite-infected mosquitoes. Am J Trop Med Hyg 33:17–20

Rossignol PA, Ribeiro JM, Spielman A (1986) Increased biting rate and reduced fertility in sporozoite-infected mosquitoes. Am J Trop Med Hyg 35:277–279

Schlenke TA, Begun DJ (2003) Natural selection drives *Drosophila* immune system evolution. Genetics 164:1471–1480

Schwartz A, Koella JC (2002) Melanization of plasmodium falciparum and C-25 sephadex beads by field-caught *Anopheles gambiae* (Diptera: Culicidae) from southern Tanzania. J Med Entomol 39:84–88

Shao F, Golstein C, Ade J, Stoutemyer M, Dixon JE, Innes RW (2003) Cleavage of Arabidopsis PBS1 by a bacterial type III effector. Science 301:1230–1233

Sinden RE (1997) Infection of Mosquitoes with Rodent Malaria. In: Crampton JM, Beard CB, Louis C (eds) The molecular biology of insect disease vectors. Chapman and Hall, London, pp 67–91

Sluiters JF, Visser PE, van der Kaay HJ (1986) The establishment of *Plasmodium berghei* in mosquitoes of a refractory and a susceptible line of *Anopheles atroparvus*. Z Parasitenkd 72:313–322

Somboon P, Prapanthadara L, Suwonkerd W (1999) Selection of *Anopheles dirus* for refractoriness and susceptibility to *Plasmodium yoelii nigeriensis*. Med Vet Entomol 13:355

Stahl EA, Dwyer G, Mauricio R, Kreitman M, Bergelson J (1999) Dynamics of disease resistance polymorphism at the Rpm1 locus of *Arabidopsis*. Nature 400:667–671

Stöven S, Silverman N, Junell A, Hedengren-Olcott M, Erturk D, Engstrom Y, Maniatis T, Hultmark D (2003) Caspase-mediated processing of the *Drosophila* NF-kappaB factor Relish. Proc Natl Acad Sci USA 100:5991–5996

Stremlau M, Owens CM, Perron MJ, Kiessling M, Autissier P, Sodroski J (2004) The cytoplasmic body component TRIM5alpha restricts HIV-1 infection in Old World monkeys. Nature 427:848–853

Swiderski MR, Innes RW (2001) The *Arabidopsis* PBS1 resistance gene encodes a member of a novel protein kinase subfamily. Plant J 26:101–112

Tahar R, Boudin C, Thiery I, Bourgouin C (2002) Immune response of *Anopheles gambiae* to the early sporogonic stages of the human malaria parasite *Plasmodium falciparum*. EMBO J 21:6673–6680

Tao Y, Xie Z, Chen W, Glazebrook J, Chang HS, Han B, Zhu T, Zou G, Katagiri F (2003) Quantitative nature of *Arabidopsis* responses during compatible and incompatible interactions with the bacterial pathogen *Pseudomonas syringae*. Plant Cell 15:317–330

Tao Y, Yuan F, Leister RT, Ausubel FM, Katagiri F (2000) Mutational analysis of the *Arabidopsis* nucleotide binding site-leucine-rich repeat resistance gene RPS2. Plant Cell 12:2541–2554

Thomasova D, Ton LQ, Copley RR, Zdobnov EM, Wang X, Hong YS, Sim C, Bork P, Kafatos FC, Collins FH (2002) Comparative genomic analysis in the region of a major *Plasmodium*-refractoriness locus of *Anopheles gambiae*. Proc Natl Acad Sci USA 99:8179–8184

Tian D, Traw MB, Chen JQ, Kreitman M, Bergelson J (2003) Fitness costs of R-gene-mediated resistance in *Arabidopsis thaliana*. Nature 423:74–77

Torres MA, Dangl JL, Jones JD (2002) Arabidopsis gp91phox homologues AtrbohD and AtrbohF are required for accumulation of reactive oxygen intermediates in the plant defense response. Proc Natl Acad Sci USA 99:517–522

Van der Biezen EA, Jones JD (1998) Plant disease-resistance proteins and the gene-for-gene concept. Trends Biochem Sci 23:454–456

van der Kaay HJ, Boorsma L (1977) A susceptible and refractive strain of *Anopheles atroparvus* van Thiel to infection with *Plasmodium berghei berghei*. Acta Leiden 45:13–19

Vaughan JA, Hensley L, Beier JC (1994) Sporogonic development of *Plasmodium yoelii* in five anopheline species. J Parasitol 80:674–681

Vernick KD (1998) Mechanisms of immunity and refractoriness in insect vectors of eukaryotic parasites. In Brey PT, Hultmark D (eds) Molecular mechanisms of immune response in insects. Chapman and Hall, London, pp 261–309

Vernick KD, Collins FH (1989) Association of a *Plasmodium*-refractory phenotype with an esterase locus in *Anopheles gambiae*. Am J Trop Med Hyg 40:593–597

Vernick KD, Collins FH, Gwadz RW (1989) A general system of resistance to malaria infection in *Anopheles gambiae* controlled by two main genetic loci. Am J Trop Med Hyg 40:585–592

Vernick KD, Fujioka H, Seeley DC, Tandler B, Aikawa M, Miller LH (1995) *Plasmodium gallinaceum*: a refractory mechanism of ookinete killing in the mosquito, *Anopheles gambiae*. Exp Parasitol 80:583–595

Vernick KD, Waters AP (2004) Genomics and malaria control. New Engl J Med 351:1901–1904

Warren RF, Merritt PM, Holub E, Innes RW (1999) Identification of three putative signal transduction genes involved in R gene-specified disease resistance in Arabidopsis. Genetics 152:401–412

Weathersby AB, McCall JW (1968) The development of *Plasmodium gallinaceum* Brumpt in the hemocoels of refractory *Culex pipiens pipiens* Linn. and susceptible *Aedes aegypti* (Linn.). J Parasitol 54:1017–1022

Weathersby AB, McCroddan DM (1982) The effects of parabiotic twinning of susceptible and refractory mosquitoes on the development of *Plasmodium gallinaceum*. J Parasitol 68:1081–1084

Wekesa JW, Copeland RS, Mwangi RW (1992) Effect of *Plasmodium falciparum* on blood feeding behavior of naturally infected *Anopheles* mosquitoes in western Kenya. Am J Trop Med Hyg 47: 484–488

Williamson KB, Zain M (1937a) A presumptive culicine host of the human malaria parasites. Transact R Soc Trop Med Hygiene 31:109–114

Williamson KB, Zain M (1937b) A presumptive culicine host of the human malaria parasites. Nature 139:714

Zhang Y, Fan W, Kinkema M, Li X, Dong X (1999) Interaction of NPR1 with basic leucine zipper protein transcription factors that bind sequences required for salicylic acid induction of the *PR-1* gene. Proc Natl Acad Sci USA 96:6523–6528

Zhang Y, Tessaro MJ, Lassner M, Li X (2003) Knockout analysis of Arabidopsis transcription factors TGA2, TGA5, and TGA6 reveals their redundant and essential roles in systemic acquired resistance. Plant Cell 15:2647–2653

Zheng L, Cornel AJ, Wang R, Erfle H, Voss H, Ansorge W, Kafatos FC, Collins FH (1997) Quantitative trait loci for refractoriness of *Anopheles gambiae* to *Plasmodium cynomolgi* B. Science 276:425–428

Zheng L, Wang S, Romans P, Zhao H, Luna C, Benedict MQ (2003) Quantitative trait loci in *Anopheles gambiae* controlling the encapsulation response against *Plasmodium cynomolgi* Ceylon. BMC Genet 4:16

Zimmerman PA, Woolley I, Masinde GL, Miller SM, McNamara DT, Hazlett F, Mgone CS, Alpers MP, Genton B, Boatin BA, Kazura JW (1999) Emergence of FY*A(null) in a *Plasmodium vivax*-endemic region of Papua New Guinea. Proc Natl Acad Sci USA 96:13973–13977

CTMI (2005) 295:417–438

Functional Proteome and Expression Analysis of Sporozoites and Hepatic Stages of Malaria Development

P. L. Blair[1] (✉) · D. J. Carucci[2]

[1]Biology Department, Earlham College, 801 National Road West, Richmond, IN 47374, USA
blairpe@earlham.edu

[2]Foundation for the National Institutes of Health, Natcher Building, 45 Center Drive (3An-44), Bethesda, MD 20892-6300, USA

Abstract An evolution in modern malaria research occurred with the completion of the *Plasmodium falciparum* genome project and the onset and application of novel post-genomic technologies. Corresponding with these technological achievements are improvements in accessing and purifying parasite material from 'hard-to-reach' stages of malaria development. Characterization of gene and protein expression in the infectious sporozoite and subsequent liver-stage parasite development is critical to identify novel pre-erythrocytic drug and vaccine targets as well as to understand the basic biology of this deadly parasite. Both transcriptional and proteomic analyses on these stages and the remaining stages of development will assist in the 'credentialing process' of the complete malaria genome.

Abbreviations

CSP	Circumsporozoite protein
EBA-175	Erythrocyte Binding Antigen-175
EBL-1	Erythrocyte Binding-Like Antigen-1
EBL	Erythrocyte Binding Like protein family
EST	Expressed sequence tags
EEF	Exoerythrocytic forms

LC	Liquid chromatography
LCM	Laser capture microdissection
MS/MS	Tandem mass spectrometry
MSP-1	Merozoite surface protein-1
MTIP	Myosin A tail domain interacting protein
MudPIT	Multidimentional protein identification technology
PFEMP1	*P. falciparum* Erythrocyte Membrane Protein-1
ORF	Open reading frame
PyHEP17	*P. yoelii* hepatocyte erythrocyte protein-17
PyHSP70	*P. yoelii* heat shock protein-70
RAP-1	Rhoptry associated protein-1
SSP2/TRAP	Sporozoite surface protein 2/thrombospondin related adhesive protein
Y2H	Yeast-two-hybrid

1
Introduction to Malaria in the Post-genomics Era

Malaria research has transcended into a new era with the near completion of several *Plasmodium* sequencing projects and the development of post-genomic technologies [1–8]. Such modern advances allow researchers to understand and characterize the underlying mechanisms represented in the Central Dogma of Molecular Biology as it pertains to malaria parasite. Therefore, scientists now have the capability of exploring the dynamic *Plasmodium falciparum* life cycle probing within its genome, transcriptome, and proteome [1, 4–7, 9–13]. The completion of the *P. falciparum* (3D7 clone) genome sequencing project in 2002 is the basis and primer for a broad range of future experimental designs in the malaria research community for years to come. However, the contributions of the complete genome sequence are imperfect in their inability to discover biological function, expression, and relevance of the more than 5,200 predicted genes and their encoded products. With improved bioinformatics and database management required to grasp genomic-scale datasets, researchers have already developed and applied technologies for analyzing malaria stage-specific gene expression/regulation, protein expression/interactions, and for identifying novel vaccine targets [2, 14–17].

Currently, most of the *P. falciparum* developmental stages in the human host (erythrocytic stages, merozoites, sporozoites and gametocytes) are accessible and have been analyzed in large-scale expression studies. Less information is available on the more 'hard-to-reach' sexual- and liver-stage forms due to the inability routinely to purify and enrich significant amounts of nucleic acid and protein material. Together the antigens expressed in the sporozoite and hepatic stages are of interest to new vaccine target hunters [18].

A comprehensive characterization of the parasite sporozoite and liver-stage transcriptome and proteome provides a directed approach towards candidate vaccine antigen discovery rather than sifting through the entire genome on a gene-by-gene basis. An efficacious and protective vaccine against these stages focuses on eliminating the parasite prior to the clinical manifestations associated with a fulminating blood-stage infection. This chapter focuses on the past and present gene and protein expression methodologies being used to study the invasive sporozoite and ensuing liver-stages of pre-erythrocytic development.

2
Advances in Genomic and Proteomic Technologies

Malaria research is now entering a juncture where high-throughput genomic technologies are beginning to compensate for the vast information generated by the various sequencing projects [2, 3, 15, 19, 20]. Since classical genetics approaches are restricted when applied to studying malaria parasites, alternative methods for the examination of gene function and expression must be sought. Prior to the final annotation and publication of the malaria genome large-scale gene expression datasets were generated using DNA microarrays and sequencing projects examining expressed sequence tags (EST), gene sequencing tags, and serial analysis of gene expression analysis [21–25]. These methods all proved successful in offering a promising glimpse of malaria gene expression focused primarily on the routinely synchronized and cultured asexual blood-stages. A shotgun DNA microarray printed with 3,648 random inserts (representing a predicted 40% of the genome) from a mung bean nuclease genomic library and probed with cDNA from trophozoite and late stage gametocytes indicated significant differences in gene expression during these two developmental stages [23]. In another study, a microarray spotted with 944 amplified EST clones (representing approximately 18% of the genome) was independently hybridized with cDNA enriched from five periodic timepoints spanning asexual development [25]. The results showed a highly coordinated stage-dependent regulation of transcripts encoding functionally similar products. Ring and trophozoite stage parasites showed an up-regulation of transcripts associated with translational machinery components as the parasite prepares for the increased rate of protein synthesis characteristic of trophozoite metabolism. Moreover, maturing schizonts increased gene expression of adhesion/ligand molecules related with the apical complex and functionally important for merozoite re-invasion into new erythrocytes. These groundbreaking studies verified and validated the extensive potential for using microarrays for expression profiling the complete *P. fal-*

ciparum life cycle as well as demonstrating the tightly controlled regulation of malaria gene expression. However, due to the inaccessibility of certain developmental stages, technological limitations, and incompleteness of the concurrent malaria genome project, missing from these initial studies were: (1) the characterization of gene expression in sporozoites, hepatic stages, and the sexual stages in the invertebrate host; (2) a complete genome-wide survey of expression; and (3) the ability to correlate transcript expression levels to that of protein expression and/or function.

Fortunately, it has not taken long to address each of these issues. Malaria research has moved to the forefront in applying post-genomic technologies on a parasitic disease partially by leveraging from advanced methods and achievements used to characterize expression profiles in yeasts. Upon completion of the *Saccharomyces cerevisiae* genome in 1996, and within a year, an Affymetrix high-density oligonucleotide array was used to profile the expression of the complete genome [26]. A proposed goal of this study was to apply this methodology to the monitoring of genome-wide expression in a more complex organism and the malaria parasite was subsequently used as a candidate.

In a similar timeframe, within a year following the completion of the *P. falciparum* 3D7 genome, an expression profile was obtained that targeted several stages of malaria development also utilizing the Affymetrix gene chip [9, 10]. The custom designed *Plasmodium* gene chip used in this study consisted of over 500,000 probes (25-nucleotide single-stranded probes) including 260,596 from predicted coding sequences and 106,630 probes from predicted noncoding regions. Designing the chip in this manner achieved approximately one probe located every 150 bp across the 22.8 Mb *P. falciparum* genome with multiple probes within each annotated open reading frame (ORF). This feature density promotes increased confidence in expression levels as each gene's expression value is generated from numerous probes and is not dependent on the binding and signal intensity of a single oligonucleotide probe. After subtracting background signal intensities, normalizing, and comparing each feature to its representative gene, the expression values were generated. In this array, low levels of expression are generally considered <100, moderate levels between 100 and 1,000, and very high expression levels >1,000. Such expression analysis determined transcript levels for more than 95% of the predicted *P. falciparum* genes by hybridizing cDNA synthesized from synchronized blood-stage, merozoite, gametocyte, and sporozoite samples. This study represents the first microarray based stage-specific expression analysis and generation of a genome-wide inclusive dataset for *Plasmodium* sporozoites. Moreover, a cluster analysis method was applied to empirically assign function to genes without known function or homology based on similar expression profiles with genes of characterized function. Expression data for

2,235 genes, the majority having a predicted unknown function, were grouped into 15 clusters using a robust k-means clustering algorithm. The approach was further validated by demonstrating that genes of known function comprising each cluster often had similar biological roles and/or functions. For example, cluster 15 included genes upregulated in late stage schizonts known to be involved with merozoite invasion and located on the surface of merozoites and within the parasite apical complex.

Microarray and gene chip technology has been shown and will continue to be exploited as an asset to the malaria research community. Recently, a long oligonucleotide microarray was used to examine thoroughly the complete genome expression over the *P. falciparum* erythrocytic cycle sampling 48-h timepoints and hybridizing each to the array [11]. Taking this vast amount of data with the large-scale EST datasets and other microarray data available gene expression profiling during *P. falciparum* blood-stages has been comprehensively explored. With the current chip technology and the advent of the next generation of custom designed chips more complex biological questions can be addressed including examining variances in gene expression across several *Plasmodium* species, strains, and clones, identifying crucial virulence loci, characterizing drug sensitive/resistant phenotypes, host response to infection, and responses to stressed growth conditions. Additionally, with improved purification and enrichment techniques the expression profiles of the currently inaccessible malaria developmental stages can be obtained and examined thus completing expression monitoring of the entire malaria life cycle.

Advances in yeast post-genomic technologies have similarly been effectively applied to malaria in the field of proteomics [27–29]. Since gene transcript levels and patterns do not always correlate with protein expression and abundance it is important to identify the protein complement of a cell, in this case, the malaria parasite. Modern advances in protein analysis were previously applied to characterize the entire yeast proteome using a novel and essentially unbiased gel-free chromatography method linked with tandem mass spectrometry. By identifying more than fivefold more *S. cerevisiae* proteins than any other previous study, this technique proved its worth [28]. The multidimensional protein identification technology (MudPIT) system combines microcapillary two-dimensional liquid chromatography (LC) coupled with tandem mass spectrometry (MS/MS). The high-throughput MudPIT method is capable of resolving proteins from complex mixtures originating from variant subcellular localizations with minimal sample handling as well as identifying post-translational modifications [30]. Additionally, MudPIT overcomes several limitations associated with 2D gel-based proteomic studies including the rare detection of proteins that are of low abundance, insoluble, and/or have an extreme pI or molecular weight. Briefly, lysed protein samples are separated into soluble and insoluble fractions and subsequently treated with selected proteases or cyanogen bromide and formic acid plus proteases, respectively. The treated samples are loaded on a microcapillary column, eluted under increasing salt gradients across voltage, and directly subject to MS/MS. The resulting MS/MS spectra are then cross-referenced to predicted in silico digested peptides from the corresponding genome database using the SEQUEST[TM] algorithm. By this means, thousands of spectra can rapidly be analyzed resulting in the identification of the protein complement represented in both the soluble and insoluble fractions.

MudPIT analysis was performed on mixed protein samples isolated from various accessible stages (sporozoite, merozoite, trophozoite, and stage III–V gametocytes) of malaria development [12]. The resulting MS/MS spectra were compared to predicted translated products in the malaria genome database. Experiments were carefully controlled to discount potential host (human and mosquito) contamination by additionally running the SEQUEST[TM] algorithm with all spectra (including noninfected controls) against the human and mosquito genome datasets. Nearly half (2,415 proteins) of the predicted expressed annotated genes present in the genome were identified. Interestingly, more than 21% (513) were proteins solely expressed in sporozoites including known cell surface proteins, proteins associated with the apical complex, and intriguingly protein families known to be involved in blood-stage antigenic variation. More in depth descriptions of the sporozoite proteome will be discussed in Sect. 3. As expected, the merozoite proteome identified numerous proteins associated with the erythrocyte invasion process and the apical complex, notably three members of the erythrocyte binding-like (EBL) family, EBA-175, BAEBL, and MAEBL. Several merozoite surface proteins (MSP-1, 2, 3, 6, and 7) were also detected with the abundant MSP-1 obtaining extensive sequence coverage. Sequence coverage is an approximate estimator of protein abundance and represents the percent of the predicted annotated protein sequence spanned by the peptides identified in MS analysis. Trophozoite peptides showed an expected abundance of proteins involved in protein synthesis and also proteinases implicated in hemoglobin degradation and metabolism. Late-stage gametocyte samples identified several previously characterized gametocyte antigens including over 64% coverage of Pfg27/25, over 48% coverage of Pfs16, in addition to several proteins linked to gametocyte-specific transcription machinery.

Similar to the results generated from gene expression profiling, the MudPIT proteome analysis determined a tightly regulated pattern of stage-specific protein expression. Less than 7% (152 proteins) of the total proteins identified were found to be constitutively expressed across all stages tested compared to 51% as resolved by the transcriptome analysis. However, whereas genome transcription profiling experiments offer expression data for each annotated gene, MS analysis identifies a representative sampling of the total protein complement and therefore is not exhaustive. It is expected that a certain number of proteins may have been missed in these initial proteomic trials and with additional experiments currently underway and by sampling from additional parasite stages a more comprehensive malaria proteome profile will be compiled. An independent *P. falciparum* proteome project using MS (nanoLC/MS/MS) identified 1,289 proteins from blood-stages, gametocytes, and gametes was published in parallel to the MudPIT analysis [1]. Examining the data from both of these novel malaria proteome projects along with acquisition of the next generations of proteome data will assist in completing the *P. falciparum* proteome profile.

Finally, advancements in yeast-two-hybrid (Y2H) technologies offer an appealing system for characterizing both malaria parasite–parasite and parasite–host protein interactions. Y2H has been successfully applied to malaria parasites to study particular parasite interactions [31–36]. Current Y2H approaches include high-throughput methods for identifying interactions with more reliability by decreasing the incidences of false positives (R. Hughes, Prolexys Pharmaceuticals, personal communication). Briefly, the construction and normalization of plasmid libraries facilitates the generation of both large-scale baits (fused with the DNA binding domain) and preys (fused with the transcriptional activation domain). A successful prey–bait protein interaction results in the mating of the two yeast haploid forms that originated from converse mating types. The interaction induces active transcription of a reporter gene resulting in selected yeast growth only in the presence of a prey–bait protein interaction. This technology is already being performed on malaria parasites to identify *P. falciparum* parasite–parasite interactions using an asexual-stage cDNA library as a prey (S. Fields, University of Washington, personal communication). Already this method has proven effective in identifying thousands of putative protein interactions (from >18,000 searches) among blood-stage encoded products. Additionally, this application provides researchers the means to discover novel parasite–host (human and mosquito) interactions that up until now have only been indirectly studied. One enticing application for this technology is to characterize interactions between sporozoites and liver-stage parasite proteins with the host liver proteins. In this manner, putative receptor–ligand interactions utilized by the sporozoite for invasion, as well as intracellular interactions occurring between the internalized parasite and its liver host cell could be identified. Furthermore, Y2H provides the opportunity for exploring each of the host–parasite interactions attributed to an invasion process: the ookinete crossing the mosquito midgut epithelium; the sporozoite entering the salivary glands; the merozoite into erythrocytes; and the aforementioned sporozoite invasion into liver cells. Y2H effectiveness could be demonstrated by constructing a high quality parasite cDNA library or by individual cloning of a selected number of prioritized malaria genes to use as baits and the creation of host cDNA libraries as potential preys.

In order to exploit Y2H efforts and other post-genomic technologies, optimized high-throughput recombinant cloning strategies will be required to clone hundreds of P. falciparum genes of interest for facilitating a rapid workflow and transition into numerous downstream genomic applications [72]. The recombinational cloning technology uses the phage recombination sites to directionally clone genes into an entry vector (BP reaction) followed by the efficient shuttling of inserts into vectors (LR reaction) always maintaining the ORF and orientation. This streamlined technology provides a high-throughput method of cloning malaria genes for downstream applications including protein expression, DNA vaccine production, yeast two hybrid studies, knockout vectors, structure analysis/crystallization, and others.

Approximately 300 P. falciparum ORFs have been cloned into the Gateway® entry vector (pDONR207 and/or pDONR/ZEO). A two-step PCR method using gene-specific primers in the initial PCR followed by a second PCR with adapter primers containing attB1 and attB2 recombination sites prepares the PCR amplicon for efficient cloning into the entry vector. The prioritized ORFs can be selected for cloning based on their stage-specific large-scale proteome and transcriptome expression profiles and through the use of bioinformatics tools available on the PlasmoDB website (http://plasmodb.org) that will be discussed in later sections. Now, in a manner of weeks it is feasible to amplify hundreds of ORFs from a DNA or cDNA template, clone them into an entry vector, shuttle inserts into several destination vectors simultaneously in a 96-well format in preparation for several functional assays. Entry clones can be made available to the research community through public resource repositories (i.e., MR4: http://www.malaria.mr4.org/). A caveat to this approach for the full complement of 5,300 genes is that the present genomic annotation may not accurately predict start and stop codons in multiple exon genes.

3
The *Plasmodium* Sporozoite Proteome

Compared with the other stages of malaria development examined, the MudPIT analysis of P. falciparum sporozoites arguably gave the most intriguing findings [12]. Prior knowledge of the protein complement of the sporozoite stage was miniscule compared with the 1,049 identified using MS. Known and well characterized sporozoite surface proteins were among those identified including the abundant circumsporozoite protein (CSP) and sporozoite surface protein 2 (SSP2/TRAP). Two peptide spectra were found for the major surface protein CSP including a peptide located at the amino-terminal end of the distinctive NANP repeat region and the other sited further downstream in the thrombospondin type 1 domain. Together the two peptides covered 10.1% of the complete CSP protein sequence. Furthermore, SSP2 gave 24 spectra matches of varying sequence and charge state with a notable 36.1% sequence coverage. The majority of the SSP2 peptide matches fell within the hydrophobic von Willebrand factor type A domain in the amino-terminal end of the annotated protein as well as a few peptide complements situated within its own thrombospondin type 1 domain. The expected presence of these two proteins assisted in validating the LC/MS/MS approach to characterizing the sporozoite proteome. However, it was the unforeseen and novel findings that have forced the malaria research community to rethink once accepted dogmas of malaria biology.

First, the sporozoite proteome identified numerous proteins associated with the parasite apical complex that were thought to be predominantly functionally active in erythrocytic stages of development. Proteins that sequester to the apical organelles are often associated with merozoite invasion into host erythrocytes with many of these proteins grouped into distinct families based on sequence homologies and putative function. Intriguingly, the sporozoite proteome includes several members of such protein families that have been described as potential blood-stage malaria vaccine candidates. For example, peptides of two EBL family members were identified representing EBL-1 and MAEBL [37]. MAEBL is known to be a unique member of the EBL family due to its chimeric gene structure, regulated expression patterns including known sporozoite stage expression, and implications in being involved with salivary gland infectivity so it is not surprising that MAEBL was found to be in the sporozoite protein complement; however little is known about EBL-1 to explain sporozoite expression [38–42]. Although screening of an EST library provided evidence that EBL family member EBA-175 is expressed in sporozoite stages, neither the proteome nor the transcriptome (expression value=0) confirmed those results [43]. The Apical Membrane Antigen-1, which shares sequence homology with MAEBL, a vaccine target due to its predicted role in the merozoite invasion process, was identified with five MS (SpecCount=12) in sporozoite samples (14.5% sequence coverage). In accordance with the proteome results, both *MAEBL* and *AMA-1* were highly expressed in the sporozoite transcriptome with values of 1513.4 and 343.8, respectively. In fact, the sporozoite microarray expression values for *MAEBL* and *AMA-1* are higher than any blood-stage level (nearly fivefold higher for *MAEBL*). These corroborating results suggest that these putative blood-stage ligand molecules may have an important functional role in sporozoites that needs further exploration [73]. Similarly, peptides matching members of both the cytoadherence-linked asexual protein and high molecular weight rhoptry (RhopH) protein families whose members are suggested to be involved in adhesion and erythrocyte selection were identified in sporozoite samples (PFC0120w, PFI1730w, PFI0265c) [44–47]. Finally, Rhoptry Associated Protein-1 (RAP-1, PF14_0102) shown to form a complex with at least two additional RAP proteins and is a vaccine target as antibodies inhibit merozoite invasion, was identified by two spectrum matches in sporozoite samples [48]. Neither RAP-2 nor RAP-3, the remaining members of the RAP complex, were found expressed in the sporozoite proteome.

It is evident that additional experiments are required to validate sporozoite expression and potential functional roles for the majority of cases mentioned above perhaps on an individual basis. This is especially the case when confirming proteins known to be functionally associated as a complex like the RAP and RhopH protein complexes. With the exception of *AMA-1* and *MAEBL*, the proteome analysis for the other proteins described does not agree with the transcriptome results since transcript expression scores in sporozoites were minimal (<30). Consequently, one might argue that since only a single (or very few) unique mass spectra were identified for many of these putative sporozoite proteins and that not all members of a known protein family or complex were identified, perhaps the results are merely artifacts. However, the MudPIT proteome methodology earns its elegance in that if a peptide is detected then the associated protein is expressed at that particular stage of parasite development with high confidence and probability. There exists no chance for contamination with other stages in the sporozoite samples and unlike when working with the transcripts, RNA turnover is not a concern or disadvantage.

Moreover, although unbiased in many regards, the LC/MS/MS proteome approach may be somewhat dependent upon the amino acid makeup of a protein in regards to frequency of trypsin cleavage sites (L. Florens, personal communication). Therefore, determining protein abundance based on number of peptide matches and/or sequence coverage may be restricted in certain cases. For example, as mentioned previously, CSP and SSP2 are two abundant sporozoite proteins with MudPIT analysis identifying only two unique peptides for CSP and a robust 24 for SSP2. Even though these identified peptides resulted in a relatively high sequence coverage for these proteins, 10.1% for CSP and 36.1% for SSP2, finding only two peptides for CSP was somewhat surprising. Perhaps these two CSP tryptic peptides might have been found numerous times thereby providing a foundation for an argument for increased CSP expression, but this is not the case. Spectral analysis of the raw proteome data indicates that only three spectra were generated that matched to these CSP peptides. In contrast, a significantly greater 114 spectra were produced to create the 24 peptides for SSP2. Only two other loci (PFL1725w and PF10_0084) in the entire sporozoite proteome resulted in higher number of spectral matches than SSP2. The sporozoite transcriptome results achieved more realistic and accepted expression values for CSP (26,684.1) and SSP2 (12,475.8) with the CSP expression value higher than any other *P. falciparum* locus in any other stage. Why then is CSP apparently under-represented in the sporozoite proteome? It can be speculated that the amino acid composition of proteins plays a role in the downstream application of using LC/MS/MS on trypsin digested fragments of *Plasmodium* proteins. Trypsin cleavage is dependent on the presence of arginine and lysine and the charge of flanking residues. *P. falciparum* 3D7 CSP appears limited in predicted cleavage sites with relatively few arginine (6) and lysine (22) residues. In comparison, SSP2 has 30 arginine and 39 lysine residues. Furthermore, the arginines and lysines that are present in CSP are flanking the internal NANP repeat region that spans 38% of the entire protein and thus may interfere with proper tryptic digestion. Therefore, the inherent amino acid makeup for CSP might interfere with the ability to quantify protein abundance from the MudPIT MS results. This latter analysis is merely an initial hypothesis in an attempt to explain the data in this specific case, where obviously further confirmation is required to substantiate this claim. Finally, it may be useful in proteomic studies to take into account an organism's amino acid composition and codon usage bias related to the number of predicted tryptic peptide cleavage sites when undergoing MudPIT analysis.

Perhaps the most striking and unexpected finding in the *P. falciparum* 3D7 sporozoite proteome was the identification of peptide spectra corresponding to proteins of the *var* and *rif* multigene families both hitherto dogmatically associated only with malaria asexual-stage development. The extensively studied *var* gene family encodes for the PFEMP1 molecule that is involved with antigenic variation, sequestration, and rosetting in blood-stage parasites. The *P. falciparum* 3D7 haploid genome contains 59 *var* gene loci predominantly localized to subtelomeric regions on all 14 of the chromosomes [4]. The sporozoite proteome identified the encoded products of 25 *var* genes and 21 different *rifin* gene products. Perhaps even more fascinating is the apparent stage-specific expression of these identified proteins. Of the 37 total *var* products identified by MudPIT analysis in the examined *P. falciparum* stages, 21 were stage specific.

Only 10 of the sporozoite expressed *var* genes and 2 of the *rifin* genes were found expressed in blood-stages. Sporozoite transcriptome analysis of *var* gene family expression correlated with sporozoite proteome data indicating 16 (out of the 25) with expression values >50. Secondly, additional confirmation of *var* gene expression in sporozoites arises from a *P. falciparum* sporozoite cDNA sequencing project (J. Aguiar, personal communication) that identified five clones representing *var* genes. Interestingly, four of these genes (PFC1120c, PF08_0107, PF13_003, and PFI0005w) were *var* genes expressed in both the proteome and the transcriptome. The remaining gene (PFL0935c) was confirmed in the transcriptome (expression value=78.1), yet not found by peptide matches in the proteome. Therefore, expression analysis of *P. falciparum* 3D7 sporozoites using three large-scale datasets (MudPIT MS, Affymetrix gene chip profiling, and cDNA library sequencing) are in agreement in regards to *var* gene and its encoded product (PFEMP1) expression.

The proteome project also introduced the concept that malaria parasites use a stage-specific mechanism of coordinated co-expression of genes adjacent to or in close proximity to each other along the chromosome [12]. The sporozoite expression of the members of the *var* multi-gene family also appears to reflect this discovery. The majority (15 out of 25) of the *var* loci found to be expressed via MudPIT analysis also had peptides identified from one or more neighboring *var* genes. For example, *P. falciparum* 3D7 chromosome 8 has two clusters of *var* gene loci, one telomeric cluster containing three *var* genes and one internal region with three *var* genes. The proteome determined that seven of these *var* genes were expressed as PFEMP1 proteins, indicating that both clusters may be co-expressed. Chromosome 4 possesses 12 *var* genes in four clusters, yet tryptic peptides only identified encoded products of three adjacent *var* genes staking a claim that this cluster is being co-expressed based on chromosome position while the remaining clusters were silenced. This pattern of expression differs from classic antigenic switching where only a single PFEMP1 molecule is expressed on the surface of infected erythrocytes. The 21 *rif* gene family members identified in the proteome project are predominantly sporozoite stage-specific with examples of co-expressed regions along the chromosome, though not as frequently as the *var* genes. Three adjacent *rif* gene loci (PF14_004, PF14_003, and PF14_004) located in the telomeric region of chromosome 14 each had peptides identified by MudPIT and significant transcript levels (expression values >50) in gene chip analysis in sporozoite samples with no peptides found in other stages of development. Functional relevance for *var* and *rif* expression and co-regulation in the nonreplicating sporozoite has not been explored at this time, but given the importance of the expressed PFEMP1 molecule and the unknown role of *rif* encoded proteins, further characterizations of these genes will be unquestionably examined in the near future.

Furthermore, the overall findings of the *P. falciparum* sporozoite proteome project will fuel the malaria scientists' 'think tanks' and future experiments for some time. The unique repertoire of protein and gene expression that characterizes the sporozoite stage of development complements the biological requirements and survival mechanisms needed to transition from mosquito to human hosts. Of the 513 proteins solely identified in sporozoite stages, the majority have no known function or sequence homology to other organisms suggesting that a prioritized and systematic approach is needed to characterize their functional role. In addition, as described above, several of the proteins of known function identified in the sporozoite proteome were originally only thought to be functionally relevant in blood-stage development. Thus, malaria researchers need to re-evaluate the functional role of these partially characterized proteins in all stages of malaria development as each stage becomes readily accessible. We now possess volumes of data on the expression profiles of the malaria sporozoite and need only harness this information and develop innovative ways to verify experimentally and assign biological function.

4
Accessing the *Plasmodium* Liver-Stage Proteome

Recent work has explored the migration and invasion process of *Plasmodium* sporozoites into hepatocytes [49–55]. However, gene and protein expression analysis during sporozoite invasion and subsequent hepatocyte development is hampered by the inaccessibility to study this stage of parasite development. Similarly, while gene and protein expression in sporozoites has now been achieved as described above, the ability to access and purify malaria liver-stages is more restricted. With intracellular liver-stage parasites it remains difficult to purify sufficient numbers of infected hepatocytes from the enormously more abundant uninfected hepatocytes in order to extract parasite nucleic acid and protein material to have adequate signal to noise ratios. Additionally, in vitro culture systems are limited in their utility since only partial development can be achieved currently [56]. However, novel techniques and observations have stimulated a push in understanding malaria liver-stage gene expression.

Current efforts in characterizing malaria liver-stage development have successfully stemmed from the use of the rodent malaria model systems of *P. yoelii* and *P. berghei* [57–59]. These model systems circumvent the need for infected human or monkey liver samples in studying human malaria liver forms. Nearly 20 years ago, cultivation of *P. berghei* in a human hepatoma (HepG2) cell line validated an in vitro method for liver stage development [59]. Briefly, *P. berghei* sporozoites were used to inoculate HepG2 cells and complete malaria exoerythrocytic development was observed resulting in successful induction of erythrocytic infection. Furthermore, ultrastructural studies using electron microscopy demonstrated that the in vitro developmental parasite morphology complemented the morphology visualized from in vivo samples [57]. Unfortunately, although *P. falciparum* and *P. vivax* sporozoites invade HepG2 cells, further development for these human malarias is prematurely arrested. We are currently optimizing a similar in vitro culturing system with *P. falciparum* using HC-04 human hepatocyte cell line. Initial studies are hopeful for increased invasion rates and complete development reaching merozoite release and subsequent red blood cell invasion. If we can feasibly and efficaciously purify infected hepatocytes from those uninfected, expression studies could be used to profile gene and protein expression from initial time points immediately after sporozoite invasion until schizont maturation just prior to merozoite release and from sporozoites attenuated by irradiation. Until then, the rodent model systems will be responsible for 'wetting our appetites' in the comprehension of mammalian malaria liver-stage development.

The first large-scale dataset for analyzing gene expression during *Plasmodium* exoerythrocytic development utilized laser capture microdissection (LCM) techniques to target and extract *P. yoelii* parasite material for the construction and subsequent sequencing of a malaria liver-stage EST library (Sacci et al., unpublished results). LCM was applied to 40-h *P. yoelii* exoerythrocytic schizonts preserved as cryosections. The liver schizonts were purified with minimal contamination from the invaded hepatocyte. Approximately 1,500 schizonts were meticulously isolated, parasite RNA was extracted, and a cDNA library was constructed. Sequencing of 2,628 of the *P. yoelii* EST clones identified 1,662 sequences with significant homologies to known sequences by BLASTN analysis. Validation of the purity of the LCM parasite sample was achieved upon the same sequence analysis where only 1% of the total sequences (16 out of 1,346) shared homology to mouse derived sequence databases. Since the 40-h schizonts used in this study are considered to be late in exoerythrocytic development, it was not surprising that there was significant overlap (40%) with known blood-stage *P. yoelii* ESTs. However, more interest will be focused on the >600 sequences not found in other parasite stage databases possibly identifying novel liver stage-specific genes. Of course, this subset of sequences may be co-expressed in additional malaria stages not examined in this study—notably gametocyte, and stages within the invertebrate host. Nonetheless, these data corroborate the findings of the proteome and transcriptome studies by indicating a tightly regulated stage-specific profile of gene expression for malaria parasites within the liver. Also in parallel with previous large-scale expression datasets, a majority (58%) of the sequences was classified as having unknown function based on sequence similarity. As expected for late-stage *P. yoelii* schizonts preparing for extensive asexual division parasites are expressing genes associated with protein synthesis and nuclear regulation. Several known liver-stage expressed and previously characterized genes were identified (*P. yoelii* hepatocyte erythrocyte protein 17–PyHEP17 and *P. yoelii* heat shock protein-70–PyHSP-70) that further supports the validity of the EST library. Additionally, several genes that flank each other in chromosomal position appear to be co-expressed in liver development supporting the concept that malaria co-regulates regions of gene expression in a stage-specific manner. Finally, by filtering and sifting through the sequences in this malaria liver-stage library and upon further characterization of genes of unknown function conceivably a better understanding of the parasite biology may be elucidated. Perhaps the most cogent place to begin in order to identify novel vaccine targets for human malaria would be the 359 *P. yoelii* sequence orthologs present in the *P. falciparum* genome.

In one recent study, evidence suggests that accession into the malaria pre-erythrocytic liver stages may be achieved in the absence of host hepatocytes [60]. In this experiment, *P. berghei* sporozoites expressing green fluorescent protein were used to infect standard HepG2 in vitro cultures. Microscopic observations noticed sporozoites transforming into exoerythrocytic forms extracellularly after 24 h in culture at 37°C. Transformation was initiated by the formation of a nucleus containing a bulbous extension followed by a 'rounding up' into a spherical shape characteristic of hepatic trophozoites. Furthermore, cell-free sporozoite cultures show transformation into exoerythrocytic forms (EEF) that were more rapid than in development within HepG2 cells, yet EEF size and morphology were similar. Additionally, temperature (37°C) and the presence of serum were proven to be essential components for the generation of EEFs. Protein expression (using CSP, TRAP, MSP-1, Hep17, HSP70 and MTIP) and transcript presence the characteristic transition from stage-specific sporozoite (S-type) to asexual (A-type) rRNA transcripts [61, 62] verified similarities between documented infected hepatocyte gene and protein expression with that of EEFs generated in this cell-free system. A similar system for other *Plasmodium* species, including *P. falciparum* has yet to be established. Large-scale characterization and profiling of gene expression via MS, microarray technology, and/or cDNA library sequencing projects on these transformed EEF parasites could enormously expand our understanding of malaria liver-stage biology and aid in identifying novel vaccine and drug targets.

5
Exploring the Proteome Using Bioinformatics Tools

The vast amount of expression and other large-scale data generated by current high-throughput technologies is a valuable resource to the malaria research community now and for years to come. The laboratories producing these data often present the published data as a tool allowing additional scientists to examine the results and plan further experiments based on personal interests. Thus, trained bioinformaticians have become a beneficial commodity to the field, yet not all malaria research laboratories have access to these individuals or the proper training themselves to exploit the extensive datasets that represent the opening of 'Pandora's Box' to the malaria research community. To this end, the resulting large-scale datasets described herein are or will be available in the constantly evolving and public *Plasmodium* on-line database, PlasmoDB (http://plasmodb.org) [63–65] at the University of Pennsylvania.

PlasmoDB allows for users to actively mine the genomic and proteomic datasets offering the ability to perform individualized queries on a user-friendly interface. Numerous additive and/or subtractive Boolean queries to isolate clusters of genes meeting the desired criteria can assist in down-selecting targets in order to answer your specific biological question. Once a query is submitted the data output is obtained routinely less than 1 s later. Therefore, in a matter of minutes and by starting from a sample size encompassing every annotated gene in the *P. falciparum* genome such queries yield manageable and prioritized gene lists suited for initial experimentation.

For example, focusing on the MS malaria proteome results, one might be interested in *P. falciparum* genes potentially encoding proteins likely to be expressed solely in sporozoites and on the surface of the parasite, a category of genes possibly involved in hepatocyte invasion. The following query would assist in down-selecting from the 1,025 annotated genes identified in the sporozoite proteome present in PlasmoDB version 4.1. Performing an initial intersecting Boolean search for genes present in the sporozoite proteome and possessing a predicted signal peptide (SignalP) yields 122 genes. However, since we are interested in sporozoite-specific expression we need to remove genes identified in both sporozoite and other developmental stages. Fortunately, PlasmoDB keeps a temporary history of queries so that you may perform numerous queries without losing information and then combine them in three ways via union, intersection, and subtraction prompts. A second Boolean query determining all genes encoding proteins in gametocyte, or merozoite, or trophozoite stages (union) found 1,888 genes. Now, by taking advantage of the query history, we can subtract common genes from our initial search from the latter search to determine sporozoite-specific genes containing signal peptides. This result produced 72 genes, a far better starting sampling size than an entire genome. Even further down-selection to 21 genes can be achieved if you add genes with predicted transmembrane domains (TMPRED, minimum=2, maximum=20) in the initial search prior to subtraction. Scanning through this list for genes previously known and expressed on the sporozoite surface often helps in validating the search. In this search, both CTRP (circumsporozoite thrombospondin related protein) and SSP2 were identified among 10 genes with predicted function, leaving 11 genes with no predicted function ready to be studied. This is just one simple example based on stage-specific protein expression for the uses of the

PlasmoDB query engine. For a list of all possible query subjects visit the following weblink: http://plasmodb.org/plasmodb/servlet/sv?page=genesb. Updated versions obtainable on CD provide access to researchers with restricted Internet access.

Bioinformatics using PlasmoDB has proven successful in jumpstarting several research endeavors and will routinely continue to do so as more laboratories harness its worth and efficacy. PlasmoDB strives to maintain a current forum for providing malaria data to the malaria public as more large-scale datasets are being produced [65]. Comparable genomics will become more feasible as *P. yoelii*, *P. vivax* and other *Plasmodium* sequencing projects are completed and pipelined into the searchable database. Following in the PlasmoDB footsteps and in a similar framework are websites devoted to other apicomplexan parasites including *Toxoplasma* (ToxoDB) and *Cryptosporidium* (CryptoDB) [66, 67].

6
Perspectives in Credentialing the *Plasmodium* Genome/Proteome

The current revolution of malaria research in the genomic and post-genomic era will undoubtedly assist in solving numerous biological questions, support designing preventative therapeutics, and improve the selection of novel vaccine targets [14, 20, 68, 69]. To augment this, however, a prioritized systems biology approach to authenticate each gene in the genome as a potential drug or vaccine target needs to be implemented. The vast expression and genome data now available proved to be the initial drive for a movement to 'credential' the malaria genome and this process has been reviewed [14]. Obtaining all of the data available for a particular gene locus including sequence, correct gene models, annotated function and features, orthologs in *Plasmodium* and other Apicomplexans, stage-specific expression, subcellular localization, protein recognition by immune sera, epitope predictions, gene knockout and transfection experiments, parasite–parasite and parasite–host protein interactions, will be vital to this mission. Knowledge of stage-specific regulation of gene expression may continue to provide–and already has provided–insight into the potential functional role of the majority of annotated genes labeled with unknown functions [9]. With further investigation, exploiting the utility of the recombinational cloning systems and standard immunofluorescence assays and immunoelectron microscopy, subcellular localization in parasite developmental stages may be achieved to narrow down potential function even further as proteins expressed in certain organelles/compartments are realized. Moreover, additional proteomic and transcriptomic data directed to profile the remaining unexamined parasite stages including pre-erythrocytic stages, gametocytogenesis, and oocysts and ookinete mosquito stages would be able to place the final pieces into the puzzle regarding expression analysis. Incorporating loss of function data generated from knockout or transgenic transfection experiments is also pivotal in characterizing malaria genes albeit, by a gene-by-gene approach. Unfortunately, large-scale gene silencing approaches using RNAi may be elusive in *Plasmodium* parasites [70, 71].

Data generated from large-scale transcriptional studies, high-throughput proteomics, functional and knockout analyses, field polymorphism determinations, protein–protein interactions, bioinformatics and others will provide the foundation for a fully credentialed genome. Researchers can focus on establishing sets of testable criteria that can be applied to this credentialed data to generate sets of genes and proteins for drug and vaccine development. In the single year since the *P. falciparum* genome was published, strategies are being employed in a systems biological approach to understand better the malaria parasite potentially leading to interventions that may reduce its burden on human health.

Acknowledgements We wish to thank our colleagues J. Aguiar, D. Bacon, F. Huang, and J. Russell for support and communication of unpublished data. We are grateful to J. Yates and L. Florens for critical reading of this review. The opinions expressed are those of the authors and do not reflect the official policy of the Department of the Navy, Department of Defense, or the US government.

References

1. Lasonder E, et al. (2002) Analysis of the *Plasmodium falciparum* proteome by high-accuracy mass spectrometry. Nature 419:537–542
2. Carucci DJ (2001) Functional genomic technologies applied to the control of the human malaria parasite, *Plasmodium falciparum*. Pharmacogenomics 2:137–142
3. Carucci DJ (2002) Technologies for the study of gene and protein expression in *Plasmodium*. Philos Trans R Soc Lond B Biol Sci 357:13–16
4. Gardner MJ, et al. (2002) Genome sequence of the human malaria parasite *Plasmodium falciparum*. Nature 419:498–511
5. Gardner MJ, et al. (2002) Sequence of *Plasmodium falciparum* chromosomes 2, 10, 11 and 14. Nature 419:531–534
6. Hall N, et al. (2002) Sequence of *Plasmodium falciparum* chromosomes 1, 3–9 and 13. Nature 419:527–531
7. Hyman RW, et al. (2002) Sequence of *Plasmodium falciparum* chromosome 12. Nature 419:534–537
8. Carlton JM, et al. (2002) Genome sequence and comparative analysis of the model rodent malaria parasite *Plasmodium yoelii yoelii*. Nature 419:512–519
9. Le Roch KG, et al. (2003) Discovery of gene function by expression profiling of the malaria parasite life cycle. Science 301:1503–1508

10. Le Roch KG, et al. (2002) Monitoring the chromosome 2 intraerythrocytic transcriptome of *Plasmodium falciparum* using oligonucleotide arrays. Am J Trop Med Hyg **67**:233–243

11. Bozdech Z, et al. (2003) The Transcriptome of the intraerythrocytic developmental cycle of *Plasmodium falciparum*. PLoS Biol 1:E5

12. Florens L, et al. (2002) A proteomic view of the *Plasmodium falciparum* life cycle. Nature 419:520–526

13. Srinivasan P, et al. (2003) Analysis of the plasmodium and anopheles transcriptomes during oocyst differentiation. J Biol Chem

14. Doolan DL, et al. (2003) Utilization of genomic sequence information to develop malaria vaccines. J Exp Biol 206:3789–3802

15. Carucci DJ (2001) Genomic tools for gene and protein discovery in malaria: toward new vaccines. Vaccine 19:2315–2318

16. Hoffman SL, Carucci DJ (2000) *Plasmodium falciparum:* from genomic sequence to vaccines and drugs. Novartis Found Symp 229:94–100; discussion 100–104

17. Hoffman SL, et al. (1998) From genomics to vaccines: malaria as a model system. Nat Med 4:1351–1353

18. Hoffman SL, et al. (2002) Protection of humans against malaria by immunization with radiation-attenuated *Plasmodium falciparum* sporozoites. J Infect Dis 185:1155–1164

19. Carucci DJ (2000) Malaria research in the post-genomic era. Parasitol Today 16:434–438

20. Waters A (2003) Comparative genomics of malaria parasites and its exploitation in a rodent malaria model. Bioinformatics 19 (Suppl 2):II245

21. Patankar S, et al. (2001) Serial analysis of gene expression in *Plasmodium falciparum* reveals the global expression profile of erythrocytic stages and the presence of anti-sense transcripts in the malarial parasite. Mol Biol Cell 12:3114–3125

22. Rathod PK, et al. (2002) DNA microarrays for malaria. Trends Parasitol 18:39–45

23. Hayward RE, et al. (2000) Shotgun DNA microarrays and stage-specific gene expression in *Plasmodium falciparum* malaria. Mol Microbiol 35:6–14

24. Kappe SH, et al. (2001) Exploring the transcriptome of the malaria sporozoite stage. Proc Natl Acad Sci USA 98:9895–9900

25. Ben Mamoun C, et al. (2001) Co-ordinated programme of gene expression during asexual intraerythrocytic development of the human malaria parasite *Plasmodium falciparum* revealed by microarray analysis. Mol Microbiol 39:26–36

26. Wodicka L, et al. (1997) Genome-wide expression monitoring in *Saccharomyces cerevisiae*. Nat Biotechnol 15:1359–1367

27. Oshiro G, et al. (2002) Parallel identification of new genes in *Saccharomyces cerevisiae*. Genome Res 12:1210–1220

28. Washburn MP, Wolters D, Yates JR 3rd (2001) Large-scale analysis of the yeast proteome by multidimensional protein identification technology. Nat Biotechnol 19:242–247

29. Washburn MP, et al. (2003) Protein pathway and complex clustering of correlated mRNA and protein expression analyses in *Saccharomyces cerevisiae*. Proc Natl Acad Sci USA 100:3107–3112

30. MacCoss MJ, et al. (2002) Shotgun identification of protein modifications from protein complexes and lens tissue. Proc Natl Acad Sci USA 99:7900–7905

31. Dessens JT, et al. (2003) SOAP, a novel malaria ookinete protein involved in mosquito midgut invasion and oocyst development. Mol Microbiol 49:319–329
32. Mello K, et al. (2002) A multigene family that interacts with the amino terminus of plasmodium MSP-1 identified using the yeast two-hybrid system. Eukaryot Cell 1:915–925
33. Cui L, Fan Q, Li J (2002) The malaria parasite *Plasmodium falciparum* encodes members of the Puf RNA-binding protein family with conserved RNA binding activity. Nucl Acids Res 30:4607–4617
34. Daly TM, Long CA, Bergman LW (2001) Interaction between two domains of the *P. yoelii* MSP-1 protein detected using the yeast two-hybrid system. Mol Biochem Parasitol 117:27–35
35. Vlachou D, et al. (2001) *Anopheles gambiae* laminin interacts with the P25 surface protein of *Plasmodium berghei* ookinetes. Mol Biochem Parasitol 112:229–237
36. Mamoun CB, et al. (1998) Identification and characterization of an unusual double serine/threonine protein phosphatase 2C in the malaria parasite *Plasmodium falciparum*. J Biol Chem 273:11241–11247
37. Adams JH, et al. (2001) An expanding ebl family of *Plasmodium falciparum*. Trends Parasitol 17:297–299
38. Preiser P, et al. (2004) Antibodies against MAEBL ligand domains M1 and M2 Inhibit sporozoite development in vitro. Infect Immun 72:3604–3608
39. Blair PL, et al. (2002) *Plasmodium falciparum* MAEBL is a unique member of the ebl family. Mol Biochem Parasitol 122:35–44
40. Kariu T, et al. (2002) MAEBL is essential for malarial sporozoite infection of the mosquito salivary gland. J Exp Med 195:1317–1323
41. Ghai M, et al. (2002) Identification, expression, and functional characterization of MAEBL, a sporozoite and asexual blood stage chimeric erythrocyte-binding protein of *Plasmodium falciparum*. Mol Biochem Parasitol 123:35–45
42. Blair PL, et al. (2002) Transcripts of developmentally regulated *Plasmodium falciparum* genes quantified by real-time RT-PCR. Nucl Acids Res 30:2224–2231
43. Gruner AC, et al. (2001) Expression of the erythrocyte-binding antigen 175 in sporozoites and in liver stages of *Plasmodium falciparum*. J Infect Dis 184:892–897
44. Sherman IW, Eda S, Winograd E (2003) Cytoadherence and sequestration in *Plasmodium falciparum*: defining the ties that bind. Microbes Infect 5:897–909
45. Craig A (2000) Malaria: a new gene family (clag) involved in adhesion. Parasitol Today 16:366–367; discussion 405
46. Kaneko O, et al. (2001) The high molecular mass rhoptry protein, RhopH1, is encoded by members of the clag multigene family in *Plasmodium falciparum* and *Plasmodium yoelii*. Mol Biochem Parasitol 118:223–231
47. Ling IT, et al. (2003) Characterisation of the rhoph2 gene of *Plasmodium falciparum* and *Plasmodium yoelii*. Mol Biochem Parasitol127:47–57
48. Baldi DL, et al. (2000) RAP1 controls rhoptry targeting of RAP2 in the malaria parasite *Plasmodium falciparum*. EMBO J 19:2435–2443
49. Ishino T, et al. (2004) Cell-passage activity is required for the malarial parasite to cross the liver sinusoidal cell layer. PLoS Biol 2:E4
50. Mota MM, et al. (2001) Migration of *Plasmodium* sporozoites through cells before infection. Science 291:141–144

51. Mota MM, Rodriguez A (2002) Invasion of mammalian host cells by *Plasmodium* sporozoites. Bioessays 24:149–156
52. Mota MM, Hafalla JC, Rodriguez A (2002) Migration through host cells activates *Plasmodium sporozoites* for infection. Nat Med 8:1318–1322
53. Matuschewski K. et al. (2002) *Plasmodium* sporozoite invasion into insect and mammalian cells is directed by the same dual binding system. EMBO J 21:1597–1606
54. Kappe SH, Kaiser K, Matuschewski K (2003) The *Plasmodium* sporozoite journey: a rite of passage. Trends Parasitol 19:135–143
55. Carrolo M, et al. (2003) Hepatocyte growth factor and its receptor are required for malaria infection. Nat Med 9:1363–1369
56. Hollingdale MR, et al. (1984) Inhibition of entry of *Plasmodium falciparum* and *P. vivax* sporozoites into cultured cells; an in vitro assay of protective antibodies. J Immunol 132:909–913
57. Aikawa M, et al. (1984) Ultrastructure of in vitro cultured exoerythrocytic stage of *Plasmodium berghei* in a hepatoma cell line. Am J Trop Med Hyg 33:792–799
58. Hollingdale MR, et al. (1983) Entry of *Plasmodium berghei* sporozoites into cultured cells, and their transformation into trophozoites. Am J Trop Med Hyg 32:685–690
59. Hollingdale MR, Leland P, Schwartz AL (1983) In vitro cultivation of the exoerythrocytic stage of *Plasmodium berghei* in a hepatoma cell line. Am J Trop Med Hyg 32:682–684
60. Kaiser K, Camargo N, Kappe SH (2003) Transformation of sporozoites into early exoerythrocytic malaria parasites does not require host cells. J Exp Med 197:1045–1050
61. Zhu JD, et al. (1990) Stage-specific ribosomal RNA expression switches during sporozoite invasion of hepatocytes. J Biol Chem 265:12740–12744
62. Gunderson JH, et al. (197) Structurally distinct, stage-specific ribosomes occur in *Plasmodium*. Science 238:933–937
63. Kissinger JC, et al. (2002) The *Plasmodium* genome database. Nature 419:490–492
64. Bahl A, et al. (2002) PlasmoDB: the *Plasmodium* genome resource. An integrated database providing tools for accessing, analyzing and mapping expression and sequence data (both finished and unfinished). Nucl Acids Res 30:87–90
65. Bahl A, et al. (2003) PlasmoDB: the *Plasmodium* genome resource. A database integrating experimental and computational data. Nucl Acids Res 31:212–215
66. Puiu D, et al. (2004) CryptoDB: the *Cryptosporidium* genome resource. Nucl Acids Res 32:D329–D331
67. Kissinger JC, et al. (2003) ToxoDB: accessing the *Toxoplasma gondii* genome. Nucl Acids Res 31:234–236
68. Wiesner J, et al. (2003) *New antimalarial drugs. Angew Chem Int Ed* Engl 42:5274–5293
69. Ellis JT, Morrison DA, Reichel MP (2003) Genomics and its impact on parasitology and the potential for development of new parasite control methods. DNA Cell Biol 22:395–403
70. Ullu E, Tschudi C, Chakraborty T (2004) RNA interference in protozoan parasites. Cell Microbiol 6:509–519
71. Aravind L, et al. (2003) *Plasmodium* biology: genomic gleanings. Cell 115:771–785

72. Aguiar JC, et al. (2004) High-throughput generation of P. falciparumfunctional molecules by recombinationalcloning. Genome Res 14(10B):2076-2082

73. Silvie O, et al (2004) A role for apical membrane antigen 1 during invasion of hepatocytes by Plasmodiumfalciparum sporozoites. J Biol Chem 279(10):9490-9496

Subject Index

Current Topics in Microbiology and Immunology

Volumes published since 1989 (and still available)

Vol. 272: **Doerfler, Walter; Böhm, Petra (Eds.):** Adenoviruses: Model and Vectors in Virus-Host Interactions. Virion and Structure, Viral Replication, Host Cell Interactions. 2003. 63 figs., approx. 280 pp. ISBN 3-540-00154-9

Vol. 273: **Doerfler, Walter; Böhm, Petra (Eds.):** Adenoviruses: Model and Vectors in VirusHost Interactions. Immune System, Oncogenesis, Gene Therapy. 2004. 35 figs., approx. 280 pp. ISBN 3-540-06851-1

Vol. 274: **Workman, Jerry L. (Ed.):** Protein Complexes that Modify Chromatin. 2003. 38 figs., XII, 296 pp. ISBN 3-540-44208-1

Vol. 275: **Fan, Hung (Ed.):** Jaagsiekte Sheep Retrovirus and Lung Cancer. 2003. 63 figs., XII, 252 pp. ISBN 3-540-44096-3

Vol. 276: **Steinkasserer, Alexander (Ed.):** Dendritic Cells and Virus Infection. 2003. 24 figs., X, 296 pp. ISBN 3-540-44290-1

Vol. 277: **Rethwilm, Axel (Ed.):** Foamy Viruses. 2003. 40 figs., X, 214 pp. ISBN 3-540-44388-6

Vol. 278: **Salomon, Daniel R.; Wilson, Carolyn (Eds.):** Xenotransplantation. 2003. 22 figs., IX, 254 pp. ISBN 3-540-00210-3

Vol. 279: **Thomas, George; Sabatini, David; Hall, Michael N. (Eds.):** TOR. 2004. 49 figs., X, 364 pp. ISBN 3-540-00534X

Vol. 280: **Heber-Katz, Ellen (Ed.):** Regeneration: Stem Cells and Beyond. 2004. 42 figs., XII, 194 pp. ISBN 3-540-02238-4

Vol. 281: **Young, John A. T. (Ed.):** Cellular Factors Involved in Early Steps of Retroviral Replication. 2003. 21 figs., IX, 240 pp. ISBN 3-540-00844-6

Vol. 282: **Stenmark, Harald (Ed.):** Phosphoinositides in Subcellular Targeting and Enzyme Activation. 2003. 20 figs., X, 210 pp. ISBN 3-540-00950-7

Vol. 283: **Kawaoka, Yoshihiro (Ed.):** Biology of Negative Strand RNA Viruses: The Power of Reverse Genetics. 2004. 24 figs., IX, 350 pp. ISBN 3-540-40661-1

Vol. 284: **Harris, David (Ed.):** Mad Cow Disease and Related Spongiform Encephalopathies. 2004. 34 figs., IX, 219 pp. ISBN 3-540-20107-6

Vol. 285: **Marsh, Mark (Ed.):** Membrane Trafficking in Viral Replication. 2004. 19 figs., IX, 259 pp. ISBN 3-540-21430-5

Vol. 286: **Madshus, Inger H. (Ed.):** Signalling from Internalized Growth Factor Receptors. 2004. 19 figs., IX, 187 pp. ISBN 3-540-21038-5

Vol. 287: **Enjuanes, Luis (Ed.):** Coronavirus Replication and Reverse Genetics. 2005. 49 figs., XI, 257 pp. ISBN 3-540-21494-1

Vol. 288: **Mahy, Brain W. J. (Ed.):** Foot-and-Mouth-Disease Virus. 2005. 16 figs., IX, 178 pp. ISBN 3-540-22419X

Vol. 289: **Griffin, Diane E. (Ed.):** Role of Apoptosis in Infection. 2005. 40 figs., IX, 294 pp. ISBN 3-540-23006-8

Vol. 290: **Singh, Harinder; Grosschedl, Rudolf (Eds.):** Molecular Analysis of B Lymphocyte Development and Activation. 2005. 28 figs., XI, 255 pp. ISBN 3-540-23090-4

Vol. 291: **Boquet, Patrice; Lemichez Emmanuel (Eds.)** Bacterial Virulence Factors and Rho GTPases. 2005. 28 figs., IX, 196 pp. ISBN 3-540-23865-4

Vol. 292: **Fu, Zhen F (Ed.):** The World of Rhabdoviruses. 2005. 27 figs., X, 210 pp. ISBN 3-540-24011-X

Vol. 293: **Kyewski, Bruno; Suri-Payer, Elisabeth (Eds.):** CD4+CD25+ Regulatory T Cells: Origin, Function and Therapeutic Potential. 2005. 22 figs., XII, 332 pp. ISBN 3-540-24444-1

Vol. 294: **Caligaris-Cappio, Federico, Dalla Favera, Ricardo (Eds.):** Chronic Lymphocytic Leukemia. 2005. 25 figs. VIII, 187 pp. ISBN 3-540-25279-7

.

Printed in Great Britain
by Amazon